VISUALIZING

THE ENVIRONMENT

VISUALIZING
THE ENVIRONMENT

Linda R. Berg, Ph.D.

Mary Catherine Hager, M.Sc.

Leslie G. Goodman, M.Sc.
University of Manitoba

Richard K. Baydack, Ph.D.
University of Manitoba

WILEY

In collaboration with

THE NATIONAL GEOGRAPHIC SOCIETY

CREDITS

VICE-PRESIDENT AND PUBLISHER Veronica Visentin

VICE-PRESIDENT, PUBLISHING SERVICES Karen Bryan

ACQUISITIONS EDITOR Rodney Burke

MARKETING MANAGER Patty Maher

CREATIVE DIRECTOR, PUBLISHING SERVICES Ian Koo

EDITORIAL MANAGER Karen Staudinger

DEVELOPMENTAL EDITOR Leanne Rancourt

EDITORIAL ASSISTANTS Laura Hwee and Sara Tinteri

CREATIVE DIRECTOR Harry Nolan

COVER DESIGN Ian Koo

INTERIOR DESIGN Vertigo Design, Adrian So

PHOTO RESEARCHERS Stacy Gold,
National Geographic Society, and Christina Beamish

COVER:

Main image: Chris Cheadle/Getty Images

Smaller images (left to right): Tom Brakefield/Thinkstock;
Rich Reid/NG Image Collection; Comstock; Jasper James/
The Image Bank; Comstock

This book was set in Times New Roman by PreMediaGlobal, printed and bound by Worldcolor.

ISBN-13 978-0-470-15798-5

Library and Archives Canada Cataloguing in Publication

Berg, Linda R.
    Visualizing the environment/Linda R. Berg... [et al.].—1st Canadian ed.

US eds. have title: Visualizing environmental science.
ISBN 978-0-470-15798-5

    1. Environmental sciences–Textbooks. I. Berg, Linda R.
II. Berg, Linda R. Visualizing environmental science.

GE70.V58 2010   333.7   C2010-902597-0

Printed in the United States of America

14 13 12 11 10 WC 5 4 3 2 1

From the Publisher

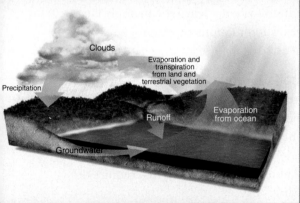

Visualizing the Environment is designed to help your students learn effectively. Created in collaboration with the National Geographic Society (NGS), *Visualizing the Environment* integrates rich visuals and media with text to direct student's attention to important information. This approach represents complex processes, organizes related pieces of information, and integrates information into clear representations. Beautifully illustrated, *Visualizing the Environment* shows your students what the discipline is all about, its main concepts and applications, while also instilling an appreciation and excitement about the richness of the subject.

Visuals, as used throughout this text, are instructional components that display facts, concepts, processes, or principles. They create the foundation for the text and do more than simply support the written or spoken word. The visuals include diagrams, graphs, maps, photographs, illustrations, schematics, animations, and videos.

Why should a textbook based on visuals be effective? Research shows that we learn better from integrated text and visuals than from either medium separately. Beginners in a subject benefit most from reading about the topic, attending class, and studying well-designed and integrated visuals. A visual, with good accompanying discussion, really can be worth a thousand words!

Well-designed visuals can also improve the efficiency with which information is processed by a learner. The more effectively we process information, the more likely it is that we will learn. This processing of information takes place in our working memory. As we learn we integrate new information in our working memory with existing knowledge in our long-term memory.

Have you ever read a paragraph or a page in a book, stopped, and said to yourself: "I don't remember one thing I just read?" This may happen when your working memory has been overloaded, and the text you read was not successfully integrated into long-term memory. Visuals don't automatically solve the problem of overload, but well-designed visuals can reduce the number of elements that working memory must process, thus aiding learning.

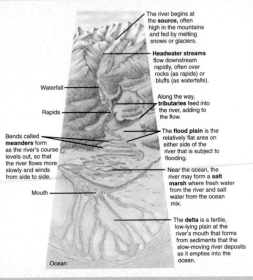

The river begins at the **source**, often high in the mountains and fed by melting snows or glaciers.

Headwater streams flow downstream rapidly, often over rocks (as rapids) or bluffs (as waterfalls).

Waterfall

Rapids

Along the way, **tributaries** feed into the river, adding to the flow.

The **flood plain** is the relatively flat area on either side of the river that is subject to flooding.

Bends called **meanders** form as the river's course levels out, so that the river flows more slowly and winds from side to side.

Mouth

Near the ocean, the river may form a **salt marsh** where fresh water from the river and salt water from the ocean mix.

The **delta** is a fertile, low-lying plain at the river's mouth that forms from sediments that the slow-moving river deposits as it empties into the ocean.

Ocean

You, as the instructor, facilitate your student's learning. Well-designed visuals, used in class, can help you in that effort. Here are six methods for using the visuals in *Visualizing the Environment* in classroom instruction.

1. **Assign students to study visuals in addition to reading the text.**

   Instead of assigning only one medium of presentation, it is important to make sure your students know that the visuals are just as essential as the written material.

2. **Use visuals during class discussions or presentations.**

   By pointing out important information as the students look at the visuals during class discussions, you can help focus students' attention on key elements of the visuals and help them begin to organize the information and develop an integrated model of understanding. The verbal explanation of important information combined with the visual representation is highly effective.

3. **Use visuals to review content knowledge.**

   Students can review key concepts, principles, processes, vocabulary, and relationships displayed visually. Better understanding results when new information in working memory is linked to prior knowledge.

4. **Use visuals for assignments or when assessing learning.**

   Visuals can be used for comprehension activities or assessments. For example, students could be asked to identify examples of concepts portrayed in visuals. Higher-level thinking activities that require critical thinking, deductive and inductive reasoning, and prediction can also be based on visuals. Visuals can be very useful for drawing inferences, for predicting, and for problem solving.

5. **Use visuals to situate learning in authentic contexts.**

   Learning is made more meaningful when a learner can apply facts, concepts, and principles to realistic situations or examples. Visuals can provide that realistic context.

6. **Use visuals to encourage collaboration.**

   Collaborative groups often are required to practise interactive processes such as giving explanations, asking questions, clarifying ideas, and arguing points of view. These interactive, face-to-face processes provide the information needed to build a verbal mental model. Learners also benefit from collaboration over visuals that require decision making or problem solving.

Visualizing the Environment not only aids student learning with extraordinary use of visuals, but it also offers an array of remarkable photos, media, and film from the National Geographic Society collections. Students using *Visualizing the Environment* also benefit from the long history and rich, fascinating resources of National Geographic.

Given all of its strengths and resources, *Visualizing the Environment* will immerse your students in the discipline, its main concepts and applications, while also instilling an appreciation and excitement about the richness of the subject area.

Additional information on learning and instructional design is provided in a special guide to using this book, *Learning from Visuals: How and Why Visuals Can Help Students Learn*, prepared by Matthew Leavitt of Arizona State University. This article is available at the Wiley website: www.wiley.com/college/visualizing. The online *Instructor's Manual* also provides guidelines and suggestions on using the text and visuals most effectively.

Visualizing the Environment also offers a rich selection of visuals in the supplementary materials that accompany the book. To complete this robust package the following materials are available: Test Bank with visuals used in assessment, PowerPoint slides, Image Gallery to provide you with the same visuals used in the text, and web-based learning materials for homework and assessment, including images, video, and media resources from National Geographic.

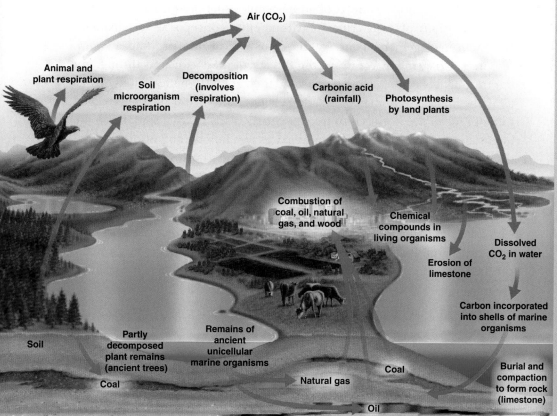

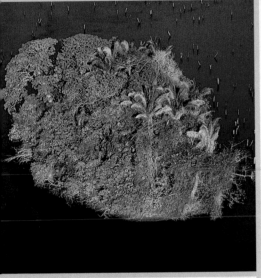

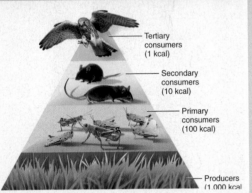

his Canadian edition of *Visualizing the Environment* offers students a valuable opportunity to identify and connect the central issues of the environment through a visual approach. As students explore the critical topics of this important field, their study of the role of humans on Earth must be interwoven with basic concepts of ecology, geography, chemistry, economics, ethics, policy, and many other disciplines. Therefore, this textbook delivers an interdisciplinary approach to studying humans and the environment, and provides students with a foundation based on environmental sustainability at the ecological, social, and economic level. *Visualizing the Environment* reinforces these interacting components and, with its premier art program, vividly illustrates the overarching role that humans play in our planet's environmental problems and successes. Most importantly, it also offers solutions for the future.

We begin *Visualizing the Environment* with an introduction of the environmental dilemmas we face in our world today, emphasizing particularly how unchecked population growth and economic inequity complicate our ability to solve these problems. We stress that solutions rest in creativity and diligence at all levels, from individual commitment to international cooperation. We revisit this theme throughout the text, offering concrete suggestions that students can adopt to make their own difference in solving environmental problems, and explaining the complications that arise when solutions are tackled on a local, regional, national, or global scale.

Yet *Visualizing the Environment* is not simply a check-list of "to do" items to save the planet. In the context of an engaging visual presentation, we offer solid discussions of such critical environmental concepts as sustainability, ecosystem management, conservation of biological diversity, and adaptive resource management. We weave the threads of these concepts throughout our treatment of ecological principles and their applications to various ecosystems, the impacts of human population change, and the problems associated with our use of the world's resources. We particularly instruct students in the importance of ecosystem goods and services to a functioning world, and the threats that restrict our planet's ability to continue to provide these services into the future.

This book is intended to serve as an introductory-level text for science-based and nonscience undergraduate students. The accessible format of *Visualizing the Environment,* coupled with our assumption that students have little prior knowledge of ecosystem ecology, allows students to easily make the transition from jumping-off points in the early chapters to the more complex concepts they encounter later. With its interdisciplinary presentation, this book is appropriate for use in one-semester environment courses offered by a variety of departments, including environmental science and studies, biology, agriculture, geography, and geology.

ORGANIZATION

Visualizing the Environment is organized around the premise that humans have accentuated many of our planet's environmental issues and now must address these issues as we use Earth's resources more efficiently so as to avoid the future disasters so often predicted in the media. The book's first four chapters lay the groundwork for creating an understanding of what environmental issues we face, how environmental sustainability and human values play a critical role in addressing these issues, how the environmental movement developed over time, how economics shapes environmental policy, and how environmental threats sometimes cause health hazards. Students also develop a richer understanding of the implications to the environment that rise from human population change.

Chapters 5 and 6 of *Visualizing the Environment* present the foundations of ecosystems, including energy flow and the cycling of matter through ecosystems, and the various ways that species interact and partition resources. Gaining familiarity with these concepts allows students to appreciate the variety of terrestrial and aquatic ecosystems that we introduce.

Chapters 7 through 12 all deal in some way with our world's resources, as humans use and manage them today, and as we assess their availability and impacts for the future. These issues cover a broad spectrum that includes biological resources, land resources, mineral and soil resources, agriculture and food, freshwater and marine resources and causes and effects of water pollution, the sources and effects of air pollution, climate and global atmospheric changes, nonrenewable and renewable energy resources, and solid and hazardous wastes.

Chapter 13 of *Visualizing the Environment* is unique to the text and offers a discussion of how to achieve our preferred future environment. Students are provided with real-world examples of how sustainable ecosystem management can help preserve the environment while also allowing for human growth and development. The environment can still be saved, and this chapter provides students with practical skills and knowledge—and most importantly, hope—of what they can do to make a difference.

ILLUSTRATED BOOK TOUR

Many visual pedagogical features have been developed specifically for *Visualizing the Environment*. Presenting the highly varied and often technical concepts woven throughout environmental science raises challenges for readers and instructors alike. The Illustrated Book Tour on the following pages provides a guide to the diverse features contributing to *Visualizing the Environment's* pedagogical plan.

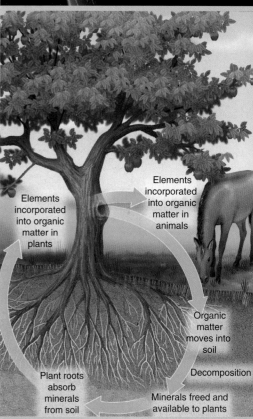

Elements incorporated into organic matter in plants

Elements incorporated into organic matter in animals

Organic matter moves into soil

Decomposition

Minerals freed and available to plants

Plant roots absorb minerals from soil

CHAPTER INTRODUCTIONS provide concise stories about some of today's most pressing environmental issues. These narratives are featured alongside striking accompanying photographs. The chapter openers also include illustrated **CHAPTER OUTLINES** that use thumbnails of illustrations from the chapter to refer visually to the content.

VISUALIZING features are specially designed multi-part visual spreads that focus on a key concept or topic in the chapter, exploring it in detail or in broader context using a combination of photos and figures.

PROCESS DIAGRAMS present a series of figures or a combination of figures and photos that describe and depict a complex process, helping students to observe, follow, and understand the process.

WHAT A SCIENTIST SEES features highlight a concept or phenomenon using photos and figures that would stand out to a professional in the field, helping students to develop observational skills.

ENVIRODISCOVERY features provide additional topical material about relevant environmental issues.

MAKING A DIFFERENCE features highlight a Canadian person or organization who is working to promote environmental sustainability or contributing to environmental knowledge through their research.

ECO CANADA CAREER FOCUS features are found at the end of each chapter and highlight the many different environmental careers available to students who want to work in the field. These features have been developed in conjunction with ECO Canada (www. eco.ca), a not-for-profit organization focused on supporting Canada's environmental industry.

The illustrated **CASE STUDIES** that cap off the text sections of each chapter offer a wide variety of in-depth examinations that address important issues in the field of environmental science.

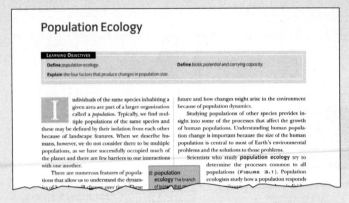

LEARNING OBJECTIVES at the beginning of each section head indicate in behavioural terms what the student must be able to do to demonstrate mastery of the material in the chapter.

CONCEPT CHECK questions at the end of each section give students the opportunity to test their comprehension of the learning objectives.

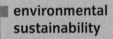

environmental sustainability

The ability to meet humanity's current needs without compromising the ability of future generations to meet their needs.

MARGINAL GLOSSARY TERMS (in green boldface) introduce each chapter's most important terms. Other important terms appear in black boldface and are defined in the text.

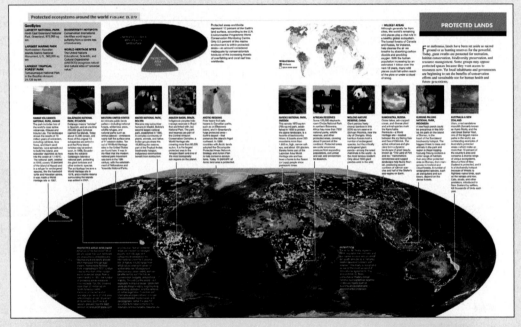

NATIONAL GEOGRAPHIC SOCIETY MAPS covering global environmental issues (Chapter 1), human population (Chapter 3), land cover types (Chapter 6), protected ecosystems (Chapter 8), water (Chapter 9), and energy (Chapter 11), and are integrated in the appropriate chapters of the book.

GLOBAL LOCATOR MAPS accompany figures addressing issues encountered in a particular geographic region. These hemispheric locator maps help students visualize where the area discussed is situated on a continent.

A "GLOBAL TO LOCAL" ICON appears in the margins of the text to reinforce the integrated theme that environmental science covers issues from the local to the global scale.

ILLUSTRATIONS AND PHOTOS support concepts covered in the text, elaborate on relevant issues, and add visual detail. Many of the photos originate from National Geographic's archives.

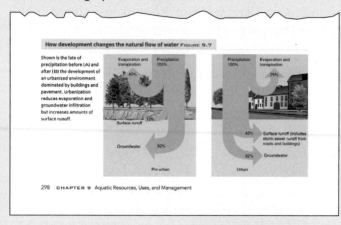

How development changes the natural flow of water FIGURE 9.7

Shown is the fate of precipitation before (A) and after (B) the development of an urbanized environment dominated by buildings and pavement. Urbanization reduces evaporation and groundwater infiltration but increases amounts of surface runoff.

298   CHAPTER 9   Aquatic Resources, Uses, and Management

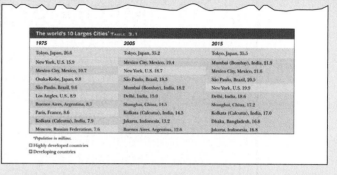

The world's 10 Larges Cities* TABLE 3.1		
1975	2005	2015
Tokyo, Japan, 26.6	Tokyo, Japan, 35.2	Tokyo, Japan, 35.5
New York, U.S. 15.9	Mexico City, Mexico, 19.4	Mumbai (Bombay), India, 21.9
Mexico City, Mexico, 10.7	New York, U.S. 18.7	Mexico City, Mexico, 21.6
Osaka-Kobe, Japan, 9.8	São Paulo, Brazil, 18.3	São Paulo, Brazil, 20.5
São Paulo, Brazil, 9.6	Mumbai (Bombay), India, 18.2	New York, U.S. 19.9
Los Angeles, U.S., 8.9	Delhi, India, 15.0	Delhi, India, 18.6
Buenos Aires, Argentina, 8.7	Shanghai, China, 14.5	Shanghai, China, 17.2
Paris, France, 8.6	Kolkata (Calcutta), India, 14.3	Kolkata (Calcutta), India, 17.0
Kolkata (Calcutta), India, 7.9	Jakarta, Indonesia, 13.2	Dhaka, Bangladesh, 16.8
Moscow, Russian Federation, 7.6	Buenos Aires, Argentina, 12.6	Jakarta, Indonesia, 16.8

*Population in millions.
□ Highly developed countries
□ Developing countries

TABLES AND GRAPHS, with data sources cited at the end of the text, summarize and organize important information.

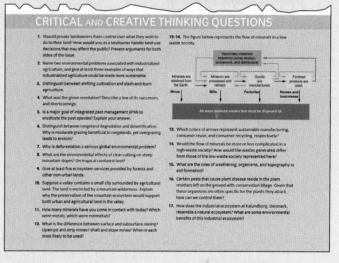

CRITICAL AND CREATIVE THINKING QUESTIONS

1. Should private landowners have control over what they wish to do to their land? How would you as a landowner handle land-use decisions that may affect the public? Present arguments for both sides of the issue.
2. Name two environmental problems associated with industrialized agriculture, and give at least three examples of ways that industrialized agriculture could be made more sustainable.
3. Distinguish between shifting cultivation and slash-and-burn agriculture.
4. What was the green revolution? Describe a few of its successes and shortcomings.
5. Is a major goal of integrated pest management (IPM) to eradicate the pest species? Explain your answer.
6. Distinguish between rangeland degradation and desertification. Why is moderate grazing beneficial to rangelands, yet overgrazing leads to erosion?
7. Why is deforestation a serious global environmental problem?
8. What are the environmental effects of clear-cutting on steep mountain slopes? On tropical rainforest land?
9. Give at least five ecosystem services provided by forests and other non-urban lands.
10. Suppose a valley contains a small city surrounded by agricultural land. The land is encircled by a mountain wilderness. Explain why the preservation of the mountain ecosystem would support both urban and agricultural land in the valley.
11. How many minerals have you come in contact with today? Which were metals; which were nonmetals?
12. What is the difference between surface and subsurface mining? Open-pit and strip mines? Shaft and slope mines? When is each most likely to be used?

13–14. The figure below represents the flow of minerals in a low-waste society.

13. Which colors of arrows represent sustainable manufacturing, consumer reuse, and consumer recycling, respectively?
14. Would the flow of minerals be more or less complicated in a high-waste society? How would the wastes generated differ from those of the low-waste society represented here?
15. What are the roles of weathering, organisms, and topography in soil formation?
16. Certain pests that cause plant disease reside in the plant residues left on the ground with conservation tillage. Given that these organisms are often specific for the plants they attack, how can we control them?
17. How does the industrial ecosystem at Kalundborg, Denmark, resemble a natural ecosystem? What are some environmental benefits of this industrial ecosystem?

CRITICAL AND CREATIVE THINKING QUESTIONS encourage critical thinking and highlight each chapter's important concepts and applications.

unrestricted use of natural resources, and increased economic growth to manage an expanding industrial base. The deep ecology worldview is an understanding of our place in the world based on harmony with nature, a spiritual respect for life, and the belief that humans and all other species have equal worth. The Aboriginal worldview is an understanding of the interconnectedness between humans and all other living organisms and the need to live in harmony with each other and with nature.

3. Environmental justice is the right of every citizen, regardless of age, race, gender, social class, or other factor, to adequate protection from environmental hazards. Environmental justice is a fundamental human right in an ethical society. A growing environmental justice movement has emerged at the grassroots level.

4 Strategies for Sustainable Living

1. Failing to confront the problem of poverty makes it impossible to manage an expanding industrial base. To stay within Earth's carrying capacity, the maximum population that can be sustained indefinitely, it will be necessary to reach a stable population and reduce excessive consumption.

2. The world's forests are being cut, burned, and seriously altered for timber and other products that the global economy requires. Also, rapid population growth and poverty are putting pressure on forests. Biological diversity, the number and variety of Earth's organisms, is declining at an alarming rate. Humans are part of Earth's web of life and are entirely dependent on that web for survival.

3. Food insecurity is the condition in which people live with chronic hunger and malnutrition. Globally, more than 800 million people lack access to the food needed for healthy productive lives.

4. The enhanced greenhouse effect is the additional warming produced by increased levels of gases that absorb infrared radiation. An increase in atmospheric CO_2, mostly produced when fossil fuels are burned, leads to climate warming. To stabilize climate, we must phase out fossil fuels in favour of renewable energy, increased energy conservation, and improved energy efficiency.

5. The air in cities in the developing world is badly polluted with exhaust from motor vehicles. Illegal squatter settlements proliferate in cities; the poorest inhabitants build dwellings using whatever materials they can scavenge. Squatter settlements have the worst water, sewage, and solid waste problems.

KEY TERMS

■ utilitarian conservationist p.000
■ environmental ethics p.000
■ environmental worldview p.000
■ Western worldview p.000

■ deep ecology worldview p.000
■ Aboriginal worldview p.000
■ environmental justice p.000
■ carrying capacity p.000

■ biolog
■ food i
■ enhar

CRITICAL AND CREATIVE THINKING QUE

The **CHAPTER SUMMARY** revisits each learning objective and redefines each marginal glossary term, featured in boldface here, and included in a list of **KEY TERMS**. Students are thus able to study vocabulary words in the context of related concepts.

What is happening in this picture ?

These hoses suck methane, which is used as fuel, from decomposing trash.

What type of renewable energy is being provided?

What would be some advantages and disadvantages of this type of energy production?

WHAT IS HAPPENING IN THIS PICTURE? is an end-of-chapter feature that presents students with a photograph relevant to chapter topics but that illustrates a situation students are not likely to have encountered previously. The photograph is paired with questions designed to stimulate creative thinking.

MEDIA AND SUPPLEMENTS

Visualizing the Environment is accompanied by a rich array of media and supplements that incorporate the visuals from the textbook extensively to form a pedagogically cohesive package. For example, a Process Diagram from the book may appear in the Instructor's Manual with suggestions for using it as a PowerPoint in the classroom; it may be the subject of a short video or an online animation; or it may appear with questions in the Test Bank.

VIDEOS

A rich collection of videos from the award-winning National Geographic Film Collection have been selected to accompany the text. The videos are available online as digitized streaming video that illustrate and expand on a concept or topic. Contextualized commentary and questions to further develop student understanding accompany each of the videos.

The videos are available in the Wiley Resource Kit.

WEB-BASED LEARNING MODULES

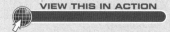

A robust suite of multimedia learning resources have been designed for *Visualizing the Environment*, again focusing on and using the visuals from the book. The content is organized into tutorial animations.

Tutorial Animations and Interactives: Animations visually support the learning of a difficult concept, process, or theory, many of them built around a specific feature such as a Process Diagram, Visualizing feature, or key visual in the chapter. The animations go beyond the content and visuals presented n the book, providing additional visual examples and descriptive narration.

BOOK COMPANION SITE (WWW.WILEY.COM/CANADA/BERG)

Instructor Resources on the book companion site include the Test Bank, Instructor's Manual, PowerPoint presentations, and all illustrations and photo in the textbook in jpeg format. Student resources include self-quizzes and flashcards.

INSTRUCTOR SUPPLEMENTS

POWERPOINT PRESENTATIONS

A complete set of visual PowerPoint presentations by Erin McCance of the University of Manitoba is available online to enhance classroom presentations. Tailored to the text's topical coverage and learning objectives, these presentations are designed to convey key text concepts, illustrated by embedded text art.

IMAGE GALLERY

All photographs, figures, maps, and other visuals from the text are online and can be used as you wish in the classroom. These online electronic files allow you to easily incorporate them into your PowerPoint presentations as you choose, or to create your own overhead transparencies and handouts.

TEST BANK

The visuals from the textbook are also included in the Test Bank by David Kemp of Lakehead University. The Test Bank has approximately 1200 test items, with at least 25 percent incorporating visuals from the book. The test items include multiple choice and essay questions testing a variety of comprehension levels. The Test Bank is available online in two formats: MS Word files and a Computerized Test Bank. The easy-to-use test-generation program fully supports graphics, print tests, student answer sheets, and answer keys. The software's advanced features allow you to create an exam to meet your exact specifications.

INSTRUCTOR'S MANUAL

The Instructor's Manual, prepared by Michael Wilson of Douglas College, begins with a special introduction on Using Visuals in the Classroom, prepared by Matthew Leavitt of the Arizona State University, in which he provides guidelines and suggestions on how to use the visuals in teaching the course. For each chapter, suggestions and directions for using web-based learning modules in the classroom are included, as well as over 50 creative ideas for in-class activities.

ALSO AVAILABLE TO PACKAGE

Environmental Science: Active Learning Laboratories and Applied Problem Sets, 2e by Travis Wagner and Robert Sanford, both of the University of Southern Main, is designed to introduce environmental science and studies students to the broad, interdisciplinary field of environmental science by presenting specific labs that use natural and social science concepts to varying degrees and by encouraging a "hands on" approach to understanding the impacts from the environmental/human interface. The laboratory and homework activities are designed to be low-cost and to reflect a sustainability approach in practice and theory. *Environmental Science: Active Learning Laboratories and Applied Problems Sets 2e* is available standalone or in a packaged with *Visualizing the Environment, Canadian Edition.* Contact your Wiley representative for more information.

EarthPulse. Utilizing full-colour imagery and National Geographic photographs, *EarthPulse* takes you on a journey of discovery covering topidcs such as *The Human Conditions, Our Relationship with Nature,* and *Our connected World.* Illustrated by specific 3examples, each section focuses on trends affecting our world today. Included are extensive full-colour world and regional maps for reference. *EarthPulse* is available only in a package with *Visualizing the Environment, Canadian Edition.* Contact your Wiley representative for more information or visit www.wiley.com/college/earthpulse.

ACKNOWLEDGMENTS

PROFESSIONAL FEEDBACK

Throughout the process of writing and developing this text and the visual pedagogy, we benefited from the comments and constructive criticism provided by the instructors listed below. We offer our sincere appreciation to these individuals for their helpful reviews:

Darren Bardati, University of Prince Edward Island/Bishop's University
Sarah Boon, University of Lethbridge
Ben Bradshaw, University of Guelph
Bill Buhay, University of Winnipeg
Steven Cooke, Carleton University
Danielle Fortin, University of Ottawa

David Kemp, Lakehead University
Michael Pidwirny, University of British Columbia, Okanagan
Silvija Stefanovic, University of Toronto, Scarborough
Michael Wilson, Douglas College
Ann Zimmerman, University of Toronto

SPECIAL THANKS

The authors of the Canadian Edition of *Visualizing the Environment* would like to firstly thank each other for the constant encouragement, inspiration, perseverance, and dedication to the cause that was essential in completing a 'Canadianized' edition of a textbook that we believe will become a mainstay in the environment field in Canadian academic institutions. Building on our experience in teaching literally thousands of students in introductory and advanced environment courses at the University of Manitoba, the text evolved over a period of years so that it now represents a synthesis of what works best for students to learn about the environment. Added to this expertise are the 'real-world' viewpoints from our extensive knowledge of the Canadian environmental and natural resources management field. The authors recognize and acknowledge that this text would not have been possible without each other.

We would not have been able to create this textbook without the tremendous work ethic and dedication put forward by Erin McCance, our hand-picked research assistant. Erin's ability to provide the detailed information needed in every chapter of the book at warp speed made our job so much easier and enjoyable. We clearly owe Erin a tremendous debt of gratitude as well as a profound acknowledgement of her unbelievable abilities in meeting deadlines even before they were set! Without Erin, *Visualizing the Environment* would not have become reality. Godspeed to you, Erin, as you move forward in your career.

We are extremely grateful to the many members of the editorial and production staff at John Wiley and Sons Canada (the Team!) who guided us through the challenging steps of developing this book, including Veronica Visentin, Publisher; Patty Maher, Marketing Manager; Karen Staudinger, Editorial Manager; Channade Fenandoe, Media Editor; and Laura Hwee and Sara Tinteri, Editorial Assistants. Their tireless enthusiasm, professional assistance, and endless patience smoothed the path as we found our way. We thank in particular Rodney Burke, our Acquisitions Editor, who would never take 'No' for an answer; Kyle Fisher, Sales Manager, for introducing both of us to the expertise that Wiley offers; Parker McDonald, our Sales Representative, for always being there, and even when we weren't, for leaving his business card under our doors; and probably most importantly, Leanne Rancourt, our Developmental Editor, for her understanding, insightful comments, belief in us, and especially for sensing when we needed a push, when she needed to pull back, when 'other things' were more important than the book, and when there was no other choice but to get it done!!! This Wiley Team was exceptional, and we hope to work with each of you again.

We wish also to acknowledge the contributions of Vertigo Design for the interior design concept, Adrian So for his design of the new Canadian features, and Ian Koo for the design of the cover. We appreciate the efforts of Christina Beamish in obtaining our text photos and permissions, and Stacy Gold, Research Editor and Account Executive at the National Geographic Image Collection, for her valuable expertise in selecting National Geographic photos. We also wish to thank Joyce Franzen, Project Editor at PreMediaGlobal, for helping to bring our book to life, and Ruth Wilson for her careful copyedit of the text.

All the writing, producing, and marketing of this book would be wasted without an energetic and dedicated sales staff. We wish to sincerely thank all the publishing representatives for their tireless efforts and good humour. It's a true pleasure to work with such a remarkable group of people.

ABOUT THE AUTHORS

Linda R. Berg is an award-winning teacher and textbook author. She received a B.Sc. in science education, M.Sc. in botany, and Ph.D. in plant physiology from the University of Maryland. Dr. Berg taught at the University of Maryland-College Park for 17 years and at St. Petersburg College in Florida for 10 years. She has taught introductory courses in environmental science, biology, and botany to thousands of students and has received numerous teaching and service awards. Dr. Berg is also the recipient of many national and regional awards, including the National Science Teachers Association Award for Innovations in College Science Teaching, the Nation's Capital Area Disabled Student Services Award, and the Washington Academy of Sciences Award in University Science Teaching. During her career as a professional science writer, Dr. Berg has authored or co-authored several leading college science textbooks. Her writing reflects her teaching style and love of science.

Mary Catherine Hager is a professional science writer and editor specializing in educational materials for life and earth sciences. She received a double-major B.A. in environmental science and biology from the University of Virginia and an M.Sc. in zoology from the University of Georgia. Ms. Hager worked as an editor for an environmental consulting firm and as a senior editor for a scientific reference publisher. For the past 15 years, Ms. Hager has published articles in environmental trade magazines, edited federal and state reports addressing wetlands conservation issues, and written components of environmental science and biology textbooks for target audiences ranging from middle school to college. Her writing and editing pursuits are a natural outcome of her scientific training and curiosity coupled with her love of reading and communicating effectively.

Leslie G. Goodman was born and raised in Calgary, Alberta, and completed a B.Sc. in zoology from the University of Calgary. She then moved to Winnipeg and completed an M.Sc. in aquatic ecology. The topic of her graduate thesis was "Phytoplankton Activity in the Red and Assiniboine Rivers as They Flow through the City of Winnipeg, Manitoba." With the creation of the Clayton H. Riddell Faculty of Environment, Earth, and Resources at the University of Manitoba, Ms. Goodman became the instructor for the Introductory Environment course. In addition, she manages a cooperative education option that works closely with students to assist in identifying career goals and employment opportunities. In total, she works with approximately 700 students annually through the Introductory Environment course and another 50 students through the cooperative education option. She has been awarded the Clayton H. Riddell Faculty of Environment, Earth, and Resources Award as Outstanding Instructor for first-year students, and has been nominated on several occasions for the University of Manitoba Outstanding Instructor for University 1 students.

Dr. Richard Baydack is a professor in the Clayton H. Riddell Faculty of Environment, Earth, and Resources at the University of Manitoba. He received a B.Sc. in zoology and a Master of Natural Resources Management from the University of Manitoba, and a Ph.D. in wildlife biology from Colorado State University. Dr. Baydack's research focus is on developing practical applications of ecosystem management for large-scale landscapes to conserve biodiversity. Recent emphasis has centred on habitat manipulation techniques to increase reproductive success of various species and determination of the effects of human activities on species survival, especially in agricultural and forested ecosystems. Several of these recent initiatives have been developed outside of North America.

He is an active member of The Wildlife Society and has been recognized as a Certified Wildlife Biologist (CWB). He is also president of the North American Grouse Partnership, a member of ECO Canada's steering committee for the National Environmental Program, and a member of the Manitoba Endangered Species Advisory Committee. He has received the prestigious Hamerstrom Award from the Prairie Grouse Technical Council for his significant contributions to the management and research on prairie grouse in North America, and has been recognized by the University of Manitoba as a recipient of the President's Outreach Award for excellence in extending his work at the university into the community.

CONTENTS *in Brief*

CONTENTS

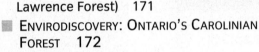

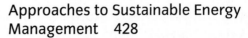

VISUALIZING FEATURES

PROCESS DIAGRAMS

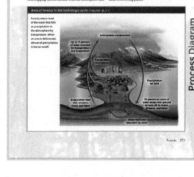

The Environment 1

DEFINING OUR ENVIRONMENT

Earth's abundant natural resources have provided the backdrop for a parade of living things to evolve. Today, millions of species inhabit the planet.

This assemblage of living and nonliving things constitutes our environment. Environments can be defined for a very minute or local area or for a broad or global scale—but in any case, the term refers to the totality of surrounding circumstances or conditions for a defined location. Environments obviously can mean different things to different people, to different cultures, to different time frames, and even to different organisms. But the set of conditions in any "environment" will be unique, and likely more complex and diverse than can be imagined!

About 100,000 years ago—a mere blip in Earth's 4.5-billion-year history—an evolutionary milestone occurred with the appearance of modern humans in Africa. Over time, our population grew, we expanded our range throughout the planet, and we increasingly impacted the environment with our presence and our technologies. Our intellectual capacity has even made it possible for us to venture into space, allowing us a view of the uniqueness of our planet in the solar system (see photograph).

In many ways these technologies have made life better, at least for those of us who live in developed nations. At the same time, much in our world indicates that we are headed for environmental changes that could harm human well-being.

The human species is a significant agent of environmental change on Earth. But the human species is creative, unique, adaptable, and capable of meeting these considerable challenges that have been placed before us.

This book introduces the major environmental problems that humans have created, but more importantly, it considers useful ways to address these issues and to minimize human impact on our planet. We cannot afford to ignore the environment because our lives and the lives of future generations depend on it. As a wise proverb says, "We have not inherited the world from our ancestors; we have only borrowed it from our children."

NATIONAL GEOGRAPHIC

3

Human Impacts on the Environment

LEARNING OBJECTIVES

Distinguish between highly developed countries, moderately developed countries, and less developed countries.

Relate human population size to natural resources and resource consumption.

Distinguish between people overpopulation and consumption overpopulation.

Describe the three factors that are most important in determining human impact on the environment.

The satellite photograph in **FIGURE 1.1** is a portrait of about 436 million people. The tiny specks of light that represent cities and large metropolitan areas, such as Toronto and Montreal in the northeast, are ablaze with light.

Earth's central environmental problem, which links all others together, is that there are many people, and the number, both in North America and worldwide, continues to grow. In 1999 the human population as a whole passed a significant milestone: 6 billion individuals. Not only is this number incomprehensibly large, but our population has grown this large in a very brief span of time. In 1960 the human population was only 3 billion (**FIGURE 1.2**). By 1975 there were 4 billion people, and by 1987 there were 5 billion. The 6.792 billion people who inhabit our planet (as of October 2009) consume vast quantities of food and water, use a great deal of energy and raw materials, and produce much waste.

Despite most countries' involvement with family planning, population growth rates will not change overnight. Several billion people will be added to the world in the 21st century, so even if we remain concerned about overpopulation and even if our solutions are very effective, the coming decades will see many problems.

poverty A condition in which people are economically unable to meet their basic needs for food, water, clothing, shelter, education, and health services.

On a global level, nearly one in four people lives in extreme **poverty**. By one measure, living in poverty is defined as having a per person income of less than $2 per day, expressed in U.S. dollars adjusted for purchasing power. About 3.5 billion people—more than half of the world's population—currently live at this level of poverty. Poverty is associated with a short life expectancy, illiteracy, and inadequate access to health services, safe water, and balanced nutrition. Although poverty is most often associated with other parts of the world, some communities in Canada, particularly in northern areas, may exhibit the above characteristics (**FIGURE 1.3**).

The world population may stabilize toward the end of the 21st century, given the family planning efforts that are currently under way. Population experts at the U.S. Population Reference Bureau have noticed a decrease in the birth rate to a current average of 2.7 children per woman, and this rate is projected to continue to decline in coming decades. Projections for the world population at the end of the 21st century have been made and range from about 7.9 billion to 10.9 billion.

No one knows whether Earth can support so many people indefinitely. Finding ways for it to do so represents one of the greatest challenges of our times. Among the tasks to be accomplished is feeding a world population considerably larger than the present one without destroying the biological communities that support us. The quality of life available to our children and grandchildren will depend to a large extent on our ability to achieve this goal.

A factor as important as population size is a population's level of **consumption**, which is the human use of material and energy. Consumption is intimately connected to a country's **economic growth**, the expansion in output of a nation's goods and services. The world's economy is growing at an enormous rate, yet this growth is unevenly distributed across the nations of the world.

Satellite view of North America at night FIGURE 1.1

This image shows most major cities and metropolitan areas in Canada, the United States, Mexico, and Central America.

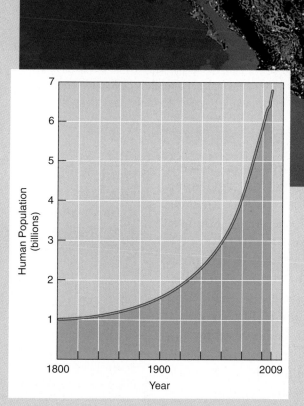

Human population numbers, 1800 to present

FIGURE 1.2

It took thousands of years for the human population to reach 1 billion (in 1800), 130 more years to reach 2 billion (1930), 30 more to reach 3 billion (1960), 15 more to reach 4 billion (1975), 12 more to reach 5 billion (1987), and 12 more again to reach 6 billion (1999).

Poverty FIGURE 1.3

The human population problem requires not only a stabilization of population numbers, but also an improvement of the economic conditions of people living in extreme poverty.

A A Japanese family from Tokyo with their possessions. People in highly developed countries consume a disproportionate share of natural resources.

B A Mexican family from Guadalajara with their possessions. Economic development in this moderately developed country has allowed many people to enjoy a middle-class lifestyle.

THE GAP BETWEEN RICH AND POOR COUNTRIES

highly developed countries Countries with complex industrialized bases, low rates of population growth, and high per person incomes.

moderately developed countries Countries with medium levels of industrialization and per person incomes lower than those of highly developed countries.

Generally speaking, countries are divided into rich (the "haves") and poor (the "have-nots"). Rich countries are known as **highly developed countries**. Canada, the United States, Japan, and most of Europe, which represent about 20 percent of the world's population, are highly developed countries (FIGURE 1.4A).

Poor countries, in which about 80 percent of the world's population live, fall into two subcategories: moderately developed and less developed. Mexico, Turkey, South Africa, and Thailand are examples of **moderately developed** countries (FIGURE 1.4B). Examples of **less developed countries** (LDCs) include Bangladesh, Mali, Ethiopia, and Laos (FIGURE 1.4C). Cheap, unskilled labour is abundant in LDCs, but capital for investment is scarce. Most economies of LDCs are agriculturally based, often on only one or a few crops. As a result, crop failure or a low world market value for that crop is catastrophic to the economy. Hunger, disease, and illiteracy are common in LDCs.

less developed countries Countries with low levels of industrialization, very high rates of population growth, very high infant mortality rates, and very low per person incomes compared with highly developed countries.

POPULATION, RESOURCES, AND THE ENVIRONMENT

People living in Canada and other highly developed countries consume many more resources per person than do people living in developing countries. This high

☐ A typical family from Kouakourou, Mali, with all their possessions. The rapidly increasing number of people in less developed countries may overwhelm their natural resources, even though individual resource requirements may be low.

rate of resource consumption affects the environment at least as much as the explosion in population that is occurring in other parts of the world.

We can make two useful generalizations about the relationships among population growth, consumption of natural resources, and environmental degradation. One, the amount of resources essential to an individual's survival is small, but rapid population growth (often found in developing countries) tends to overwhelm and deplete a country's soils, forests, and other natural resources. Two, in highly developed nations, individual demands on natural resources are far greater than the requirements for mere survival. To satisfy their desires as well as their basic needs, many people in more affluent nations deplete resources and degrade the global environment by high consumption of items such as televisions, jet skis, and cell phones.

Types of resources

Natural resources (economically referred to as land or raw materials) are naturally forming substances that are considered valuable in their relatively unmodified or "natural" form. When examining the effects of population on the environment, it is important to distinguish between nonrenewable and renewable natural resources.

Nonrenewable resources include minerals (such as aluminum, tin, and copper) and fossil fuels (coal, oil, and natural gas). Natural processes do not replenish nonrenewable resources within a reasonable duration on the human time scale. Fossil fuels, for example, take millions of years to form.

In addition to a nation's population and its level of resource use, several other factors affect how nonrenewable resources are used—including how efficiently the resource is extracted and processed and how much of it is required or consumed. Nonetheless, the inescapable fact is that Earth has a finite supply of nonrenewable resources that sooner or later will be exhausted. In time, technological advances may help find or develop substitutes for nonrenewable resources. Slowing the rate of population growth and resource consumption will help us buy time to develop such alternatives.

Some examples of **renewable resources** are trees, fish, fertile agricultural soil, and fresh water. Nature replaces these resources fairly rapidly. Forests, fisheries, and agricultural land are particularly important renewable resources in developing countries because they provide food. Indeed, many people in developing countries and elsewhere are subsistence farmers who harvest just enough food for their families to survive.

Rapid population growth can cause renewable resources to be overexploited, for example when large numbers of people must grow crops on land that is inappropriate for farming—such as mountain slopes or tropical rain forests. Although this practice may provide a short-term solution to the need for food, it causes

natural resources Naturally forming substances that are considered valuable in their relatively unmodified or "natural" form.

nonrenewable resources Natural resources that are present in limited supplies and are depleted as they are used.

renewable resources Resources that are replaced by natural processes and that can be used forever, provided they are not overexploited in the short term.

Refugee camp

A young Iraqi girl cries in front of a tent. This photograph, taken in 2007, shows the al-Husseiniya district in Baghdad, where people have fled from the violence in Diyala province.

Global Locator

Sometimes economic, political, or environmental problems force people to leave their homelands en masse. According to United Nations High Commissioner for Refugees (UNHCR), in 2007 there were more than 21 million international refugees and internally displaced persons. *International refugees* are people who moved from one country to another, whereas *internally displaced persons* are people who were forced to move but did not cross a border into another country.

Environmental degradation is recognized as a major cause of refugee flight. Over the past several decades, deterioration of agricultural lands, extended droughts, and other natural disasters have displaced perhaps 30 million people worldwide. The large number of refugees strains the economies and natural resources of the regions into which the refugees migrate.

Many people attempting to flee their homelands and move elsewhere are not welcomed with open arms. The industrial world's reluctance to embrace refugees stems from several sources:

- Domestic economic problems
- The high costs of processing asylum applications
- Scepticism about whether refugees are actually fleeing political persecution and are not simply "economic migrants"

Failure to address the high demand for asylum leads to a lower quality of life for prospective refugees:

- Refugee camps are growing more violent, with sexual and gender-based violence against women refugees particularly pronounced.
- Human trafficking (charging asylum-seekers money to illegally transport them across borders) is on the rise, with its accompanying financial exploitation, human rights abuses, and violence.
- Less developed nations take their cue from highly developed ones, refusing to accept refugees when more prosperous nations will not.

problems in the long run because when these lands are cleared for farming, their agricultural productivity declines rapidly, and severe environmental deterioration occurs. Renewable resources, then, are *potentially* renewable. They must be used in a manner that gives them time to replace or replenish themselves.

The effects of population growth on natural resources are particularly critical in developing countries. The economic growth of developing countries is frequently tied to the exploitation of their natural resources, often for export to highly developed countries. Developing countries are faced with the difficult choice of exploiting natural resources to provide for their expanding populations in the short term (that is, to pay for food or to cover debts) or conserving those resources for future generations. It is instructive to note that some of the economic growth and development in highly developed nations came about through the exploitation—and in some cases the destruction—of their resources. Continued growth and development in highly developed countries is increasingly reliant on the importation of these resources from less developed countries.

Poverty is tied to the effects of population pressures on natural resources and the environment. Poor people in developing countries may find themselves trapped in a vicious cycle of poverty. They use environmental resources unwisely for short-term gain (that is, to survive), but this exploitation degrades the resources and diminishes long-term prospects of economic development.

Population size and resource consumption
Resource issues are clearly related to population size—intuitively, we know that more people use more resources. But an equally important factor is a population's resource consumption. Consumption is both an economic and a social act. Consumption provides the consumer with a sense of identity as well as status among peers. The advertising industry, especially in highly developed countries, promotes consumption as a way to achieve happiness. We are encouraged to spend, to consume.

People in highly developed countries can be extravagant and wasteful consumers; their use of resources is greatly out of proportion to their numbers. A single child

■ **people overpopulation**
A situation in which there are too many people in a given geographic area.

■ **consumption overpopulation**
A situation in which each individual in a population consumes too large a share of resources.

born in a highly developed country has a greater impact on the environment and on resource depletion than perhaps 20 children born in a developing country. Many natural resources are needed to provide the automobiles, air conditioners, disposable diapers, cell phones, DVD players, computers, clothes, newspapers, athletic shoes, furniture, boats, and other "comforts" of life in highly developed nations. Thus, the disproportionately large consumption of resources by people in highly developed countries affects natural resources and the environment as much as or more than the population explosion in the developing world.

People overpopulation and consumption overpopulation A country is overpopulated if the level of demand on its resource base results in damage to the environment. In comparing human impact on the environment in developing and highly developed countries, we see that a country can be overpopulated in two ways. **People overpopulation** occurs when the environment is worsening because there are too many people, even if those people consume few resources per person. People overpopulation is the current problem in many developing nations. In contrast, **consumption overpopulation** results from the consumption-oriented lifestyles that exist in highly developed countries. The effects of consumption overpopulation on the environment are the same as those of people overpopulation—pollution, resource depletion, and degradation of the environment.

Many affluent, highly developed nations, including Canada, suffer from consumption overpopulation. Highly developed nations represent only 20 percent of the world's population, yet they consume significantly more than half of its resources. According to the Worldwatch Institute, highly developed nations account for the lion's share of total resources consumed:

- 86 percent of aluminum used
- 76 percent of timber harvested
- 68 percent of energy produced
- 61 percent of meat eaten
- 42 percent of the fresh water consumed

Consumption overpopulation FIGURE 1.5

North American consumption is actively promoted in advertisements like those shown here in Dundas Square in Toronto. Highly developed nations like Canada consume more than 50 percent of the world's resources, produce 75 percent of its pollution and waste, and represent only 19 percent of the total population.

These nations also generate 75 percent of the world's pollution and waste (**FIGURE 1.5**).

Ecological footprints The **ecological footprint** has emerged as the world's premier measure of humanity's or an individual's demand on nature. It measures how much land and water area a human population or being requires to produce the resources it consumes and to absorb its wastes. Conceived in 1990 by Mathis Wackernagel and William Rees at the University of British Columbia, the ecological footprint is now widely used by scientists, businesses, governments, individuals, and institutions working to monitor ecological resource use and advance sustainable development. The concept is especially useful in enabling individuals to gauge the impact of their lifestyle on the resources of our planet, thus empowering each of us to share the responsibility of Earth's well-being.

The World Wildlife Fund's Living Planet Report regularly compares the ecological footprint for all countries in the world. The comparison is based on the average amount of productive land and ocean needed to supply a person in every country with food, energy, water, housing, transportation, and waste disposal. Data from 2005 show that in developing nations, such as India and Nigeria, the ecological footprint is about 1 hectare. In

Canada, the ecological footprint is about 7.1 hectares (and it is even higher in the United States at 9.4 hectares). If all 6.7 billion people in the world had the same lifestyle and level of consumption as an average person in Canada, and assuming no changes in technology, we would need three additional planets the size of Earth to accommodate us all (**FIGURE 1.6**).

In 2005, the global ecological footprint was 2.7 gha per person (a global hectare—gha—is a hectare with world-average ability to produce resources and absorb wastes). On the supply side, the total productive area, or biocapacity, was 2.1 gha per person. Humanity's footprint first exceeded the Earth's total biocapacity in the 1980s; this overshoot has been increasing since then. In 2005, demand was 30 percent greater than supply.

New consumers in developing countries As developing nations increase their economic growth and improve their standard of living, more and more people in those countries purchase consumer goods. By the early 2000s, more new cars were sold annually in Asia than in North America and western Europe combined. These new consumers may not consume at the high level of someone in a highly developed nation, but their consumption has increasingly adverse effects on the environment. For example, air pollution

Ecological footprints FIGURE 1.6

A In LDCs such as India, about 1 hectare meets the resource requirements of an average person; the ecological footprint of each individual in a highly developed country such as Canada is about 7 hectares.

India   Canada

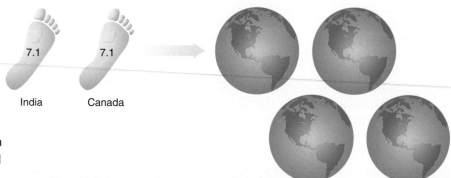

India   Canada

B If everyone in the world had the same level of consumption as an average person in Canada, it would take the resources and land area of four Earths.

from traffic in urban centres in developing countries is bad and getting worse every year. Millions of dollars are lost to health problems caused by air pollution in these cities.

Population, consumption, and environmental impact When you turn on the tap to brush your teeth in the morning, you probably do not think about where the water comes from or about the environmental consequences of removing it from a river or the ground. Likewise, most North Americans do not think about where the energy comes from when they flip on a light switch or start their car, van, or truck. Many of us may not pay attention to the obvious fact that all the materials that make up the products we use every day come from Earth, nor do we grasp that these materials eventually are returned to Earth, mainly in sanitary landfills.

Such human impacts on the environment are difficult to assess. One way to estimate them is to use the three factors most important in determining environmental impact (I):

- The number of people (P)
- The affluence per person, which is a measure of the consumption, or amount of resources used per person (A)

- The environmental effects (resources needed and wastes produced) of the technologies used to obtain and consume the resources (T)

This method of assessment is usually referred to as the IPAT equation: $I = P \times A \times T$.

Biologist Paul R. Ehrlich and physicist John P. Holdren first proposed the IPAT model in the 1970s. It shows the mathematical relationship between environmental impacts and the forces that drive them. To determine the environmental impact of carbon dioxide (CO_2) emissions from motor vehicles, for example, multiply the population by the number of cars per person (affluence or consumption per person) by the average annual CO_2 emissions per year (technological impact). This model demonstrates that although improving motor vehicle efficiency and developing cleaner technologies will reduce pollution and environmental degradation, a larger reduction will result if population and per person consumption are also controlled.

Although useful, the IPAT equation must be interpreted with care, in part because we often do not understand all of the environmental impacts of a particular technology. Motor vehicles, for example, are linked not only to global warming from CO_2 emissions but also to air pollution (tailpipe exhaust), water pollution (improperly

The NIMBY response is an unavoidable complication in addressing many environmental issues. NIMBY stands for "not in my backyard." As soon as people hear that a power plant, an incinerator, or a hazardous waste disposal site may be situated nearby, the NIMBY response rears its head. Part of the reason NIMBYism is so prevalent is that, despite the assurances experts give that a site will be safe, no one can guarantee complete safety and no possibility of an accident.

A sister response to NIMBY is NIMTOO, which stands for "not in my term of office." Politicians who wish to get re-elected are sensitive to their constituents' concerns and are not likely to support the construction of power plants or waste disposal sites in their districts.

Given human nature, the NIMBY and NIMTOO responses are not surprising. But emotional reactions, however reasonable they may be, do little to constructively solve complex environmental problems. Consider the siting and disposal of radioactive waste from nuclear power plants. There is universal agreement that we need to safely locate a nuclear plant and isolate radioactive waste until it decays enough to cause little danger. But NIMBY and NIMTOO supporters, with their associated demonstrations, lawsuits, and administrative hearings, prevent us from effectively dealing with radioactive waste disposal. Every potential plant location or disposal site is near someone's home, in some politician's riding. Most people agree that our generation has the responsibility to dispose of hazardous waste, but the problem is that we want to put it in some other province, in someone else's backyard. Arguing against any disposal scheme that is proposed will simply result in letting the waste remain where it is now. Although this may be the only politically acceptable solution, it is unacceptable from a safety viewpoint.

Steam rises from two of the cooling towers of a nuclear power plant. Many nuclear power plants store highly radioactive spent fuel on site because there is currently no place to safely dispose of it.

disposed motor oil and antifreeze), stratospheric ozone depletion (from leakage of air conditioner coolants), and solid waste (disposal of automobiles in sanitary landfills). There are currently about 550 million cars on the planet, and the number is rising.

The three factors in the IPAT equation are always changing in relation to each other. For example, consumption of a particular resource may increase, but technological advances may decrease the environmental impact of the increased consumption. Consumer trends and choices also affect environmental impact. Because of such trends and uncertainties, the IPAT equation is of limited use when making long-term predictions.

But the IPAT equation is valuable in helping to identify what we don't know or understand about consumption and its environmental impact. For example, which kinds of consumption have the greatest destructive impact on the environment? Which regions of a country are responsible for the greatest environmental disruption? How can we alter the activities in these areas to achieve a greater societal good? It will take years to address such questions, but the answers should help decision makers in business and government formulate policies that will alter consumption patterns in an environmentally responsible way. The ultimate goal should be to make consumption sustainable so that humanity's current practices do not compromise the ability of future generations to use and enjoy the riches of our planet.

To summarize, as human numbers and consumption increase worldwide, so does humanity's impact on Earth, posing new challenges to us all. Success in achieving sustainability in population size and consumption will require the cooperation of all the world's peoples.

CONCEPT CHECK **STOP**

How is human population growth related to natural resource depletion and environmental degradation?

What is the difference between people overpopulation and consumption overpopulation?

What are the three most important factors in determining human impact on the environment?

Environmental Sustainability and Sustainable Development

LEARNING OBJECTIVES

Define *environmental sustainability* and *sustainable development*.

Identify human behaviours that threaten environmental sustainability.

Outline some of the complexities associated with the concept of sustainable consumption.

Only by understanding the environment and how it works can we make the necessary decisions to protect it. Only by valuing all our precious natural and human resources can we hope to build a sustainable future.

—Kofi Anan,
former Secretary-General of the United Nations

One of the most important concepts in this text is **environmental sustainability**. *Sustainability* implies that the environment will function indefinitely without going into a decline from the stresses that human society imposes on natural systems (such as fertile soil, water, and air). Environmental sustainability applies to many levels, including the individual, communal, regional, national, and global levels.

environmental sustainability
The ability to meet humanity's current needs without compromising the ability of future generations to meet their needs.

DAVID SUZUKI

There are many advocates of environmental sustainability out there, but few have had the impact that David Suzuki has had and continues to have on influencing individuals, businesses, and governments to make sustainability a priority. A world-renowned scientist, environmentalist, and broadcaster, Suzuki has been educating the public on nature and the environment for over 40 years. His unique ability to communicate complex scientific concepts to the public in a clear and engaging manner has made his contributions to environmental education and conservation both in Canada and around the world matched by few.

In 1990, he co-founded the David Suzuki Foundation, a non-profit organization committed to educating and researching sustainable ways to protect nature's diversity and to working toward a balance between human needs and the Earth's ability to sustain life. Suzuki has authored over 40 books and has created several television series (*A Planet for the Taking, The Secret of Life,* and *The Brain*), each winning critical acclaim. Over and above these accomplishments, Suzuki is well known as the host of *The Nature of Things,* the scientific television program produced by the Canadian Broadcasting Corporation (CBC), and for his long-running CBC radio series *Quirks and Quarks.* Suzuki is the winner of numerous awards, including the prestigious Tory Peterson Memorial Medal from Harvard University, a UNESCO prize for science, a United Nations Environment Programme medal, and an induction as a Companion of the Order of Canada.

Suzuki's outspoken passion and commitment to addressing serious environmental concerns have brought forth social and political awareness and change. His contributions have placed Canada on the world stage in scientific leadership and have positioned him as a recognized world leader in sustainable ecology.

For more information on David Suzuki and the David Suzuki Foundation, please go to www.davidsuzuki.org.

Environmental sustainability is based upon the following ideas:

- We must consider the effects of our actions on the well-being of the natural environment, including all living things.
- Earth's resources are not present in infinite supply. We must live within limits that let renewable resources regenerate for future needs.
- We must understand all the costs to the environment and to society of products we consume.
- We must each share in the responsibility for environmental sustainability.

Many environmental experts think that human society is not operating sustainably because of the following human behaviours (see **FIGURE 1.7** and **FIGURE 1.8** on pages 16–17):

- We are using nonrenewable resources such as fossil fuels as if they were present in unlimited supplies.

- We are using renewable resources such as fresh water and forests faster than they are replenished naturally.
- We are polluting the environment—the land, rivers, ocean, and atmosphere—with toxins as if the capacity of the environment to absorb them were limitless.
- Our numbers continue to grow despite Earth's finite ability to feed and sustain us and to absorb our wastes.

If left unchecked, these activities may threaten the life-support systems of Earth to the extent that recovery is impossible. Our first goal should be to critically evaluate which changes our society is willing to make.

SUSTAINABLE DEVELOPMENT

At first glance, the issues may seem simple. Why don't we just stop the overconsumption, population growth, and pollution? The solutions are more complex than they

A Cod off the east coast of Canada are overexploited.

C Annual world oil consumption, 1950 to 2006. This increase is unsustainable.

B A clear-cut forest near Hinton, Alberta. Forest harvest is unsustainable in many parts of the world.

D An oil refinery produces noxious emissions. Air pollution control devices reduce these toxic emissions but often are not used because they are expensive. Photographed in Fort McMurray, Alberta.

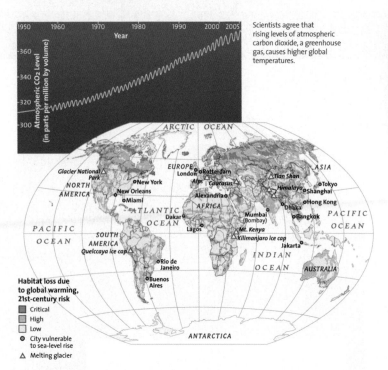

Scientists agree that rising levels of atmospheric carbon dioxide, a greenhouse gas, causes higher global temperatures.

Clear-cutting large tracts of timber, without sustainable replanting, contributes to deforestation, erosion, and loss of habitat.

Habitat loss due to global warming, 21st-century risk
- ▢ Critical
- ▢ High
- ▢ Low
- ⊙ City vulnerable to sea-level rise
- △ Melting glacier

Vanishing forest
- ▢ Frontier forest (large, mostly virgin forest)
- ▢ Degraded forest
- ▢ Frontier forest 8000 years ago

▲ GLOBAL WARMING

Temperatures across the world are increasing at a rate not seen at any other time in the last 10,000 years. Although climate variation is a natural phenomenon, human activities that release carbon dioxide and other greenhouse gases into the atmosphere—industrial processes, fossil fuel consumption, deforestation, and land use change—are contributing to this warming trend. Scientists predict that if this trend continues, one-third of plant and animal habitats will be dramatically altered and more than one million species will be threatened with extinction in the next 50 years. And even small increases in global temperatures can melt glaciers and polar ice sheets, raising sea levels and flooding coastal cities and towns.

▲ DEFORESTATION

Of the 13 million hectares of forest lost each year, mostly to make room for agriculture, more than half are in South America and Africa, where many of the world's tropical rainforests and terrestrial plant and animal species can be found. Loss of habitat in such species-rich areas takes a toll on the world's biodiversity. Deforested areas also release, instead of absorb, carbon dioxide into the atmosphere, contributing to global climate change. Deforestation can also affect local climates by reducing evaporative cooling, leading to decreased rainfall and higher temperatures.

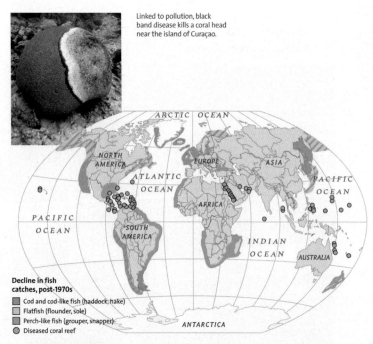

Linked to pollution, black band disease kills a coral head near the island of Curaçao.

In China's Tengger Desert, sand threatens to engulf nearby railroad lines despite a grid of straw meant to help stabilize the drifts.

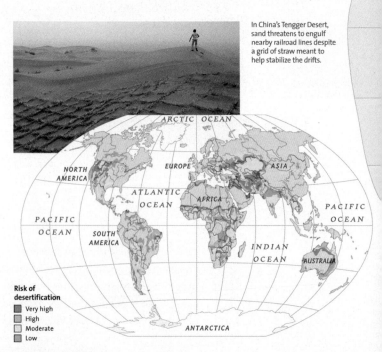

Decline in fish catches, post-1970s
- ▢ Cod and cod-like fish (haddock, hake)
- ▢ Flatfish (flounder, sole)
- ▢ Perch-like fish (grouper, snapper)
- ⊙ Diseased coral reef

Risk of desertification
- ▢ Very high
- ▢ High
- ▢ Moderate
- ▢ Low

▲ THREATENED OCEANS

Oceans cover more than two-thirds of the Earth's surface and are home to at least half of the world's biodiversity, yet they are the least understood ecosystems. The combined stresses of overfishing, pollution, increased carbon dioxide emissions, global climate change, and coastal development are having a serious impact on the health of oceans and ocean species. Over 70% of the world's fish species are depleted or nearing depletion, and 50% of coral reefs worldwide are threatened by human activities.

▲ DESERTIFICATION

Climate variability and human activities, such as grazing and conversion of natural areas to agricultural use, are leading causes of desertification, the degradation of land in arid, semi-arid, and dry subhumid areas. The environmental consequences of desertification are great—loss of topsoil, increased soil salinity, damaged vegetation, regional climate change, and a decline in biodiversity. Equally critical are the social consequences—more than 2 billion people live in and make a living off these dryland areas, covering about 41% of Earth's surface.

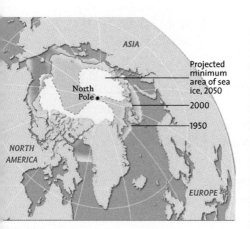

POLAR ICE CAP

ver the last 50 years, the xtent of polar sea ice has oticeably decreased. Since 970 alone, an area larger han Norway, Sweden, and Denmark combined has melted. This trend is predicted to accelerate as temperatures rise in the Arctic and across the globe.

Labels on map: ASIA, North Pole, NORTH AMERICA, EUROPE, Projected minimum area of sea ice, 2050, 2000, 1950

GeoBytes

ACIDIFYING OCEANS
Oceans are absorbing an unprecedented 20 to 25 million tonnes of carbon dioxide each day, increasing the water's acidity.

ENDANGERED REEFS
Some 95% of coral reefs in Southeast Asia have been destroyed or are threatened.

RECORD TEMPERATURES
The 1990s were the warmest decade on record in the last century.

WARMING ARCTIC
While the world as a whole has warmed nearly 0.6°C over the last hundred years, parts of the Arctic have warmed 4 to 5 times as much in only the last 50 years.

DISAPPEARING RAINFORESTS
Scientists predict that the world's rainforests will disappear within the next one hundred years if the current rate of deforestation continues.

OIL POLLUTION
Nearly 1.3 million tonnes of oil seep into the world's oceans each year from the combined sources of natural seepage, extraction, transportation, and consumption.

ACCIDENTAL DROWNINGS
Entanglement in fishing gear is one of the greatest threats to marine mammals.

A FAREWELL TO FROGS?
Worldwide, almost half of the 5700 named amphibian species are in decline.

ENVIRONMENT

With the growth of scientific record keeping, observation, modelling, and analysis, our understanding of Earth's environment is improving. Yet even as we deepen our insight into environmental processes, we are changing what we are studying. At no other time in history have humans altered their environment with such speed and force. Nothing occurs in isolation, and stress in one area has impacts elsewhere. Our agricultural and fishing practices, industrial processes, extraction of resources, and transportation methods are leading to extinctions, destroying habitats, devastating fish stocks, disturbing the soil, and polluting the oceans and the air. As a result, biodiversity is declining, global temperatures are rising, polar ice is shrinking, and the ozone layer continues to thin.

▼ **POLLUTION**
No corner of the Earth is immune to pollution, be it in the air, soil, or water. Concentrations of pollution can be found in the industrial centres of North America, Europe, and, increasingly, Asia—and areas downwind or downstream from them. Shipping routes are sources of pollution, from oil spills to garbage dumpings.

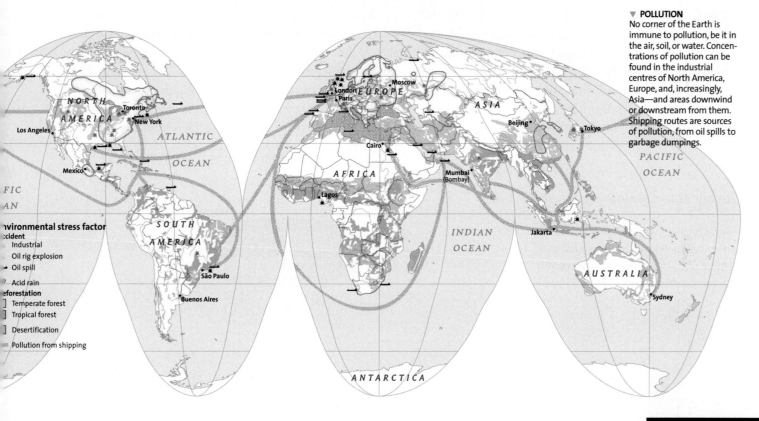

nvironmental stress factor
ccident
▪ Industrial
 Oil rig explosion
✦ Oil spill
 Acid rain
eforestation
▪ Temperate forest
▪ Tropical forest
▪ Desertification
 Pollution from shipping

Map labels: NORTH AMERICA, Los Angeles, Toronto, New York, Mexico, ATLANTIC OCEAN, SOUTH AMERICA, São Paulo, Buenos Aires, London, Paris, EUROPE, Moscow, Cairo, AFRICA, Lagos, Mumbai (Bombay), Beijing, Tokyo, ASIA, Jakarta, INDIAN OCEAN, PACIFIC OCEAN, AUSTRALIA, Sydney, ANTARCTICA, FIC AN

OZONE DEPLETION

rst noted in the mid-1980s, e springtime "ozone hole" ver the Antarctic continues grow. With sustained forts to restrict chlorofluo- carbons (CFCs) and other zone-depleting chemicals, ientists have begun to e what they hope is a velling off in the rate of depletion. Stratospheric ozone shields the Earth from the sun's ultraviolet radiation. Thinning of this protective layer puts people at risk for skin cancer and cataracts. It can also have devastating effects on the Earth's biological functions.

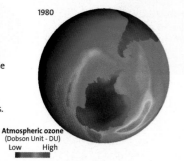
1980

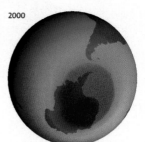
2000

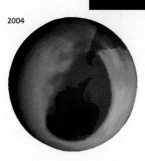
2004

Atmospheric ozone
(Dobson Unit - DU)

Low   High

may initially seem, in part because of various interacting ecological, societal, and economic factors.

Environmental sustainability is a concept that people have discussed for many years. As early as the 1970s, "sustainability" was used to describe an economy in equilibrium with basic ecological support systems. Some ecologists have pointed to the "limits of growth" and presented the alternative of a "steady state economy" in order to address environmental concerns. This concept uses the notion of valuing ecological goods and services—the benefits arising from the ecological functions of healthy ecosystems. Such benefits accrue to all living organisms, including animals and plants, rather than to humans alone. Recently, there has been a growing recognition of the importance to society that ecological goods and services provide for health, social, cultural, and economic needs.

Our Common Future, the 1987 report of the UN World Commission on Environment and Development, (also known as the Brundtland Report and the Brundtland Commission, named after the Chairperson of the UN Commission and former Prime Minister of Norway, Gro Harlem Brundtland) presented the closely related concept of **sustainable development** (FIGURE 1.9).

The report linked the environment's ability to meet present and future needs to the state of technology and social organization existing at a given time and in a given place. The number of people, their degree of affluence (that is, their level of consumption), and their choices of technology all interact to produce the total effect of a given society, or of society at large, on the sustainability of the environment.

Even using the best technologies imaginable, Earth's productivity still has its limits, and our use of it cannot be expanded indefinitely. *Sustainable development can occur only within the limits of the environment.* To live within these limits, population growth must be held at a level that we can sustain, and the wealthy must first stabilize their use of natural resources and then reduce this use to a level that can be maintained. The world does not contain nearly enough resources to sustain everyone at the level of consumption that is enjoyed in, for example, Canada or Japan. Suitable strategies, however, do exist to reduce these levels of consumption without concurrently reducing the real quality of life.

■ **sustainable development**
Economic growth that meets the needs of the present without compromising the ability of future generations to meet their own needs.

Sustainable development FIGURE 1.9

Three factors—environmentally sound decisions, economically viable decisions, and socially equitable decisions—interact to promote sustainable development.

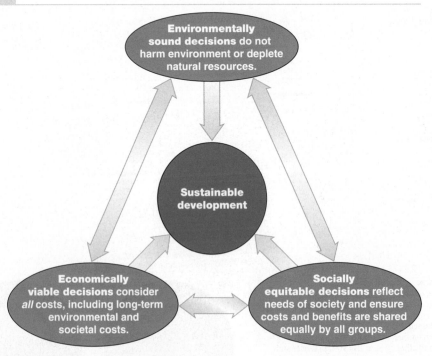

Environmentally **sound decisions** do not harm environment or deplete natural resources.

Sustainable development

Economically **viable decisions** consider *all* costs, including long-term environmental and societal costs.

Socially **equitable decisions** reflect needs of society and ensure costs and benefits are shared equally by all groups.

SUSTAINABLE CONSUMPTION

As we discussed earlier, consumption overpopulation is pollution and degradation of the environment that occurs when each individual in a population consumes too large a share of resources. Consumption overpopulation stems from the lifestyles of people living in highly developed nations. *Lifestyle* is interpreted broadly to include goods and services bought for food, clothing, housing, travel, recreation, and entertainment. In evaluating consumption overpopulation, all aspects of the production, use, and disposal of these goods and services are taken into account, including environmental costs. Such an analysis provides a sense of what it means to consume sustainably versus unsustainably.

Sustainable consumption, like sustainable development, forces us to address whether our present actions undermine the long-term ability of the environment to meet the needs of future generations. Factors that affect sustainable consumption include population, economic activities, technology choices, social values, and government policies.

At the global level, sustainable consumption requires the eradication of poverty, which in turn requires that poor people in developing countries *increase* their consumption of certain essential resources (**FIGURE 1.10**). For their increased consumption to be sustainable, however, the consumption patterns of people in highly developed countries must change.

Widespread adoption of sustainable consumption will not be easy. It will require major changes in the

> **sustainable consumption** The use of goods and services in a way that satisfies basic human needs and improves the quality of life but that also minimizes resource use and preserves resources for the use of future generations.

The challenge of eradicating poverty in developing countries FIGURE 1.10

Global Locator INDIA

A Squatters live in dilapidated shacks under a water pipe in Delhi, India. Leaks in the pipe provide water for these people.

B Desperately poor pavement people eat, sleep, and raise their families on a street in Calcutta, India.

Car sharing FIGURE 1.11

More than 10,000 Torontonians have signed up for AutoShare, Toronto's longest-running car-sharing program.

consumption patterns and lifestyles of most people in highly developed countries. Some examples of promoting sustainable consumption include switching from motor vehicles to public transport and bicycles and developing durable, repairable, recyclable products.

People in highly developed nations are being introduced to a type of sustainable consumption known as simplicity, which recognizes that individual happiness and quality of life are not necessarily linked to the accumulation of material goods. People who embrace this notion of **voluntary simplicity** recognize that a person's values and character define that individual more than how many things he or she owns. This belief may require a change in behaviour as people purchase and use fewer items than they might have formerly. It is a commitment at the individual level to saving the planet for future generations.

One example of voluntary simplicity is daily use of public transportation or bicycles rather than automobiles. Increasingly, Canadian urban centres are developing more efficient public transportation systems to ease stress on the environment. Reducing the use of personal automobiles may also reduce the numbers of cars manufactured and the number of kilometres that an individual drives (FIGURE 1.11).

As people adopt new lifestyles, they must be educated so that they understand the reasons for changing practices that may be highly ingrained or traditional. Both formal and informal education are important in bringing about change and in contributing to sustainable consumption. If people understand the way the natural world functions, they can appreciate their own place in it and value sustainable actions.

Any long-term involvement in the condition of the world must start with individuals—our values, attitudes, and practices. Each of us makes a difference, and it is ultimately our collective activities that make the world what it is.

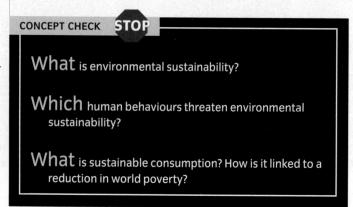

CONCEPT CHECK STOP

What is environmental sustainability?

Which human behaviours threaten environmental sustainability?

What is sustainable consumption? How is it linked to a reduction in world poverty?

Environmental Economics

NATIONAL GEOGRAPHIC

LEARNING OBJECTIVES

Explain how economics is related to natural capital. Make sure you include sources and sinks.

Give two reasons why the national income accounts are incomplete estimates of national economic performance.

Distinguish between the following economic terms: marginal cost of pollution, marginal cost of pollution abatement, and optimum amount of pollution.

Economics is the study of how people use their limited resources to try to satisfy their unlimited wants. Economists try to understand the consequences of the ways in which people, businesses, and governments allocate their resources. Seen through an economist's eyes, the world is one large marketplace, where resources are allocated to a variety of uses, and where goods (a car, a pair of shoes, a hog) and services (a haircut, a museum tour, an education) are consumed and paid for. In a free market, supply and demand determine the price of a good. If something in great demand is in short supply, its price will be high. High prices encourage suppliers to produce more of a good or service, as long as the selling price is higher than the cost of producing the good or service.

Economies depend on the natural environment and its processes as *sources* for raw materials and *sinks* for waste products (**FIGURE 1.12**). Both sources and sinks contribute to **natural capital**. According to economists, the environment provides natural capital for human production and consumption. Resource degradation and pollution represent the overuse of natural capital. *Resource degradation* is the overuse of sources, and *pollution* is the overuse of sinks; both threaten our long-term economic future.

> **natural capital**
> Earth's resources and processes that sustain living organisms, including humans; includes minerals, forests, soils, groundwater, clean air, wildlife, and fisheries.

Visualizing

Economics and the environment FIGURE 1.12

Economies Depend on Natural Capital Sources for Raw Materials and Sinks for Waste Products

Natural Capital: Sources are the part of the environment from which materials move.

Raw Materials

Economy Products

Production — Products and money flow between production and consumption. — Consumption

Money

Natural Capital: Sinks are the part of the environment that receives input of materials.

Waste Products

Approximately 3.3 billion cubic metres of wood is harvested annually; 17 percent of that is used for making paper.

Per person annual consumption of paper in Canada is 263 kg. In 2006, net earnings of the world's 100 largest forest and paper companies totalled $15 billion.

Paper and paper products account for more than one-third of Canada's total waste. Only one-quarter of Canada's paper waste is recycled.

NATIONAL INCOME ACCOUNTS AND THE ENVIRONMENT

Much of our economic well-being flows from natural assets—such as land, rivers, the ocean, oil, timber, and the air we breathe—rather than human-made assets.

national income accounts A measure of the total income of a nation's goods and services for a given year.

Ideally, for the purposes of economic and environmental planning, the **national income accounts** should include both the use and misuse of natural resources and the environment. Two measures used in national income accounting are *gross domestic product (GDP)* and *net domestic product (NDP)*. Both GDP and NDP provide estimates of national economic performance that are used to make important policy decisions.

Unfortunately, current national income accounting practices provide an incomplete or inaccurate measure of income because they do not incorporate environmental factors. Two important conceptual problems exist with the way the national income accounts currently handle the economic use of natural resources and the environment, particularly natural resource depletion and the costs and benefits of pollution control. Better accounting for environmental quality would help address whether, for any given activity, the benefits (both economic and environmental) exceed the costs.

Natural resource depletion

If a manufacturing firm produces some product (output) but in the process wears out a portion of its plant and equipment, that firm's output is counted as part of GDP, but the depreciation of capital is subtracted in the calculation of NDP. Thus NDP is a measure of the net production of the economy after a deduction for used-up capital. In contrast, when an oil company drains oil from an underground field, the value of the oil produced is counted as part of the nation's GDP, but no offsetting deduction to NDP is made to account for the nonrenewable resources used up (**FIGURE 1.13**).

AN ECONOMIST'S VIEW OF POLLUTION

An important aspect of the operation of a free-market system is that the person consuming a product should pay for all the cost of producing it. However, production

or consumption of a product often has an **external cost**.

A product's market price does not usually reflect an external cost—that is, the buyer or seller doesn't pay for the external cost. As a result, the market system generally does not operate in the most efficient way.

external cost A harmful environmental or social cost that is borne by people not directly involved in selling or buying a product.

Consider the following example of an external cost. Suppose an industry makes a product and, in doing so, also releases a pollutant into the environment. The price of the product when it is sold reflects the cost of making it but not the cost of the pollutant's damage to the environment. This damage is the external cost of the product. Because this damage is not included in the product's price and because the consumer may not know that the pollution exists or that it harms the environment, the cost of the pollution has no impact on the consumer's decision to buy the product. As a result, consumers of the product may buy more of it than they would if its true cost, including the cost of pollution, were known or reflected in the selling price.

The failure to add the price of environmental damage to the cost of products generates a market force that increases pollution. From the perspective of economics, then, one of the causes of the world's pollution problem

The Hibernia oil platform on the Grand Banks in the Atlantic Ocean FIGURE 1.13
The oil field being drained under the sea floor is part of the GDP. The fact that it will be drained dry someday is not taken into account.

is the failure to consider negative external costs in the pricing of goods.

How much pollution is acceptable?

To assign a proper price to pollution, economists first try to answer a basic question: How much pollution should we allow in our environment? Imagine two environmental extremes: a wilderness in which no pollution is produced but neither are goods, and a "sewer" that is completely polluted from excess production of goods. In our world, a move toward a better environment almost always entails a cost in terms of goods.

How do we, as individuals, as a country, and as part of the larger international community, decide where we want to be between the two extremes of a wilderness and a sewer? Economists analyze the marginal costs of environmental quality and of other goods to answer such questions. A **marginal cost** is the additional cost associated with one more unit of something. An example of marginal costs associated with pollution is the effect of pollution on human health and on organisms in the natural environment. (See **FIGURE 1.14** on page 24.)

The trade-off between environmental quality and more goods involves balancing marginal costs of two kinds: (1) the cost, in terms of environmental damage, of more pollution (the marginal cost of pollution) and (2) the cost, in terms of giving up goods, of eliminating pollution (the marginal cost of pollution abatement).

Determining the **marginal cost of pollution** involves assessing the risks associated with the pollution—for example, damage to health, property, or agriculture. Let's consider a simple example involving the marginal cost of sulphur dioxide, a type of air pollution produced during the combustion of fuels containing sulphur. Sulphur dioxide is removed from the atmosphere as acid rain, which causes damage to the environment, particularly aquatic ecosystems. Economists add up the harm of each additional unit of pollution—in this example, each tonne of sulphur dioxide added to the atmosphere. As the total amount of pollution increases, the harm of each additional unit usually also increases, and as a

marginal cost of pollution The added cost of an additional unit of pollution.

marginal cost of pollution abatement The added cost of reducing one unit of a given type of pollution.

cost-benefit diagram A diagram that helps policy makers make decisions about costs of a particular action and benefits that would occur if that action were implemented.

optimum amount of pollution The amount of pollution that is economically most desirable.

result, the curve showing the marginal cost of pollution slopes upward.

The **marginal cost of pollution abatement** tends to rise as the level of pollution declines. It is relatively inexpensive to reduce automobile exhaust emissions by half, but costly devices are required to reduce the remaining emissions by half again. For this reason, the curve showing the marginal cost of pollution abatement slopes downward.

A graph that shows marginal-cost curves plotted together is called a **cost-benefit diagram**. Economists use this type of diagram to identify the point at which the marginal cost of pollution equals the marginal cost of abatement—that is, the point where the two curves intersect. As far as economics is concerned, this point represents an **optimum amount of pollution**. At this optimum, the cost to society of having less pollution is offset by the benefits to society of the activity creating the pollution.

There are two major flaws in the economist's concept of optimum pollution. First, it is difficult to determine the true cost of environmental damage caused by pollution. Usually, there are many polluters and many affected individuals. Second, when economists add up pollution costs, they do not take into account the possible disruption or destruction of the environment. The web of relationships within the environment is extremely intricate and may be more vulnerable to pollution damage than is initially obvious. This is truly a case where the whole is much greater than the sum of its parts, and it is inappropriate for economists to simply add up the costs of lost elements in a polluted environment.

CONCEPT CHECK **STOP**

What is natural capital? How is economics related to natural capital?

Why are national income accounts incomplete estimates of total national economic performance?

Economists can analyze the marginal costs of pollution and other environmental issues to determine the ideal spot on the continuum between pristine wilderness (A) and extreme pollution, like smog (B). One example of a marginal cost of pollution is its effect on human health (C).

NATIONAL
GEOGRAPHIC

Environmental Science and Studies

Our inadequate scientific understanding of how the environment works and how different human choices affect the environment is a major reason that problems of environmental sustainability are difficult to resolve. Even for established environmental problems, political and social controversy often prevent widespread acceptance that an environmental threat is real. Because the effects of many interactions between the environment and humans are unknown or difficult to predict, we generally do not know what corrective actions should be taken until our scientific understanding is more complete. But since decisions respecting the environment can often not wait until more or "better" knowledge is attained, the environment is increasingly being thought of in non-scientific ways that consider wide-ranging implications from the perspective of diverse disciplines.

Environmental science and environmental studies are newly developing, interdisciplinary fields of investigation that combine information from any discipline that is related to the environment, such as biology, geography, chemistry, geology, physics, economics, sociology, cultural anthropology, natural resource management, agriculture, engineering, law, politics, ethics, architecture, and management.

Ecology, the discipline of biology that studies the interrelationships between organisms and their environment, is a basic tool of environmental science and studies. Environmental scientists, planners, analysts, and managers examine the relationships among human populations, determine various uses of

environmental science and environmental studies The interdisciplinary fields of humanity's relationship with other organisms and the physical environment. Environmental science relies more heavily on the scientific method. Environmental studies deals more with social issues, which often are value-laden. But both "science" and "studies" deal with the all-encompassing and interdisciplinary aspects of the environment.

natural resources and their associated impacts, and develop approaches to achieve environmental sustainability. There are many and varied processes that are used in dealing with environmental matters, and advances in this field continue on a daily basis. The key to dealing with the environment lies in the need to generate increasing levels of knowledge about its complexity, diversity, and well-being, while at the same time making the best decisions possible based on available information and societal wants, needs, and aspirations. Adapting these decisions over time as more knowledge becomes available is also an important feature of environmental science and studies.

GOALS FOR ENVIRONMENTAL INVESTIGATIONS

Humanity has been establishing general principles about how the natural world functions over many generations. More recently, the concepts of *sustainable development* and *environmental sustainability* have become the focus for environmental decision making on a global scale. We are increasingly using these principles to develop viable solutions to environmental problems—solutions that are based as much as possible on scientific knowledge.

Environmental problems are generally complex, however, and scientific understanding of them is often less complete than we would like. Many times, incomplete knowledge exists on a particular subject, so recommendations for solutions are often based on probabilities rather than on precise answers.

The environmental problems and issues considered in this book are serious ones that society must address into the future. But we are not simply providing a "doom and gloom" listing of problems, coupled with predictions of a bleak future. On the contrary, the focus of environmental science and studies, and our focus as individuals and as world citizens, is on identifying,

A botanist compares a cotton leaf from an experimental plant (*left*) to a control plant (*right*). The two plants were raised under identical conditions, except that the experimental plant was genetically engineered to resist insect pests.

understanding, and solving problems that we as a society have generated. (See Chapter 13 for more on achieving our preferred future environment.) It is encouraging that individuals, businesses, and governments are already doing a great deal, although more must be done to address the problems of today's world.

SCIENCE AS A PROCESS

The key to successfully solving any environmental problem is rigorous scientific evaluation. It is important to understand clearly just what science is, as well as what it is not. Most people think of **science** as a body of knowledge—a collection of facts about the natural world. However, science is also a dynamic *process*, a systematic way to investigate the natural world. Science seeks to reduce the apparent complexity of our world to general principles, which are then used to make predictions, solve problems, or provide new insights.

Scientists collect objective **data** (singular, *datum*), the information with which science works. Data are collected through observation and experimentation and then analyzed or interpreted (**FIGURE 1.15**). Scientific conclusions are inferred from the available data and are not based on faith, emotion, or intuition. Scientists publish their findings in scientific journals, and other scientists examine and critique their work. A requirement of science is repeatability—that is, observations and experiments must produce consistent data when they are repeated. This scrutiny by other scientists reveals any inconsistencies in results or interpretation. These errors are discussed openly, and ways to eliminate them are developed.

There is no absolute certainty or universal agreement about anything in science. Science is an ongoing enterprise, and scientific concepts must be re-evaluated in light of newly discovered data. Thus, scientists never claim to know the final answer about anything because scientific understanding changes. Instead, science aims to discover and better understand the general principles that govern the operation of the natural world.

Several areas of human endeavour are not scientific, but are equally important in reaching optimal decisions for the environment. For example, ethical principles often have a religious or cultural foundation, and political principles reflect social systems. The social sciences comprise academic disciplines concerned with the study of the social life of human groups and individuals. Although perhaps not as directed by process as "pure science," the study of humans and their activities is a necessary and required condition for effective environmental investigation. Humans form a part of the environment; hence their viewpoints, feelings, perceptions, beliefs, attitudes, and behaviours are important considerations for any environmental issue.

The scientific method

The established processes that scientists use to answer questions or solve problems are collectively called the **scientific method**. Although there are many variations of the scientific method, it basically involves five steps:

scientific method
The way a scientist approaches a problem, by formulating a hypothesis and then testing it.

1. Recognize a question or an unexplained occurrence in the natural world.

2. Develop a *hypothesis*, or an educated guess, to explain the problem.

3. Design and perform an experiment to test the hypothesis.

4. Analyze and interpret the data to reach a conclusion.

5. Share new knowledge with the scientific community.

Although the scientific method is often portrayed as a linear sequence of events, science is rarely as straightforward or tidy as the scientific method implies (see What a Scientist Sees on page 28). Good science involves creativity, not only in recognizing questions and developing hypotheses but also in designing experiments. Because scientists try to expand our current knowledge, their work is in the realm of the unknown. Many creative ideas end up as dead ends, and there are often temporary setbacks or reversals of direction as scientific knowledge progresses. Scientific knowledge often expands in surprising ways, with the "big picture" emerging slowly from confusing and sometimes contradictory details.

Scientific discoveries are often incorrectly portrayed in the media as "new facts" that have just come to light. Then later, additional "new facts" that question the validity of the original study may be reported. If you were to read the scientific papers on which such media reports are based, however, you would find that all the scientists involved probably made very tentative conclusions based on their data. Science progresses from uncertainty to less uncertainty, not from certainty to greater certainty. Thus, science self-corrects over time, even though it never actually "proves" anything, in the strictest sense of the word.

The importance of prediction

A scientific **hypothesis** needs to be useful—it needs to tell you something you want to know. A hypothesis is most useful when it makes predictions because the predictions provide a way to test the validity of the hypothesis. If your experiment refutes your prediction, then you must carefully recheck the entire experiment. If the prediction is still refuted, then you must reject the hypothesis. The more verifiable predictions a hypothesis makes, the more valid that hypothesis is.

Each of the many factors that influence a process is called a **variable**. To evaluate alternative hypotheses about a specific variable, it is necessary to hold all other variables constant so that they are not misleading or confusing. To test a hypothesis about a variable, we carry out two forms of the experiment in parallel. In the **experimental group**, the chosen variable is altered in a known way. In the **control group**, that variable isn't altered. In all other respects, the experimental group and the control group are the same. We then ask, What is the difference, if any, between the outcomes for the two groups? Any difference must be due to the influence of the variable we changed because all other variables remained the same. Much of the challenge of science lies in designing control groups and in successfully isolating a single variable from all other variables.

Theories

A **theory** is an integrated explanation of numerous hypotheses, each of which is supported by a large body of observations and experiments. A theory condenses and simplifies many data that previously appeared to be unrelated. Because a theory demonstrates the relationships among different data, it simplifies and clarifies our understanding of the natural world. A good theory grows as additional information becomes known. It predicts new data and suggests new relationships among a range of natural phenomena.

Theories are the solid ground of science, the explanations of which we are most sure. This definition contrasts sharply with the general public's use of the word *theory*, which implies lack of knowledge or a guess. In this book, the word *theory* is always used in its scientific sense, to refer to a broadly conceived, logically coherent, and well-supported explanation.

Although theories are generally accepted, remember that there is no absolute truth in science, only varying degrees of uncertainty. Science is continually evolving as new evidence comes to light, and therefore its conclusions are always provisional or uncertain. It is always possible

The Scientific Method

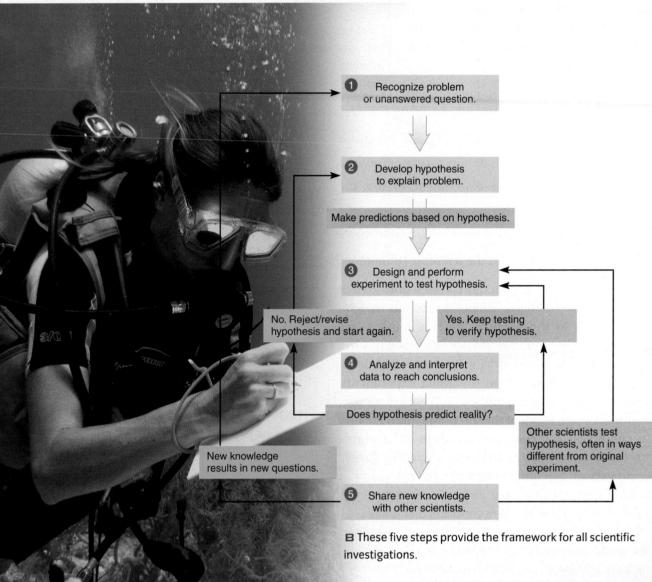

1. Recognize problem or unanswered question.

2. Develop hypothesis to explain problem.

Make predictions based on hypothesis.

3. Design and perform experiment to test hypothesis.

No. Reject/revise hypothesis and start again.

Yes. Keep testing to verify hypothesis.

4. Analyze and interpret data to reach conclusions.

Does hypothesis predict reality?

New knowledge results in new questions.

Other scientists test hypothesis, often in ways different from original experiment.

5. Share new knowledge with other scientists.

B These five steps provide the framework for all scientific investigations.

A A field scientist makes observations critical to understanding damage to coral reefs from global climate change. Photographed at Turneffe Atoll, Belize.

C Many scientists present their research during poster sessions at scientific meetings. This allows their work to be critically assessed by others in the scientific community.

that the results of a future experiment will contradict a prevailing theory and show at least one aspect of it to be false.

Uncertainty, however, does not mean that scientific conclusions are invalid. For example, overwhelming evidence links cigarette smoking to incidence of lung cancer. We can't state with absolute certainty that every smoker will be diagnosed with lung cancer, but this uncertainty does not mean that there is no correlation between smoking and lung cancer. On the basis of the available evidence, we say that people who smoke have an increased risk of developing lung cancer.

In conclusion, the aim of science is to increase human comprehension by explaining the processes and events of nature. Scientists work under the assumption that all phenomena in the natural world have natural causes, and they formulate theories to explain these phenomena. The process of science as a human endeavour has shaped the world we live in and transformed our views of the universe and how it works.

CONCEPT CHECK **STOP**

What is environmental science and studies? What are some of the disciplines involved in environmental science and studies?

What are the five steps of the scientific method? Why is each important?

ECO CANADA   CAREER FOCUS

ENVIRONMENTAL COMMUNICATIONS OFFICERS

Environmental communications officers perform an important role in gathering and disseminating environmental information.

Imagine you are standing at the back of a crowded press room, a sea of television cameras, spotlights, microphones, and notepads between you and your boss at the front. You are an environmental communications officer for a local conservation agency and your boss, the agency's president, is about to make an official statement on the recent municipal decision to halt development near a sensitive wetland. You know exactly what the statement will say because you are the person who wrote it.

As the conservation agency's environmental communications officer, you began working on this controversial wetlands development almost two years ago, when you first learned that a group of land developers proposed to drain and build on the wetlands. Your conservation agency was well aware of how sensitive and valuable the wetlands are, so you drafted a communications strategy to ensure that the agency's opposition to the proposed development was well understood and widely heard. You organized a series of informational evenings open to the public and made certain the media covered these events. You also ensured that all staff members within your own office were well informed on the issue and knew how to properly answer questions on the proposed development from the media and general public. Part of your communications strategy included educating local businesses about the environmental impacts of such a development. You also made certain your organization attended industry events and municipal council meetings to communicate your concerns. Months and months of your hard work have paid off, with the town council agreeing to protect the wetlands area.

Environmental communications officers disseminate information on environmental issues and events on behalf of their organizations. They can be involved in long-term activities, such as public information and education campaigns, as well as short-term activities, such as responding to a toxic spill. Environmental communications officers use their skills in writing, design, media relations, and networking to educate the public and encourage environmental protection and conservation.

For more information or to look up other environmental careers go to www.eco.ca.

THE FLOOD OF THE CENTURY: RED RIVER VALLEY, MANITOBA

Manitobans are used to floods: Ever since First Nations residents told stories of huge floods covering the entire Red River basin, and ever since the disastrous flood that almost washed away the entire colony of the first Scottish settlers, Manitobans have built dikes and piled sandbags almost every year. But in spring 1997 residents knew they were in for the Big One.

That year, between November 1996 and April 1997, the Red River Valley received almost double its average precipitation. And when a blizzard struck the region in early April, dumping 50 centimetres of snow, the Red River turned into the "Red Sea," covering some 2000 square kilometres, an area equivalent to the size of Prince Edward Island.

There was unanimous agreement that what saved Winnipeg from complete flooding and evacuation was the Red River Floodway—a massive ditch built in the 1960s to divert the Red River runoff from the city. But despite the preventive measures, two communities were flooded out: Ste. Agathe, population 400, and Grande Pointe, population 450. In all, some 800 homes between the U.S. border and Winnipeg were damaged, 100 of them irreparably. Total damages have been estimated at more than $150 million.

The Governments of Canada and the United States asked the International Joint Commission (IJC) to investigate the causes and effects of flooding in the Red River basin and to provide recommendations to improve the preparedness for future flooding. In 2000, the IJC submitted a report, *Living with the Red*. The report found that the first settlers in the basin established their communities along the Red River—the principal transport route into the region. These settlement patterns set the stage for most of today's flooding problems. The relatively flat topography of the basin places most of its residents on or near the flood plain of the river, and for economic and social reasons this is unlikely to change.

The IJC report also indicated that several "natural" environmental factors in the Red River basin need to be re-established wherever possible, including:

- Creation of large reservoirs for storage of meltwater that would lessen the flow in spring along the entire river

- Use of micro-storage options in local areas to reduce flow from agricultural fields

- Restoration of natural wetlands along the length of the basin to provide their necessary functions of retaining flood waters and reducing peak flows or total flood volumes or both

- Reduction of artificial drainage of agricultural lands to reduce flows into the Red River system.

Even though the Flood of the Century was caused primarily by unusual weather conditions in the Red River area over a two-year period, later investigations revealed that human alterations to the environment served as a catalyst and contributed to the severity of the impact and damage that occurred. Consequently, recommendations for mitigation approaches have been made, and are now being implemented, to ensure that the natural environment and its associated interactions are preserved as much as possible.

The Red River floods on a regular basis, but the Flood of the Century occurred in 1997.

SUMMARY

1 Human Impacts on the Environment

1. **Highly developed countries** are countries that have complex industrialized bases, low rates of population growth, and high per person incomes. **Moderately developed countries** are developing countries that have medium levels of industrialization and average per person incomes lower than those of highly developed countries. **Less developed countries (LDCs)** are developing countries with low levels of industrialization, very high rates of population growth, very high infant mortality rates, and very low per person incomes (relative to highly developed countries). **Poverty**, which is common in LDCs, is a condition in which people are unable to meet their basic needs for food, clothing, shelter, education, or health.

2. The increasing global population is placing stresses on the environment as humans consume ever-increasing quantities of food and water, use more energy and raw materials, and produce enormous amounts of waste and pollution. **Nonrenewable resources** are natural resources that are present in limited supplies and are depleted as they are used. **Renewable resources** are resources that natural processes replace and that therefore can be used forever, provided that they are not overexploited in the short term.

3. **People overpopulation** is a situation in which too many people live in a given geographic area. Developing countries have people overpopulation. **Consumption overpopulation** is a situation that occurs when each individual in a population consumes too large a share of resources. Highly developed countries have consumption overpopulation.

4. Environmental impact and the forces that drive it can be modelled by the IPAT equation: $I = P \times A \times T$. Environmental impact (I) has three factors: the number of people (P); the affluence per person (A), which is a measure of the consumption, or amount of resources used per person; and the environmental effect of the technologies used to obtain and consume those resources (T).

2 Environmental Sustainability and Sustainable Development

1. **Environmental sustainability** is the ability to meet humanity's current needs without compromising the ability of future generations to meet their needs. Sustainability implies that the environment can function indefinitely without going into a decline from the stresses that human society imposes on natural systems. **Sustainable development** is economic growth that meets the needs of the present without compromising the ability of future generations to meet their own needs. Environmentally sound decisions, economically viable decisions, and socially equitable decisions interact to promote sustainable development. Sustainable development is the mechanism by which sustainability can be achieved.

2. Human behaviours that threaten environmental sustainability include overuse of renewable and nonrenewable resources, pollution, and overpopulation.

3. **Sustainable consumption** is the use of goods and services that satisfy basic human needs and improve the quality of life but that also minimize the use of resources so they are available for future use. Voluntary simplicity recognizes that individual happiness and quality of life are not necessarily linked to the accumulation of material goods.

3 Environmental Economics

1. Economics is the study of how people use their limited resources to try to satisfy their unlimited wants. Economies depend on the natural environment and its processes as sources for raw materials and sinks for waste products. Both sources and sinks contribute to **natural capital**, which is Earth's resources and processes that sustain living organisms, including humans. Natural capital includes minerals, forests, soils, groundwater, clean air, wildlife, and fisheries.

2. **National income accounts** are a measure of the total income of a nation's goods and services for a given year. An **external cost** is a harmful environmental or social cost that is borne by people not directly involved in buying or selling a product. National income accounts are incomplete estimates of national

economic performance because they do not include both natural resource depletion and the environmental costs of economic activities. Many economists, government planners, and scientists support more comprehensive income accounting that includes these estimates.

3. From an economic point of view, the appropriate amount of pollution is a trade-off between harm to the environment and inhibition of development. The **marginal cost of pollution** is the added cost of an additional unit of pollution. The **marginal cost of pollution abatement** is the added cost of reducing one unit of a given type of pollution. Economists think the use of resources for pollution abatement should increase only until the cost of abatement equals the cost of the pollution damage. This results in the **optimum amount of pollution**—the amount of pollution that is economically most desirable and acceptable from society's perspective.

4 Environmental Science and Studies

1. **Environmental science and environmental studies** are newly developing, interdisciplinary fields of investigation that combine information from any discipline that is related to the environment, such as biology, geography, chemistry, geology, physics, economics, sociology, cultural anthropology, natural resources management, agriculture, engineering, law, politics, ethics, architecture, and management. Science relies heavily on use of the scientific method, whereas studies focuses on social science and its inherent value judgements.

2. The **scientific method** is the way a scientist approaches a problem, by formulating a hypothesis and then testing it by means of an experiment. (1) A scientist recognizes and states the problem or unanswered question. (2) The scientist develops a hypothesis, or an educated guess, to explain the problem. (3) An experiment is designed and performed to test the hypothesis. (4) Data, the results obtained from the experiment, are analyzed and interpreted to reach a conclusion. (5) The conclusion is shared with the scientific community.

KEY TERMS

- **poverty** p. 4
- **highly developed countries** p. 6
- **moderately developed countries** p. 6
- **less developed countries** p. 6
- **natural resources** p. 7
- **nonrenewable resources** p. 7
- **renewable resources** p. 7
- **people overpopulation** p. 9

- **consumption overpopulation** p. 9
- **environmental sustainability** p. 13
- **sustainable development** p. 18
- **sustainable consumption** p. 19
- **natural capital** p. 21
- **national income accounts** p. 22
- **external cost** p. 22
- **marginal cost of pollution** p. 23

- **marginal cost of pollution abatement** p. 23
- **cost-benefit diagram** p. 23
- **optimum amount of pollution** p. 23
- **environmental science and environmental studies** p. 25
- **scientific method** p. 27

CRITICAL AND CREATIVE THINKING QUESTIONS

1. Criticize the following statement: "Population growth in developing countries is of much more concern than is population growth in highly developed countries."

2. Give at least two examples and discuss what you can do as an individual to promote environmental sustainability.

3. People want scientists to give them precise, definitive answers to environmental problems. Explain why this is not possible.

4. Make a three-column list of the material items you own or would like to own. Column 1 consists of items you have that meet your basic needs; column 2 contains items you have that you wanted

but didn't need; and column 3 contains items you would like to possess in the near future. Which of the items in columns 2 and 3 would you be willing to give up to reduce your personal impact on the environment?

5. Write a brief paragraph describing two or three ways your personal quality of life could be improved. Do any of these involve material items? If so, explain how these items would improve your life.

6. How are sustainable consumption and voluntary simplicity related?

7. Development is sometimes equated with economic growth. Explain the difference between sustainable development and development as an indicator of economic growth, using the figure shown to the right.

8. How do the three factors shown in the figure to the right interact to promote sustainable development?

9–11. Your throat feels scratchy, and you think you're coming down with a cold. You take a couple of vitamin C pills and feel better. You conclude that vitamin C helps prevent colds.

9. Is your conclusion valid from a scientific standpoint? Why or why not?

10. Develop a testable hypothesis to answer the question, Does taking vitamin C pills help prevent colds?

11. Describe how you would conduct a controlled experiment to test your hypothesis.

12–14. Examine the graph shown to the right, which shows an estimate of the discrepancy between the wealth of the world's poorest countries and that of the richest countries.

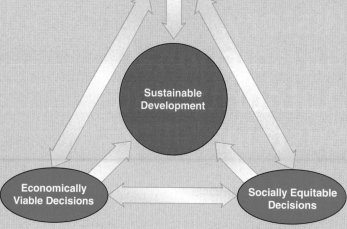

12. How has the distribution of wealth changed from the 1800s to the present? How would you explain this difference?

13. Based on the trend evident in this graph, predict what the graph might look like in 100 years.

14. Some economists think that our current path of economic growth is unsustainable. Do the data in this graph support or refute this idea? Explain your answer.

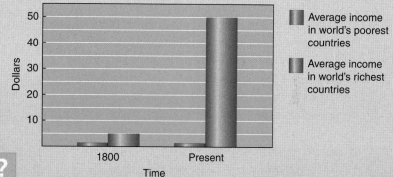

What is happening in this picture ?

The Great Bear Rainforest Campaign began in 1995 to stop destructive logging practices in BC's coastal rainforest.

What are some of the environmental, social, and economic factors associated with this issue?

Why would local residents and environmental groups protest against logging in this area?

Environmental History, Jurisdictional Authority, and Sustainability

2

THE GLOBAL COMMONS

Ecologist Garrett Hardin (1915–2003) is best known for his 1968 essay "The Tragedy of the Commons." In it he contended that our inability to solve complex environmental problems is the result of a struggle between short-term individual welfare and long-term *environmental sustainability* (the ability to meet humanity's current needs without compromising the ability of future generations to meet their needs).

Hardin used the commons to illustrate this struggle. In medieval Europe, the inhabitants of a village shared pastureland, called the commons, and each herder could bring animals onto the commons to graze. The more animals a herder brought onto the commons, the greater the advantage to that individual. When every herder in the village brought as many animals onto the commons as possible, however, the plants were killed from overgrazing.

In today's world, Hardin's parable has particular relevance at the global level. The commons are those parts of our environment that are available to everyone but for which no single individual has responsibility: the atmosphere, water, wildlife, fisheries, and forests (see photograph). These modern-day commons, sometimes collectively called the *global commons,* are experiencing increasing environmental stress. The world needs effective legal and economic policies to prevent the degradation of our global commons. Clearly, all people, businesses, and governments must foster a strong sense of *stewardship,* or shared responsibility, for the sustainable care of our planet.

This chapter examines the role of ethics, legislation, and values in environmental issues.

CHAPTER OUTLINE

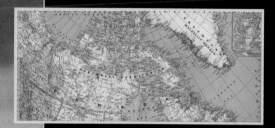

35

History of the Environmental Movement

Since the first European contacts with North America, beginning with the Vikings in about 1000 and Christopher Columbus in 1492, the first two centuries of this continent's history were a time of widespread environmental destruction. Land, timber, wildlife, rich soil, clean water, and other resources were cheap and seemingly inexhaustible. The European settlers did not dream that the bountiful natural resources of North America would one day become scarce. During the 1700s and most of the 1800s, many North Americans had a **frontier attitude**, a desire to conquer nature and put its resources to use in the most lucrative manner possible.

EARLY FOREST CONSERVATION

It is not known the exact makeup of the great forests of North America that existed prior to settlement by Europeans. We know that Aboriginal peoples had impacts on forests across the continent, the most extensive one likely being the use of fire, both accidental and deliberate. They, of course, also cut down trees to use for shelter, tools, boats, and fuel. But Aboriginal peoples' impacts on forests were negligible, since their use of forests represents sustainable consumption; forests would have regenerated naturally far quicker than Aboriginal peoples could have cut them down.

However, in the early 1500s the first significant use of North American forests occurred with the establishment of the European fishing and fur industries, which also had a profound effect on the wildlife that were being hunted and fished. Large European fleets made annual expeditions to inshore fishing grounds to catch cod, and shore installations were established to process the fish for shipment to Europe. These installations included docks, residential buildings, warehouses, drying racks, and other facilities, all of which used large amounts of timber growing in the coastal regions (FIGURE 2.1). Perhaps surprisingly, regulations were adopted to control this usage, and these constituted the first forest laws of the continent.

Actual colonization in Canada did not begin until the 17th century when settlements began to be established along coastal areas and major river systems, resulting in an increasing demand for forest products. Following the usual practice in the New England colonies of the U.S., the French and British in Canada instituted what were known as "broad arrow" policies. This meant that trees suitable for shipbuilding were reserved for the Crown and marked with arrows burned into their trunks.

Early coastal settlement in Hopedale, Labrador FIGURE 2.1

The earliest established communities in Canada required lots of forest products to build homes, ships, and other facilities.

Officials were appointed to enforce the laws, and the first of these (the Surveyor General of His Majesty's Woods) was appointed in Nova Scotia in 1728. Severe penalties were imposed on anyone cutting these trees without permission.

Settlement of North America throughout the 1700s and 1800s created serious and mostly negative impacts on continental forests and wildlife. A vast array of forest products were used for these developments, and their usage resulted in severe alterations to the natural environment, including wildlife habitat destruction.

During the 19th century, many naturalists began to voice concerns about conserving natural resources. John James Audubon (1785–1851) painted lifelike portraits of birds and other animals in their natural surroundings that aroused widespread public interest in the wildlife of North America (**FIGURE 2.2**). Henry David Thoreau (1817–1862), a prominent American writer, lived for two years on the shore of Walden Pond near Concord, Massachusetts. There he observed nature and contemplated how people could economize and simplify their lives to live in harmony with the natural world. George Perkins Marsh (1801–1882) was a farmer, linguist, and diplomat at various times during his life. Today he is most remembered for his book *Man and Nature,* published in 1864, which provided one of the first discussions of humans as agents of global environmental change.

In the United States, President Theodore Roosevelt, an avid outdoorsman and hunter, appointed Gifford Pinchot (1865–1946) the first head of the U.S. Forest Service. Both Roosevelt and Pinchot were **utilitarian conservationists** who viewed forests in terms of their usefulness to people—such as in providing jobs and renewable resources. Pinchot supported expanding the nation's forest reserves and managing them scientifically (for instance, harvesting trees only at the rate at which they regrow). In Canada, a similar approach was taken with the establishment of the Canadian Forest Service in 1899. Clifford Sifton (1861–1929), who was Minister of the Interior under Prime Minister Wilfrid Laurier, was responsible for placing forests under federal control and organized the Canadian Forestry Association.

utilitarian conservationist
A person who values natural resources because of their usefulness to humans but uses them sensibly and carefully.

This portrayal is one of 500 engravings in Audubon's classic *The Birds of America*, completed in 1844. Shown are two male Louisiana tanagers (also called western tanagers, top) and male and female scarlet tanagers (bottom).

THE NORTH AMERICAN WILDLIFE CONSERVATION MODEL

With the settlement of the west, the North American wildlife populations were diminished by overhunting and habitat loss. Many species came close to extinction by the 19th century, a notable example being the devastation brought to over 40 million buffalo. Perhaps surprisingly, the first push to conserve and protect wild-life and their habitats came from hunters and fishers. They realized that wildlife is an exhaustible resource and limits were needed to protect their rapidly disappearing numbers. They pushed for hunting regulations and established conservation groups to protect habitat, and in the process raised social recognition that animal populations are vulnerable to commercial exploitation, to local depletions, and to national extinctions. By the 1860s, the commitment of hunters and anglers to preserving wild nature and restoring depleted resources was well organized and strong with nearly 500 associations of various kinds dedicated to these efforts.

The efforts of these early conservationists evolved into the **North American Wildlife Conservation Model**. This model is based on two basic principles: (1) that fish and wildlife belong to all North American citizens, and (2) resources should be managed sustainably. Well over a century later, these doctrines and efforts persist, supported by hunting and fishing licence fees, equipment taxes, and state/provincial game and non-game management programs. Without the foresight of these early conservationists, we would today face a continent without white-tailed deer, pronghorn, antelope, elk, wild turkeys, wood ducks, and many more species.

CONSERVATION IN THE MID-20TH CENTURY

The 20th century brought a new awareness of the environment and our impact on it. Many people began to realize that our unfettered use of forests and wildlife would have severe consequences in the future, and the public started to take notice.

Furthermore, the 20th century witnessed events that had significant impacts on the environment. During the droughts of the 1930s, windstorms carried away much of the topsoil in parts of the Great Plains (**FIGURE 2.3**), forcing many farmers to abandon their farms and search for work elsewhere. This dust bowl alerted governments to the need for soil conservation, and led to the establishment of the Prairie Farm Rehabilitation Administration in 1935. Other conservation groups also emerged at this time. Ducks Unlimited Canada, founded in 1938,

MAKING A DIFFERENCE

SHANE MAHONEY

Native Newfoundlander Shane Mahoney is an internationally known biologist, writer, and lecturer on environmental and resource-conservation issues. Mahoney has devoted over 26 years to investigating the diverse dynamics of species such as seabirds, waterfowl, black bears, lynx, and woodland caribou. He has authored or co-authored over 120 scientific and popular articles and reports and has appeared and/or assisted with films for the CBC, BBC, Turner Broadcasting, and National Geographic.

Mahoney is widely recognized for his intense and deep personal commitment to understanding humans' place in nature and is considered the leading authority on the North American Wildlife Conservation Model. Mahoney captivates audiences with his evolutionary, spiritual, and historical writings, and challenges society to strive for a better understanding of human civilizations and their unique place within the biodiversity we must conserve and protect. His internationally recognized vision and commitment to sustainable development have made him a Canadian leader in forward-thinking conservation.

Mahoney is currently the Executive Director of Sustainable Development and Strategic Science of the Newfoundland and Labrador Department of Tourism, Culture and Recreation.

Check out *Opportunity for All*, a documentary on the history of the North American Model of Wildlife Conservation, created by the Rocky Mountain Elk Foundation and Shane Mahoney: https://www.rmef.org/Hunting/HuntersConservation/DVD/.

The dust bowl FIGURE 2.3

Drought during the 1930s resulted in windstorms that carried away much of the topsoil on many Prairie farms.

was formed by a group of conservation-minded sportsmen who wanted to stop the destruction and neglect of Canada's wetlands, which was happening because of drought, agriculture, and urban expansion.

Aldo Leopold (1886–1948) was a wildlife biologist and environmental visionary who was extremely influential in the conservation movement of the mid- to late-20th century (**FIGURE 2.4**). His textbook *Game Management,* published in 1933, supported the passage of a 1937 act in which new taxes on sporting weapons and ammunition funded wildlife management and research. Leopold also wrote philosophically about humanity's relationship with nature and about the need to conserve wilderness areas in *A Sand County Almanac,* published in 1949. Leopold argued persuasively for a land ethic and the sacrifices that such an ethic requires. Albert Hochbaum (1911–1988), a well-known conservationist in Canada, studied under Leopold and brought Leopold's conservation principles to Canada.

Leopold had a profound influence on many American thinkers and writers, including Wallace Stegner (1909–1993), who penned his famous "Wilderness Essay" in 1962. Stegner's essay, written to a commission that was conducting a national inventory of wilderness lands, helped create support for the passage of the Wilderness Act of 1964. Stegner wrote:

> *Something will have gone out of us as a people if we ever let the remaining wilderness be destroyed; if we permit the last virgin forests to be turned into comic books and plastic*

Aldo Leopold FIGURE 2.4

Leopold's *A Sand County Almanac* is widely considered an environmental classic.

cigarette cases; if we drive the few remaining members of the wild species into zoos or to extinction; if we pollute the last clean air and dirty the last clean streams and push our paved roads through the last of the silence, so that never again will Americans be free in their own country from the noise, the exhausts, the stinks of human and automotive waste. . .

We simply need that wild country available to us, even if we never do more than drive to its edge and look in. For it can be a means of reassuring ourselves of our sanity as creatures, a part of the geography of hope.

During the 1960s, public concern about pollution and resource quality began to increase, in large part due to the work of marine biologist Rachel Carson (1907–1964). Carson wrote about interrelationships among living organisms, including humans, and the natural environment (**FIGURE 2.5**). Her most famous

Rachel Carson FIGURE 2.5

Carson's book *Silent Spring* heralded the beginning of the environmental movement.

work, *Silent Spring*, was published in 1962. In it Carson wrote against the indiscriminate use of pesticides:

Pesticide sprays, dusts, and aerosols are now applied almost universally to farms, gardens, forests, and homes— nonselective chemicals that have the power to kill every insect, the "good" and the "bad," to still the song of birds and the leaping of fish in the streams, to coat the leaves with a deadly film, and to linger on in soil—all this though the intended target may be only a few weeds or insects. Can anyone believe it is possible to lay down such a barrage of poisons on the surface of the earth without making it unfit for all life? They should not be called "insecticides," but "biocides."

Silent Spring heightened public awareness and concern about the dangers of uncontrolled use of DDT and other pesticides, including poisoning birds and other wildlife and contaminating human food supplies. Ultimately, the book led to restrictions on the use of certain pesticides. Around this time, the media began to increase its coverage of environmental incidents, such as hundreds of deaths in New York City from air pollution (1963) and closed beaches and fish kills in Lake Erie from water pollution (1965).

In 1968, when the population of Earth was "only" 3.5 billion people, ecologist **Paul R. Ehrlich** published *The Population Bomb*. In it he described the damage occurring to Earth's life support system because it was supporting such a huge population, including the depletion of essential resources such as fertile soil, groundwater, and other living organisms. Ehrlich's book raised the public's awareness of the dangers of overpopulation and triggered debates about how to deal effectively with population issues.

THE ENVIRONMENTAL MOVEMENT

Until 1970 the voice of **environmentalists**, people concerned about the environment, was heard primarily through societies such as the Sierra Club and the Canadian Wildlife Federation. There was no generally perceived **environmental movement** until the spring of 1970, when Gaylord Nelson, former senator of Wisconsin, urged Harvard graduate student Denis Hayes to organize the first nationally celebrated Earth Day. This event awakened environmental consciousness to population growth, overuse of resources, and pollution and degradation of the environment. On Earth Day 1970, an estimated 20 million Americans planted trees, cleaned roadsides and riverbanks, and marched in parades to demonstrate their support of improvements in resource conservation and environmental quality.

In the years that followed the first Earth Day, environmental awareness and the belief that individual actions could repair the damage humans were doing to Earth became a pervasive popular movement. Musicians and other celebrities popularized environmental concerns. Many of the world's religions—such as Christianity, Judaism, Islam, Hinduism, Buddhism, Taoism, Shintoism, Confucianism, and Jainism—embraced environmental themes such as protecting endangered species and controlling global climate change.

By Earth Day 1990, the movement had spread around the world, signalling the rapid growth in environmental consciousness. An estimated 200 million people in 141 nations, including Canada, demonstrated to increase public awareness of the importance of individual efforts ("Think globally, act locally") (**FIGURE 2.6**). The theme of Earth Day 2000, "Clean Energy Now," reflected the dangers of global climate change and what individuals and communities could do: replace fossil fuel energy sources, which produce greenhouse gases, with solar electricity, wind power, and the like. However, by 2000 many environmental activists had begun to think that the individual actions Earth Day espouses, while collectively important, are not as important as pressuring governments and large corporations to make environmentally friendly decisions. In 2007, global environmental concern was expressed in the efforts of Live Earth, a series of concerts held worldwide to launch programs to combat climate change.

Since the first Earth Day celebrations, the environmental movement has taken many different routes and continues to evolve to this day. One notable environmental organization that has emerged in the last 40 years is Greenpeace, a non-governmental organization for the protection and conservation of the environment, which was founded as the Greenpeace

Earth Day 1991 in Ottawa FIGURE 2.6

Foundation in Vancouver in 1971. Another is one founded in 1990 by the most famous environmentalist here in Canada, David Suzuki. The David Suzuki Foundation is an environmental charity that works toward balancing human needs with the Earth's ability to sustain life. The Foundation's goal is to find and communicate practical ways to achieve that balance. (For more on David Suzuki and his contributions to the environmental movement, see the Making a Difference feature in Chapter 1.) **FIGURE 2.7** on page 42 shows a timeline of selected environmental events since Earth Day 1970.

CONCEPT CHECK STOP

What role did each of the following have in North American environmental history: protecting forests, conservation in the mid-20th century, and the environmental movement of the late-20th century?

What was the environmental contribution of Rachel Carson?

List three environmental organizations or movements that have emerged in North America over the last 40 years.

Timeline of selected environmental events, from 1970 to the present FIGURE 2.7

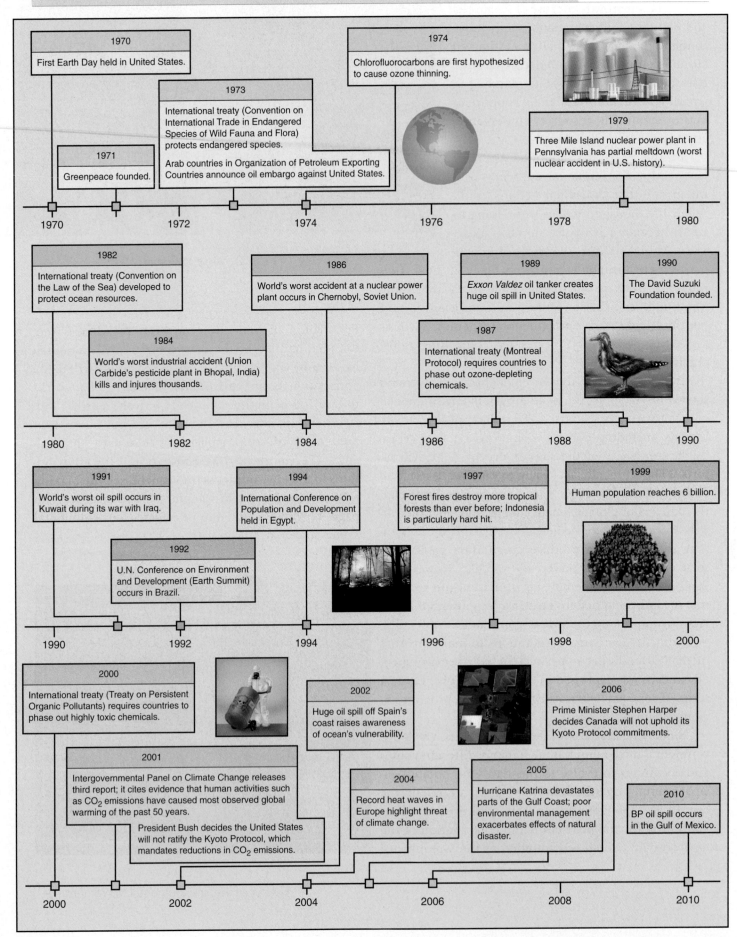

1970
First Earth Day held in United States.

1971
Greenpeace founded.

1973
International treaty (Convention on International Trade in Endangered Species of Wild Fauna and Flora) protects endangered species.

Arab countries in Organization of Petroleum Exporting Countries announce oil embargo against United States.

1974
Chlorofluorocarbons are first hypothesized to cause ozone thinning.

1979
Three Mile Island nuclear power plant in Pennsylvania has partial meltdown (worst nuclear accident in U.S. history).

1970 1972 1974 1976 1978 1980

1982
International treaty (Convention on the Law of the Sea) developed to protect ocean resources.

1984
World's worst industrial accident (Union Carbide's pesticide plant in Bhopal, India) kills and injures thousands.

1986
World's worst accident at a nuclear power plant occurs in Chernobyl, Soviet Union.

1987
International treaty (Montreal Protocol) requires countries to phase out ozone-depleting chemicals.

1989
Exxon Valdez oil tanker creates huge oil spill in United States.

1990
The David Suzuki Foundation founded.

1980 1982 1984 1986 1988 1990

1991
World's worst oil spill occurs in Kuwait during its war with Iraq.

1992
U.N. Conference on Environment and Development (Earth Summit) occurs in Brazil.

1994
International Conference on Population and Development held in Egypt.

1997
Forest fires destroy more tropical forests than ever before; Indonesia is particularly hard hit.

1999
Human population reaches 6 billion.

1990 1992 1994 1996 1998 2000

2000
International treaty (Treaty on Persistent Organic Pollutants) requires countries to phase out highly toxic chemicals.

2001
Intergovernmental Panel on Climate Change releases third report; it cites evidence that human activities such as CO_2 emissions have caused most observed global warming of the past 50 years.

President Bush decides the United States will not ratify the Kyoto Protocol, which mandates reductions in CO_2 emissions.

2002
Huge oil spill off Spain's coast raises awareness of ocean's vulnerability.

2004
Record heat waves in Europe highlight threat of climate change.

2005
Hurricane Katrina devastates parts of the Gulf Coast; poor environmental management exacerbates effects of natural disaster.

2006
Prime Minister Stephen Harper decides Canada will not uphold its Kyoto Protocol commitments.

2010
BP oil spill occurs in the Gulf of Mexico.

2000 2002 2004 2006 2008 2010

Jurisdictional Authority and Environmental Legislation

The environment in many parts of the world is subject to a diverse array of rules and regulations as determined by the local society. This is because the environment is composed of the "commons" or public goods. In today's world, the commons are those parts of our environment that are available to everyone but for which no single individual has responsibility—for example, the atmosphere, water, wildlife, fisheries, and forests. Or put another way, the commons are those parts of our environment that everyone bears responsibility for since we are all a part of it. In many countries of the world, people have determined that governments will represent societal needs, wants, and aspirations when it comes to the environment, which means that each one of us has the right to help determine how the environment is used, managed, and maintained for the future.

JURISDICTIONAL AUTHORITY

Common law In Canada, society has determined that the federal and provincial governments will largely have responsibility for the environment. This jurisdictional authority is based largely on the principles of British common law (except in Quebec, where the Napoleonic Code from France is applied). The foundation of British common law is legal precedents that have been established by the court system and serve as examples of how the law should be applied in the future.

Many of these precedents now form the basis for environmental rules and regulations.

An important principle of common law is that property owners have the right to use their land as they wish, generally including all of the resources found within their property boundaries. This means that owners of property in effect have control over the environment that they own. In general, then, people can do whatever they want on their property, unless government rules and regulations provide restrictions. Thus, common law has little influence on protecting the environment; rather it focuses on societal use of the environment.

Another important principle is that property owners are not required to show that their use of their property is not detrimental to the use of others. Hence, any person who believes that his or her enjoyment of the use of common property is affected by another has the right to object and seek a remedy through the court system. However, no one has the right, including the government, to determine how someone uses his or her own property if the use is contained to only that property.

Statute law The legal precedents established under common law eventually become statute law (that is, an act passed by the government) if it is deemed beneficial to society. Which government has control over the environmental rules and regulations in its jurisdiction was established in the British North America Act (BNA Act) of 1867. When Canada became a country, its leaders created a series of responsibilities for the federal and provincial governments through statute law that remain in effect to this day (**FIGURE 2.8**).

The BNA Act divided responsibilities between the federal and provincial governments. The sections of the act that relate to the environment are 91, 92, and 92A. Section 91 gives the federal government jurisdiction over any situation that might affect "the Peace, Order, and Good Government of Canada..." which is known as the POGG component of the act. Essentially, section 91 gives the federal government authority in any matter

Canada's jurisdictional boundaries FIGURE 2.8

This map shows the jurisdictional boundaries in Canada. Although only four provinces initially joined together to form Canada in 1867, the same jurisdictional responsibilities set forth in the BNA Act were also granted to the other provinces as they joined.

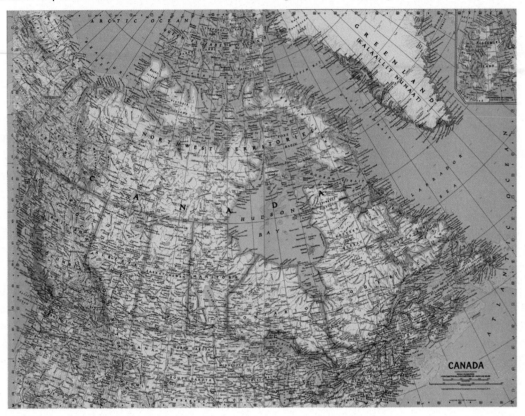

that might affect the citizenry of the country. Clearly, environmental matters could fall into this category and they often do since environmental matters can have significant impacts on all Canadians.

Sections 92 and 92A of the BNA Act set the authority for provincial governments to establish regulations for several issues, many of which relate to the environment:

- Taxation
- Public lands
- Municipal institutions
- Local works and undertakings
- Exploration for nonrenewable natural resources
- Development, conservation, and management of nonrenewable natural resources and forestry resources
- Development, conservation, and management of sites and facilities for the generation and production of electrical energy

These stipulations provide the provincial governments in Canada with the jurisdictional authority to make rules and regulations governing the environment for land resources, water, fisheries, wildlife, forestry, minerals, energy, and any other matters that are purely provincial. Some exceptions apply in relation to any federal lands (national parks, Indian reserves, etc.), any matters that are interprovincial or international (water flowage across boundaries, migratory wildlife, etc.), or any matter deemed to enable the POGG clause (section 91 of the Act).

These various jurisdictional agreements and rules have been revised and modified over time to ensure that proper emphasis is being placed on the respective societal goals and objectives of the day. As society desires and demands different things from the environment, jurisdictional matters need to be reconsidered and resolved. The varying emphasis of federal and provincial authority waxes and wanes to some extent with the wishes of the majority of the people in various jurisdictions.

Increasingly, environmental non-government organizations (ENGOs), and business and industry themselves, are taking on additional responsibility for protecting of the environment. This tendency to "go green" is sometimes seen as a marketing strategy, but nonetheless can lead to significant environmental benefits that complement the work of government agencies. As part of the sustainable development paradigm, every segment of society is expected to take a leadership role and make a difference in terms of the environment, and this trend has become more commonplace over time.

ENVIRONMENTAL LEGISLATION

By the late 1960s, in response to a substantial increase in population growth post–World War II, and stemming from the recognition of the damage resulting from centuries of exploitation of our natural resources, the need to protect and conserve Canada's natural environment became clear. On June 11, 1971, the Canadian government responded by legally incorporating Environment Canada as the department responsible for preserving and enhancing the nation's natural environment. The department consolidates all environmental efforts under one administrative unit with the mandate to oversee the creation and management of environmental policies and programs. The range of the department's responsibilities include the protection and conservation of air, water, land, fish, wildlife, and forests.

Environmental impact assessment In 1992, the **Canadian Environmental Assessment Act** was passed. The introduction of this act initiated a formal review process, at federal and provincial levels, for any major project with potential impact on the environment. This important legislation requires project developers to submit an **environmental impact assessment (EIA)** to the appropriate government agency (**FIGURE 2.9**). The project review process varies depending on the jurisdiction within which the development is set to take place, although it rarely occurs that a development project involves only one government agency. The federal government, for example, must approve any development project that requires federal funding or involves the use of federal property.

Enforcement In 1999, the **Canadian Environmental Protection Act (CEPA)** was passed, further solidifying Canada's commitment to the protection of the environment and sustainable development. Prior to that, the federal Environment Assessment and Review Process

Environmental impact assessments FIGURE 2.9

Detailed environmental impact assessments help federal and provincial agencies consider the environmental impacts of proposed activities. When the environmental impacts are judged too severe, alternative actions are considered.

(EARP) served to address Canada's environmental protection commitments.

CEPA is considered one of the most important pieces of federal environmental legislation. The act applies to all elements of the environment including air, water, land, layers of the atmosphere, organic and inorganic matter, and all living organisms. In addition, CEPA legislation addresses toxic contamination of Canada's air, water, and land, and gives the Canadian government the ability to enforce strict penalties on polluters and those responsible for polluting activities such as ocean dumping and hazardous waste spills.

The responsibility to protect and conserve Canada's environment is a shared management effort between the federal and provincial/territorial governments. All provinces and territories have general environmental protection laws that address specific environmental protection relating to concerns such as air pollution, water pollution, water conservation, waste management, and the transportation of hazardous materials.

According to CEPA, the Environment Canada Enforcement Branch is responsible for ensuring compliance with federal environmental statutes. Two types of enforcement officers, environmental enforcement officers and wildlife enforcement officers, are employed to ensure social compliance with CEPA legislation. The environmental enforcement officers administer regulations from CEPA and pollution provisions from the Fisheries Act. The wildlife enforcement officers ensure compliance with the Migratory Birds Convention Act, Canada Wildlife Act, and the Species at Risk Act (see Chapter 7), among others.

On June 18, 2009, the Environmental Enforcement Act was passed. This act increases the enforcement capabilities of Environment Canada, calling for larger fines and penalties for individuals and corporations failing to comply with CEPA legislation. It recognizes the severity of the damage caused by environmental offences and the need to correspondingly increase the severity of the associated fines and penalties.

Accomplishments of environmental legislation

Today, Environment Canada administers over two dozen acts and assists with the administration of many others. These acts address a wide range of environmental issues, such as endangered species, clean water, clean air, energy conservation, hazardous wastes, and pesticides. Despite many challenges, environmental legislation has had an overall positive effect on environmental protection in Canada. Here are some highlights of what has been accomplished with the help of environmental legislation:

- Twenty-three national parks and three national marine conservation areas have been established since 1970. National parks are located in every one of the nation's 13 provinces and territories. (See Chapter 8.)

- Many endangered species are recovering with the help of environmental legislation such as the whooping crane, peregrine falcon, swift fox, white pelican, wood bison, and ferruginous hawk.

- The amount of strictly protected area in Canada has increased from over 36 million hectares in 1992 to over 60 million hectares in 2001. Protected land has emerged as the crucial instrument to preserving national biodiversity. Despite the increase in strictly protected land area in Canada, there is still more work to be done. Of the 194 terrestrial ecoregions in Canada, 113 have some strictly protected area; however, this leaves 81 ecoregions in Canada with little or no protection.

- The total volume of toxic chemicals released into Canadian surface waters dropped by 63 percent between 1994 and 1999, although toxic chemicals are still causing significant water pollution through agricultural runoff, urban runoff, and airborne deposits of toxic chemicals.

- The percentage of the municipal population receiving some form of sewage treatment in Canada has risen from 72 percent of the population in 1983 to 97 percent in 1999. Treatment plants are important for reducing water pollution since they can remove significant levels of pollutants from municipal urban wastewater, which represents one of the largest sources of pollutants to Canadian waters.

- There was a measured overall improvement in air quality in major cities across Canada between 1979 and 1992. Environment Canada, through the Clean Air Act of 1969 and National Air Pollution Surveillance (the provincial–federal program launched to measure air quality in Canada), continues to establish limits on common air pollutants to protect human health and the environment.

CONCEPT CHECK **STOP**

What authorities did the BNA Act give to the provinces that relate to the environment?

What are environmental impact assessments and why are they important?

What are the key pieces of environmental legislation in Canada?

Human Values and Environmental Sustainability

LEARNING OBJECTIVES

Define *environmental ethics.*

Discuss distinguishing features of the Western, deep ecology, and Aboriginal worldviews.

Define *environmental justice.*

We now shift our attention to the views of different individuals and societies and how those views affect our ability to understand and solve sustainability problems. **Ethics** is the branch of philosophy that is derived through the logical application of human **values**. These values are the principles that an individual or a society considers important or worthwhile. Values are not static entities but change as societal, cultural, political, and economic priorities change. Ethics help us determine which forms of conduct are morally acceptable and unacceptable, right and wrong. Ethics play a role in any type of human activity that involves intelligent judgement and voluntary action. Whenever alternative, conflicting values occur, ethics help us choose which value is better, or worthier, than other values.

Environmental ethics examines moral values to determine how humans should relate to the natural environment. Environmental ethicists consider such issues as what role we should play in determining the fate of

environmental ethics A field of applied ethics that considers the moral basis of environmental responsibility.

Earth's resources, including other species, or how we might develop an environmental ethic that is acceptable in the short term for us as individuals and also in the long term for our species and the planet. These issues and others like them are difficult intellectual questions that involve political, economic, societal, and individual trade-offs.

Environmental ethics considers not only the rights of people living today, both individually and collectively, but also the rights of future generations (**FIGURE 2.10** on page 48). This aspect of environmental ethics is critical because the impacts of today's activities and technologies are changing the environment. In some cases these impacts may be felt for hundreds or even thousands of years. Addressing issues of environmental ethics puts us in a better position to use science and technology for long-term environmental sustainability.

WORLDVIEWS

Each of us has a particular **worldview**—that is, a perspective based on a collection of our basic values that helps us make sense of the world, understand our place and purpose in it, and determine right and wrong behaviours. These worldviews lead to behaviours and lifestyles that may or may not be compatible with environmental sustainability.

Tomorrow's generation FIGURE 2.10

The choices made today will determine whether future generations, such as these students will inherit a sustainable world.

Western worldview FIGURE 2.11

A Early logging operation using oxen on Cowichan Lake, Vancouver Island.

Global Locator

B The Western worldview in action today. These logs were obtained illegally from a protected Indonesian forest.

environmental worldview

A worldview based on how the environment works, our place in the environment, and right and wrong environmental behaviours.

Western worldview

A worldview based on human superiority over nature, the unrestricted use of natural resources, and economic growth to manage an expanding industrial base.

Two extreme, competing **environmental worldviews** are the Western worldview and the deep ecology worldview. These two worldviews, admittedly broad generalizations, are at nearly opposite ends of a spectrum of worldviews relevant to global sustainability problems, and each approaches environmental responsibility in a radically different way.

The traditional **Western worldview**, also known as the *expansionist worldview,* is human centred and utilitarian. It mirrors the beliefs of the 19th-century frontier attitude, a desire to conquer and exploit nature as quickly as possible (**FIGURE 2.11**). The Western worldview also advocates the inherent rights of individuals, accumulation of wealth, and unlimited consumption of goods and services to provide material comforts. According to the Western worldview, humans have a primary obligation to humans and are therefore responsible for managing natural resources to benefit human society. Thus, any concerns about the environment are derived from human interests.

The **deep ecology worldview** is a diverse set of viewpoints that dates from the 1970s and is based on the work of Arne Naess, a Norwegian philosopher, and others, including ecologist Bill Devall and philosopher George Sessions. The principles of deep ecology, as expressed by Naess in *Ecology, Community and Lifestyle* (1989), include the following:

> **deep ecology worldview** A worldview based on harmony with nature, a spiritual respect for life, and the belief that humans and all other species have an equal worth.

1. Both human and nonhuman life has intrinsic value. The value of nonhuman life forms is independent of the usefulness they may have for narrow human purposes.

2. Richness and diversity of life forms contribute to the flourishing of human and nonhuman life on Earth (**FIGURE 2.12**).

3. Humans have no right to reduce this richness and diversity except to satisfy vital needs.

4. Present human interference with the nonhuman world is excessive, and the situation is rapidly worsening.

5. The flourishing of human life and cultures is compatible with a substantial decrease in the human population. The flourishing of nonhuman life requires such a decrease.

6. Significant change of life conditions for the better requires changes in economic, technological, and ideological structures.

7. The ideological change is mainly that of appreciating life quality rather than adhering to a high standard of living.

8. Those who subscribe to the foregoing points have an obligation to participate in the attempt to implement the necessary changes.

Preservation of biological diversity is an important part of the deep ecology worldview FIGURE 2.12

A Western hemlock trees, ferns, mosses, and other plants in Olympic National Park, Washington.

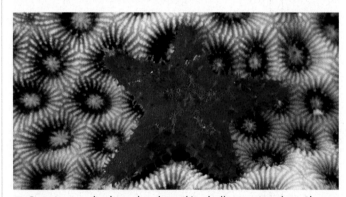

B Sea star perched on a hard coral in shallow water along the coast of Vanuatu in the South Pacific Ocean.

C An African black leopard, also known as a black panther, in a tree. Leopards often hide their food in trees to avoid sharing it with lions and hyenas.

Compared with the Western worldview, the deep ecology worldview represents a radical shift in how humans relate to the environment. The deep ecology worldview stresses that all forms of life have the right to exist and that humans are not different or separate from other organisms. Humans have an obligation to themselves and to the environment. The deep ecology worldview advocates sharply curbing human population growth. It does not advocate returning to a society free of today's technological advances, but instead proposes a significant rethinking of our use of current technologies and alternatives. It asks individuals and societies to share an inner spirituality connected to the natural world.

Most people today do not fully embrace either the Western worldview or the deep ecology worldview. The Western worldview is **anthropocentric** and emphasizes the importance of humans as the overriding concern in the grand scheme of things. In contrast, the deep ecology worldview is **biocentric** and views humans as one species among others. The planet's natural resources could not support its more than 6.7 billion humans if each consumed the high level of goods and services sanctioned by the Western worldview. On the other hand, the world as envisioned by the deep ecology worldview could support only a fraction of the existing human population (**Figure 2.13**).

In addition to the Western and deep ecology worldviews, Aboriginal peoples in Canada and throughout the world have also developed their own worldview based on their cultural and spiritual values. The **Aboriginal worldview**, passed from generation to generation through traditional stories, rituals, and spiritual beliefs, is an important component of modern Aboriginal beliefs and lifestyles. Although many different Aboriginal peoples are found in Canada and throughout the world, several common beliefs are evident in their various cultures. One of the most prevalent is the interconnectedness among all living and nonliving things, which leads to the belief that

> **Aboriginal worldview** A worldview that emphasizes the interconnectedness among all living and nonliving things. The belief that humans need to live in harmony with each other and with nature is integral to this worldview.

Embracing deep ecology FIGURE 2.13

At one time or another, most of us yearn for the simpler life that the tenets of deep ecology advocate. However, there are far too many people and far too little land for us all to embrace this lifestyle. Photographed on Gotland Island, Sweden.

humans need to live in harmony with each other and with nature.

Since many Aboriginal peoples believe that humans are connected with nature and everyone has a responsibility to respect others, including animals, birds, plants, and inanimate objects, this worldview teaches honour

The Aboriginal worldview maintains a deep connection between humans and nature. Symbols of animals and nature are often incorporated into traditional dress, ceremonies, and spiritual objects.

and respect for the animals that are hunted, the fish that are caught, the plants and their products that are eaten (**FIGURE 2.14**). To ensure future survival and continued existence, the Aboriginal worldview requires that respect for the land and all of its benefits must be preserved. The sum total of all of the beliefs and values of Aboriginal peoples is often referred to as **traditional knowledge**.

More and more often in Canada, regulations are requiring government and private sector officials to recognize the traditional knowledge of Aboriginal peoples. When decisions are made about resource development, the possible impacts as identified by the Aboriginal worldview are becoming increasingly important. In many ways, the Aboriginal worldview reflects current concerns about the environment as expressed by various features of the Western worldview as well as the deep ecology worldview.

All three of these worldviews are useful to keep in mind as you examine various environmental issues in later chapters. In the meantime, you should think about your own worldview and discuss it with others—whose worldviews will probably be different from your own. Thinking leads to actions, and actions lead to consequences. What are the short-term and long-term consequences of your particular worldview? We must develop and incorporate into our culture a long-lasting, environmentally sensitive worldview if the environment is to be sustainable for us, for other living organisms, and for future generations.

ENVIRONMENTAL JUSTICE

In the early 1970s, the Board of National Ministries of the American Baptist Churches coined the term *eco-justice* to link social and environmental ethics. At the local level, eco-justice encompasses environmental inequities faced by low-income minority communities. Many studies indicate that low-income communities and/or communities of visible minorities are more likely than others to have chemical plants, hazardous waste facilities, sanitary landfills, sewage treatment

Poor minority neighbourhoods often have the most polluted and degraded environments.

plants, and incinerators (**FIGURE 2.15**). A 1990 study at Clark Atlanta University, for example, found that six of eight incinerators in Houston, Texas, were located in predominantly black neighbourhoods. Such communities often have limited involvement in the political process and may not even be aware of their exposure to increased levels of pollutants.

This increased exposure can lead to significant health concerns. The high incidence of asthma in many minority communities may be caused or exacerbated by exposure to environmental pollutants. Few studies, except those documenting lead contamination, have examined how environmental pollutants interact with other socioeconomic factors to cause health problems.

Currently, we have little scientific evidence showing to what extent a polluted environment is responsible for the disproportionate health problems of poor and minority communities. In First Nations communities of northwestern Ontario, for example, high concentrations of mercury in fish that were caused by effluents from pulp and paper mills lead to increased incidence of mercury poisoning in area residents. Recommended restrictions on consumption of fish in the diet of local residents were provided by public health agencies as a result.

Environmental justice and ethical issues

There is an increasing awareness that environmental decisions, such as where to locate a hazardous waste

environmental justice The right of every citizen to adequate protection from environmental hazards.

landfill, have important ethical dimensions. The most basic ethical dilemma centres on the rights of the poor and disenfranchised versus the rights of the rich and powerful. Whose rights should have priority in these decisions? The challenge is to find and adopt solutions that respect all social groups, including those yet to be born. **Environmental justice** is a fundamental human right in an ethical society. Although we may never completely eliminate past environmental injustices, we have a moral imperative to prevent them today so that their negative effects do not disproportionately affect any particular segment of society.

In response to these concerns, a growing environmental justice movement has emerged at the grassroots

level as a strong motivator for change. Advocates are calling for special efforts to clean up hazardous sites in low-income neighbourhoods, from inner-city streets to Aboriginal reserves. Many advocates cite the need for more research on human diseases that environmental pollutants may influence.

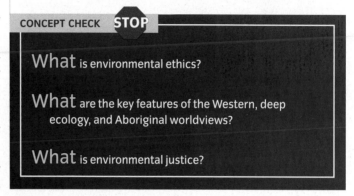

CONCEPT CHECK STOP

What is environmental ethics?

What are the key features of the Western, deep ecology, and Aboriginal worldviews?

What is environmental justice?

Strategies for Sustainable Living

LEARNING OBJECTIVES

Relate poverty and population growth to carrying capacity and global sustainability.

Discuss problems related to loss of forests and declining biological diversity.

Describe the extent of food insecurity.

Define *enhanced greenhouse effect* and explain how stabilizing climate is related to energy use.

Describe at least two problems in cities in the developing world.

There is no shortage of suggestions for ways to address the world's many environmental problems. We have organized this section around the five recommendations for sustainable living presented in the 2006 book *Plan B 2.0: Rescuing a Planet Under Stress and a Civilization in Trouble* by Lester R. Brown. If we as individuals and collectively as governments were to focus our efforts and financial support on Brown's plan, we think the quality of human life would be much improved. Brown's five recommendations for sustainable living are as follows:

1. Eliminate poverty and stabilize the human population.

2. Protect and restore Earth's resources.

3. Provide adequate food for all people.

4. Mitigate climate change.

5. Design sustainable cities.

Seriously addressing these recommendations offers hope for the kind of future we want for our children and grandchildren (**FIGURE 2.16** on page 54).

A plan for sustainable living FIGURE 2.16

A **Family planning in Egypt** Women at a health clinic learn about family planning and birth control.

RECOMMENDATION 1: ▶ ELIMINATE POVERTY AND STABILIZE THE HUMAN POPULATION.

B **Young trees in Scotland** These evergreens are being cultivated as part of a reforestation project on land unsuitable for growing crops.

RECOMMENDATION 2: ▶

PROTECT AND RESTORE EARTH'S RESOURCES.

VIEW THIS IN ACTION

D **Apartment buildings with solar panels in the modern city Orot, Israel.**

RECOMMENDATION 4: ▶ MITIGATE CLIMATE CHANGE.

C **Fish breeding pens off the coast of Norway.** Fish farming is expanding at an unprecedented scale, in part because wild fisheries are being over fished.

RECOMMENDATION 3: ▶

PROVIDE ADEQUATE FOOD FOR ALL PEOPLE.

E **Bicycles at a train station in Amsterdam.** Each resident in the Netherlands rides a bicycle an average of 917 kilometres per year.

RECOMMENDATION 5: ▶ DESIGN SUSTAINABLE CITIES.

NATIONAL GEOGRAPHIC

RECOMMENDATION 1: ELIMINATE POVERTY AND STABILIZE THE HUMAN POPULATION

The ultimate goal of economic development is to make it possible for humans throughout the world to enjoy long, healthy lives. A serious complication lies in the fact that the distribution of the world's resources is unequal. We who live in North America are collectively the wealthiest people who have ever existed, with the highest standard of living (shared with a few other rich countries). Canada, with less than 1 percent of the world's people, typically ranks among the top 10 wealthiest countries but depends on other nations for this prosperity. Yet we often seem unaware of this relationship and tend to underestimate our effects on the environment that supports us.

Failing to confront the problem of poverty around the world makes it impossible to attain global sustainability (**FIGURE 2.17**). For example, most people would find it unacceptable that about 29,000 infants and children under the age of five die each day (2006 data from UN Children's Fund). Most of these deaths could have been prevented through access to adequate food and basic medical techniques and supplies. For us to allow so many to go hungry and to live in poverty threatens the global ecosystem that sustains us all. Everyone must have a reasonable share of Earth's productivity.

Raising the standard of living for poor countries requires the universal education of

carrying capacity
The maximum population that can be sustained by a given environment or by the world as a whole.

children and the elimination of illiteracy. Improving the status of women is crucial because women are often disproportionately disadvantaged in poor countries. In many developing countries, women have few rights and little legal ability to protect their property, their rights to their children, and their income.

We have entered an era of global trade, within which we must establish guidelines for national, corporate, and individual behaviours. For example, the flow of money from developing countries to highly developed countries has exceeded the flow in the other direction for many years. Former West German Chancellor Willy Brandt termed this phenomenon "a blood transfusion from the sick to the healthy." A world that values social justice and environmental sustainability must reverse this flow. Debts from the poorest countries should be forgiven more readily than they are now, and international development assistance should be enhanced.

Population growth rates are generally highest where poverty is most intense. If we pay consistent attention to overpopulation and devote the resources necessary to make family planning available for everyone, the human population will stabilize. If we do not continue to emphasize family planning measures, we simply will not achieve population stability.

To stay within Earth's **carrying capacity** we must reach and sustain a stable population and reduce excessive consumption. These goals must be coupled with educational programs everywhere, so that people understand that Earth's carrying capacity is not unlimited. There is no hope for a peaceful world without overall population stability, and there is no hope for regional economic sustainability without regional population stability.

RECOMMENDATION 2: PROTECT AND RESTORE EARTH'S RESOURCES

To build a sustainable society, we must preserve the natural systems that support us. The conservation of nonrenewable resources, such as oil and minerals, is obvious, although discoveries of new supplies of nonrenewable resources sometimes give the illusion that they are inexhaustible. Renewable resources such as forests, biodiversity, soils, fresh water, and fisheries must

Child at work FIGURE 2.17

This child works as a labourer at a quarry in Accra, Ghana. He was blinded from an accident at the quarry.

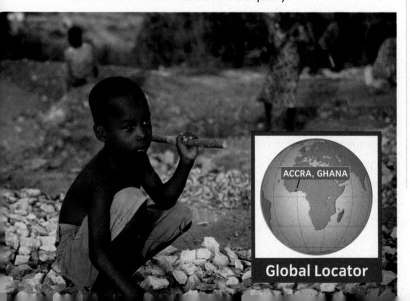

ACCRA, GHANA

Global Locator

be used in ways that ensure their long-term productivity. Their capacity for renewal must be understood and respected. However, renewable resources have been badly damaged over the past 200 years. Until environmental sustainability becomes a part of economic calculations, susceptible natural resources will continue to be consumed unsustainably, driven by short-term economics.

The world's forests

Many of the world's forests are being cut, burned, or seriously altered at a frightening rate. For example, logging in British Columbia, Oregon, Washington, Alaska, and Siberia is destroying old-growth forests. Tropical forests are also being overexploited. Many products come to the industrialized world from the tropics (hardwoods; foods such as beef, bananas, coffee, and tea; medicines). As trees are destroyed, only a small fraction of them are replanted.

The pressure of rapid population growth and widespread poverty also harms the world's forests. In many developing countries, forests have traditionally served as a "safety valve" for the poor, who, by consuming small tracts of forest on a one-time basis and moving on, find a source of food, shelter, and clothing. But now the numbers of people in developing countries are too great for their forests to support. Tropical rainforests—biologically the world's richest terrestrial areas—have been reduced to less than half their original area. Methods of forest clearing that were suitable when population levels were lower and forests had time to recover from temporary disturbances simply do not work any longer. They convert a potentially renewable resource into an unsustainable one.

Loss of biodiversity

We have a clear interest in protecting Earth's **biological diversity** and managing it

biological diversity The number and variety of Earth's organisms.

sustainably because we obtain from living organisms all our food, most medicines, many building and clothing materials, biomass for energy, and numerous other products. In addition, organisms and the natural environment provide an array of **ecosystem services** without which we would not survive. These services include the protection of watersheds and soils, the development of fertile agricultural lands, the determination of both local climate and global climate, and the maintenance of habitats for animals and plants.

Over the next few decades, we can expect human activities to cause the rate of extinction to increase to perhaps hundreds of species a day. How big a loss is this? Unfortunately, we still have a limited understanding about the world's biological diversity. An estimated five-sixths of all species have not yet been scientifically described. Some 80 percent of the species of plants, animals, fungi, and microorganisms on which we depend are found in developing countries. How will these relatively poor countries sustainably manage and conserve these precious resources? Biological diversity is an intrinsically local problem, and each nation must address it for the sake of its own people's future, as well as for the world at large. Like most other challenges of sustainable development, biological diversity can be addressed adequately only if we provide international assistance where needed, including help in training scientists and engineers from developing countries.

Biological diversity and human cultural diversity are intertwined: They are, in fact, two sides of the same coin. **Cultural diversity** is Earth's variety of human communities, each with its individual languages, traditions, and identities (**FIGURE 2.18**). Cultural diversity enriches the collective human experience. For that reason, the United Nations Educational, Scientific, and Cultural Organization (UNESCO) supports the protection of minorities in the context of cultural diversity.

RECOMMENDATION 3: PROVIDE ADEQUATE FOOD FOR ALL PEOPLE

Globally, more than 800 million people lack access to the food needed for healthy, productive lives. This estimate, according to a 2004 report by the UN Food and Agriculture Organization, includes a high percentage of children. Children are particularly susceptible to food deficiencies because their brains and bodies cannot develop properly without adequate nutrition. Most malnourished people live in rural areas of the poorest developing nations. The link between poverty and **food insecurity** is inescapable.

Improving agriculture is one of the highest priorities for achieving global sustainability. In general, grain production per person has kept pace with human population growth over

food insecurity The condition in which people live with chronic hunger and malnutrition.

Humans are part of the web of life FIGURE 2.18

Yanomami children in Brazil enjoy a photographer's camera. Intrusion into isolated areas such as the Amazon basin threatens both biological diversity and the cultures of indigenous people who have lived in harmony with nature for hundreds of generations.

Global Locator

NORTH AMERICA
BRAZIL
SOUTH AMERICA

the past 50 years. However, expanded agricultural productivity has taken place at high environmental costs. Moreover, the global population continues to expand, putting additional pressure on food production.

Worldwide, little additional land that is not currently under cultivation is suitable for agriculture. One way to increase the productivity of agricultural land is through **multi-cropping**, or growing more than one crop per year. For example, winter wheat and summer soybean crops are grown in some areas of North America. However, multi-cropping can be accomplished only in regions where water supplies are adequate for irrigation. Also, care must be taken to prevent a decline in soil fertility from such intensive use.

The negative environmental effects of agriculture, including loss of soil fertility, soil erosion, aquifer depletion, soil and water pollution, and air pollution, must be brought under control (**FIGURE 2.19**). Many strategies exist to retard the loss of topsoil, conserve water and energy, and reduce the use of agricultural chemicals. For example, in **conservation tillage**, residues from previous crops are left in the soil, partially covering it and helping to hold topsoil in place (see Chapter 8 for more on agricultural land uses).

We must develop sustainable agricultural systems that provide improved dietary standards, such as the inclusion of high-quality protein in diets in developing countries. China's expanding use of **aquaculture** is an

Damage to soil resources FIGURE 2.19

Extensive withdrawal of water contributed to erosion in this once-fertile area of Kenya. Careful stewardship of the land prevents such damage.

AFRICA
KENYA

Global Locator

example of efficient protein production. The carp that are raised in Chinese aquaculture are efficient at converting food into high-quality protein. In China, fish production by aquaculture now exceeds poultry production. However, aquaculture, like all human endeavours, has negative environmental effects that must be addressed for it to be sustainable on a large scale.

RECOMMENDATION 4: MITIGATE CLIMATE CHANGE

A widely discussed human effect on the environment is climate change caused by the **enhanced greenhouse effect**. The most important greenhouse gas, carbon dioxide or CO_2, is produced when we burn fossil fuels—coal, oil, and natural gas.

enhanced greenhouse effect The additional warming produced by increased levels of gases that absorb infrared radiation.

Although Earth's climate has been relatively stable during the past 10,000 years, human activities are causing it to change. The average global temperature increased by almost 1° C during the 20th century; more than half of that warming occurred during the past 30 years. Climate scientists generally agree that Earth's climate will continue to change rapidly during the 21st century.

We must address climate change in an aggressive and coordinated fashion, but how do we get all nations of the world to adopt the necessary approaches? Many policy makers say that we should wait until scientific knowledge of climate change is complete. This reasoning is flawed because Earth's climate system is extremely complex, and we may never completely understand it.

For example, we often say that an increase in atmospheric CO_2 leads to climate warming. However, the increase in CO_2, like other human impacts, is not a simple cause-and-effect relationship but instead a cascade of interacting responses that ripple through the environment (**Figure 2.20**) We cannot begin to predict how these changes will affect humans or other organisms.

Stabilizing the climate requires a comprehensive energy plan to include phasing out fossil fuels in favour of renewable energy (such as solar and wind power), increasing energy conservation, and improving energy efficiency. Many national and local governments as well as corporations and environmentally aware individuals are setting goals to cut carbon emissions. Other nations, however, have not recognized the urgency of the global climate problem. We need a global consensus to address climate change. Chapter 10 discusses the enhanced greenhouse effect and global climate change in detail.

RECOMMENDATION 5: DESIGN SUSTAINABLE CITIES

At the beginning of the Industrial Revolution, in approximately 1800, only 3 percent of the world's people lived in cities, and 97 percent were rural, living on farms or in small towns. In the two centuries since then, population distribution has changed radically—toward the cities. More people live in Mexico City today than were living in all the cities of the world 200 years ago. This is a staggering difference in the way people live. Almost 50 percent of the world's population now lives in cities, and the percentage continues to grow. In industrialized countries such as the United States and Canada, almost 80 percent of the people live in cities.

City planners around the world are trying a variety of approaches to make cities more livable. Many cities are developing urban transportation systems to reduce the use of cars and the problems associated with them, such as congested roads, large areas devoted to parking, and air pollution. Urban transportation ranges from mass transit subways and light rails to pedestrian and bicycle pathways.

Investing in urban transportation in ways other than building more highways encourages commuters to use forms of transportation other than automobiles. When a city is built around people instead of cars—such as establishing parks and open spaces instead of highways and parking lots—urban residents have an improved quality of life. Air pollution, including emission of climate-warming CO_2, is substantially reduced.

Water scarcity is a major issue for many cities of the world. Some city planners think that innovative approaches must be adopted where water resources are scarce. For example, some cities, such as Singapore, recycle some of their wastewater after it has been treated.

Effectively dealing with the problems in squatter settlements is an urgent need. Evicting squatters does not

Cascading responses of increased carbon dioxide through the environment FIGURE 2.20

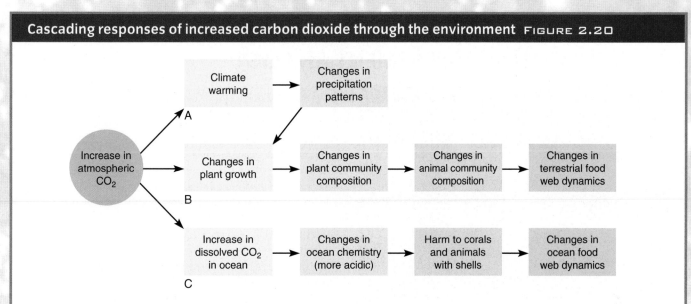

A Most people know that an increase in atmospheric CO_2 leads to global warming. But this phenomenon is far from a simple cause-and-effect relationship. Increasing CO_2 may cause a cascade of interacting responses throughout the Earth system. **B** Effects of increased atmospheric CO_2 on land plants. **C** Effects of increased atmospheric CO_2 on the ocean. Both **B** and **C** lead to increased extinctions.

Squatter settlement FIGURE 2.21

Manila, in the Philippines, is a city of contrasts, with gleaming modern skyscrapers and abjectly poor squatter settlements.

PHILIPPINES

Global Locator

address the underlying problem of poverty. Instead, cities should incorporate some sort of plan for the eventual improvement of squatter settlements (**FIGURE 2.21**). Providing basic services—such as clean water to drink, transportation (so people can find gainful employment), and garbage pickup—would help improve the quality of life for the poorest of the poor.

CONCEPT CHECK STOP

What is the global extent of poverty?

What are two ecosystem services that natural resources such as forests and biological diversity provide?

What is food insecurity?

How is stabilizing climate related to energy use?

What are two serious problems in urban environments?

Strategies for Sustainable Living   59

ENVIRONMENTAL ASSESSMENT ANALYST

Environmental assessment analysts ensure that the environmental effects of proposed projects are identified, assessed, and mitigated.

Imagine you are standing in the quiet shade of an undisturbed forested area in the northern part of the province. Until a few years ago, there wasn't much interest in this area, but the discovery of a large natural gas reserve lying underneath the forest has the potential to change all that. An energy corporation is proposing to tap this natural gas field and build an underground pipeline to carry the natural gas to southern markets. You are an environmental assessment analyst and you and your team are visiting the site as part of an environmental assessment on the proposed project. Your job is to ensure the project's potential environmental effects are identified, assessed, and mitigated, and that accurate information is provided to decision makers to decide whether the project should proceed.

As the lead environmental assessment analyst for this project, you must decide what kind of assessment the project needs and what provincial, federal, and other environmental legislation applies. Since the pipeline is a large-scale project, federal funds and regulator approvals are required, and because the area is very sensitive to human disturbance, it has been decided that a comprehensive study level of assessment is required. A comprehensive study is an intensive environmental assessment under federal environmental assessment legislation designed to identify, assess, and mitigate adverse environmental effects and evaluate the significance of the residual effects of a proposed development.

You and your team will spend months gathering data and information from the site and reviewing case studies from similar developments. You will also spend time consulting with area residents and members of a local Aboriginal community to gather their comments on the pipeline, as well as posting assessment information on the Internet for additional feedback from the public.

Once you have all required information, you will prepare an environmental assessment report that outlines the potential environmental consequences of the development and provides conclusions on the significance of the residual effects on the environment. If the project is approved, you may also be involved in following up on the assessment, for example, monitoring during pipeline construction to ensure that mitigation measures are implemented and effective and the residual effects remain insignificant.

Environmental assessment analysts research and analyze environmental data and information for the preparation of environmental assessment reports in accordance with federal (i.e., Canadian Environmental Assessment Act) and provincial environmental assessment legislation. Environmental assessment analysts evaluate proposed projects and provide factual information for effective planning and decision making that promotes public awareness, environmental protection and management, and sustainable development.

For more information or to look up other environmental careers go to www.eco.ca.

JAKARTA, INDONESIA

There is an urgent need to improve the environment and quality of life in cities, particularly in rapidly growing cities of developing countries. Consider Jakarta, Indonesia, which had a population of 16.9 million in 2006 and has continued to grow since then. (This number includes several nearby cities that have merged with Jakarta.)

Jakarta is plagued with many of the problems found in other rapidly growing cities in the developing world. The air in Jakarta is badly polluted with the exhaust from cars, buses, and motorbikes that transport about two million commuters into the city each day. Water pollution is also a critical problem. At least 95 percent of human wastes produced in the city are not cleaned up at sewage treatment plants. Instead, human sewage and garbage are dumped directly into nearby rivers. Jakarta's municipal water supply is so polluted that piped water must be boiled before people can drink it.

As groundwater has been depleted to meet the city's needs, parts of Jakarta have subsided so that many areas are increasingly flood prone, particularly during the rainy season (see photograph). Illegal squatter settlements proliferate; the poorest inhabitants build dwellings on vacant land using whatever materials they can scavenge. As in other cities around the world, in Jakarta, squatter settlements have the worst water, sewage, and solid waste problems.

Although we have painted a grim picture of Jakarta, it should be noted that improvements are slowly beginning to occur. As part of an ambitious long-term transportation upgrade, Jakarta has developed a busway with dedicated bus lines that reduce commuter times and encourage people to commute by bus rather than by automobile. A monorail and subway system are also planned.

NATIONAL GEOGRAPHIC

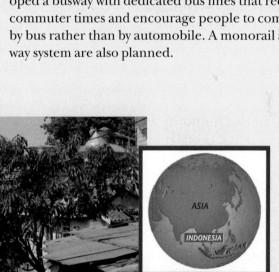

Global Locator

Poorly constructed houses lining a canal in Jakarta

SUMMARY

1 History of the Environmental Movement

1. The first two centuries of North American history were a time of widespread environmental destruction. During the 1700s and early 1800s, most North Americans had a desire to conquer and exploit nature as quickly as possible. During the 19th century, many naturalists became concerned about conserving natural resources. The earliest conservation legislation revolved around protecting land—forests, parks, and monuments. By the late 20th century, environmental awareness had become a pervasive popular movement.

2. Many writers, artists, and hunters during the 19th and 20th centuries had a profound impact on the way people viewed the environment. John James Audubon's art aroused widespread interest in the wildlife of North America. Henry David Thoreau wrote about living in harmony with the natural world. George Perkins Marsh wrote about humans as agents of global environmental change. Hunters and fishers were the backbone of the North American Model of Wildlife Conservation, whose two basic principles—that our fish and wildlife belong to all North American citizens and the need for sustainable management—are still persistent environmental doctrines today. In *A Sand County Almanac,* Aldo Leopold wrote about humanity's relationship with nature. Rachel Carson published *Silent Spring,* alerting the public about the dangers of uncontrolled pesticide use. Paul Ehrlich published *The Population Bomb,* which raised the public's awareness of the dangers of overpopulation.

3. The last 40 years has seen dramatic growth in the environmental movement. The first Earth Day was held on April 22, 1970, and has since become a worldwide phenomenon. Greenpeace has had a profound impact on lobbying governments and bringing environmental issues to the public's attention. The David Suzuki Foundation has found and communicated practical ways for individuals to make changes that will achieve balance between human needs and Earth's capacity to meet those needs.

2 Jurisdictional Authority and Environmental Legislation

1. In Canada, jurisdictional responsibility between the federal and provincial governments was established under the 1867 British North America Act. The federal government has jurisdiction over anything that might affect all Canadians, which clearly would include the environment under many circumstances. The provinces have jurisdiction over renewable and nonrenewable resources. Since 1971, with the incorporation of Environment Canada as the department responsible for preserving and enhancing the nation's natural environment, the federal government has addressed many environmental problems.

2. The Canadian Environmental Assessment Act was passed in 1992 and established the need for appropriate environmental impact assessments (EIAs) for all federal projects. These assessments help the government consider the environmental impacts of all proposed projects, and if the environmental costs are considered too severe, alternative actions are considered.

3. In 1999, the Canadian Environmental Protection Act (CEPA) was passed, which initiated serious environmental protection in Canada. CEPA addresses toxic contamination of Canada's air, water, and land, and gives the government the ability to enforce strict penalties on polluters.

3 Human Values and Environmental Sustainability

1. **Environmental ethics** is a field of applied ethics that considers the moral basis of environmental responsibility and how far this responsibility extends. Environmental ethicists consider how humans should relate to the natural environment.

2. An **environmental worldview** is a worldview that helps us make sense of how the environment works, our place in the environment, and right and wrong environmental behaviours.

The **Western worldview** is an understanding of our place in the world based on human superiority and dominance over nature, the unrestricted use of natural resources, and increased economic growth to manage an expanding industrial base. The **deep ecology worldview** is an understanding of our place in the world based on harmony with nature, a spiritual respect for life, and the belief that humans and all other species have equal worth. The **Aboriginal worldview** is an understanding of the interconnectedness between humans and all other living organisms and the need to live in harmony with each other and with nature.

3. **Environmental justice** is the right of every citizen, regardless of age, race, gender, social class, or other factor, to adequate protection from environmental hazards. Environmental justice is a fundamental human right in an ethical society. A growing environmental justice movement has emerged at the grassroots level.

4 Strategies for Sustainable Living

1. Failing to confront the problem of poverty makes it impossible to attain global sustainability. To stay within Earth's **carrying capacity**, the maximum population that can be sustained indefinitely, it will be necessary to reach a stable population and reduce excessive consumption.

2. The world's forests are being cut, burned, and seriously altered for timber and other products that the global economy requires. Also, rapid population growth and poverty are putting pressure on forests. **Biological diversity**, the number and variety of Earth's organisms, is declining at an alarming rate. Humans are part of Earth's web of life and are entirely dependent on that web for survival.

3. **Food insecurity** is the condition in which people live with chronic hunger and malnutrition. Globally, more than 800 million people lack access to the food needed for healthy, productive lives.

4. The **enhanced greenhouse effect** is the additional warming produced by increased levels of gases that absorb infrared radiation. An increase in atmospheric CO_2, mostly produced when fossil fuels are burned, leads to climate warming. To stabilize climate, we must phase out fossil fuels in favour of renewable energy, increased energy conservation, and improved energy efficiency.

5. The air in cities in the developing world is badly polluted with exhaust from motor vehicles. Illegal squatter settlements proliferate in cities; the poorest inhabitants build dwellings using whatever materials they can scavenge. Squatter settlements have the worst water, sewage, and solid waste problems.

KEY TERMS

1. Describe how the North American Wildlife Conservation Model came about. How did this model impact Canadian wildlife?

2. What are some of the accomplishments achieved by the environmental legislation established in the last quarter of the 20th century? Take a moment to think about the current level of environmental protection that exists in Canada. Is this level of protection suitable? What are some the challenges that still exist with environmental protection?

3. What is the message of this cartoon? What environmental issue probably inspired it, and when was it likely published?

4. What social groups generally suffer the most from environmental pollution and degradation? What social groups generally benefit from this situation?

5. In its broadest sense, how does environmental justice relate to highly developed countries and less developed countries?

6. What is the role of education in changing personal attitudes and practices that affect the environment?

7. State whether each of the following statements reflects the Western worldview, the deep ecology worldview, or both. Explain your answers.
   a. Species exist to be used by humans.
   b. All organisms, humans included, are interconnected and interdependent.
   c. There is a unity between humans and nature.
   d. Humans are a superior species capable of dominating other organisms.
   e. Humans should protect the environment.
   f. Nature should be used, not preserved.
   g. Economic growth will help Earth manage an expanding human population.
   h. Humans have the right to modify the environment to benefit society.
   i. All forms of life are intrinsically valuable and therefore have the right to exist.

8. Write a paragraph that briefly describes your environmental worldview.

9. Write a one-page essay describing what kind of world you want to leave for your children.

10. The graphs below show a computer simulation by the U.S. National Climate Assessment. In the first graph (top), the level of atmospheric CO_2 is projected for the 21st century. As a result of increasing levels of CO_2 in the atmosphere, more CO_2 dissolves in ocean water, where it forms carbonic acid. The increasing acidity dissolves and weakens coral skeletons, which are composed of calcium carbonate (shown in the bottom graph). (Values in parts A and B are midrange projections.)

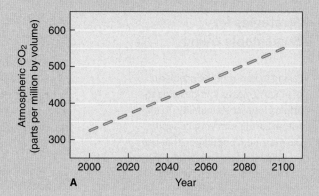

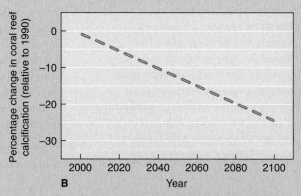

   a. Why could rising CO_2 levels in the atmosphere be catastrophic to corals and other shell-forming organisms?
   b. How do these graphs relate to Figure 2.20 (page 59)?

11. Many conservationists today think that human health and well-being should be an important part of conservation efforts. Explain why humans and the natural world are interconnected, using tropical rainforests and indigenous people as an example.

12. Are an improved standard of living and a reduction in the level of consumption mutually exclusive? Give an example that supports your answer.

These fishermen have seen their catches decline in recent years. Why do you think fish shortages are occurring around the world?

How does Garrett Hardin's description of the tragedy of the commons in medieval Europe relate to the ocean's fisheries today?

Name an additional example of a global commons other than the one shown in this photograph.

Human Populations

NATIONAL GEOGRAPHIC

SLOWING POPULATION GROWTH IN CHINA

Of all the countries in the world, China, with an estimated population of 1.3 billion people, has the largest population. Recognizing that its rate of population growth had to decrease or the quality of life for everyone in China would be compromised, in 1971 the Chinese government began to pursue birth control policies seriously. It urged couples to marry later, increase spacing between children, and have fewer children.

In 1979 China instigated a more aggressive plan to reduce the birth rate, announcing incentives to promote later marriages and one-child families (see inset). A couple who signed a pledge to limit themselves to a single child might be eligible for incentives like medical care and schooling for the child, cash bonuses, preferential housing, and retirement funds. Penalties were instituted, including fines and the surrender of all of these privileges, if a second child was born.

China's aggressive plan brought about the most rapid and drastic reduction in fertility in the world, from 5.8 births per woman in 1970 to 1.6 births per woman today. However, the plan was controversial and unpopular because it compromised individual freedom of choice. In some instances, social pressures caused women who were pregnant with a second child to get an abortion.

In China, sons are valued more highly than daughters because sons carry on the family name and traditionally provide old-age security for their parents. A disproportionate number of male babies have been born in recent years, suggesting that some expectant parents determine the sex of their fetus and abort it if it is female. In the past, parents required to conform to the one-baby policy abandoned or killed thousands of newborn baby girls because they wanted a boy.

In 1984 the one-child family policy was relaxed in rural China, where 70 percent of all Chinese live. China's recent population control program has relied on education, publicity campaigns (see photo), and fewer penalties to achieve its goals.

NATIONAL GEOGRAPHIC

Population Ecology

Individuals of the same species inhabiting a given area are part of a larger organization called a *population.* Typically, we find multiple populations of the same species and these may be defined by their isolation from each other because of landscape features. When we describe humans, however, we do not consider there to be multiple populations, as we have successfully occupied much of the planet and there are few barriers to our interactions with one another.

There are numerous features of populations that allow us to understand the dynamics of how they will change over time. These population characteristics are distinct from those of the individuals found within them. Some of these features include birth and death rates, growth rates, population density, and age structure. By having an understanding of these features, population ecologists can predict changes that might arise in the future and how changes might arise in the environment because of population dynamics.

Studying populations of other species provides insight into some of the processes that affect the growth of human populations. Understanding human population change is important because the size of the human population is central to most of Earth's environmental problems and the solutions to those problems.

Scientists who study **population ecology** try to determine the processes common to all populations (**FIGURE 3.1**). Population ecologists study how a population responds to its environment—such as how individuals in a given population compete for food or other resources, and how predation, disease, and other environmental pressures affect that population. Environmental pressures such as these prevent populations—whether of bacteria or maple trees or giraffes—from increasing indefinitely.

population ecology The branch of biology that deals with the number of individuals of a particular species found in an area and why those numbers increase or decrease over time.

What we learn about one population helps us make predictions about other populations FIGURE 3.1

At first glance, the two populations shown here appear to have little in common, but they share many characteristics.

A A cluster of soft corals. Photographed in the Western Pacific Ocean.

B A herd of impala in a woodland meadow. Photographed in Rwanda.

HOW DO POPULATIONS CHANGE IN SIZE?

Populations of organisms, whether sunflowers, eagles, or humans, change over time. On a global scale, this change is due to two factors: the rate at which individual organisms produce offspring (the birth rate) and the rate at which individual organisms die (the death rate) (**FIGURE 3.2A**). In humans, the birth rate (b) is usually expressed as the number of births per 1000 people per year, and the death rate (d) as the number of deaths per 1000 people per year. The **growth rate (r)** of a population is the birth rate (b) minus the death rate (d), or $r = b - d$. Growth rate is also referred to as natural increase in populations. It is common to express birth and death rates as measures of the number of births and deaths per 1000 individuals for a given time period. We refer to these rates as the *crude birth rate* or *crude death rate*, respectively.

If organisms in the population are born faster than they die, the growth rate is more than zero, and population size increases. If organisms in the population die faster than they are born, the growth rate is less than zero, and population size decreases. If the growth rate is equal to zero, births and deaths match, and population size is stationary, despite continued reproduction and death.

In addition to birth and death rates, **dispersal**—movement from one region or country to another—affects local populations. There are two types of dispersal: **immigration** (i), in which individuals enter a population and increase its size, and **emigration** (e), in which individuals leave a population and decrease its size. The growth rate (r) of a local population must take into account birth rate (b), death rate (d), immigration (i), and emigration (e) (**FIGURE 3.2B**). The formula for calculating growth rate taking into account these factors is as follows:

> **growth rate (r)**
> The rate of change (increase or decrease) of a population's size, which is typically expressed in percentage per year.

Growth rate = [(Birth rate + Immigration rate)
 − (Death rate + Emigration rate)]

For example, if a population has a birth rate of 20 per 1000, a death rate of 10 per 1000, an immigration rate of 7 per 1000, and an emigration rate of 3 per 1000, then this population would have a growth rate of 14 per 1000, calculated as follows:

$$\left(\frac{20}{1000} + \frac{7}{1000} \right) - \left(\frac{10}{1000} + \frac{3}{1000} \right) = \frac{14}{1000}$$

This number tells us the net change in the population's size per 1000 individuals. This means that if this population has 1000 people in year 1, it will increase to 1014 in year 2, 1028 in year 3, and so on. These population increases are often expressed as percentages, which we can calculate using the following formula:

Growth rate × 100%

Thus, a growth rate of $\frac{14}{1000}$ would be expressed as:

$$\frac{14}{1000} \times 100\% = 1.4\%$$

Factors that interact to change population size FIGURE 3.2

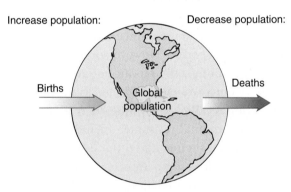

A On a global scale, the change in a population is due to the number of births and deaths.

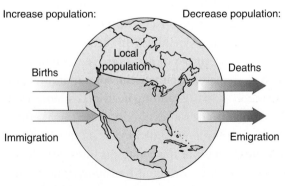

B In local populations, such as the population of Canada, the number of births, deaths, immigrants, and emigrants affect population size.

Using percentages to measure population growth allows scientists to compare increases and decreases in species that have far different population sizes. They can also calculate changes that will occur in the population over longer periods.

Population density

Population density describes the number of individuals within a population per unit area. Typically, the larger the organism, the lower its population density will be. This is because larger organisms require more resources—and thus more area—to survive.

High population densities have both benefits and drawbacks for organisms. High density means that it will be easier to find mates and group together, but it can also lead to competition if there is too little space or food, or if there are too few mates. In addition, constant close contact increases the likelihood of transmission of infectious diseases. For these reasons, many organisms will leave an area if it becomes too densely populated. Low population densities mean that organisms have more space and resources, but it may be harder to find a mate or other companions.

MAXIMUM POPULATION GROWTH

Different species have different abilities to reproduce and create offspring. This is described as the **biotic potential**. Several factors influence the biotic potential of a species: the age at which reproduction begins, the fraction of the lifespan during which an individual can reproduce, the number of reproductive periods per lifetime, and the number of offspring produced during each period of reproduction. These factors, called *life history characteristics,* determine whether a particular species has a large or a small biotic potential.

Generally, larger organisms like blue whales have the smallest biotic potentials, whereas microorganisms have the greatest biotic potentials. Under ideal conditions (that is, in an environment with unlimited resources), certain bacteria reproduce by dividing in half every 30 minutes (**Figure 3.3A**). At this rate of growth, a single

exponential population growth The population growth that occurs when environmental resources are not limited and there is a constant rate of reproduction. With exponential growth, the birth rate alone controls how fast (or slow) the population grows.

biotic potential The maximum rate at which a population could increase under ideal conditions.

bacterium increases to a population of more than 1 million in just 10 hours and exceeds 1 billion in 15 hours. If you plot population number versus time, the graph takes on the characteristic J shape of **exponential population growth** (**Figure 3.3B** and **C**). When a population grows exponentially, the larger the population gets, the faster it grows. Regardless of species, whenever a population grows at a constant rate, population size plotted versus time gives the same J-shaped curve. The J-shaped curve may be increased in slope as the growth rate increases. The increase can continue under ideal conditions until the population growth achieves the biotic potential, at which point the growth rate will not increase any further.

ENVIRONMENTAL RESISTANCE AND CARRYING CAPACITY

Certain populations may exhibit exponential population growth for a short period. However, organisms do not reproduce indefinitely at their biotic potentials because the surrounding environment sets limits to further growth. This limitation is called **environmental resistance**. Examples of environmental resistance include limited food, water, shelter, and other essential resources, as well as temperature, space, increased disease, and predation.

Using the earlier example, we find that bacteria never reproduce unchecked for an indefinite period because they run out of food and living space and poisonous body wastes accumulate in their vicinity. With crowding, bacteria become more susceptible to parasites (remember that high population densities facilitate the spread of infectious organisms such as viruses among individuals) and predators (high population densities also increase the likelihood of a predator catching an individual). As the environment deteriorates, bacterial birth rate declines and death rate increases. The environmental conditions might worsen to a point where the death rate exceeds the birth rate, and as a result, the population decreases. Thus, the environment controls population size, because as the population

Exponential population growth FIGURE 3.3

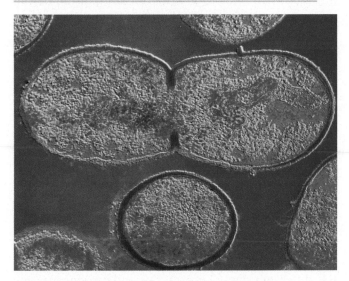

A *Streptococcus* bacterium in the process of dividing.

Time (hours)	Number of bacteria
0	1
1	4
2	16
3	64
4	256
5	1024
6	4096
7	16,384
8	65,536
9	262,144
10	1,048,576

B When bacteria divide at a constant rate, their number increases exponentially.

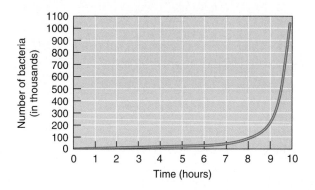

C When the population numbers are graphed against time, the curve of exponential population growth has a characteristic J shape. The maximum rate of growth is observed when the biotic potential is achieved.

increases, so does environmental resistance, which limits population growth. Furthermore, the interaction between an organism's biotic potential and the environmental resistance to its population growth helps determine the fate of the population.

Over longer periods, the rate of population growth may decrease to nearly zero. This levelling off occurs at or near the environment's **carrying capacity (K)** and represents the number of individuals in the population that are able to exist over time given the limitations dictated by the environment. In nature, carrying capacity is dynamic and changes in response to alterations in the environment. A particularly cold, high snowpack winter, for example, might create harsh conditions and decrease the availability of vegetation in an area. This change, in turn, would lower the carrying capacity for deer and other herbivores in that environment.

carrying capacity (K) The largest population that can be sustained over the long term given that there are no changes in the environment.

When a population influenced by environmental resistance is graphed over a long period (**FIGURE 3.4** on page 72), the curve has an S shape. The curve shows the population's initial exponential increase (note the curve's J shape at the start, when environmental resistance is low). Then the population size levels out as it approaches the carrying capacity of the environment. Although the S curve is an oversimplification of how most populations change over time, it fits some populations studied in the laboratory, as well as a few studied in nature.

A population rarely stabilizes at K (carrying capacity), as shown in Figure 3.4, but its size may temporarily rise higher than K. It will then drop back to, or below, the carrying capacity. Sometimes a population that overshoots K will experience a *population crash*, an abrupt decline from high to low population density when resources are exhausted. Such an abrupt change is commonly observed in bacterial cultures, zooplankton, and other populations whose resources are exhausted.

The availability of winter forage largely determines the carrying capacity for reindeer, which live in cold

Population growth as carrying capacity is approached FIGURE 3.4

In many laboratory studies population growth produces a curve with a characteristic S shape. Carrying capacity is found by identifying the population number when a zero population growth rate is observed.

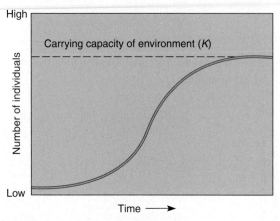

northern habitats. In 1910, a small herd of reindeer was introduced on one of the Pribilof Islands in the Bering Sea (FIGURE 3.5A). The herd's population increased exponentially for about 25 years until there were many more reindeer than the island could support, particularly in winter. The reindeer overgrazed the vegetation until the plant life was almost wiped out. Then, in slightly over a decade, as reindeer died from starvation, the number of reindeer plunged to less than 1 percent of the population at its peak (FIGURE 3.5B). If reindeer overgraze the vegetation, it takes 15 to 20 years for it to recover. During that period, the carrying capacity for reindeer is greatly reduced.

To understand exponential growth more fully, go to www.otherwise.com/population/exponent.html and complete an exponential growth population simulation. Use different values of growth rate and carrying capacity to identify changes arising in the population.

A population crash FIGURE 3.5

A A herd of reindeer on one of the Pribilof Islands off the coast of Alaska.

B Graph of the reindeer population originally introduced to one of the Pribilof Islands in 1911. Note the population crash, which followed the peak population attained in 1935.

The lesser snow goose, also known as the blue goose, is native to North America. Over the past four decades, the snow goose population has tripled and it continues to grow at a rate of 5 percent each year. The breeding population of the lesser snow goose is nearing 5 million birds, the highest population levels ever recorded.

Snow geese are migratory birds, and in North America they travel through the central flyway across some of the most productive agricultural land in America. Traditionally, snow geese migrated south for the winter and used their strong bills to dig up roots or marsh grasses along coastal wetlands. However, the transition of the landscape along the central flyway to cultivated farming fields provided new, high-quality food options for these migrating birds, and now snow geese feed on any field grain they encounter. Snow geese also take advantage of fall-seeded grains, such as winter wheat.

The end result is a boosted survival rate in a population of birds that are able to benefit from a ready supply of high-quality feed from September through April or May. Many biologists believe that this change to their winter feeding habits is the contributing factor to their population boom. The increase in population means that more snow geese, in better physical condition, are returning to breed in the Arctic each year, degrading the tundra and severely affecting the future sustainability of the Arctic coastal marsh ecosystem. Nearly one-third of the coastal salt marsh habitat along the west coast of the Hudson Bay has been destroyed while another third is badly damaged. Adding to this dilemma is the long timeframe necessary for habitats to recover in Arctic climates. Even if no

further damage occurred, recovery of the already damaged coastal and marshland habitats would take decades or longer.

Overabundant snow geese populations present challenges for wildlife managers and government leaders faced with degrading ecosystems and agriculture. In 1986, under the North American Waterfowl Management Plan (NAWMP), the Arctic Goose Joint Venture (AGJV) was established through the efforts of the governments of Canada, the United States, and Mexico. The AGJV promotes continent-wide research to investigate the problem of overabundant goose populations. Wildlife agencies have introduced several strategies to help reduce the snow goose population. A spring hunt was introduced in some areas, calling for a modification to be made to the Canada–U.S. Migratory Birds Convention Act, with the hopes that it may help reduce goose population numbers. Subsistence hunting was increased for northern Aboriginal communities, electronic calls are now allowed for use by hunters, and some provinces have approved a Sunday hunt.

Continued research and snow goose surveys will be required to determine whether these initiatives will be successful in reducing the snow goose population.

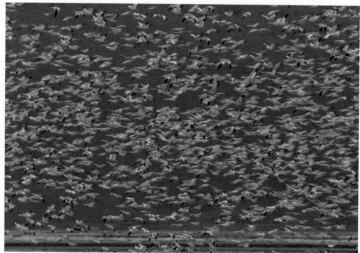

Each spring, millions of birds arrive in the Arctic to breed and immediately begin to feed on freshwater and salt marsh habitats, pulling up grasses by the root and leaving the ground bare behind them. With the sheer volume of birds, an entire area is quickly stripped bare.

CONCEPT CHECK 🛑 **STOP**

How do each of the following affect population size: birth rate, death rate, immigration, and emigration?

What is biotic potential? Carrying capacity?

How do biotic potential and/or carrying capacity produce the J-shaped and S-shaped population growth curves?

Human Population Patterns and Demographics

Now that you have examined some of the basic concepts of population ecology, let's apply those concepts to the human population. **FIGURE 3.6** shows the increase in human population since 1800. The characteristic J curve of exponential population growth reflects the decreasing amount of time it has taken to add each additional billion people to our numbers. It took thousands of years for the human population to reach 1 billion, a milestone that took place around 1800. The United Nations projects that the human population will reach 7 billion by 2013. Population experts predict that the population will level out during the 21st century, possibly forming an S curve as observed in other species.

One of the first people to recognize that the human population cannot increase indefinitely was **Thomas Malthus** (1766–1834), a British economist. Noting that human population can increase faster than its food supply, he warned that the inevitable consequences of population growth would be famine, disease, and war. Since Malthus's time, the human population has increased from about 1 billion to more than 6 billion. On the surface, it seems that Malthus was wrong. Our population has grown dramatically because scientific advances have allowed food production to keep pace with population growth. Malthus's ideas may ultimately be proven correct, however, because we don't know whether this increased food production is sustainable.

Our world population was 6.7 billion in 2008, an increase of about 80 million from 2007. This increase isn't due to a rise in the birth rate (b). In fact, the world birth rate has declined slightly during the past 200 years. The population growth is due instead to a dramatic decrease in the death rate (d) because greater food production, better medical care, and improvements in water quality and sanitation practices have increased life expectancy for a great majority of the global population (**FIGURE 3.7**).

PROJECTING FUTURE POPULATION NUMBERS

The human population has reached a turning point. Although our numbers continue to increase, the world growth rate (r) has declined slightly over the past several

Human population numbers, 1800 to present FIGURE 3.6

Until recently, the human population has been increasing exponentially. There are now indications that the human population is beginning to level out, forming an S curve.

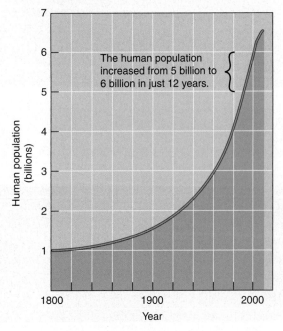

The human population increased from 5 billion to 6 billion in just 12 years.

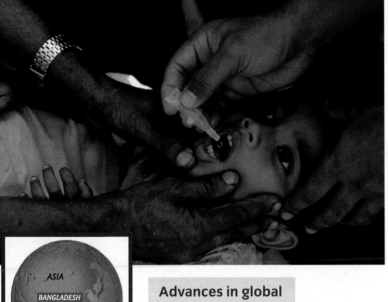

ASIA
BANGLADESH

Global Locator

Advances in global health FIGURE 3.7

A child in Bangladesh receives a dose of oral polio vaccine. At one time, polio killed or crippled millions of children each year.

Population projections to 2050 FIGURE 3.8

In 2004, the United Nations made three projections of world population in 2050, each based on different fertility rates.

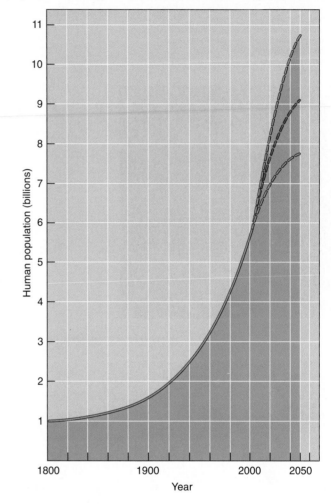

■ **zero population growth** The state in which the population remains the same size because the birth rate equals the death rate.

years, from a peak of 2.2 percent per year in the mid-1960s to the current growth rate of 1.2 percent per year. Population experts at the United Nations and the World Bank project that the growth rate will continue to decrease slowly until **zero population growth** is attained toward the end of the 21st century.

The United Nations periodically publishes population projections for the 21st century. The latest (2004) U.N. figures forecast that the human population will total between 7.7 billion (their low projection) and 10.6 billion (their high projection) in the year 2050, with 9.1 billion thought to be most likely (**FIGURE 3.8**). These estimates vary widely depending on what assumptions are made about standard of living, resource consumption, technological innovations, and waste generation. If we want all people to have a high level of material well-being equivalent to the lifestyles in highly developed countries, then Earth will support far fewer humans than if everyone lives just above the subsistence level.

The main unknown factor in any population growth scenario is Earth's carrying capacity. Most published estimates of how many people Earth can support range from 4 billion to 16 billion. What will happen to the

human population when it approaches Earth's carrying capacity? Optimists suggest that a decrease in the birth rate will stabilize the human population. Some experts take a more pessimistic view and predict that our ever-expanding numbers will cause widespread environmental degradation and make Earth uninhabitable for humans as well as other species. This doesn't mean we will go extinct as a species, but it projects severe hardship for many people. Other experts think the human population has already exceeded the carrying capacity of the environment, a potentially dangerous situation that threatens our long-term survival as a species.

Human population trends are summarized in **FIGURE 3.9** on pages 76 and 77.

Human Population Patterns and Demographics   75

The human population FIGURE 3.9

Human population issues are both local and global in scope.

▶ **POPULATION DENSITY**

Population density can be measured as the average number of people per square unit in a given area (arithmetic density). Populations, however, are not evenly distributed. Often, they're gathered around arable land. Comparing populations to farmland (physiologic density), can make statistics more meaningful.

Egypt, for example, where nearly 90% of the citizens are clustered in the Nile Valley, has a modest overall density of 74 people per sq km but a physiologic density of 3089 people per agricultural sq km, among the world's highest.

Population growth rate as a percentage of total population, 2005 estimate
(1 block=1 million people)

- More than 3%
- 2.0%–2.9%
- 1.0%–1.9%
- 0%–0.9%
- Negative growth

2005 population (millions) in parentheses

Not all countries or territories shown

▲ **POPULATION CARTOGRAM**

The world appears quite different when countries are sized proportionally to their populations and mapped. Underpopulated Canada, the world's second-largest country, is reduced to a small strip above the United States, while small, crowded Japan looms large. India and China become the global giants.

▶ **POPULATION PYRAMIDS**

When population is expressed in bars representing age and gender and stacked up (males left, females right), country profiles emerge that have ramifications for the future. Countries with high birth rates and high percentages of young, such as Nigeria, look like pyramids. Countries such as Italy, whose birth rate is below the replacement fertility level of 2.1 children per couple, show bulges in the higher age brackets. The United States clearly shows the "baby boom" of children born in the years after World War II.

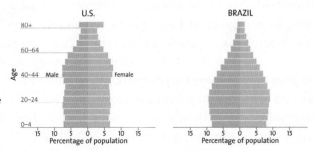

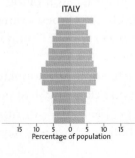

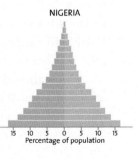

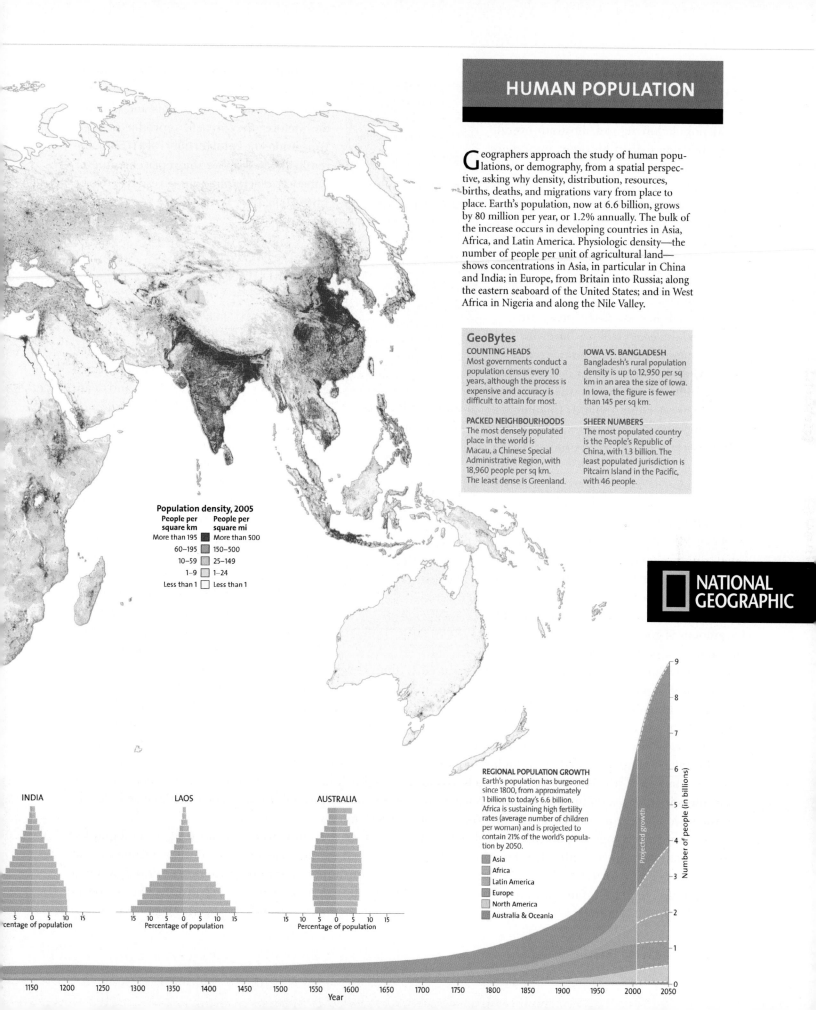

HUMAN POPULATION

Geographers approach the study of human populations, or demography, from a spatial perspective, asking why density, distribution, resources, births, deaths, and migrations vary from place to place. Earth's population, now at 6.6 billion, grows by 80 million per year, or 1.2% annually. The bulk of the increase occurs in developing countries in Asia, Africa, and Latin America. Physiologic density—the number of people per unit of agricultural land—shows concentrations in Asia, in particular in China and India; in Europe, from Britain into Russia; along the eastern seaboard of the United States; and in West Africa in Nigeria and along the Nile Valley.

GeoBytes

COUNTING HEADS
Most governments conduct a population census every 10 years, although the process is expensive and accuracy is difficult to attain for most.

IOWA VS. BANGLADESH
Bangladesh's rural population density is up to 12,950 per sq km in an area the size of Iowa. In Iowa, the figure is fewer than 145 per sq km.

PACKED NEIGHBOURHOODS
The most densely populated place in the world is Macau, a Chinese Special Administrative Region, with 18,960 people per sq km. The least dense is Greenland.

SHEER NUMBERS
The most populated country is the People's Republic of China, with 1.3 billion. The least populated jurisdiction is Pitcairn Island in the Pacific, with 46 people.

Population density, 2005

People per square km	People per square mi
More than 195	More than 500
60–195	150–500
10–59	25–149
1–9	1–24
Less than 1	Less than 1

NATIONAL GEOGRAPHIC

REGIONAL POPULATION GROWTH
Earth's population has burgeoned since 1800, from approximately 1 billion to today's 6.6 billion. Africa is sustaining high fertility rates (average number of children per woman) and is projected to contain 21% of the world's population by 2050.

- Asia
- Africa
- Latin America
- Europe
- North America
- Australia & Oceania

Projected growth

Number of people (in billions)

INDIA

LAOS

AUSTRALIA

Percentage of population

Year

DEMOGRAPHICS OF COUNTRIES

World population figures illustrate overall trends but don't describe other important aspects of the human population story, such as population differences from country to country. **Demographics** provides interesting information on the populations of various countries. Recall from Chapter 1 that countries are classified into two groups—highly developed and developing—based on population growth rates, degree of industrialization, and relative prosperity.

Highly developed countries have the lowest birth rates in the world. Some countries, such as Germany, have birth rates just below what is needed to sustain their populations and are declining slightly in numbers. Highly developed countries also have low **infant mortality rates** (FIGURE 3.10A) and longer life expectancies. The infant mortality rate of Canada, for example, is 5.4, compared with a world rate of 49, and life expectancy in Canada is 80 years versus 68 years worldwide. In addition, highly developed countries tend to have high per person GNI PPP, which is a country's gross national income (GNI) in purchasing power parity (PPP) divided by its population. It indicates the amount of goods and services an average citizen of that particular country could buy in the United States (which is the baseline currency used for comparative purposes). For example, the per person GNI PPP in Canada is $36,170, which is very high compared with the worldwide figure of $7439.

In *moderately developed countries*, birth rates and infant mortality rates are higher than those of highly developed countries, but they are declining. Moderately developed countries have a medium level of industrialization, and their average per person GNI PPPs are lower than those of highly developed countries. Less developed countries have the shortest life expectancies, the lowest average per person GNI PPPs, the highest birth rates, and the highest infant mortality rates in the world (FIGURE 3.10B).

demographics The applied branch of sociology that deals with population statistics.

infant mortality rate The number of deaths of infants under age 1 per 1000 live births.

replacement-level fertility The number of children a couple must produce to "replace" themselves.

total fertility rate (TFR) The average number of children born to each woman.

demographic transition The process whereby a country moves from relatively high birth and death rates to relatively low birth and death rates.

Replacement-level fertility is usually given as 2.1 children. The number is greater than 2.0 because some infants and children die before they reach reproductive age. Worldwide, the **total fertility rate (TFR)** is currently 2.6, well above the replacement level. Canada's total fertility rate was estimated to be 1.6 in 2008.

THE DEMOGRAPHIC TRANSITION

Demographers recognize four demographic stages based on their observations of Europe as it became industrialized and urbanized (FIGURE 3.11). These stages converted Europe from relatively high birth and death rates to relatively low birth and death rates. All highly developed and moderately developed countries with more advanced economies have gone through this **demographic transition**, and demographers assume that the same progression will occur in less developed countries as they industrialize.

Why has the population stabilized in more than 30 highly developed countries in the fourth (postindustrial) demographic stage? The reasons are complex. Declining birth rate is associated with an improvement in living standards. It is difficult to say whether improved socioeconomic conditions have resulted in a decrease in birth rate or whether a decrease in birth rate has resulted in improved socioeconomic conditions. Perhaps both are true. Another reason for the decline in birth rate in highly developed countries is the increased availability of family planning services. Other socioeconomic factors that influence birth rate are increased education, particularly of women, and urbanization of society (discussed later in this chapter).

Once a country reaches the fourth demographic stage, is it correct to assume that it will continue to have a low birth rate indefinitely? We don't know. Low birth rates may be a permanent response to the socioeconomic factors of an industrialized, urbanized society.

Infant mortality rates FIGURE 3.10

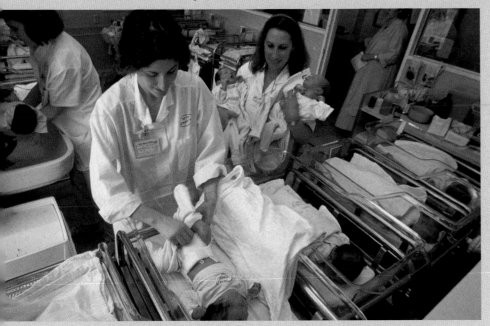

A Nurses care for newborn infants. In Canada, the infant mortality rate is 5 per 1000 live births.

B This infant, just born in the less developed country of Bangladesh, is underweight and therefore at risk. The infant mortality rate in Bangladesh is 34 per 1000 live births.

The demographic transition FIGURE 3.11

Demographers have identified four stages through which a population progresses as its society becomes industrialized.

VIEW THIS IN ACTION

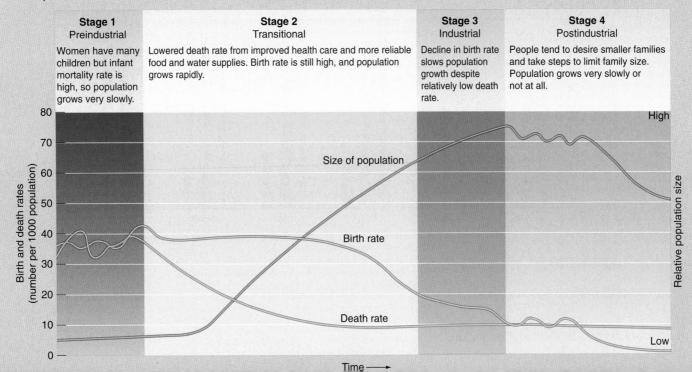

Stage 1	Stage 2	Stage 3	Stage 4
Preindustrial	Transitional	Industrial	Postindustrial
Women have many children but infant mortality rate is high, so population grows very slowly.	Lowered death rate from improved health care and more reliable food and water supplies. Birth rate is still high, and population grows rapidly.	Decline in birth rate slows population growth despite relatively low death rate.	People tend to desire smaller families and take steps to limit family size. Population grows very slowly or not at all.

Birth and death rates (number per 1000 population)

Relative population size

Size of population

Birth rate

Death rate

High

Low

Time ⟶

On the other hand, low birth rates may be a response to socioeconomic factors, such as the changing roles of women in highly developed countries. Unforeseen changes in the socioeconomic status of women and men in the future may again change birth rates. No one knows for sure.

The population in many developing countries is beginning to approach stabilization (**FIGURE 3.12**). For example, the TFR in Brazil in 1960 was 6.7 children per woman; today it is 2.3. Worldwide, the TFR in developing countries has decreased from an average of 6.1 children per woman in 1970 to 2.7 today.

Although fertility rates in these countries have declined, many still exceed replacement-level fertility. Consequently, populations in these countries are still increasing. Even when fertility rates equal replacement-level fertility, population growth will still continue for some time. To understand why this is so, let's examine the age structure of various countries.

AGE STRUCTURE OF COUNTRIES

A population's **age structure** helps predict future population growth. The number of males and the number of females at

> **age structure**
> The number and proportion of people at each age in a population.

each age, from birth to death, are represented in an age structure diagram. Each diagram is divided vertically in half, the left side representing the males in a population and the right side the females. The bottom third of each diagram represents pre-reproductive humans (between 0 and 14 years of age); the middle third, reproductive humans (15 to 44 years); and the top third, postreproductive humans (45 years and older). The widths of these segments are proportional to the population sizes: A wider segment implies a larger population. The overall shape of an age structure diagram indicates whether the population is increasing, stable, or shrinking.

The age structure diagram of a country with a high growth rate, based on a high fertility rate—for example, Nigeria or Bolivia—is shaped like a pyramid (**FIGURE 3.13A**). The largest percentage of the population is in the pre-reproductive age group (0 to 14 years of age), so the probability of future population growth is great. A positive **population growth momentum** exists because when all these children mature, they will become the parents of the next generation, and this group of parents will be larger than the previous group. Even if the fertility rate of such a country has declined to replacement level (couples are having smaller families than their parents did), the population will

Fertility changes in selected developing countries FIGURE 3.12

Since the 1960s, fertility levels have dropped dramatically in many developing countries.

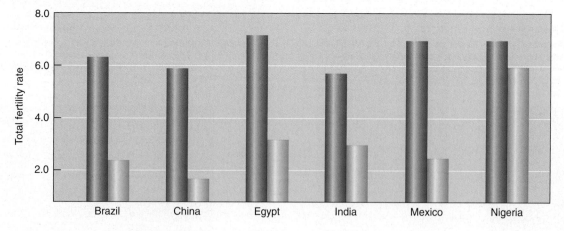

■ 1960–1965

▮ 2007

Age structure diagrams FIGURE 3.13

Shown are countries with (A) rapid (Nigeria), (B) slow (United States), and (C) declining (Germany) population growth.

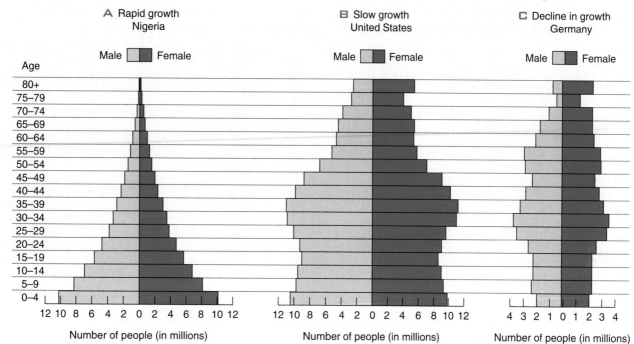

continue to grow for some time. Population growth momentum explains how a population's present age distribution affects its future growth.

The more tapered bases of the age structure diagrams of countries with slowly growing, stable, or declining populations indicate that a smaller proportion of the population will become the parents of the next generation (FIGURE 3.13B and C). The age structure diagram of a stable population (neither growing nor shrinking) demonstrates that the numbers of people at pre-reproductive and reproductive ages are approximately the same. A larger percentage of the population is older—that is, postreproductive—than in a rapidly increasing population. Many countries in Europe have stable populations.

In a shrinking population, the pre-reproductive age group is smaller than either the reproductive or postreproductive age group. Russia, Bulgaria, and Germany are examples of countries with slowly shrinking populations.

Worldwide, 28 percent of the human population is under age 15 (FIGURE 3.14A on page 82). When these people enter their reproductive years, they have

the potential to cause a large increase in the growth rate. Even if the birth rate doesn't increase, the growth rate will increase simply because there are more people reproducing.

Most of the world population increase since 1950 has taken place in developing countries as a result of the younger age structure and the higher-than-replacement-level fertility rates of their populations. In 1950, 67 percent of the world's population was in developing countries in Africa, Asia (minus Japan), and Latin America. Between 1950 and 2000, the world's population more than doubled in size, but most growth occurred in developing countries. This trend is reflected in the current number of people in developing countries, which has increased to 82 percent of the world population. Most of the population increase during the 21st century will take place in developing countries, largely as a result of their younger age structures. It is anticipated that these countries will have economic difficulty supporting such growth.

Declining fertility rates have profound social and economic implications because as fertility rates drop, the percentage of the population that is elderly increases

A Percentages of the population under age 15 in 2007. The higher this percentage, the greater the potential for population growth.

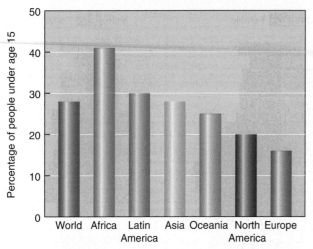

B Percentages of the population older than 65 in 2007. Lower fertility rates lead to aging populations.

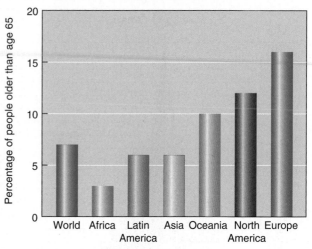

(FIGURE 3.14B). An aging population has a higher percentage of people who are chronically ill or disabled, and these people require more health care and other social services. An aging population also reduces a country's productive workforce, increases its tax burden, and strains its social security, health, and pension systems. Consequently, governments with growing elderly populations may offer incentives to the elderly to work longer before retiring.

POPULATION DYNAMICS IN CANADA

Let's take a look at the demographics of Canada. The Great Depression of the 1930s in Canada was a time of economic hardship, and people had fewer children. Following the end of World War II, however, the economy grew and the birth rate in Canada increased. The generation born in the period from 1946–1966 is known as the *baby boomers*. This period saw a large bulge emerge on the age structure pyramid for Canada with the increase in the number of children born. You can follow this bulge on the age structure pyramid shown in FIGURE 3.15A, which is now moving toward the older demographic segment for Canada since the oldest of the baby boomers are now in their sixties. Canada is getting older! In 2006, seniors accounted for 13.2 percent of the country's

population, almost double the 7.2 percent of the population that seniors accounted for at the start of the baby boom in 1946.

In 2011, the first baby boomers will turn 65 and more and more workers will retire, potentially creating a shortage of workers in Canada unless immigrant workers step in to fill the void. Immigration in Canada accounts for close to two-thirds of Canada's population growth. Within the next 10 years, close to half of the populations in large cities such as Toronto and Vancouver will be immigrants.

Not all of Canada exhibits the same age structure as that shown in Figure 3.15A. FIGURE 3.15B shows the age structure pyramid for Nunavut. On the national pyramid you can see that the natural birth rate for Canada is contracting, illustrating that the population growth resulting from natural births in the country is declining. The current birth rate in Canada is just over 1.5 children for every woman, whereas the replacement level, above which the aging process would stop, is 2.1 children per woman (during the baby boom, the fertility rate reached as high as 3.9 children per woman). In Nunavut, however, the birth rate contrasts with national trends. The pyramid for Nunavut has a broad base, indicating a high proportion of the population is in their pre-reproductive stage of life. This implies that the population in Nunavut will see a rapid rate of

Canada's demographics FIGURE 3.15

Shown are age structure pyramids for Canada as a whole (A) and Nunavut (B) in 2006. Compare these pyramids with those in Figure 3.13. What similarities do you see between Canada's pyramid and that of the United States? What about Nunavut and Nigeria? What conclusions can you draw from these similarities?

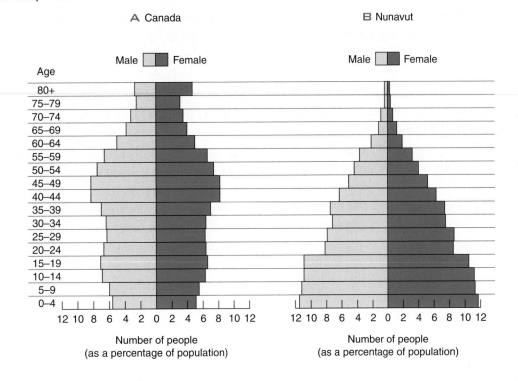

A Canada

B Nunavut

population growth and a low proportion of older people. The pyramid shape also shows a compression in the number of individuals in their late teens and early twenties, showing a tendency for people to move out of the area at this age. This burst in population growth in Nunavut will be accompanied with an increase in economic, social, and environmental needs.

If fertility and immigration trends in Canada continue as they have, projections indicate that the Canadian population will stabilize between 35 and 50 million people by the end of the century. Projections also suggest that the Aboriginal population in Canada will continue to grow twice as fast as the rest of the population. (For more information on population dynamics in Canada and for animations of population changes, please visit Statistics Canada's website at www.statcan.gc.ca/kits-trousses/animat/edu06a_0000-eng.htm.)

CONCEPT CHECK STOP

How would you describe human population growth for the past 200 years?

When determining Earth's carrying capacity for humans, why is it not enough to just consider human numbers? What else must be considered?

What is infant mortality rate? How does it affect life expectancy?

If all the women in the world suddenly started bearing children at replacement-level fertility rates, would the population stop increasing immediately? Why or why not?

Stabilizing World Population

LEARNING OBJECTIVES

Relate total fertility rates to each of the following: cultural values, social and economic status of women, the availability of family planning services, and government policies.

Explain the link between education and total fertility rates.

Dispersal used to be a solution for overpopulation, but not today. As a species, we humans have expanded our range throughout Earth, and few habitable areas remain with the resources to adequately support a major increase in human population. As well, increasing the death rate is not an acceptable means of regulating population size. Clearly, reducing the birth rate is the way to control our expanding population. Cultural traditions, women's social and economic status, family planning, and government policies all influence total fertility rate (TFR).

CULTURE AND FERTILITY

The values and norms of a society—what is considered right and important and what is expected of a person—constitute a part of that society's **culture.** Gender—that is, varying roles men and women are expected to fill—is an important part of culture. Different societies have different gender expectations (**FIGURE 3.16**). With respect to fertility and culture, a couple may be expected to have the number of children traditional in their society.

Varying roles of men and women FIGURE 3.16

A In many societies, men do the agricultural work. This Argentinian man is harvesting grapes.

B In sub-Saharan Africa, women do most of the agricultural work in addition to caring for their children. Photographed in Mali.

High TFRs are traditional in many societies. The motivations for having many babies vary from one region to another, but a major reason for high TFRs is often correlated with high infant and child mortality rates. For a society to endure, it must produce enough children who survive to reproductive age. If infant and child mortality rates are high, TFRs must be high to compensate. Although infant and child mortality rates are decreasing, it will take longer for TFRs to decline. Parents must have confidence that the children they already have will survive before they stop having additional babies. As well, the traditional mindset to have large families usually takes a longer period of time to be altered.

Higher TFRs in some developing countries are also due to the important economic and societal roles of children. In some societies, children usually work in family enterprises such as farming or commerce, contributing to the family's livelihood. The International Labour Organization estimates that, worldwide, about 218 million children under the age of 15 work full time; more than 95 percent of these children live in developing countries (**FIGURE 3.17**). When these children become adults, they provide support for their aging parents. In contrast, children in highly developed countries have less value as a source of labour because they attend school and because less human labour is required in an industrialized society. Furthermore, highly developed countries provide many social services for the elderly, so the burden of their care doesn't fall entirely on offspring.

Many cultures place a higher value on male children than on female children. In these societies, a woman who bears many sons achieves a high status; thus, the social pressure to have male children keeps the TFR high.

Religious values are another aspect of culture that affects TFRs. Several studies done in the United States point to differences in TFRs among Catholics, Protestants, and Jews. In general, Catholic women have a higher TFR than either Protestant or Jewish women, and women who don't follow any religion have the lowest TFRs of all. The observed differences in TFRs may not be the result of religious differences alone. Other variables, such as ethnicity (certain religions are associated with particular ethnic groups) and residence (certain religions are associated with urban or with rural living), complicate any generalizations that might be made.

THE SOCIAL AND ECONOMIC STATUS OF WOMEN

Gender inequality exists in most societies; in these societies, women don't have the same rights, opportunities, or privileges as men. Because sons are more highly

Working children FIGURE 3.17

These Indonesian children are making yarn for the textile industry.

ASIA

INDONESIA

Global Locator

valued than daughters, girls are often kept at home to work rather than being sent to school. In most developing countries, a higher percentage of women are illiterate than men. Fewer women than men attend secondary school. In some African countries only 2 to 5 percent of girls are enrolled in secondary school. Worldwide, some 90 million girls aren't given the opportunity to receive a primary (elementary school) education. Laws, customs, and lack of education often limit women to low-skilled, low-paying jobs. In such societies, marriage is usually the only way for a woman to achieve social influence and economic security.

Evidence suggests that the single most important factor affecting high TFRs may be the low status of women in many societies (FIGURE 3.18). A significant way to tackle population growth, then, is to improve the social and economic status of women.

Marriage age and educational opportunities, especially for women, also affect fertility. The average age at which women marry affects the TFR; in turn, the laws and customs of a given society affect marriage age. Women who marry are more apt to bear children than women who don't marry, and the earlier a woman marries, the more children she is likely to have. Consider Sri Lanka and Bangladesh, two developing countries in South Central Asia. In Sri Lanka the average age at marriage is 25, and the average number of children born per woman is 2.0. In contrast, in Bangladesh the average age at marriage is 17, and the average number of children born per woman is 3.0.

In nearly all societies women with more education tend to marry later and have fewer children. Providing women with educational opportunities delays their first childbirth, thereby reducing the number of childbearing years and increasing the amount of time between generations. Education provides greater career opportunities and may change women's lifetime aspirations. In North America, it isn't uncommon for a woman to give birth to her first child in her thirties or forties, after establishing a career.

Education increases the probability that women will know how to control their fertility. It also provides knowledge to improve the health of the women's families, which results in a decrease in infant and child mortality.

Gender discrimination FIGURE 3.18

Ethiopia is traditionally a male-dominated society in which women and girls have a lower status than men. These young girls are attending an elementary school. Overall, only 17 percent of Ethiopian girls have access to a secondary education, as compared with 28.3 percent of boys.

ETHIOPIA

Global Locator

A study in Kenya showed that 10.9 percent of children born to women with no education died by age 5, as compared with 7.2 percent of children born to women with a primary education, and 6.4 percent of children born to women with a secondary education. Education also increases women's career options and provides ways of achieving status besides having babies.

Education may also have an indirect effect on TFR. Children who are educated have a greater chance of improving their living standards, partly because they have more employment opportunities. Parents who recognize this may be more willing to invest in the education of a few children than in the birth of many children whom they can't afford to educate. The ability of better-educated people to earn more money may be one reason smaller family size is associated with increased family income.

FAMILY PLANNING SERVICES

Socioeconomic factors may encourage people to want smaller families, but fertility reduction won't become a reality without the availability of health and family planning services. The governments of most countries recognize the importance of educating people about basic maternal and child health care. Developing countries that have significantly lowered their TFRs credit the results to effective family planning programs. Prenatal care and proper birth spacing make women healthier. In turn, healthier women give birth to healthier babies, leading to fewer infant deaths.

Family planning services provide information on reproductive physiology and contraceptives, as well as the actual contraceptive devices, to people who wish to control the number of children they have or to space out the time between their children's births. Family planning programs are most effective when they are designed with sensitivity to local social and cultural beliefs. Family planning services don't try to force people to limit their family sizes; rather, they attempt to convince people that small families (and the contraceptives that promote small families) are acceptable and desirable.

Contraceptive use is strongly linked to lower TFRs. Research has shown that 90 percent of the decrease in fertility in 31 developing countries was a direct result of increased knowledge and availability of contraceptives (FIGURE 3.19). In highly developed countries, where TFRs are at replacement levels or lower, an average of 68 percent of married women of reproductive age use contraceptives. Fertility declines are occurring in developing countries where contraceptives are readily available. During the 1970s, 1980s, and 1990s, use of

Access to contraceptives
FIGURE 3.19

A birth control vendor explains condoms to women at the Adjame market in Côte d'Ivoire (Ivory Coast), a West African country with a TFR of 5.0. Currently, only 7 percent of Ivory Coast women aged 15 to 49 use modern methods of contraception.

Microcredit programs extend small loans ($50 to $500) to very poor people to help them establish businesses that generate income. The people use these loans for a variety of projects. Some have purchased used sewing machines to make clothing faster than sewing by hand. Others have opened small grocery stores after purchasing used refrigerators to store food so that it does not spoil.

The **Foundation for International Community Assistance (FINCA)** is a not-for-profit agency that administers a global network of microcredit banks. FINCA uses *village banking,* in which a group of very poor neighbours guarantees one another's loans, administers group lending and saving activities, and provides mutual support. These village banks give autonomy to local people.

FINCA primarily targets women because an estimated 70 percent of the world's poorest people are women. FINCA believes that the best way to alleviate the effects of poverty and hunger on children is to provide their mothers with a means of self-employment. A woman's status in the community is raised as she begins earning income from her business.

contraceptives in East Asia and many areas of Latin America increased significantly, and these regions experienced a corresponding decline in birth rate. In areas where contraceptive use remained low, such as parts of Africa, little or no decline in birth rate took place.

Family planning centres provide information and services primarily to women. As a result, in the male-dominated societies of many developing countries, such services may not be as effective as they could otherwise be. Polls of women in developing countries reveal that many who say they don't want additional children still don't practise any form of birth control. When asked why they don't use birth control, these women frequently respond that their husbands or in-laws want additional children.

GOVERNMENT POLICIES AND FERTILITY

The involvement of governments in childbearing and childrearing is well established. Laws determine the minimum age at which people may marry and the amount of compulsory education they receive. Governments may allot portions of their budgets to family planning services, education, health care, old-age security, or incentives for smaller or larger family size. The tax structure, including additional charges or allowances based on family size, also influences fertility.

In recent years, the governments of at least 78 developing countries—41 in Africa, 19 in Asia, and 18 in Latin America and the Caribbean—have recognized that they must limit population growth. These countries have formulated policies, such as economic rewards and penalties, to achieve this goal. Most countries sponsor family planning projects, which are integrated with health care, education, economic development, and efforts to improve women's status.

CONCEPT CHECK STOP

What is the relationship between fertility rate and marriage age? Between fertility and educational opportunities for women?

What is family planning? Is family planning effective in reducing fertility rates?

Urbanization

LEARNING OBJECTIVES

Define *urbanization* and describe trends in the distribution of people in rural and urban areas.

Describe some of the problems associated with the rapid growth rates of large urban areas.

Explain how compact development makes a city more livable.

The geographic distribution of people in rural areas, towns, and cities significantly influences the social, environmental, and economic aspects of population growth. During recent history, the human population has become increasingly urbanized. **Urbanization** involves the movement of people from rural to urban areas as well as the transformation of rural areas into urban areas. When Europeans first settled in North America, the majority of the population consisted of farmers in rural areas. Today, approximately 80 percent of the Canadian population lives in cities.

urbanization
A process whereby people move from rural areas to densely populated cities.

One important distinction between rural and urban areas isn't how many people live there but how people make a living. Most people residing in rural areas have occupations that involve harvesting natural resources—such as fishing, logging, and farming. In urban areas, most people have jobs that are not connected directly with natural resources. Cities have traditionally provided more jobs than rural areas because cities are sites of industry, economic development, and educational and cultural opportunities.

Every city is unique in terms of its size, climate, culture, and economic development (**FIGURE 3.20**). Although there is no such thing as a typical city, certain

Vancouver, British Columbia FIGURE 3.20

Vancouver is Canada's most important port city on the Pacific coast. Noted for its lovely parks and gardens, Vancouver has mild weather and a wet winter season. The Greater Vancouver region has a population of more than 2 million, including a large immigrant population.

Brownfield of vacant warehouses and stores FIGURE 3.21

traits are common to city populations in general. One basic characteristic of city populations is their far greater heterogeneity with respect to race, ethnicity, religion, and socioeconomic status compared with rural populations. People living in urban areas tend to be younger than those living in the surrounding countryside. The young age structure of cities is due to the influx of many young adults from rural areas.

Urban and rural areas often have different proportions of males and females, and cities in developing nations tend to have more males. In cities in Africa, for example, males migrate to the city in search of employment, whereas females tend to remain in the country and tend their farms and children. Cities in highly developed countries often have a higher ratio of females to males. Women in rural areas often have little chance of employment after they graduate from high school, so they move to urban areas.

ENVIRONMENTAL PROBLEMS OF URBAN AREAS

Growing urban areas affect land-use patterns. Suburban sprawl that encroaches into former forest, wetland, desert, or agricultural land destroys or fragments wildlife habitat. Portions of Calgary, for example, are former wetlands. Most cities have blocks and blocks of **brownfields**—areas of abandoned, vacant factories, warehouses, and residential sites that may be contaminated from past uses (**FIGURE 3.21**). Meanwhile, the suburbs continue to expand outward, swallowing natural areas and farmland.

Reuse of brownfields is complicated because many have environmental contaminants that must be cleaned up before redevelopment can proceed. Nonetheless, brownfields represent an important potential land resource. Halifax is particularly known for its redevelopment of brownfields. Residential and commercial sites there now occupy several of these former brownfields (**FIGURE 3.22**).

In addition, most workers in cities have to commute many kilometres through traffic-congested streets from the suburbs where they live to downtown areas where they work. Because development is so spread out in the suburbs, automobiles are a necessity to accomplish everyday chores. This dependence on motor vehicles as our primary means of transportation increases air pollution and causes other environmental problems.

The high density of automobiles, factories, and commercial enterprises in urban areas causes a buildup of airborne emissions, including particulate matter (dust), sulphur oxides, carbon oxides, nitrogen oxides, and volatile organic compounds. Urban areas in developing nations have the worst air pollution in the world. Although we have made progress in reducing air

Brownfield redevelopment in Pittsburgh Figure 3.22

Brownfield redevelopment projects, like the one shown here in Pittsburgh, often feature upscale housing, recreation, and commerce.

pollution in highly developed nations, the atmosphere in many cities often contains higher levels of pollutants than are acceptable based on health standards.

Cities also affect water flow because they cover the rainfall-absorbing soil with buildings and paved roads. Storm sewage systems are built to handle the runoff from rainfall, which is polluted with organic wastes (garbage, animal droppings, and so on), motor oil, lawn fertilizers, and heavy metals. In most cities across Canada, urban runoff is treated in sewage treatment plants before being discharged into nearby waterways. In many cities, however, high levels of precipitation can overwhelm the sewage treatment plants, resulting in sewer overflows and the release of untreated urban runoff directly into rivers and streams. When this occurs, the polluted runoff contaminates water far beyond the boundaries of the city.

Highland Creek in Toronto is a good example of the effects of urbanization on waterways. The urban expansion and development of the Scarborough area, where the river runs through down to Lake Ontario, has resulted in natural plant cover being replaced by buildings, roadways, parking lots, and sidewalks. These changes to the landscape mean that significantly less water is absorbed into the soil and a far greater amount runs off into storm drains and waterways. This has substantially increased the amount of water that has historically travelled through the river, increasing erosion and changing water turbidity. Artificial rock cages and diversions have been built to try to reduce the erosion damage. Small dams have been constructed to try to even out the river water flow. The urban runoff also carries with it a number of pollutants such as fertilizers, garbage, oil, gas, and pesticides. The quality of the water used for drinking, aquatic life, wildlife, vegetation, and human recreation is reduced.

ENVIRONMENTAL BENEFITS OF URBANIZATION

Although the previous discussion may suggest that the concentration of people in cities has a harmful effect overall on the environment, urbanization does have the potential to provide tangible environmental benefits that in many cases outweigh its environmental problems. A well-planned city actually benefits the environment by reducing pollution and preserving rural areas.

One solution to urban growth is **compact development**, which uses land efficiently. Dependence on motor vehicles and their associated pollution are reduced as people walk, cycle, or take public transit such as buses or light rails to work and shopping districts (**FIGURE 3.23**). With compact development, fewer parking lots

| compact development |
| Design of cities in which tall, multiple-unit residential buildings are close to shopping and jobs, and all are connected by public transportation. |

and highways are needed, leaving more room for parks, open space, housing, and businesses. Compact development makes a city more livable and attractive to people.

Portland, Oregon, provides a good example of compact development. Although Portland is still grappling with many issues, the city government has developed effective land-use policies that dictate where and how growth will occur. The city looks inward to brownfields rather than outward to the suburbs for new development sites. Since 1975, Portland's population has grown 50 percent,

Compact development in Canada FIGURE 3.23

A Tip Top Lofts is an example of compact development in Canada. Located in Toronto, Ontario, this project converted a former 1920s industrial building into condominium lofts.

B Vancouver's Pacific Place is a brownfield redevelopment project that has received national attention by blending housing for approximately 15,000 people with more than 50 acres of public parks and three kilometres of seawall walkway.

from 0.9 million to 1.8 million (in 2005), yet the urbanized area increased by only about 2 percent. In contrast, from 1975 to 2005, the population of Chicago grew 22 percent, yet its urbanized area increased more than 50 percent due to sprawl.

Although the automobile is still the primary means of transportation in Portland, the city's public transportation system is an important part of its regional master plan. Public transportation incorporates light-rail lines, bus routes (many of which feature buses arriving every 15 minutes), bicycle lanes, and walkways. Employers are encouraged to provide bus passes to their employees instead of paying for parking. The emphasis on public transportation has encouraged commercial and residential growth along light rail and bus stops instead of in suburbs.

URBANIZATION TRENDS

Urbanization is a worldwide phenomenon. Currently, 49 percent of the world population lives in urban areas with populations of 2000 or greater. The percentage of people living in cities compared with rural settings is greater in highly developed countries than in developing countries. In 2007, urban inhabitants made up 75 percent of the total population of highly developed countries but only 43 percent of the total population of developing countries.

Although proportionately more people still live in rural settings in developing countries, urbanization is increasing rapidly. Currently, most urban growth in the world is occurring in developing countries, whereas highly developed countries are experiencing little urban growth. As a result of the greater urban growth of developing nations, most of the world's largest cities are in developing countries. In 1975, four of the world's 10 largest cities—Mexico City, São Paulo, Buenos Aires, and Kolkata (Calcutta)—were in developing countries. In 2005, eight of the world's 10 largest cities—Mexico City, São Paulo, Mumbai (Bombay), Delhi, Shanghai, Kolkata (Calcutta), Jakarta, and Buenos Aires (**TABLE 3.1**)—were in developing countries. By 2015, eight of the world's 10 largest cities will still be from developing countries, although Dhaka, Bangladesh, will have made the list, replacing Buenos Aires (**FIGURE 3.24** on page 94).

According to the United Nations, almost 400 cities worldwide have a population of at least 1 million inhabitants, and 284 of these cities are in developing countries. The number and size of **megacities** (cities with more than 10 million inhabitants) has also increased. In some places, separate urban areas have merged into

The world's 10 largest cities* TABLE 3.1		
1975	*2005*	*2015*
Tokyo, Japan, 26.6	Tokyo, Japan, 35.2	Tokyo, Japan, 35.5
New York, U.S. 15.9	Mexico City, Mexico, 19.4	Mumbai (Bombay), India, 21.9
Mexico City, Mexico, 10.7	New York, U.S. 18.7	Mexico City, Mexico, 21.6
Osaka-Kobe, Japan, 9.8	São Paulo, Brazil, 18.3	São Paulo, Brazil, 20.5
São Paulo, Brazil, 9.6	Mumbai (Bombay), India, 18.2	New York, U.S. 19.9
Los Angeles, U.S., 8.9	Delhi, India, 15.0	Delhi, India, 18.6
Buenos Aires, Argentina, 8.7	Shanghai, China, 14.5	Shanghai, China, 17.2
Paris, France, 8.6	Kolkata (Calcutta), India, 14.3	Kolkata (Calcutta), India, 17.0
Kolkata (Calcutta), India, 7.9	Jakarta, Indonesia, 13.2	Dhaka, Bangladesh, 16.8
Moscow, Russian Federation, 7.6	Buenos Aires, Argentina, 12.6	Jakarta, Indonesia, 16.8

Population in millions.

☐ Highly developed countries
☐ Developing countries

In 2015, eight of the 10 largest cities will be in developing countries: Mumbai (Bombay), Mexico City, São Paulo, Delhi, Shanghai, Kolkata (Calcutta), Dhaka, and Jakarta.

2015 projections

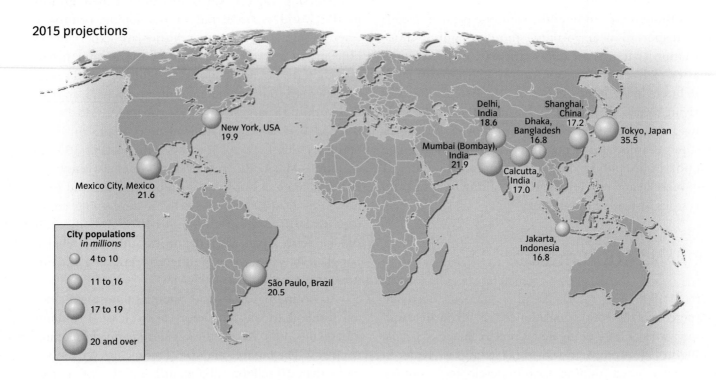

urban agglomerations, urbanized core regions, each of which consists of several adjacent cities or megacities and their surrounding developed suburbs. An example is the Tokyo-Yokohama-Osaka-Kobe agglomeration in Japan, which is home to about 50 million people. However, according to the United Nations Population Division, most of the world's urban population still lives in small or medium-sized cities with populations of less than 1 million.

The fast pace of urban growth in developing nations has outstripped the limited capacity of many cities to provide basic services. It has overwhelmed their economic growth (although cities still offer more job possibilities than rural areas). Consequently, cities in developing nations are generally faced with more serious challenges than cities in highly developed countries. These challenges include substandard housing (slums and squatter settlements); poverty; exceptionally high unemployment; heavy pollution; and inadequate or nonexistent water, sewage, and waste disposal. Rapid urban growth also strains school, medical, and transportation systems.

CONCEPT CHECK STOP

What is urbanization?

What are some of the problems caused by rapid urban growth in developing countries?

How does compact development help combat some of the problems of urbanization?

ENVIRONMENTAL PLANNER

Environmental planners are responsible for balancing the needs of urban living with sustaining the environment.

Imagine standing at the front of a small boardroom, patiently checking your equipment and glancing through your notes one last time while you wait for everyone to arrive. You are an environmental planner and today is a big day for you. You will present the first draft of your report on a proposed highway to a group of city aldermen, an environmental advocate, and community representatives. This proposed highway is supposed to skirt the southern edge of the city limits and alleviate traffic congestion on inner-city roads, but the proposed route also crosses a few environmentally sensitive wetlands. As one of the city's chief environmental planners, you've been tasked with finding a way to build the highway without threatening the local environment.

As an environmental planner, you balance the economic demands of the city's growth with the environmental concerns associated with urban expansion. When you first reviewed the preliminary route for the highway, the easiest solution to protecting the wetlands seemed to be moving the route to go around these areas. In some sections, this idea worked, so you redrew the route a little closer to the city limits. But in other sections, the wetlands were too big to go around, so now you must come up with a plan for constructing the highway through the wetlands with minimal environmental impact.

You start by consulting experts in the field, including wildlife biologists and wetland ecologists, for recommendations on how to move the wetlands. Perhaps by enlarging the wetlands outside of the highway, the birds and other inhabitants will nest and burrow far enough away from the road so they won't be affected by the traffic. Your plan will also include the construction of barriers to minimize the impact of noise pollution, as well as drainage maps to prevent contamination from vehicles and the road. Your report will address all these considerations and outline how the new highway can be built without threatening the sensitive environment.

Environmental planners are responsible for developing short- and long-term plans for land use in urban and rural areas while balancing considerations such as social, economic, and environmental issues. They also contribute to environmental impact assessments. Environmental planners can be involved in a range of fields, including strategic, commercial, and industrial development, as well as heritage, tourism, and integrated resource planning. Environmental planners work on a range of scales, from local planning to regional and national strategies.

For more information or to look up other environmental careers go to www.eco.ca.

URBAN PLANNING IN CURITIBA, BRAZIL

Livable cities aren't restricted to highly developed countries. Curitiba, a Brazilian city of more than 2.9 million people, provides a good example of *compact development* in a moderately developed country. Curitiba's city officials and planners have had notable successes in public transportation, traffic management, land-use planning, waste reduction and recycling, and community livability.

The city developed an inexpensive, efficient mass transit system that uses clean, modern buses that run in high-speed bus lanes. High-density development was largely restricted to areas along the bus lines, encouraging population growth where public transportation was already available. About 72 percent of commuters use mass transportation (**Figure A**). Since the 1970s, Curitiba's population has more than doubled, yet traffic has declined by 30 percent. Curitiba doesn't rely on automobiles as much as other comparably sized cities do, so it has less traffic congestion and significantly cleaner air, both of which are major goals of compact development. Instead of streets crowded with vehicular traffic, the centre of Curitiba is a *calcadao*, or "big sidewalk," that consists of 49 downtown blocks of pedestrian walkways connected to bus stations, parks, and bicycle paths (**Figure B**).

Over several decades, Curitiba purchased and converted flood-prone properties along rivers in the city to a series of interconnected parks crisscrossed with bicycle paths. This move reduced flood damage and increased the per person amount of "green space" from 0.5 m^2 (5.4 ft^2) in 1950 to 50 m^2 (538 ft^2) today, a significant accomplishment considering Curitiba's rapid population growth during the same period.

Another example of Curitiba's creativity is its labour-intensive garbage purchase program, in which poor people exchange filled garbage bags for bus tokens, surplus food (eggs, butter, rice, and beans), or school notebooks. This program encourages garbage pickup from the unplanned shantytowns (which garbage trucks can't access) that surround the city. Curitiba supplies more services to these unplanned settlements than most cities do. It tries to provide water, sewer, and bus service for them.

These changes didn't happen overnight. Urban planners can carefully reshape most cities over several decades to make better use of space and to reduce dependence on motor vehicles. City planners and local and regional governments are increasingly adopting measures to provide the benefits of compact development in the future.

Curitiba, Brazil

A Bus passengers pay their fares in advance in the tubular bus stations and then walk directly onto the bus as soon as it arrives.

B The downtown area of Curitiba is filled with open terraces lined with shops and restaurants.

SUMMARY

1 Population Ecology

1. **Population ecology** is the branch of biology that deals with the number of individuals of a particular species found in an area and how and why those numbers change over time.

2. The **growth rate (r)** is the rate of change (increase or decrease) of a population's size, expressed in percentage per year. On a global scale, growth rate is due to the birth rate (b) and the death rate (d): $r = b - d$. Emigration (e), the number of individuals leaving an area, and immigration (i), the number of individuals entering an area, also affect a local population's growth rate.

3. **Biotic potential** is the maximum rate a population could increase under ideal conditions. **Exponential population growth** is the accelerating population growth that occurs when optimal conditions allow a constant reproductive rate for limited periods. Eventually, the growth rate decreases to around zero or becomes negative because of environmental resistance, unfavourable environmental conditions that prevent organisms from

reproducing indefinitely at their biotic potential. The **carrying capacity (K)** is the largest population a particular environment can support sustainably (long term) if there are no changes in that environment.

2 Human Population Patterns and Demographics

1. It took thousands of years for the human population to reach 1 billion (around 1800). Since then, the population has grown exponentially. The United Nations projects that the population will reach 7 billion by 2013. Although our numbers continue to increase, the growth rate (r) has declined slightly over the past several years. The population should reach **zero population growth**, in which it remains the same size because the birth rate equals the death rate, toward the end of the 21st century.

2. Estimates of Earth's carrying capacity for humans vary widely depending on what assumptions are made about standard of living, resource consumption, technological innovations, and waste generation. In addition to natural environmental constraints, human choices and values determine Earth's carrying capacity for humans.

3. **Demographics** is the applied branch of sociology that deals with population statistics. As a country becomes industrialized, it goes through a **demographic transition** as it moves from relatively high birth and death rates to relatively low birth and death rates.

4. The **infant mortality rate** is the number of deaths of infants under age 1 per 1000 live births. The **total fertility rate (TFR)** is the average number of children born to each woman. **Replacement-level fertility** is the number of children a couple must produce to "replace" themselves. **Age structure** is the number and proportion of people at each age in a population. A country can have replacement-level fertility and still experience population growth if the largest percentage of the population is in the pre-reproductive years. In contrast to developing countries, highly developed countries have low infant mortality rates, low total fertility rates, and an age structure in which the largest percentage of the population isn't in the pre-reproductive years.

3 Stabilizing World Population

1. Four factors are most responsible for high total fertility rates: high infant and child mortality rates, the important economic and societal roles of children in some cultures, the low status of women in many societies, and a lack of health and family planning services. The single most important factor affecting high TFRs is the low status of women. The governments of many developing countries are trying to limit population growth.

2. Education of women decreases the total fertility rate, in part by delaying the first childbirth. Education increases the likelihood that women will know how to control their fertility. Education also increases women's career options, which provide ways of achieving status besides having babies.

4 Urbanization

1. **Urbanization** is the process whereby people move from rural areas to densely populated cities. In developing nations, most people live in rural settings, but their rates of urbanization are rapidly increasing.

2. Rapid urbanization makes it difficult to provide city dwellers with basic services such as housing, water, sewage, and transportation systems.

3. **Compact development** is the design of cities so that tall, multiple-unit residential buildings are close to shopping and jobs, and all are connected by public transportation.

KEY TERMS

- **population ecology** p. 68
- **growth rate (r)** p. 69
- **biotic potential** p. 70
- **exponential population growth** p. 70
- **carrying capacity (K)** p. 71

- **zero population growth** p. 75
- **demographics** p. 78
- **infant mortality rate** p. 78
- **replacement-level fertility** p. 78
- **total fertility rate (TFR)** p. 78

- **demographic transition** p. 78
- **age structure** p. 80
- **urbanization** p. 89
- **compact development** p. 92

CRITICAL AND CREATIVE THINKING QUESTIONS

1. Draw a graph to represent the long-term growth of a population of bacteria cultured in a test tube that contains a nutrient medium that isn't replenished.

2. In Bolivia, 38 percent of the population is younger than 15, and 4 percent is older than 65. In Austria, 16 percent of the population is younger than 15, and 17 percent is older than 65. On the basis of this information, which of these countries will have the higher population growth momentum over the next two decades? Why?

3. Should the rapid increase in world population be of concern to the average citizen in Canada? Why or why not?

4. How is human population growth related to natural resource depletion and environmental degradation?

5. Although the United States ranks third in population number, some experts contend that it is the most overpopulated country in the world. Explain why.

6. Explain the rationale behind this statement: It is better for highly developed countries to spend millions of dollars on family planning in developing countries now than to have to spend billions of dollars on relief efforts later.

7. Discuss this statement: The current human population crisis causes or exacerbates all environmental problems.

8. Discuss some of the ethical issues associated with overpopulation. Is it ethical to have more than two children? Is it ethical to consume so much in the way of material possessions? Is it ethical to try to influence a couple's decision about family size?

9. Tanzania, Argentina, and Poland have about the same population sizes (38 million to 39 million each). The current total fertility rates of these countries are Tanzania, 5.4; Argentina, 2.5; and Poland, 1.3. Assuming that fertility rates decline to 2.0 by 2050 in Tanzania and Argentina, will there be a difference in population among the three countries? If so, which country will have the highest population in 2050? Explain your answer.

10. If you were to draw an age structure diagram for Poland, with a total fertility rate of 1.3, which of the following overall shapes would the diagram have? Explain why a country like Poland faces a population decline even if its fertility rate were to start increasing today.

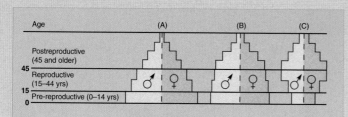

The following graph shows the 10 countries with the largest populations in 2007. Use it to answer questions 11 and 12.

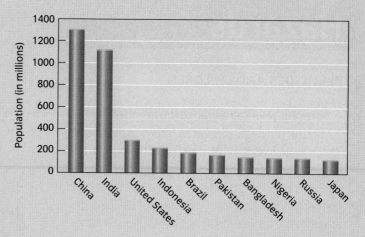

11. Population experts project that in 2050 the Democratic Republic of Congo and Ethiopia will replace Russia and Japan in the top 10. Do Russia and Japan have high, medium, or low fertility rates? Do the Democratic Republic of Congo and Ethiopia have high, medium, or low fertility rates? Explain your answers.

12. Experts project that the United States will have 420 million people in 2050. Given that the U.S. fertility rate is relatively low and stable (2.1), why do you suppose the United States has one of the highest population growth rates in the developed world?

What is happening in this picture ?

This photo shows people—mainly displaced rural workers—picking through trash at the Smoky Mountain garbage dump in Manila, Philippines. Most of these people moved to Manila looking for work. Why didn't they find better employment?

They are looking mainly for scraps of plastic and metal, which they can sell. What valuable environmental service does such scavenging provide?

Human Health and Environmental Hazards

PESTICIDES AND CHILDREN

In recent years, attention to the health effects of household pesticides on children has increased. These pesticides appear to be a greater threat to children than to adults because children tend to play on floors and lawns where they are exposed to greater concentrations of pesticide residues. Also, children are probably more sensitive to pesticides because their bodies are still developing. Several preliminary studies suggest that exposure to household pesticides may cause brain cancer and leukemia in children, but scientists must do more research before they can reach any firm conclusions.

Research supports the hypothesis that exposure to pesticides may affect the development of intelligence and motor skills in young children. One study, published in *Environmental Health Perspectives* in 1998, compared two groups of rural Yaqui Indian preschoolers in Mexico. These two nearly identical groups differed mainly in their exposure to pesticides: One group lived in a farming community where pesticides were used frequently (see photograph), and the other lived in an area where pesticides were rarely used. When asked to draw a person, most of the 17 children from the low-pesticide area drew recognizable stick figures (see part a of inset), whereas most of the 34 children from the high-pesticide area drew meaningless lines and circles (see part b). Additional tests of simple mental and physical skills revealed similar striking differences between the two groups of children.

Drawings of a person
(by 4-year-olds)

Foothills 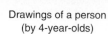 Valley

(a) 54 mo.
female

(b) 54 mo.
female

Risk Assessment

LEARNING OBJECTIVES

Define *risk* and *risk assessment.*

Explain how the four steps of risk assessment help determine adverse health effects.

Risk is inherent in all our actions and in everything in our environment. All of us take risks every day of our lives. Walking on stairs involves a small risk, but a risk nonetheless because sometimes people die from falls on stairs. Using household appliances is slightly risky because sometimes people die from electrocution when they operate appliances with faulty wiring or use appliances in an unsafe manner. Driving or riding in a car or flying in a jet has risks that are easier for most of us to recognize. Yet few of us hesitate to get in a car or board a plane because of the associated risk. It is important to have an adequate understanding of the nature and size of risks before deciding what actions are appropriate to avoid them (**FIGURE 4.1**).

Each of us uses our intuition, habit, and experience to make many decisions regarding risk every day. However, environmental and health risks often affect many individuals, and

the best choices cannot always be made on an intuitive level. **Risk management** is the process of identifying, assessing, and reducing risks. The four steps involved in **risk assessment** for

> **risk** The probability of harm (such as injury, disease, death, or environmental damage) occurring under certain circumstances.

> **risk assessment** The use of statistical methods to quantify risks so they can be compared and contrasted.

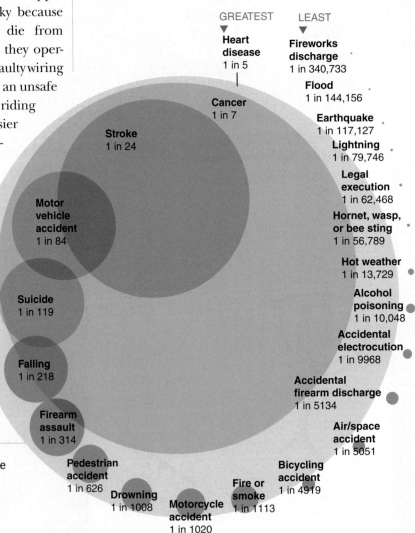

Lifetime probability of death by selected causes FIGURE 4.1

These 2003 data from the National Safety Council are for U.S. residents.

GREATEST ▼   LEAST ▼

Heart disease 1 in 5

Cancer 1 in 7

Stroke 1 in 24

Motor vehicle accident 1 in 84

Suicide 1 in 119

Falling 1 in 218

Firearm assault 1 in 314

Pedestrian accident 1 in 626

Drowning 1 in 1008

Motorcycle accident 1 in 1020

Fire or smoke 1 in 1113

Bicycling accident 1 in 4919

Air/space accident 1 in 5051

Accidental firearm discharge 1 in 5134

Accidental electrocution 1 in 9968

Alcohol poisoning 1 in 10,048

Hot weather 1 in 13,729

Hornet, wasp, or bee sting 1 in 56,789

Legal execution 1 in 62,468

Lightning 1 in 79,746

Earthquake 1 in 117,127

Flood 1 in 144,156

Fireworks discharge 1 in 340,733

The four steps of risk assessment for adverse health effects FIGURE 4.2

① Hazard identification
Does exposure to substance cause increased likelihood of adverse health effects such as cancer or birth defects?

② Dose–response assessment
What is the relationship between amount of exposure (dose) and seriousness of adverse health effect? A person exposed to a low dose may have no symptoms, whereas exposure to a high dose may result in illness.

③ Exposure assessment
How much, how often, and how long are humans exposed to substance in question? Where humans live relative to emissions is also considered.

④ Risk characterization
What is the probability of an individual or population having adverse health effects? Risk characterization evaluates data from dose–reponse assessment and exposure assessment (steps 2 and 3). Risk characterization indicates that agricultural workers, for example, are more vulnerable to pesticide exposure than other groups.

Agricultural workers have a greater than average exposure to chemicals such as pesticides.

adverse health effects are summarized in **FIGURE 4.2**. Once a risk assessment is performed, its results are combined with relevant political, social, and economic considerations to determine whether we should reduce or eliminate a particular risk and, if so, what we should do. This evaluation includes the development and implementation of laws to regulate hazardous substances.

Risk assessment helps estimate the probability that an event will occur and lets us set priorities and manage risks in an appropriate way. As an example, consider a person who smokes a pack of cigarettes a day and drinks well water containing traces of the cancer-causing chemical trichloroethylene (in acceptable amounts, as established by Health Canada). Without knowledge of risk assessment, this person might buy bottled water in an attempt to reduce his or her chances of getting cancer. Based on risk assessment calculations, the annual risk from smoking is 0.00059, or 5.9×10^{-1}, whereas the annual risk from drinking water with accepted levels of trichloroethylene is 0.000000002, or 2.0×10^{-4}. This means that this person is almost *300,000 times* more likely to get cancer from smoking than to get it from ingesting such low levels of trichloroethylene (**FIGURE 4.3** on page 104). Knowing this, the person in our example would, we hope, stop smoking.

One of the most perplexing dilemmas of risk assessment is that people often ignore substantial risks but get extremely upset about minor risks. The average life expectancy of smokers is more than eight years less than that of nonsmokers, and almost one-third of all smokers die from diseases that the habit causes or exacerbates. Yet many people get more upset over a one-in-a-million chance of getting cancer from pesticide residues on food than they do over the relationship between smoking and cancer. Perhaps part of the reason for this attitude is that behaviours such as diet, smoking, and exercise are parts of our lives that we can change if we choose to (TABLES 4.1 and 4.2). Risks over which most of us have no control, such as pesticide residues or nuclear wastes, tend to evoke more fearful responses.

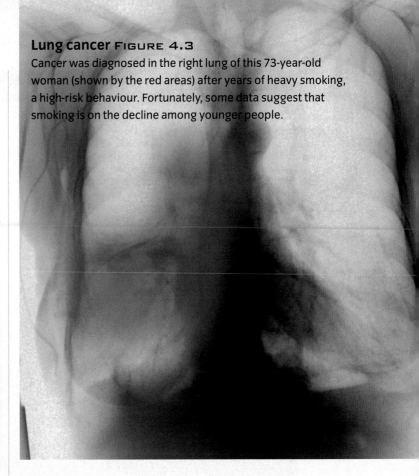

Lung cancer FIGURE 4.3
Cancer was diagnosed in the right lung of this 73-year-old woman (shown by the red areas) after years of heavy smoking, a high-risk behaviour. Fortunately, some data suggest that smoking is on the decline among younger people.

Smoking in Canada TABLE 4.1

	Percent who smoke
Men	22
Women	18
Adolescent boys (15–19 yrs)	16
Adolescent girls (15–19 yrs)	19
Young men (20–24 yrs)	31
Young women (20–24 yrs)	28
Aboriginal Canadians	
First Nations	56
Metis	57
Inuit	72
Statistics by Province	
Newfoundland and Labrador	23
Prince Edward Island	23
Nova Scotia	22
New Brunswick	24
Quebec	23
Ontario	19
Manitoba	21
Saskatchewan	21
Alberta	21
British Columbia	16

Preventable causes of death in Canada TABLE 4.2

	Number of deaths/year in Canada
Smoking	45000
Alcohol	6700
Traffic accidents	1900
AIDS	1300
Homicide	550

CONCEPT CHECK STOP

What are risk and risk assessment?

What are the four steps of risk assessment?

Environmental Health Hazards

LEARNING OBJECTIVES

Define *toxicology* and distinguish between acute toxicity and chronic toxicity

Explain why public water supplies are monitored for fecal coliform bacteria even though most strains of *E. coli* do not cause disease.

Describe the link between environmental changes and emerging diseases, such as the swine and avian strains of influenza.

Threats or hazards to our health, particularly from toxic chemicals in the environment, make for big news. Many of these stories are more sensational than factual. In fact, human health is generally better today than at any previous time in our history, and our life expectancy continues to increase rather than decline. This does not mean that you should ignore chemicals that humans introduce into the environment. Nor does it mean you should discount the stories that the news media sometimes sensationalize. These stories serve an important role in getting the regulatory wheels of the government moving to protect us as much as possible from the dangers of our technological and industrialized world.

Scientists use the term **environmental hazard** to describe situations and events that pose threats to the surrounding environment. For example, the human body may be exposed to both natural and synthetic chemicals in the air we breathe, the water we drink, and the food we eat. All chemicals, even "safe" chemicals such as sodium chloride (table salt), are toxic if exposure is high enough. For example, a 1-year-old child will die from ingesting about 2 tablespoons of table salt; table salt is also harmful to people with heart or kidney disease. Chemicals with adverse effects are known as **toxicants**.

Toxicology involves (a) studying the effects of toxicants on living organisms, (b) studying the mechanisms that cause toxicity, and (c) developing ways to prevent or

environmental hazard Situations and events that pose threats to the surrounding environment.

toxicology The study of toxicants, chemicals with adverse effects on health.

acute toxicity Adverse effects that occur within a short period after high-level exposure to a toxicant.

chronic toxicity Adverse effects that occur after a long period of low-level exposure to a toxicant.

minimize adverse effects. (Developing appropriate handling or exposure guidelines for specific toxicants is one of these ways.)

The effects of toxicants following exposure can be immediate *(acute toxicity)* or prolonged *(chronic toxicity)*. Symptoms of **acute toxicity** range from dizziness and nausea to death. Acute toxicity occurs immediately to several days following a single exposure. In comparison, **chronic toxicity** generally produces damage to vital organs, such as the kidneys or liver, following long-term, low-level exposure to a toxicant. Human diseases that are the result of chronic toxicity are *noninfectious*—that is, they are not transmitted from one human host to another. Toxicologists know far less about chronic toxicity than they do about acute toxicity, partly because the symptoms of chronic toxicity often mimic those of other chronic diseases associated with risky lifestyle patterns, poor nutrition, and aging. Also, it is difficult to isolate a causative agent from among the multiple toxicants we are routinely exposed to.

DISEASE-CAUSING AGENTS IN THE ENVIRONMENT

Disease-causing agents are *infectious* organisms, such as bacteria, viruses, protozoa, and parasitic worms that cause diseases. Typhoid, cholera, bacterial dysentery, polio, and infectious hepatitis are some of the most common bacterial or viral diseases that are transmissible through contaminated food and water. Diseases such as these are considered environmental health hazards. Other human diseases, such as acquired immunodeficiency syndrome (AIDS), are not transmissible through the environment and aren't discussed here.

The vulnerability of our public water supplies to waterborne disease-causing agents was dramatically demonstrated in 2000 when the first waterborne outbreak in North America of a deadly strain of *Escherichia coli* occurred in Walkerton, Ontario (**FIGURE 4.4**). In May 2000, many residents in the small town of 5000 people began to experience bloody diarrhea, gastrointestinal infections, and other symptoms of the *E. coli* infection and the highly dangerous O157:H7 strain of *E. coli* bacteria. Seven people died from *E. coli* contamination and several thousand became sick.

The inquiry into the outbreak found numerous inadequacies at the Walkerton Public Utilities Commission, including failing to use adequate doses of chlorine, failing to monitor chlorine residuals, creating false entries on daily operating reports, and recording misleading information about where microbiological samples were taken. The Ontario government was held responsible for not monitoring water quality and ensuring public

> **pathogen**
> An agent (usually a microorganism) that causes disease.

health safety. The inquiry's recommendations have been incorporated into new provincial legislation and have influenced provincial policies across Canada.

Monitoring and enforcing accurate testing, such as should have been done and accurately reported in the Walkerton water contamination case, are essential for indicating the possible presence of various disease-causing agents. Although most strains of coliform bacteria found in sewage do not cause disease, testing for these bacteria is a reliable way to indicate the likely presence of **pathogens** in water. Although the Walkerton Public Utilities Commission insisted the water was clean, it was eventually proven that reports had been falsified.

In addition to *E. coli*, there are numerous other parasitic and disease-causing agents transmitted by polluted water. Some of these are further described in **TABLE 4.3**, and some are more common to Canada than others, including giardiasis (commonly called beaver fever) and cryptosporidiosis. Effective water treatment technologies have reduced the risk of these waterborne pathogens and are described further in Chapter 9.

Drinking water is regularly monitored for the presence of *E. coli* and typically involves **fecal coliform tests** (**FIGURE 4.5**). A small sample of water is passed through a filter to trap the bacteria, which are then transferred to a petri dish that contains nutrients. After an incubation period, the number of greenish colonies present indicates the number of *E. coli*. Safe drinking water should contain no more than one coliform bacterium per 100 mL of water (about 1/2 cup); safe swimming water should have no more than 200 per 100 mL of water; and general recreational water (for boating) should have no more than 2000 per 100 mL. In contrast, raw sewage may contain several million coliform bacteria per 100 mL of water. Water pollution and purification are discussed further in Chapter 9.

ENVIRONMENTAL CHANGES AND EMERGING DISEASES

Human health has improved significantly over the past several decades, but environmental factors remain a significant cause of human disease in many areas of the world. **Epidemiologists**, scientists who investigate the outbreaks of both infectious and noninfectious diseases

Clean drinking water FIGURE 4.4

In Canada, we generally assume our drinking water is safe. The Walkerton tragedy of 2000 proved that this is not always the case.

Some human diseases transmitted by polluted water TABLE 4.3

Disease	Type of organism	Symptoms
Cholera	Bacterium	Severe diarrhea, vomiting; fluid loss of as much as 19 litres per day causes cramps and collapse
Dysentery	Bacterium	Infection of the colon causes painful diarrhea with mucus and blood in the stools; abdominal pain
Enteritis	Bacterium	Inflammation of the small intestine causes general discomfort, loss of appetite, abdominal cramps, and diarrhea
Typhoid	Bacterium	Early symptoms include headache, loss of energy, fever; later, a pink rash appears, along with (sometimes) hemorrhaging in the intestines
Giardiasis	Protozoan	Majority of those affected develop gradual abdominal cramps, diarrhea, soft stools, vomiting, fever, and upper GI symptoms including nausea, bloating, and acid indigestion
Infectious hepatitis	Virus	Inflammation of liver causes jaundice, fever, headache, nausea, vomiting, severe loss of appetite, muscle aches, and general discomfort
Poliomyelitis	Virus	Early symptoms include sore throat, fever, diarrhea, and aching in limbs and back; when infection spreads to spinal cord, paralysis and atrophy of muscles occur
Cryptosporidiosis	Protozoon	Diarrhea and cramps that last up to 22 days
Amoebic dysentery	Protozoon	Infection of the colon causes painful diarrhea with mucus and blood in the stools; abdominal pain
Schistosomiasis	Fluke	Tropical disorder of the liver and bladder causes blood in urine, diarrhea, weakness, lack of energy, repeated attacks of abdominal pain
Ancylostomiasis	Hookworm	Severe anemia, sometimes symptoms of bronchitis

Fecal coliform test FIGURE 4.5

This test indicates the likely presence of disease-causing agents in water. A water sample is first passed through a filtering apparatus. **A** The filter disk is then placed on a medium that supports coliform bacteria for 24 hours. **B** After incubation, the number of bacterial colonies is counted. Each colony of *Escherichia coli* arose from a single coliform bacterium in the original water sample.

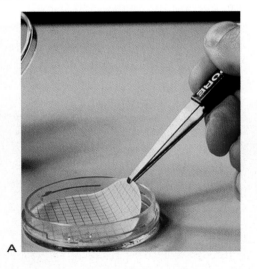

A

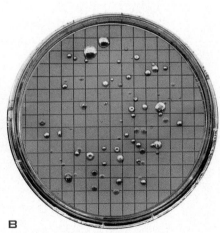

B

in a population, have established links between human health and human activities that alter the environment. A World Health Organization report released in 1997 concluded that about 25 percent of diseases and injuries worldwide are related to human activities that cause environmental changes. The environmental component of human health is sometimes direct and obvious, as when people drink unsanitary water and contract dysentery, a waterborne disease that causes diarrhea. Diarrhea causes 4 million deaths worldwide each year, mostly in children.

The health effects of many human activities are complex and often indirect. The disruption of natural environments may give disease-causing agents an opportunity to thrive. Development activities such as cutting down forests, building dams, and expanding agriculture may bring more humans into contact with new or previously rare disease-causing agents. Such projects may increase the population and distribution of disease-carrying organisms such as mosquitoes, thereby increasing the spread of disease (**Figure 4.6**). Social factors may also contribute to disease epidemics. Highly concentrated urban populations promote the rapid spread of infectious organisms among large numbers of people (**Figure 4.7**). Global travel also has the potential to contribute to the rapid spread of disease as infected individuals move easily from one place to another.

Consider malaria, a disease that mosquitoes transmit to humans. Each year, between 300 million and 500 million people worldwide contract malaria, and it causes more than 1 million deaths. About 60 different species of *Anopheles* mosquito transmit the parasites that cause malaria. Each mosquito species thrives in its own unique combination of environmental conditions (such as elevation, amount of precipitation, temperature, relative humidity, and availability of surface water).

Another recent concern is the possibility of a pandemic of avian (bird) and swine (H1N1) influenza or flu. A **pandemic** is a disease that reaches nearly every part of the world and has the potential to infect almost every person. Avian influenza is a strain of influenza virus that is common in birds (**Figure 4.8**). It tends to be difficult for humans to contract because it is usually transferred from birds to humans but not from human to human. It is extremely potent once contracted, however, and has a high fatality rate. Of concern for scientists is the possible evolution of a strain easily transferred from human to human. Such a strain, which might involve a single mutation in the viral genes, could kill millions of people in a single year.

Understanding and controlling an avian influenza pandemic requires study of the environment that allows the virus to survive and travel, as well as cooperation

Road clearing in the Amazon rainforest FIGURE 4.6

The drainage ditches that will be added to each side of the road will hold standing water where mosquito larvae thrive.

Global Locator

Crowds on a street in Hong Kong, China FIGURE 4.7

The development of cities, and the concentration of people living in them, permit the rapid spread of infectious disease-causing agents.

among many governments and individuals. The virus often originates in areas that have dense populations of domestic birds, especially chickens, raised in small cages. In the past several years, large numbers of domestic poultry have been killed and burned to prevent or stop disease outbreaks. Avian flu is now endemic in domestic and wild birds in Asia, Europe, and Northern Africa.

Dead swan FIGURE 4.8

Deaths of wild swans provided early evidence that avian influenza had migrated from Asia to Europe.

In April 2009, a new strain of the A (H1N1) influenza virus, commonly referred to as swine flu, was identified. Swine influenza viruses do not normally infect humans; however, from time to time, human infections do occur resulting in the human swine flu.

CONCEPT CHECK STOP

What is the difference between acute and chronic toxicity?

Why is the fecal coliform test performed on public drinking water supplies?

How is the incidence of swine and avian strains of influenza related to human activities that alter the environment?

Movement and Fate of Toxicants

Some chemically stable toxicants are particularly dangerous because they resist degradation and readily move around in the environment. These include certain pesticides, radioactive isotopes, heavy metals such as mercury, flame retardants such as PBDEs (polybrominated diphenylethers), and industrial chemicals such as PCBs (polychlorinated biphenyls).

The effects of the pesticide **DDT (dichlorodiphenyltrichloroethane)** on many bird species demonstrate the problem. Falcons, pelicans, bald eagles, and many other birds are sensitive to traces of DDT in their tissues. Substantial evidence indicates that DDT causes these birds to lay eggs with thin, fragile shells that usually break during incubation, causing the chicks' deaths. Canada banned DDT in 1985, following the United States' ban in 1972. Since that time there has been a marked recovery in bird populations previously affected by DDT exposure.

The impact of DDT on birds is the result of (1) its persistence, (2) bioaccumulation, and (3) biological magnification. **Persistence** means that the substance is extremely stable and may take many years to break down into a less toxic form. When an organism can't metabolize (break down) or excrete a toxicant, it is simply stored, usually in fatty tissues. Over time, the organism may accumulate high concentrations of the toxicant. The buildup of a persistent toxicant in an organism is called **bioaccumulation** (**FIGURE 4.9**).

Organisms at the top of the food chain tend to store greater concentrations of bioaccumulated toxicants in their bodies than those lower on the food chain. As an

biological magnification
The increase in toxicant concentrations as a toxicant passes through successive levels of the food chain.

example of **biological magnification**, consider a food chain studied in a salt marsh that was sprayed with DDT over several years for mosquito control: algae and plankton → shrimp → American eel → Atlantic needlefish → ring-billed gull (**FIGURE 4.10**).

Another example is mercury, which is a heavy metal that is also shown to exhibit bioaccumulation and biological magnification. While mercury is found in small amounts naturally, the concentration in the environment has increased dramatically since the Industrial Revolution. With this, there has been an observed bioaccumulation of mercury in its various forms in natural ecosystems. For example, mercury may be found in extremely low concentration in sea water, where it is absorbed by algae. In successive trophic levels of the food chain, the mercury levels build, such as in zooplankton, small nekton, and then successively in larger fish. Anything that consumes these fish also consumes the higher levels of mercury that the fish have accumulated. Predatory fish such as swordfish and sharks and predatory birds such as osprey and eagles have higher concentrations of mercury in their tissue than could be accounted for by direct exposure to mercury alone. Herring contains mercury at 0.01 ppm (parts per million), whereas a shark contains mercury at greater than 1 ppm. All top carnivores, from fish to humans, are at risk of health problems from biological magnification. Scientists therefore test pesticides to ensure that they do not persist and accumulate in the environment. The EnviroDiscovery box on page 112 discusses the impact mercury contamination had on the English-Wabigoon river system in Ontario.

Effect of DDT on birds FIGURE 4.9

A A bald eagle feeds its chick.

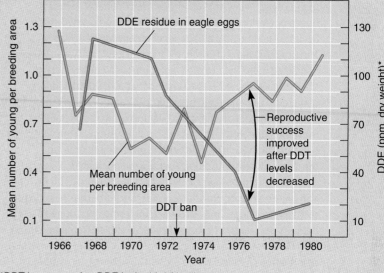

*DDT is converted to DDE in the birds' bodies

B A comparison of the number of successful bald eagle offspring with the level of DDT residues in their eggs.

Biological magnification of DDT FIGURE 4.10

A A ring-billed gull stands at the edge of a lake.

B Note how the level of DDT, expressed as parts per million, ▶ increased in the tissues of various organisms as DDT moved through the food chain from producers to consumers (bottom to top of figure). The ring-billed gull at the top of the food chain had approximately 1.5 million times more DDT in its tissues than the water contained.

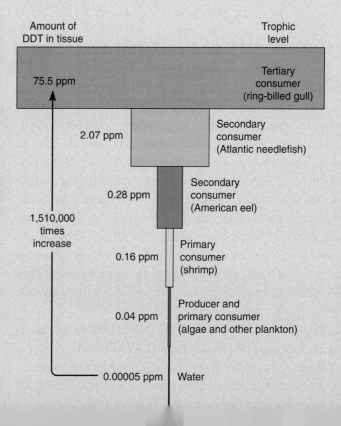

In 1970, mercury contamination was discovered in the English-Wabigoon river system, located in northwestern Ontario. The Wabigoon River flows from Wabigoon Lake in Dryden, Ontario, to the English River. In 1962, the Dryden Chemical Company began operations, producing chlorine and sodium hydroxide used by a nearby pulp and paper company for bleaching pulp. The Dryden Chemical Company dumped its waste water, which contained mercury, into the Wabigoon River, contaminating the entire watershed. It is estimated that over 9000 kilograms of mercury were dumped between 1962 and 1970, and the English-Wabigoon river system became known as one of the most mercury-contaminated freshwater systems in the world.

On March 26, 1970, the Ontario government ordered the Dryden Chemical Company to stop dumping mercury into the river system; however, it did not place any restrictions on the company's airborne emissions of mercury. Airborne emissions of mercury continued until 1975 when the company stopped using mercury cells, and in 1976, the Dryden Chemical Company closed.

However, the damage had been done. This illegal dumping of industrial chemical waste caused large amounts of mercury to enter into the food chain, accumulating mainly in fish, a substantial food source for Aboriginal peoples living in nearby communities. The consumption of the mercury-contaminated fish led to serious health effects such as birth defects, sensory disturbances, and a severe neurological disorder known as Ontario Minamata disease. The mercury contamination also severely impacted the area's economy. The main sources of jobs, especially for Aboriginal residents, were commercial fishing and guiding sport fishermen, meaning a loss of economic security for many.

The Grassy Narrows and Whitedog First Nations communities wanted compensation for the health, economic, and social effects of the mercury contamination. In 1985, a memorandum of agreement was signed committing the federal and provincial governments, along with the two parent companies involved (Reed Limited and Great Lakes Forest Products Ltd.) to pay out $16.67 million in compensation, although community members have seen little of this money as it remains tied up in bureaucratic processes.

MOBILITY IN THE ENVIRONMENT

Persistent toxicants tend to move through the soil, water, and air, sometimes travelling long distances from their original source. For example, pesticides applied to agricultural lands may be washed into rivers and streams by rain, harming aquatic life (**FIGURE 4.11**). If the pesticide level in their aquatic ecosystem is high enough, plants and animals may die. At lower pesticide levels, aquatic life may still suffer from symptoms of chronic toxicity, such as bone degeneration in fishes. These symptoms may, for example, decrease fishes' competitiveness or increase their chances of being eaten by predators. Mobility of persistent toxicants is also a risk for humans.

Long-range transport of pollutants
The **long-range transport of air pollutants (LRTAP)** is the transport of atmospheric pollutants within a moving air mass for a distance over 100 kilometres. For example, the Atlantic regions of Canada receive mercury and other harmful substances from industrial sites in the northeastern United States, the Great Lakes Basin, as well as from other continents. Long-range transport of chemicals can occur by water and air. Atmospheric wind and ocean currents are the primary forces behind long-range distribution of chemicals.

The incidence of pesticides and other industrial pollutants found in remote areas of the world, such as the Arctic, where no industrial pollutants are found, gives rise to international concerns of the long-range transport of harmful substances. Main air currents take industrial air pollution from the major industrial areas of the northern hemisphere, circulate them, carry them long distances, and deposit them in the Arctic (**FIGURE 4.12**). The end result is an increase of pollutants in the biosphere of the Arctic. Arctic ecosystems are particularly vulnerable to this pollution because of their cold temperature and highly specialized food chains that have high biomagnification potential. The Arctic may be experiencing the deposition of harmful chemicals from industrial areas as far away as Asia and Europe.

Mobility of pesticides in the environment FIGURE 4.11

The intended pathway of pesticides in the environment is shown in the green band across the top of the figure, and the actual pathways are shown across the bottom.

VIEW THIS IN ACTION

Intended pathway for pesticide

Aerial spraying of pesticide, evaporation → Air → Gravitational settling and precipitation → Crop → Target pest

Actual pathways of pesticide in the environment

Agricultural soil

Harvest

Precipitation

Erosion, leaching

Animals

Food

Runoff and seepage

Freshwater and aquatic organisms

Especially groundwater

Humans

Food

Ocean, marine organisms, and ocean sediments

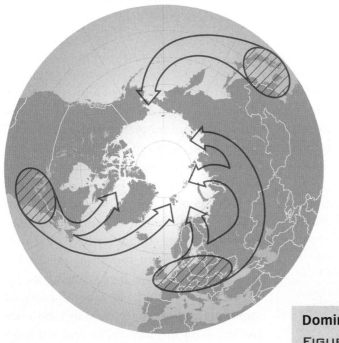

Central industrial areas

OASIS-Canada, a subsector of the larger OASIS-International, is a research program that is studying the long-range transport of these harmful chemicals. OASIS-Canada hopes to determine how these harmful chemicals travelled to the Arctic, if these chemicals end up in the arctic food chain, and the impact they have on arctic ecosystems.

Researchers can no longer rely on contamination levels near industrial sites to determine pollution intensity. Everything is connected and the impacts of the emission of harmful chemicals from cities, factories, and other pollutant sources are felt in the most remote areas of the globe. The EnviroDiscovery box on page 114 discusses one project that is underway to study the effects of long-range transport of mercury on vulnerable ecosystems.

Dominating air currents and the pollution of the Arctic
FIGURE 4.12

You would think that arctic ecosystems, located far from common pollution-creating sources, would be a pristine environment; however, this is not the case. Many harmful chemicals are found in arctic ecosystems.

Mercury is emitted from human activities, contaminated soils, water, and some natural sources. Mercury can be transported through river systems in sediment loads and also through the atmosphere. Once mercury is released to the atmosphere it can be deposited in the vicinity of the mercury source or it can be carried long-range distances by air masses. The deposition of mercury is variable and depends on a number of factors such as meteorology, temperature, solar radiation, and humidity. The Canadian Arctic is a vulnerable ecosystem because of this long-range mercury transport. In the Canadian Arctic and Greenland, the levels of mercury in arctic ringed seals and beluga whales have increased two to four times over the last 25 years. The arctic aquatic food chain is long with many levels, and as a result top predators have high concentrations of biomagnified mercury.

The tropic ecosystems are also vulnerable. Large quantities of mercury are released into the waters of the Amazon from the gold mining operations located there. The effects of the mercury in the water are felt locally but also across a much larger floodplain reaching Brazil, Bolivia, and Paraguay. Other sources of mercury in the tropical rainforests are from agricultural slash-and-burn clearing of the land for agriculture and mining. These activities cause the mercury already stored in the soil to be exposed.

Canada is doing its part to contribute toward an understanding of ecosystem impacts from mercury loading. The Experimental Lakes Area, situated in northwestern Ontario,

is conducting a whole ecosystem research project called METAALICUS (Mercury Experiment to Assess Atmospheric Loading in Canada and the United States) to assess the impacts of atmospheric deposition of mercury on ecosystem processes and biological magnification (specifically fish mercury accumulation). These results will assist to establish regulations on emission levels of mercury to the atmosphere by electrical utilities and others.

For more information go to Environment Canada's website, www.ec.gc.ca, and search mercury and the environment.

Experimental Lakes Area

THE GLOBAL BAN OF PERSISTENT ORGANIC POLLUTANTS

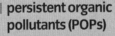

The **Stockholm Convention on Persistent Organic Pollutants**, which was adopted in 2001, is an important U.N. treaty that seeks to protect human health and the environment from the 12 most toxic **persistent organic pollutants**, or **POPs**, on Earth (**TABLE 4.4**). Some POPs disrupt the endocrine system (discussed later in this chapter), cause cancer, or adversely affect the developmental processes of organisms.

There are several characteristic features associated with these persistent organic pollutants that have been contributing factors to the establishment of the treaty. POPs are carbon-based molecules and often contain highly reactive

persistent organic pollutants (POPs)

Persistent toxicants that bioaccumulate in organisms and travel through air and water to contaminate sites far from their source.

chlorine that is harmful to human health and the environment. POPs may be ingested by organisms and stored in living tissues and therefore undergo biological magnification. Most of the POPs that are found in the environment tend to be manufactured by humans and therefore are synthetic in nature.

For example, polychlorinated biphenyls (PCBs) are industrial chemicals that were synthesized and commercialized in North America in 1929 for use in manufacturing electrical equipment, heat exchangers, hydraulic systems, and for other specialized applications. PCBs were used in various capacities until the late 1970s, and although they were never manufactured in Canada, they were widely used throughout the country. In 1977, in response to the recognition of the environmental and health effects of PCBs, the Canadian government

Persistent organic pollutants: The "dirty dozen" TABLE 4.4	
Persistent organic pollutant	**Use**
Aldrin	Insecticide
Chlordane	Insecticide
DDT (dichlorodiphenyl-trichloroethane)	Insecticide
Dieldrin	Insecticide
Endrin	Rodenticide and insecticide
Heptachlor	Fungicide
Hexachlorobenzene	Insecticide; fire retardant
Mirex™	Insecticide
Toxaphene™	Insecticide
PCBs (polychlorinated biphenyls)	Industrial chemicals
Dioxins	By-products of certain manufacturing processes
Furans (dibenzofurans)	By-products of certain manufacturing processes

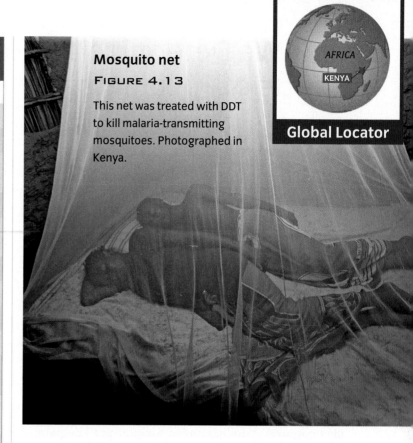

Mosquito net

FIGURE 4.13

This net was treated with DDT to kill malaria-transmitting mosquitoes. Photographed in Kenya.

Global Locator

AFRICA
KENYA

banned the import, manufacture, and sale of PCBs, making them illegal, and the release of PCBs into the environment was made illegal in 1985.

Despite these efforts, the legislation did not address owners of PCB equipment, allowing the equipment to be used until the end of its service capabilities. PCBs continued to contaminate the environment through spills and fires. It was clear that the action to "phase out" PCBs had to speed up. In 2008, Environment Canada reversed the earlier Chlorobiphenyl Regulations and the Storage of PCB Material Regulations and placed specific dates and regulations for the destruction of PCBs in service and storage, reorganizing the Canadian Environmental Protection Act (CEPA) of 1999. The new legislation addresses the risks posed by the use, storage, and release of PCBs to the environment and human health, and hopes to reduce these risks by accelerating the elimination of these harmful substances.

The Stockholm Convention requires countries to develop plans to eliminate the production and use of intentionally produced POPs. A notable exception to this requirement is that DDT is still produced and used to control malaria-carrying mosquitoes in countries where no affordable alternatives exist (FIGURE 4.13).

CONCEPT CHECK STOP

What is meant by a persistent toxicant?

What problems are associated with bioaccumulation and biological magnification?

What is long-range transport of toxins? Why is it important to examine arctic ecosystems for pollutants?

What are POPs, and why has the international community banned them?

How We Determine the Health Effects of Pollutants

LEARNING OBJECTIVES

Describe how a dose-response curve is used to determine the health effects of environmental pollutants.

Describe the most common method of determining whether a chemical causes cancer.

Distinguish among additive, synergistic, and antagonistic interactions in chemical mixtures.

Explain why children are particularly susceptible to toxicants.

We measure toxicity by the dose at which adverse effects are produced. A **dose** of a toxicant is the amount that enters the body of an exposed organism. The **response** is the type and amount of damage that exposure to a particular dose causes. A dose may cause death (*lethal dose*) or cause harm but not death (*sublethal dose*). Lethal doses, which are usually expressed in milligrams of toxicant per kilogram of body weight, vary depending on the organism's age, sex, health, and metabolism, and on how the dose was administered (all at once or over a period of time). The lethal doses for humans of many toxicants are known through records of homicides and accidental poisonings.

One way to determine acute toxicity is to administer different-sized doses to populations of laboratory animals, measure the responses, and use these data to predict the chemical effects on humans (**FIGURE 4.14**). The dose that is lethal to 50 percent of a population of test animals is called the **lethal dose–50 percent**, or **LD$_{50}$**.

It is usually reported in milligrams of chemical toxicant per kilogram of body weight. An inverse relationship exists between the size of the LD$_{50}$ and the acute toxicity of a chemical: the smaller the LD$_{50}$, the more toxic the chemical, and conversely, the greater the LD$_{50}$, the less toxic the chemical (**TABLE 4.5**). The LD$_{50}$ is determined for new synthetic chemicals—thousands are produced each year—as a way of estimating their toxic potential. It is generally assumed that a chemical with a low LD$_{50}$ for several species of test animals is also very toxic in humans. Using laboratory animals for testing is increasingly becoming less common as alternatives are being devised by the pharmaceutical industry.

Laboratory rat FIGURE 4.14

The results of chemicals administered orally to laboratory animals are extrapolated to humans.

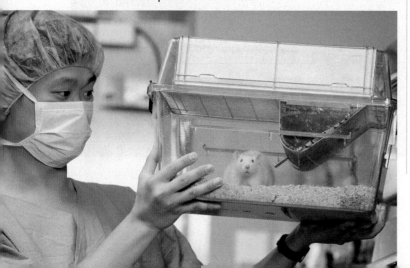

LD$_{50}$ values for selected chemicals	TABLE 4.5
Chemical	**LD$_{50}$(mg/kg)***
Aspirin	1750.0
Ethanol	1000.0
Morphine	500.0
Caffeine	200.0
Heroin	150.0
Lead	20.0
Cocaine	17.5
Sodium cyanide	10.0
Nicotine	2.0
Strychnine	0.8

**Administered orally to rats.*

The **effective dose–50 percent**, or **ED$_{50}$**, measures a wide range of biological responses, such as stunted development in the offspring of a pregnant animal, reduced enzyme activity, or onset of hair loss. The ED$_{50}$ is the dose that causes 50 percent of a population to exhibit whatever response is under study. For example, we could determine the effective dose of a pain killer such as Tylenol or Aspirin that would relieve a headache in 50 percent of the population.

> ■ **dose–response curve** In toxicology, a graph that shows the effects of different doses on a population of test organisms.

To develop a **dose–response curve**, scientists first test the effects of high doses and then work their way down to a **threshold** level, the maximum dose that has no measurable effect (or, alternatively, the minimum dose that produces a measurable effect) (**FIGURE 4.15**). Scientists assume that doses lower than the threshold level are safe.

A growing body of evidence, however, suggests that for certain toxicants there is no safe dose. A threshold does not exist for these chemicals, and even the smallest amount causes a measurable response.

CANCER-CAUSING SUBSTANCES

Because cancer is so feared, for many years it was the only disease evaluated in chemical risk assessment. Environmental contaminants are linked to many serious health concerns, including other diseases, birth defects, damage to the immune system, reproductive problems, and damage to the nervous system or other body systems. We focus here on risk assessment as it relates to cancer, but noncancer hazards are assessed in similar ways.

The most common method of determining whether a chemical causes cancer is to expose laboratory animals, such as rats, to extremely large doses of it and see whether they develop cancer. This method is indirect and uncertain, however. For one thing, although humans and rats are both mammals, they are different organisms and may respond differently to exposure to the same chemical. (Even rats and mice often respond differently to the same toxicant.)

Another problem is that laboratory rats are exposed to massive doses of the suspected **carcinogen** relative to their body size, whereas humans are

> ■ **carcinogen** Any substance (for example, chemical, radiation, virus) that causes cancer.

Dose–response curves FIGURE 4.15

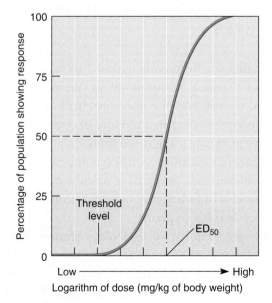

A This hypothetical dose–response curve demonstrates two assumptions: first, that the biological response increases as the dose is increased; second, that harmful responses occur only above a certain threshold level.

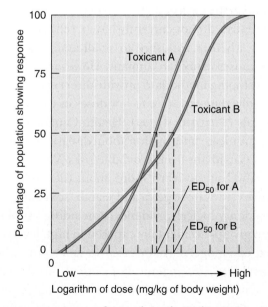

B Dose–response curves for two hypothetical toxicants, A and B. In this example, A has a lower effective dose–50 percent (ED$_{50}$) than B. However, at lower doses, B is more toxic than A.

usually exposed to much lower amounts. Researchers must use large doses to cause cancer in a small group of laboratory animals within a reasonable amount of time. Otherwise, such tests would take years, require thousands of test animals, and be prohibitively expensive to produce enough data to have statistically significant results.

Risk assessment assumes that you can extrapolate (work backward) from the huge doses of chemicals and the high rates of cancer they cause in rats to determine the expected rates of cancer in humans exposed to lower amounts of the same chemicals. However, little evidence exists to indicate that extrapolating backward is scientifically sound. Even if you are reasonably sure that exposure to high doses of a chemical causes the same effects for the same reasons in both rats and humans, you cannot assume that these same mechanisms work at low doses in humans. The body metabolizes small and large doses of a chemical in different ways. For example, enzymes in the liver may break down carcinogens in small quantities, but an excessive amount of carcinogen might overwhelm the liver enzymes.

In short, extrapolating from one species to another and from high doses to low doses is uncertain and may overestimate or underestimate a toxicant's danger. However, animal carcinogen studies provide valuable information: A toxicant that does not cause cancer in laboratory animals at high doses is not likely to cause cancer in humans at levels found in the environment or in occupational settings.

Scientists do not currently have a reliable way to determine whether exposure to small amounts of a substance causes cancer in humans. However, toxicologists are developing methods to provide direct evidence of the risk involved in exposure to low doses of cancer-causing chemicals (**Figure 4.16**). Health Canada is working to communicate information about disease prevention to protect Canadians from avoidable risks. Visit the Health Canada website to learn more about their educational initiatives (www.hc-sc.gc.ca and search "toxicology").

Epidemiological evidence, including studies of human groups accidentally exposed to high levels of suspected carcinogens, is also used to determine whether chemicals are carcinogenic. For example, in 1989 epidemiologists in Germany established a direct link between cancer and a group of persistent organic pollutants called *dioxins* (see Table 4.4 on page 115). They observed the incidence of cancer in workers exposed to high

Measuring low doses of a toxicant (dioxin) in human blood serum Figure 4.16

Scientists are developing increasingly sophisticated methods of biomonitoring to analyze human tissues and fluids. Photographed at the Centers for Disease Control and Prevention's Environmental Health Laboratory.

concentrations of dioxins during an accident at a chemical plant in 1953 and found unexpectedly high levels of cancers in their digestive and respiratory tracts.

RISK ASSESSMENT OF CHEMICAL MIXTURES

We humans are frequently exposed to various combinations of chemical compounds, in the air we breathe, the food we eat, and the water we drink. For example, cigarette smoke contains a mixture of chemicals, as does automobile exhaust. Cigarette smoke is a mixture of air pollutants that includes hydrocarbons, carbon dioxide, carbon monoxide, particulate matter, cyanide, and a small amount of radioactive materials from the fertilizer used to grow tobacco plants.

The vast majority of toxicology studies have been performed on single chemicals rather than chemical mixtures, and for good reason. Mixtures of chemicals interact in a variety of ways, increasing the level of complexity in risk assessment, a field already complicated by many uncertainties. Moreover, there are simply too many chemical mixtures to evaluate.

Chemical mixtures interact by **additivity**, **synergy**, or **antagonism**. When a chemical mixture is *additive*, the effect is exactly what you would expect, given the individual effects of each component of the mixture. If a chemical with a toxicity level of 1 is mixed with a different chemical, also with a toxicity level of 1, the combined effect of exposure to the mixture is 2. A chemical mixture that is *synergistic* has a greater combined effect than expected; two chemicals, each with a toxicity level of 1, might have a combined toxicity of 3. An *antagonistic* interaction in a chemical mixture results in a smaller combined effect than expected; for example, two chemicals, each with a toxicity level of 1, might have a combined effect of 1.3.

If toxicological studies of chemical mixtures are lacking, how is risk assessment for chemical mixtures assigned? Toxicologists use *additivity* to assign risk to mixtures—that is, they add the known effects of each compound in the mixture. Such an approach sometimes overestimates or underestimates the actual risk involved, but it is the best method currently available. The alternative—waiting for years or decades until numerous studies have been designed, funded, and completed—is unreasonable.

CHILDREN AND CHEMICAL EXPOSURE

Because children weigh less than adults, they are more susceptible to the effects of chemicals. Consider a toxicant with an LD_{50} of 100 mg/kg. A potentially lethal dose for a child who weighs 11.3 kg (25 lb) is $100 \times 11.3 = 1130$ mg, which is equal to a scant 1/4 teaspoon if the chemical is a liquid. In comparison, the potentially lethal dose for an adult who weighs 68 kg (150 lb) is 6800 mg, or about 2 teaspoons. This exercise demonstrates that we must protect children from exposure to environmental chemicals because lethal doses are smaller for children than for adults.

Children and pollution

Consider the toxicants in air pollution. Air pollution is a greater health threat to children than it is to adults (**FIGURE 4.17**). The lungs continue to develop throughout childhood, and air pollution restricts lung development. In addition, a child has a higher metabolic rate than an adult and therefore

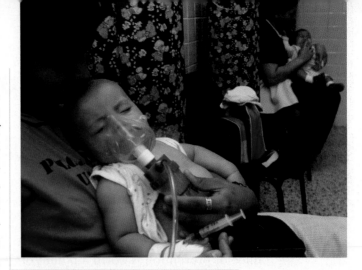

Air pollution and respiratory disease in children FIGURE 4.17

A Honduran mother gives oxygen to her baby, who suffers from environmentally linked respiratory disease. Farmers nearby burn land to prepare for the planting season; the resulting smoke triggers breathing problems, mostly in children and the elderly.

needs more oxygen. To obtain this oxygen, a child breathes more air—about two times as much air per half-kilogram of body weight as an adult. This means that a child also breathes more air pollutants into the lungs. A 1990 study in which autopsies were performed on 100 Los Angeles children who died for reasons unrelated to respiratory problems found that more than 80 percent had subclinical lung damage, which is lung disease in its early stages, before clinical symptoms appear. (Los Angeles has some of the worst air quality in the world.)

NATIONAL GEOGRAPHIC

CONCEPT CHECK STOP

What is a dose–response curve?

What is one way that scientists determine whether a chemical causes cancer? What are two problems with this method?

What are the three ways that chemical mixtures interact?

Why are children particularly susceptible to toxicants?

The Precautionary Principle

LEARNING OBJECTIVES

Discuss the precautionary principle as it relates to the introduction of new technologies or products.

You've probably heard the expression "An ounce of prevention is worth a pound of cure." This statement is the heart of the **precautionary principle** that many politicians and environmental activists advocate. According to the precautionary principle, we should not introduce new technology or chemicals until it is demonstrated that (a) the risks are small and (b) the benefits outweigh the risks. The precautionary principle puts the burden of proof on the developers of the new technology or substance. It asks them to prove that a product is safe beyond a reasonable doubt instead of society discovering its harmful effects after it has been introduced.

> **precautionary principle** A practice that involves making decisions about adopting a new technology or chemical product by assigning the burden of proof of its safety to its developers.

The precautionary principle is also applied to existing technologies when new evidence suggests that they are more dangerous than originally thought. For example, when observations and experiments suggested that lead added to gasoline as an anti-knock ingredient was contaminating soil, particularly in inner cities near major highways, the precautionary principle led to the phase-out of leaded gasoline. While the principle is commonly assigned to environmental hazards, it is also applicable to the management and conservation of species at risk and sustainable development.

To many people the precautionary principle is just common sense, given that science and risk assessment often cannot provide definitive answers to policy makers dealing with environmental and public health problems. However, the precautionary principle does not require that developers provide absolute proof that their product is safe; such proof would be impossible to provide.

In 2000, the Canadian government, through multidepartmental working groups, began working on a federal framework for consistent and coherent application of the precautionary principle into nationwide science-based programs and risk decision making. The federal framework's main objectives were

- to improve predictability, credibility, and consistency of federal precautionary approaches ensuring adequate, reasonable, and cost-effective approaches;

- to increase public and private stakeholder confidence in the federal precautionary decision-making process;

- to minimize crisis and capitalize on opportunities; and

- to increase Canada's ability to influence international standards and applications of the precautionary approach.

The Canadian government created a discussion paper titled "Guiding Principles," which outlined the federal framework and objectives. The government distributed these materials to stakeholders throughout various agencies soliciting reactions to the concepts and principles of the framework and calling for their feedback.

The precautionary principle has generated much controversy. Some scientists fear that the precautionary principle challenges the role of science and endorses making decisions without the input of science. Some critics contend that its imprecise definition reduces trade and limits technological innovations. For example, several European countries made precautionary decisions to ban beef from Canada and the United States because these countries use growth hormones to make the cattle grow faster. Europeans contend that the growth

Is Europe's ban of Canadian beef the result of its concern over the safety of hormone-treated beef or an excuse to support the European beef industry?

hormone might harm humans eating the beef, but the ban, in effect since 1989, is widely viewed as protecting the European beef industry (**FIGURE 4.18**). Another international controversy in which the precautionary principle is involved is the cultivation of genetically modified foods (discussed further in Chapter 8).

CONCEPT CHECK STOP

What is the precautionary principle?

What are two criticisms of the precautionary principle?

ECO CANADA **CAREER FOCUS**

ECOTOXICOLOGIST

Imagine you are kneeling on the bank of a cold glacier-fed stream with your arm in the water taking a grab sample. You are an ecotoxicologist and you are taking samples of the creek because there is concern that a potentially harmful chemical has polluted the water. Biologists studying fish downstream have noticed the fish population is almost entirely mature adults, with very few young fish. The absence of younger fish indicates that the population is not reproducing, which indicates some form of toxicant in the water. You have been asked to investigate the situation, determine a cause, and find a way to reverse the effects.

As an ecotoxicologist, you must find out why the creek's fish have stopped reproducing and if something can be done about it. The first step toward finding these answers is to visit the site itself, both to gather water samples and to see if there is obvious evidence of the cause in the surrounding environment, for example, dumped chemical containers or a spill site.

Even when the cause seems apparent, you must investigate a little deeper. You will bring the water samples to the lab for analysis, to get a better idea of the different chemicals present in the

Ecotoxicologists study the causes and effects of environmental toxins that affect living organisms.

water. From the list of chemicals, you will look at each one to see if it is responsible for stopping reproduction.

But you aren't finished once you have found the culprit or culprits; you will also study the mechanism by which the contaminant acts on the fish, which will be the key to reversing its effects. It is also important that you find the source of the toxicant so you can prevent similar incidences in the future. It's a big task, but your specialty is answering the questions of what, how, and where disturbances are in the ecosystem.

Environmental toxicology is a specialty of toxicology that studies the harmful effects of chemical, biological, and physical agents on living organisms, including humans. Environmental toxicologists, or ecotoxicologists, draw on a variety of scientific disciplines to predict, measure, and explain the frequency and severity of adverse effects of environmental toxins on living organisms. Environmental toxicologists strive to improve environmental protection through a better understanding of the hazards and risks to which organisms are exposed.

For more information or to look up other environmental careers go to www.eco.ca.

ENDOCRINE DISRUPTERS

Mounting evidence suggests that dozens of widely used industrial and agricultural chemicals are **endocrine disrupters**, which interfere with the normal actions of the endocrine system (the body's hormones) in humans and animals. These chemicals include chlorine-containing industrial compounds known as PCBs and dioxins, the heavy metals lead and mercury, pesticides such as DDT, and certain plastics and plastic additives.

Hormones are chemical messengers that organisms produce to regulate their growth, reproduction, and other important biological functions. Some endocrine disrupters mimic the *estrogens,* a class of female sex hormones. Other endocrine disrupters mimic *androgens* (male hormones such as testosterone) or *thyroid hormones.* Like hormones, endocrine disrupters are active at very low concentrations and therefore may cause significant health effects at relatively low doses.

Many endocrine disrupters appear to alter the reproductive development of various animal species. A chemical spill in 1980 contaminated Lake Apopka, Florida's third-largest lake, with DDT and other agricultural chemicals that have known estrogenic properties. In the years following the spill, male alligators had low levels of testosterone and elevated levels of estrogen. The mortality rate for eggs in this lake was extremely high, which reduced the alligator population for many years (see photo).

Humans may also be at risk from endocrine disrupters. The number of reproductive disorders, infertility cases, and hormonally related cancers (such as testicular cancer and breast cancer) appears to be increasing. However, we cannot make definite connections between environmental endocrine disrupters and human health problems at this time because of the limited number of human studies. Complicating such assessments is the fact that humans are also exposed to *natural* hormone-mimicking substances in the plants we eat. For example, soy-based foods such as bean curd and soymilk contain natural estrogens.

Environment Canada has made endocrine disrupting substance (EDS) research a top priority to gather the information necessary to guide sound policy and regulatory decisions. Canadian scientists are among world leaders in studying the effects of endocrine disruptors on wildlife populations, specifically fish populations in the Great Lakes. Environment Canada is working closely with international organizations such as the Organisation for Economic Co-operation and Development (OECD) and the United Nations Environment Programme (UNEP), among others, to address the global issue of an alarming number of substances that result in endocrine disruption.

Environment Canada, in partnership with Health Canada, also manages the Toxic Substance Research Initiative, which provides support and funding for continued EDS research. In each of the Regional Ecosystem Initiatives, Environment Canada has included an EDS research component, establishing a multidisciplinary research strategy, collaborating knowledge and efforts from provincial and territorial governments, industry, universities, and the federal government. Through these initiatives, the federal government will gather information on the impacts of EDS on Canadians and the environment aiding them in their formation of effective regulations and controls.

Lake Apopka alligators

A young American alligator hatches from eggs that University of Florida researchers took from Lake Apopka, Florida. Many of the young alligators that hatch have abnormalities in their reproductive systems. This young alligator may not leave any offspring.

SUMMARY

1 Risk Assessment

1. A **risk** is the probability of harm (such as injury, disease, death, or environmental damage) occurring under certain circumstances. **Risk assessment** is the use of statistical methods to quantify the risks of an action so they can be compared with other risks.

2. The four steps of risk assessment are hazard identification (Does exposure cause an increased risk of an adverse health effect?), dose–response assessment (What is the relationship between the dose and the extent of the adverse health effect?), exposure assessment (How often and how long are humans exposed to the substance?), and risk characterization (What is the probability of an adverse effect?).

2 Environmental Health Hazards

1. **Toxicology** is the study of **toxicants**, chemicals that have adverse effects on health. **Acute toxicity** is adverse effects that occur within a short period after high-level exposure to a toxicant. **Chronic toxicity** is adverse effects that occur after a long period of low-level exposure to a toxicant.

2. A **pathogen** is an agent (usually a microorganism) that causes disease. Although most strains of coliform bacteria do not cause disease, the fecal coliform test is a reliable way to indicate the likely presence of pathogens in water.

3. About 25 percent of disease and injury worldwide is related to human-caused environmental changes. The environmental component of human

health is sometimes direct, as when people drink unsanitary water and contract a waterborne disease. The health effects of other human activities are complex and indirect, as when the disruption of natural environments gives disease-causing agents an opportunity to break out of isolation.

3 Movement and Fate of Toxicants

1. Some toxicants exhibit persistence—they are extremely stable in the environment and may take many years to break down into less toxic forms. Bioaccumulation is the buildup of a persistent toxicant in an organism's body. **Biological magnification** is the increase in toxicant concentration as

a toxicant passes through successive levels of the food chain.

2. Persistent toxicants do not stay where they are applied but tend to move through the soil, water, and air, sometimes over long distances. For example, pesticides applied to agricultural lands may wash into rivers and streams, harming fishes, and industrial air pollutants from major industrial centres in the northern hemisphere travel on air currents to the Arctic.

3. The Stockholm Convention on Persistent Organic Pollutants requires countries to eliminate the production and use of the 12 worst **persistent organic pollutants (POPs)**. POPs are a group of persistent toxicants that bioaccumulate in organisms and travel thousands of kilometres through air and water, contaminating sites far removed from their source.

4 How We Determine the Health Effects of Pollutants

1. A **dose–response curve** is a graph that shows the effect of different doses on a population of test organisms. Scientists test the effects of high doses and work their way down to a threshold level, the maximum dose that has no measurable effect. It is assumed that doses lower than the threshold level will not have an effect on the organism and are safe.

2. A **carcinogen** is any substance (for example, chemical, radiation, virus) that causes cancer. The most common method of determining whether a chemical is carcinogenic is to expose laboratory animals such as rats to extremely large doses of that chemical and see if they develop cancer. It is assumed that we can extrapolate from the huge doses of chemicals and the high rates of cancer they cause in rats to determine the rates of cancer

expected in humans exposed to lower amounts of the same chemicals.

3. When a chemical mixture is additive, the effect is exactly what you would expect, given the individual effects of each component of the mixture. A chemical mixture that is synergistic has a greater combined effect than expected. An antagonistic interaction in a chemical mixture results in a smaller combined effect than expected.

4. Because children weigh less than adults, they are more susceptible to chemicals; the potentially lethal dose for a child is considerably less than the potentially lethal dose for an adult. Air pollution restricts a child's lung development. Also, a child has a higher metabolic rate than an adult and therefore breathes more air and more air pollutants into the lungs.

5 The Precautionary Principle

1. The **precautionary principle** argues for making decisions about adopting a new technology or chemical product by assigning the burden of proof of its safety to the developers of that technology or product.

KEY TERMS

- **risk** p. 102
- **risk assessment** p. 102
- **environmental hazard** p. 105
- **toxicology** p. 105
- **acute toxicity** p. 105
- **chronic toxicity** p. 105
- **pathogen** p. 106
- **biological magnification** p. 110
- **persistent organic pollutants (POPs)** p. 114
- **dose–response curve** p. 117
- **carcinogen** p. 117
- **precautionary principle** p. 120

CRITICAL AND CREATIVE THINKING QUESTIONS

1. How do acute and chronic toxicity differ?

2. Distinguish among persistence, bioaccumulation, and biological magnification.

3–5. The figure to the right shows the organisms sampled in the salt marsh study of DDT discussed in the text (see **Figure 4.10** on page 111).

3. If DDT is sprayed on land to control insects, how does it get into the bodies of aquatic species?

4. Why does the Atlantic needlefish (5) contain more DDT in its body than an American eel (4)?

5. How does high concentration of DDT cause reproductive failure in birds at the top of the food chain?

6. Describe how a persistent pesticide might move around in the environment.

7. What is the Stockholm Convention on Persistent Organic Pollutants?

8. Why is air pollution a greater threat to children than it is to adults?

9. Which risk—an extremely small amount of a cancer-causing chemical in drinking water or smoking cigarettes—tends to generate the greatest public concern? Explain why this view is counterproductive.

10. Should public policy makers be more concerned with public risk perception or with actual risks? Explain your answer.

11. Describe the four steps of risk assessment for adverse health effects.

12. Is the absence of scientific certainty about the health effects of an environmental pollutant synonymous with the absence of risk? Explain your answer.

13. Is risk assessment important in toxicology? Explain.

14. Would you support Canada adopting the precautionary policy in all of its legislation? Why or why not?

What is happening in this picture ?

This mouse developed cancer after exposure to high levels of a toxicant. What uncertainties are associated with extrapolating this result to low levels of exposure of that toxicant in humans?

Why do we use animal testing to determine whether a new pesticide causes cancer?

How Ecosystems Work

5

LAKE VICTORIA'S ECOLOGICAL IMBALANCE

Africa's Lake Victoria, the world's second-largest freshwater lake, was once home to about 400 species of small, colourful fishes known as cichlids (pronounced "SICK'-lids"). Some of these cichlids grazed on algae; others consumed dead organic material at the lake bottom; still others ate insects, shrimp, or other cichlids. They thrived throughout the lake ecosystem, providing much-needed protein to 30 million humans living near the lake.

Today, more than half of the cichlids and other native fish species in Lake Victoria are extinct. Because most of the algae-eating cichlids have disappeared, the algal population has increased explosively. When these algae die, their decomposition uses up the dissolved oxygen in the water. The bottom zone of the lake, once filled with cichlids, is empty because it contains too little dissolved oxygen. Any fishes venturing into the anoxic (no oxygen) zone suffocate. Local fishermen, who once caught and ate hundreds of types of fishes, now catch only a few types.

How did this happen? In the early 1960s, the Nile perch (see inset) was introduced into the lake to stimulate the local economy. For about 20 years, as its population slowly increased, the Nile perch didn't have an appreciable effect on the lake. But in 1980, fishermen noticed increasing catches of Nile perch and decreasing harvests of native fishes. By 1985, most of the catch was Nile perch, which was increasing in number because it had an abundant food supply—the cichlids.

In 1988 the water hyacinth, a South American plant, invaded Lake Victoria, adding to the ecological havoc (see large photo). Efforts to release hyacinth-eating weevils into the lake supplemented by chemical and mechanical treatment eventually provided effective control of the weed.

Ecological imbalances such as that in Lake Victoria can occur with disruptions large and small, from the clearing of a small forest to the pollution of a continent's largest river.

NATIONAL GEOGRAPHIC

Defining Ecology and the Ecosystem Concept

LEARNING OBJECTIVES

Define *ecology, ecosystem,* and *feedback loops.*

Distinguish between the following ecological levels: population, community, landscape, and biosphere.

In the 19th century the German biologist Ernst Haeckel first developed the concept of ecology and devised its name—*eco* from the Greek word for "house" and *logy* from the Greek word for "study." Thus, **ecology** literally means "the study of one's house." The environment—one's house—consists of two parts: the *biotic* (living) environment, which includes all organisms, and the *abiotic* (nonliving, or physical) environment, which includes physical factors such as living space, temperature, sunlight, soil, wind, and precipitation.

Ecology is the broadest field within the biological sciences. We tend to think of ecology as the study of the interactions that occur between organisms and their environment. It is linked to every other biological discipline and to other fields as well. Geology and earth science are extremely important to ecology, especially when ecologists examine the physical environment of Earth. Chemistry and physics are also important. Humans are biological organisms, and all our activities have a bearing on ecology. Even economics and politics have profound ecological implications, as discussed in Chapter 2.

Ecologists are most interested in the levels of biological organization including and above the level of the individual organism. Individuals of the same species occur in **populations.** (A *species* is a group of similar organisms whose members freely interbreed with one another in the wild to produce fertile offspring; members of one species generally don't interbreed with other species of organisms.) A population ecologist might study a population of walruses or a population of marsh grass.

Populations are organized into communities. The number and kinds of species that live within a **community**, along with their relationships with one another, characterize the community. A community ecologist might study how organisms interact with one another—including feeding relationships (who eats whom)—in a coral reef community or in an alpine meadow community (**FIGURE 5.1**).

Ecosystem is a more inclusive term than community. An ecosystem includes all the biotic interactions of a community as well as the interactions between organisms and their abiotic environment in a defined geographic area. In an ecosystem, all the biological, physical, and chemical components of an area form an extremely complicated interacting network of energy flow and materials cycling. An ecosystem ecologist might examine how energy, nutrient composition, or water affects the organisms living in a tundra community or a coastal bay ecosystem. Note that humans are considered to be a component of an ecosystem, and that humans generally "draw the line" indicating where an ecosystem begins and ends. For example, we might designate a lake as one ecosystem and the adjacent forest as another. In reality, these two regions are interacting with one another.

There is no right or wrong way to delimit an ecosystem, as long as a working definition of its constitution is provided. Ecosystems can be based on political jurisdictions, administrative divisions, or ideally, ecological parameters. Combinations of these categories are also possible. But it is necessary to describe the basis and rationale for delineating the area described as an ecosystem. And if an ecosystem crosses political boundaries, it is important to

ecology The study of the interactions among organisms and between organisms and their abiotic environment.

population A group of organisms of the same species that live and interact together in the same area at the same time.

community A natural association that consists of all the populations of different species that live and interact together within an area at the same time.

ecosystem Biotic communities and their associated abiotic components that interact in a defined geographic area.

An alpine meadow community FIGURE 5.1

Alpine flowers put on a colourful show during their brief blooming period. Photographed in Berchtesgaden National Park, Germany. (*Inset*) The golden-mantled ground squirrel, which is larger than a chipmunk, is a burrowing mammal common in alpine meadows in western North America.

recognize that additional considerations may be necessary to ensure efficient and effective management. This may result in interprovincial as well as international negotiations, as is the case for Lake Winnipeg where the watershed crosses four provinces and three states in the United States (see the Case Study in Chapter 9 for a more in-depth discussion of Lake Winnipeg).

Another important feature of ecosystems and landscapes is that they are interconnected and will respond in a cause-and-effect relationship. Consider, for example, the parameters of precipitation and forest growth as components of a simplified ecosystem. As precipitation patterns are altered (perhaps as a consequence of climate change) and drought conditions prevail for extended periods, some tree species may be unable to withstand the stressful conditions. With the loss of these trees, less water percolates into the ground and this further reduces overall moisture for parched roots. Overall, there is a trend toward fewer trees and less groundwater.

This situation describes the concept of **feedback loops**. Feedback loops are defined as circular processes where the output in a system serves as input to that same system. We can further define these feedback loops in terms of their overall effect on the ecosystem. If the feedback loop promotes the change and this becomes increasingly pronounced in the ecosystem over time, we describe this as a **positive feedback loop**. Note that "positive" refers to the direction of change rather than a desired outcome. The example above describes positive feedback: a loss of precipitation results in a loss of forest, which in turn further promotes an additional loss of available moisture.

In contrast, **negative feedback loops** reduce or inhibit the change and over time help stabilize the system.

> **feedback loop**
> A chain of cause-and-effect responses where the output from an event influences the same event in the future to form a loop.

A Grizzly bears raise their cubs and spend much of their time in forests that are close by streams where they can catch salmon.

A connection between two ecosystems within a landscape FIGURE 5.2

B A grizzly bear fishes for salmon

Populations of predators and prey represent a classic example of negative feedback. As the population of prey (e.g., rabbits) increase in an ecosystem, predator populations also increase due to the availability of more food. Increased predation causes a decline in the prey population that will later adjust the predator population downward as food becomes scarce. While there may be slight changes in the populations of the predator and prey, overall the trend is to maintain and stabilize these populations. The climate system has a number of positive and negative feedback loops that are examined in greater detail in Chapter 10.

The ultimate goal of ecologists is to understand how ecosystems function. They study how ecosystem processes collectively regulate the global cycles of water, carbon, nitrogen, and oxygen that are essential to the survival of humans and all other organisms. As humans increasingly alter ecosystems for their own uses, the natural functioning of ecosystems is changed, and we must learn whether these changes will affect the *sustainability* of our life-support system.

Landscape ecology is a subdiscipline in ecology that studies the connections among ecosystems. Consider, for example a **landscape** consisting of a forest ecosystem and a coastal ecosystem. One possible connection between these two ecosystems is the grizzly bear (**FIGURE 5.2**). Grizzly bears play a highly complex role in the ecosystem and have been called **ecosystem engineers** because of their ability to modify habitat. Grizzly bears congregate along coastal water ecosystems, feeding on salmon, but often raise their cubs and spend their time in the forests.

Grizzly bears will transfer salmon-derived nitrogen, carbon, and phosphorus from the coastal ecosystem into the forest ecosystem when they move these fish onto dry land and eat them, disperse nutrient-rich feces, and leave partially eaten fish carcasses in the forest. In addition, grizzlies disperse seeds over large areas of land, often across various ecosystems, contributing to nutrient cycling. Landscapes, then, are based on larger land areas that may include several ecosystems.

The organisms of the **biosphere**—Earth's communities, ecosystems, and landscapes—depend on one another and on the other realms of Earth's physical environment: the atmosphere, hydrosphere, and lithosphere. The *atmosphere* is the gaseous envelope

> **landscape** A region that may include several interacting ecosystems.
>
> **biosphere** The layer of Earth that contains all living organisms.

surrounding Earth; the *hydrosphere* is Earth's supply of water—liquid and frozen, fresh and salty; and the *lithosphere* is the soil and rock of Earth's crust. Ecologists who study the biosphere examine the global interrelationships of Earth's atmosphere, land, water, and organisms.

The biosphere teems with life. Where do these organisms get the energy to live? And how do they harness this energy? We now examine the importance of energy to organisms, which survive only as long as the environment continuously supplies them with energy. We will revisit energy as it relates to human endeavours in many chapters throughout this text.

CONCEPT CHECK **STOP**

Give some examples of a working definition of an ecosystem.

What is the difference between a community and an ecosystem? Between an ecosystem and a landscape?

Energy Flow in Ecosystems

LEARNING OBJECTIVES

Define *energy*, and state the first and second laws of thermodynamics.

Distinguish between producers, consumers, and decomposers.

Summarize how energy flows through a food web.

nergy is the capacity or ability to do work. Organisms require energy to grow, move, reproduce, and maintain and repair damaged tissues. Energy exists as stored energy—called **potential energy**—and as **kinetic energy**, the energy of motion. We can think of potential energy as an arrow on a drawn bow (**FIGURE 5.3**). When the string is released, this potential energy is converted to kinetic energy as the motion of the bow propels the arrow. Similarly, the grass a bison eats has chemical potential energy, some of which is converted to kinetic energy and heat as the bison runs across the prairie. Thus, energy changes from one form to another.

Another example of potential energy is water dammed for hydroelectric power generation. A dam is used to block water flow and generate electricity when water is forced by gravity to move toward and through turbines. The work being done by the water to turn the turbines generates mechanical energy that powers the dam's generators and is converted into electricity.

energy The capacity or ability to do work.

The human body is yet another example of energy transformation. Food that we consume is stored in the body as carbohydrates, proteins, or fats and oils as potential energy. This potential energy is stored as glucose until blood glucose levels drop below optimum levels. When we exercise, we call on this stored potential energy, which is converted into energy for muscle movement and then transformed into kinetic energy.

THE FIRST AND SECOND LAWS OF THERMODYNAMICS

Thermodynamics is the study of energy and its transformations. Two laws about energy apply to all things in the universe: the first and second laws of thermodynamics. According to the **first law of thermodynamics**, an

first law of thermodynamics A physical law which states that energy cannot be created or destroyed, although it can change from one form to another.

Potential and kinetic energy FIGURE 5.3

Potential energy is stored in the drawn bow (**A**) and is converted to kinetic energy (**B**) as the arrow speeds toward its target. Photographed in Athens, Greece, during the 2004 Summer Olympics.

organism may absorb energy from its surroundings, or it may give up some energy into its surroundings, but the total energy content of the organism and its surroundings is always the same. An organism cannot create the energy it requires to live. Instead, it must capture energy from the environment to use for biological work, a process that involves the transformation of energy from one form to another. In **photosynthesis**, for example, plants absorb the radiant energy of the sun and convert it into the chemical energy contained in the bonds of sugar molecules (**FIGURE 5.4**). Later, an animal that eats the plant may transform some of the chemical energy into the mechanical energy of muscle contraction, enabling the animal to walk, run, slither, fly, or swim.

photosynthesis
The biological process that captures solar energy and transforms it into the chemical energy of organic molecules, which are manufactured from carbon dioxide and water.

As each energy transformation occurs, much of the energy is changed to heat energy that is released into the cooler surroundings. No organism can ever use this energy again for biological work; it is "lost" from the biological point of view. However, it isn't gone from a thermodynamic point of view because it still exists in the surrounding physical environment. The food you eat that enables you to walk or run doesn't destroy the chemical energy that was once present in the food molecules. After you have performed the task of walking or running, the energy still exists in your surroundings as heat energy.

According to the **second law of thermodynamics**, the amount of useable energy available to do work in the universe decreases over time. The second law of thermodynamics is consistent with the first law—that is, the total amount of energy in the universe is not decreasing with time. However, the total amount of energy in the universe available to do biological work is decreasing over time.

second law of thermodynamics
A physical law which states that when energy is converted from one form to another, some of it is degraded into heat, a less useable form that disperses into the environment.

A The sun powers photosynthesis, producing chemical energy stored in the leaves and seeds of this umbrella tree. Photographed in Hanging Rock State Park, North Carolina.

B The chemical energy stored in flowers and leaves transfers to this juvenile hare as it eats.

Less useable energy is more diffuse, or disorganized, than more useable energy. *Entropy* is a measure of this disorder or randomness. Organized, useable energy has low entropy, whereas disorganized energy such as heat has high entropy. Another way to explain the second law of thermodynamics is that entropy, or disorder, in a system tends to increase over time. As a result of the second law of thermodynamics, no process that requires an energy conversion is ever 100 percent efficient because much of the energy is dispersed as heat, resulting in an increase in entropy. For example, an automobile engine, which converts the chemical energy of gasoline to mechanical energy, is between 20 and 30 percent efficient: only 20 to 30 percent of the original energy stored in the chemical bonds of the gasoline molecules is actually transformed into mechanical energy, or work.

Burning firewood provides an example of how energy is transformed from potential to kinetic energy and while doing so increases entropy in the system. Heat energy is released as the wood ignites. The energy initially stored in the wood is in a "more ordered" state with low entropy. Once the wood has been burned, this energy generates tremendous heat and leads to a "less ordered state" with increasing disorder and higher entropy.

PRODUCERS, CONSUMERS, AND DECOMPOSERS

The organisms of an ecosystem are divided into three categories based on how they obtain nourishment: producers, consumers, and decomposers. Virtually all ecosystems contain representatives of all three groups, which interact extensively with one another, both directly and indirectly.

Green plants, algae, and cyanobacteria (blue-green algae) use the sun's radiation directly to produce their own food. These organisms are referred to as **autotrophs** (self-feeders) or **producers**. Through photosynthesis, these plants are able to manufacture carbohydrates (or sugars) from simple inorganic substances, generally carbon dioxide and water, using the energy of sunlight. The reaction of photosynthesis can be described by the following equation:

> **producers**
> An organism that manufactures large organic molecules from a simple inorganic substance.

$$6CO_2 + 12H_2O + \text{Sun's Energy} \rightarrow C_6H_{12}O_6 + 6O_2 + 6H_2O$$

Organisms that consume other organisms are known as **consumers**, or **heterotrophs**. These can be further described by their feeding relationships. Consumers that eat producers are *primary consumers,* or *herbivores.* Grasshoppers, deer, moose, and rabbits are examples of primary consumers (**FIGURE 5.5A**). *Secondary consumers* eat primary consumers, whereas *tertiary consumers* eat secondary consumers. Both secondary and tertiary consumers are *carnivores* that eat other animals. Lions, spiders, and lizards are examples of carnivores (**FIGURE 5.5B**). Other consumers, called *omnivores,* eat a variety of organisms. Bears, pigs, and humans are examples of omnivores.

Some consumers, called detritus feeders or *detritivores,* consume **detritus**, organic matter that includes animal carcasses, leaf litter, and feces (**FIGURE 5.5C**). Detritus feeders, such as snails, crabs, clams, and worms, are especially abundant in aquatic environments, where they consume the organic matter in the bottom muck. Earthworms are terrestrial (land-dwelling) detritus feeders, as are termites, beetles, snails, and millipedes. Detritus feeders work together with microbial decomposers to destroy dead organisms and waste products.

Bacteria and fungi are important examples of **decomposers**, organisms that break down dead organisms and waste products (**FIGURE 5.5D**). Decomposers release simple inorganic molecules, such as carbon dioxide and mineral salts, which producers can then reuse.

consumers
Organisms that can't make their own food and use the bodies of other organisms as a source of energy and bodybuilding materials.

decomposers
Microorganisms that break down dead organic material and use the decomposition products to supply themselves with energy.

VIEW THIS IN ACTION

Consumers and decomposers FIGURE 5.5

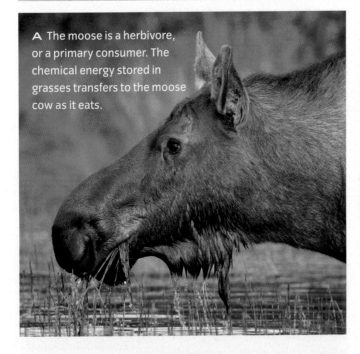

A The moose is a herbivore, or a primary consumer. The chemical energy stored in grasses transfers to the moose cow as it eats.

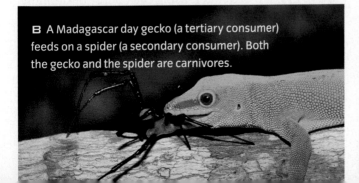

B A Madagascar day gecko (a tertiary consumer) feeds on a spider (a secondary consumer). Both the gecko and the spider are carnivores.

C A crab forages for detritus along stones near a stream. Photographed in the Philippines.

D These mushrooms are growing on a dead beech tree in Ostmuritz/Serrahn National Park, Germany. The mushrooms are reproductive structures; the invisible branching, threadlike body of the mushroom grows underground, decomposing dead organic material.

THE PATH OF ENERGY FLOW IN ECOSYSTEMS

In an ecosystem, **energy flow** occurs in a series of feeding relationships we call **food chains** and **food webs** (**FIGURE 5.6**). Each level, or "link," in a food chain is a **trophic level** (the Greek *tropho* means "nourishment"). Autotrophs, the primary producers, represent the largest group at the first trophic level; herbivores, the primary consumers, form the second trophic level; and carnivores, the secondary consumers, form the third trophic level. The top predators are found at the fourth trophic level.

At every step in a food chain are detritivores and decomposers, which respire while they digest and decompose the dead organisms and fecal wastes into smaller compounds. While this decomposition takes place, the remaining energy is exploited and carbon dioxide is produced. Inorganic chemicals are made available for uptake by plants in the constant recycling of matter.

The first trophic level of the food chain contains the largest amount of energy. At each trophic level within the food chain, most of the energy that organisms use is lost as waste (or latent) heat through respiration. Therefore, only a small amount of energy is actually transferred to the second trophic level through herbivory and then to the third trophic level. As a general rule, the higher trophic level will only obtain 10 percent of the original energy available from the trophic level below it. Therefore, the highest trophic levels contain the least energy. Although exceptions can be found, this energy transfer and loss limits ecosystems to as few as four trophic levels.

energy flow
The passage of energy in a one-way direction through an ecosystem.

Energy flow through a food chain FIGURE 5.6

Much of the energy acquired by a given level of a food chain is used and escapes into surrounding environment as heat. This energy, as the second law of thermodynamics stipulates, is unavailable to the next level of the food chain.

VIEW THIS IN ACTION

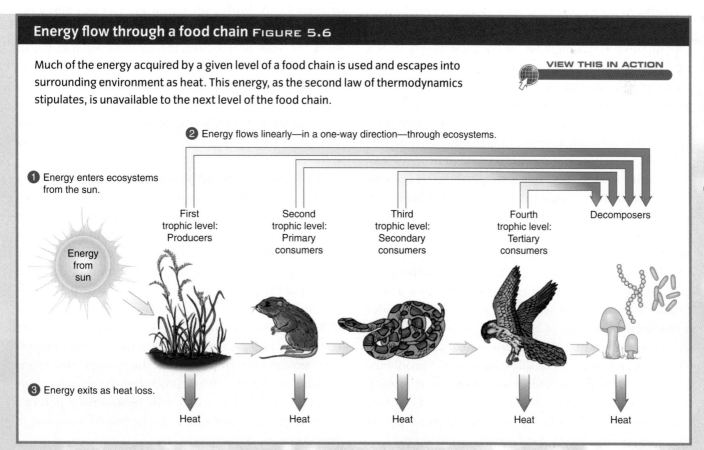

1 Energy enters ecosystems from the sun.

2 Energy flows linearly—in a one-way direction—through ecosystems.

3 Energy exits as heat loss.

Energy from sun

First trophic level: Producers

Second trophic level: Primary consumers

Third trophic level: Secondary consumers

Fourth trophic level: Tertiary consumers

Decomposers

Heat | Heat | Heat | Heat | Heat

Process Diagram

Simple food chains rarely occur in nature because few organisms eat just one kind of organism. The flow of energy through an ecosystem typically takes place in accordance with a range of food choices for each organism involved. In an ecosystem of average complexity, numerous alternative pathways are possible. An owl eating a rabbit is a different energy pathway than an owl eating a snake. A food web, a complex network of interconnected food chains in an ecosystem, is a more realistic model of the flow of energy and materials through ecosystems (**FIGURE 5.7**).

BIOLOGICAL PRODUCTION

We measure the flow of energy in ecosystems through an assessment of biological production. **Net primary production (NPP)** is a measure of the amount of plant mass that is first generated through photosynthesis but also remains after cellular respiration has taken place. This is an important piece of information for us to consider because all of humanity's food, fibre, and fuel is derived from plant material.

Biological primary production is determined over a time scale from days to years and can be measured in one of three ways: biomass, energy stored, or carbon uptake through measurements of photosynthesis. The most common and easily measured is biomass. Envision an ecologist taking a swath of grasses and their buried roots, drying them to rid the grasses of any moisture, and placing them on a metric scale to obtain a measurement of dry weight and you get a sense of how one can determine biomass. **Biomass** is simply the total amount of organic matter (or amount of fixed energy) measured at a point in time. The units of biomass are expressed as mass per unit area (e.g., g/m^2). To obtain a measure of NPP, we track the changes in biomass (B) over time ($NPP = B2 - B1$). Increases in biomass indicate an increase in NPP, while negative results point to a loss of productivity in the ecosystem. This approach provides a crude measurement of biological production since it overlooks energy that is lost as latent heat and respired. Therefore, we typically consider this to be net productivity (rather than gross). Other techniques used to measure biological production include tracing ^{14}C carbon radioisotope uptake during photosynthesis in phytoplankton too small to take measurements of weight.

By taking repeated measurements of biomass or productivity over various time intervals, ecologists are able to track and monitor the amount of energy within a particular system and determine whether and how it changes over time. This is an important estimation because ecosystems may be found to have a decline in production (e.g., a forest may be overharvested and not given enough time to restore to its natural state) or an increase in production (e.g., a new protected area may be created where hunting is no longer permitted, allowing a wildlife species to grow in numbers).

When we examine the flow of energy through an ecosystem along the feeding relationships established as food chains and webs, we can consider changes in the size of the trophic level in terms of the number of individuals and biomass. When we do so, we realize that there is an attenuation from the base of the food chain to the highest trophic level that depicts a *pyramid*. These relationships are referred to as a pyramid of energy, biomass, or numbers depending on what we're considering. At the bottom of the pyramid, the first trophic level has the largest number and biomass of autotrophs. With the significant loss of energy between the first and second trophic levels, there is a pronounced reduction in both the biomass and number of herbivores. The energy has been lost to the surrounding ecosystem as biologically unuseable latent heat. The top predators, typically the tertiary consumers, are only a restricted number of individuals, and their biomass represents a very small proportion of that found in the ecosystem.

We can use trophic-level efficiency to quantify this energy transfer in an ecosystem. This is done by comparing the biomass of one trophic level with that of the next higher trophic level. For example, we can determine the efficiency of transfer of cows grazing in a pasture. Let's say the average biomass of cows for the size of the pasture was determined to be $0.5 \text{ kg}/m^2$ and the grasses they were grazing on had a biomass of $3.5 \text{ kg}/m^2$. The trophic level efficiency would be 0.5/3.5, or 14.3 percent. From studies in other ecosystems, we can take 10 percent as the average efficiency, or "rule of thumb," from producers to primary consumers. This efficiency may only be around 1 percent between primary consumers to secondary consumers (herbivores to carnivores).

Food web at the edge of an eastern Canadian deciduous forest FIGURE 5.7

This food web is greatly simplified compared with what actually happens in nature. Many species are not included, and numerous links in the web are not shown.

First trophic level
1. White pine
2. White oak
3. Blackberry
4. Red clover
5. Acorns
6. Grasses

Second trophic level
7. Grey squirrel
8. White-tailed deer
9. Eastern chipmunk
10. Eastern cottontail
11. Insects
12. Worms and ants
13. Moths

Third trophic level
14. Eastern bluebird
15. Red-winged blackbird
16. American robin
17. Red-headed woodpecker
18. Deer mouse

Fourth trophic level
19. Barred owl
20. Red-tailed hawk
21. Red fox

Detritivores/Decomposers
22. Spiders
23. Fungi
24. Bacteria and fungi

What proportion of your diet consists of animal products? Eating at lower trophic levels decreases your ecological footprint because you are essentially bypassing the considerable amount of plant material used to move up the trophic pyramid. What are some ways you could reduce your ecological footprint through the food choices you make?

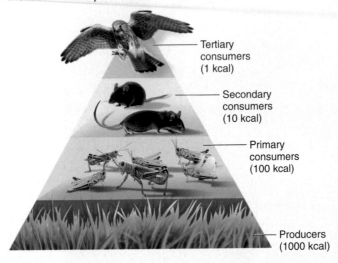

Tertiary consumers (1 kcal)

Secondary consumers (10 kcal)

Primary consumers (100 kcal)

Producers (1000 kcal)

FIGURE 5.8 is an illustration of a pyramid of energy that represents the energy flow in ecosystems. The food chain includes an assemblage of plants, grasshoppers, rodents, and hawks. For every hawk found, many rodents, still more grasshoppers, and an immense number/biomass of plants are required to provide the necessary energy and to support the hawk's growth and vitality.

This relationship illustrates why meat consumption is a factor in assessing ecological footprints. Consuming more meat requires more plants and crops than would be required if we became vegetarians. For every one kilogram of beef we consume, 17 kilograms of grass biomass had to be ingested. The trend, however, is toward greater production of beef, pork, and poultry to support our growing appetites. We consider this impact in the case study on human appropriation of net primary production at the end of this chapter.

CONCEPT CHECK **STOP**

What is the first law of thermodynamics? What is the second?

Why is a balanced ecosystem unlikely to contain only producers and consumers? Or only consumers and decomposers? Explain your answers.

What is trophic-level efficiency and what does this mean in relation to the food webs found in ecosystems?

The Cycling of Matter in Ecosystems

LEARNING OBJECTIVE

Diagram and explain the carbon, hydrologic, nitrogen, and phosphorus cycles.

I n contrast to energy flow, matter, the material of which organisms are composed, moves in numerous cycles from one part of an ecosystem to another—from one organism to another and from living organisms to the abiotic environment and back again. We call these cycles of matter **biogeochemical cycles** because they involve biological, geological, and chemical interactions. Four different biogeochemical cycles of matter—carbon, hydrologic (water), nitrogen, and phosphorus—are considered in our discussion in this chapter.

biogeochemical cycles The various cycles that move matter from one part of an ecosystem to another. They include the carbon, hydrologic, nitrogen, and phosphorus cycles.

Scientists recognize that organisms have profoundly altered the way chemicals interact and move in the environment. For example, plants undergo photosynthesis and form glucose as an end product of the reaction. Indeed, without plants, organic carbon would not be created and life could not exist as we know it today. We recognize that the dependency of life on these chemicals represents essential nutrients or food to drive the biosphere. **Macronutrients** are those chemicals needed by all life in large quantities and include carbon, hydrogen, oxygen, phosphorus, nitrogen, and sulphur; plants also require large amounts of potassium, calcium, and magnesium. Conversely, **micronutrients** are chemicals required only in small quantities and include boron, copper, iron, chloride, manganese, molybdenum, and zinc.

Biogeochemical cycles include reservoirs and pathways. **Reservoirs** represent regions where chemicals pool and reside for a period of time. These reservoirs can be found in the biosphere (organisms), the hydrosphere (water), the atmosphere (air), and the lithosphere (rocks). We further describe these reservoirs as sources and sinks. Sources represent the reservoir supplying the chemical whereas sinks represent the reservoir receiving the chemical. **Pathways** facilitate the movement of chemicals from the source to the sink reservoirs.

Biogeochemists, researchers who investigate the biogeochemical cycles and their interactions, are able to measure both the rates at which chemicals move in pathways from one reservoir to another and the size of the reservoir itself. With this information we gain a better understanding of how and to what extent humans impact the movement of chemicals. Modelling allows researchers to analyze human influence on the flow of these chemicals and helps us predict the consequences of these changes to our environment. Climate change associated with the carbon cycle and water pollution resulting from alterations to the nitrogen and/or phosphorus cycles are examples of human-induced alterations of the chemical cycles in ecosystems. These are discussed in greater detail in upcoming chapters.

THE CARBON CYCLE

Carbon, the fourth most abundant element in the universe, is an essential chemical element of all living matter. Carbon has the unique ability to form bonds with four other atoms or form double bonds with itself, which allows carbon to combine to form compounds that may be solids (such as limestone, wood, plastic, diamond, and graphite), liquids, or gases at normal temperatures (such as the gas carbon dioxide). Carbon is a macronutrient and essential to the formation of proteins, carbohydrates, and other molecules that are essential to living organisms. Carbon makes up approximately 0.04 percent of the atmosphere as a gas, CO_2. It is present in the ocean in several chemical forms depending on the pH. Bicarbonate (HCO_3^-) accounts for the largest proportion of carbon in oceans while smaller proportions of carbonate (CO_3^{2-}) and CO_2 are also found. In rocks, carbon may be found as limestone. The global movement of carbon between the abiotic environment, including the atmosphere, oceans, and lithosphere, and the living biosphere is defined as the **carbon cycle** (**FIGURE 5.9** on page 140).

The carbon cycle can be described in terms of biological processes and abiotic or geological activities. In the biological pathways, photosynthesis fixes CO_2 from either the atmosphere or water into sugars like glucose (recall the equation for photosynthesis was given on page 133). The organic carbon is transferred from one trophic level to the next through feeding relationships defined in the food chain. Carbon dioxide is then returned to the atmosphere when living organisms—producers, consumers, and decomposers—respire. During respiration, sugars are broken down to carbon dioxide, and the potential energy is transformed into kinetic energy available for work.

Sometimes the carbon in biological molecules is not recycled back to the atmosphere. Instead, it may be sequestered or stored in a reservoir for a relatively long period of time. For example, a large amount of carbon is stored in the wood of trees, where it may stay for several hundred years or even longer. Productive topsoils contain large quantities of organic carbon that only gradually decompose to release carbon dioxide to the atmosphere. Coal, oil, and natural gas, called *fossil fuels* because they formed from the remains of ancient organisms, are vast deposits of organic carbon compounds—the end products of photosynthesis that occurred millions of years ago. In *combustion*, organic molecules in wood, coal, oil, and natural gas are burned, with accompanying releases of heat, light, and carbon dioxide.

The carbon cycle FIGURE 5.9

The movement of carbon between the abiotic environment (the atmosphere, lithosphere, and ocean) and living organisms (biosphere) is a process known as the carbon cycle. Because proteins, carbohydrates, and other molecules contain large amounts of carbon, the process is essential to life.

VIEW THIS IN ACTION

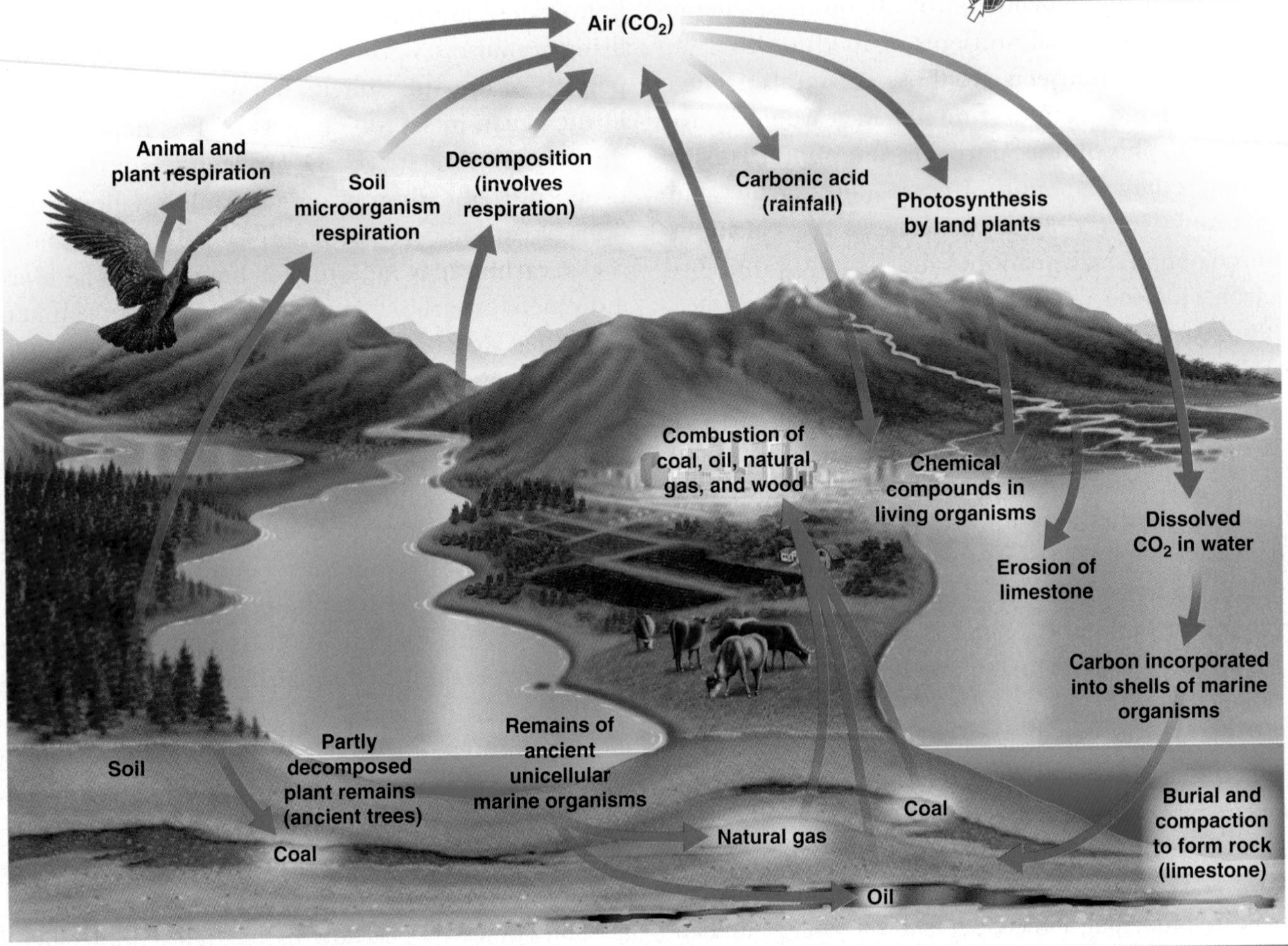

Air (CO_2)

Animal and plant respiration

Soil microorganism respiration

Decomposition (involves respiration)

Carbonic acid (rainfall)

Photosynthesis by land plants

Combustion of coal, oil, natural gas, and wood

Chemical compounds in living organisms

Dissolved CO_2 in water

Erosion of limestone

Carbon incorporated into shells of marine organisms

Soil

Partly decomposed plant remains (ancient trees)

Remains of ancient unicellular marine organisms

Coal

Burial and compaction to form rock (limestone)

Coal

Natural gas

Oil

The **geologic carbon cycle** is the component of the carbon cycle that interacts with the rock cycle in the process of weathering, volcanism, and precipitation and burial of minerals. Carbonic acid (H_2CO_3) is formed when carbon dioxide (CO_2) combines with water. This carbonic acid reaches the lithosphere when it rains and reacts with minerals at the surface, causing ions to dissolve in rainwater through a process called weathering. This is then carried in rivers and streams until eventually reaching the ocean where it settles to the ocean floor as minerals like calcite ($CaCO_3$). The continual deposition and burial of calcite sediment eventually forms sedimentary rock called limestone. Limestone is also created by the carbon contained in the shells of marine organisms, which settle to the ocean floor and are eventually cemented together. This cycle continues until these minerals are drawn into the mantle of the Earth by subduction, a process in which lithospheric plates collide as one descends on another. Carbon is then returned into the atmosphere as carbon dioxide during a volcanic eruption. Over hundreds of millions of years, these interactions work to stabilize atmospheric carbon dioxide concentrations.

Scientists like Charles Keeling, an oceanographer with Scripps Institution of Oceanography, have studied atmospheric levels of CO_2. Keeling's data, commonly known as the "Keeling curve," contains the longest continuous record of atmospheric CO_2 and substantiates reports that human activities have significantly altered atmospheric CO_2 levels from the natural carbon cycle.

Since the beginning of the Industrial Revolution, the burning of fossil fuels as well as deforestation have contributed to a long-term rise in atmospheric CO_2. Burning oil and coal has released previously sequestered carbon into the atmosphere at a rate far quicker than it can be removed due to a loss of forests that would otherwise take in carbon through a process of photosynthesis. Other human-induced changes have altered the rates of input to the atmosphere as well as the removal to the biosphere. Consequently, atmospheric concentrations of CO_2 continue to rise. Because CO_2 increases the atmosphere's ability to hold heat, it has been called a "greenhouse gas" (see Chapter 10 for a more in-depth discussion of greenhouse gases and the enhanced greenhouse effect).

The elevated concentrations of atmospheric CO_2 combine with water vapour and ocean waters to form carbonic acid. Scientists have already reported a change in oceanic pH that is expected to increase the acidity of the oceans, which in turn is expected to have profound influences on the marine community. The reduced availability of shell-building carbonates could result in collapsing the populations of bivalves such as clams and scallops, softening the shells of lobster and other crustaceans, and further stressing coral reefs.

The Mauna Loa Observatory in Hawaii (see picture) is a premier atmospheric research facility. The undisturbed air, remote location, and minimal influences of vegetation and human activity make this observatory an ideal place for scientists to gather data related to atmospheric change.

THE HYDROLOGIC CYCLE

You are probably most familiar with the **hydrologic cycle**. Water continuously circulates from the ocean to the atmosphere to the land and back to the ocean (**FIGURE 5.10** on page 142).

Through the hydrologic cycle, a renewable supply of purified water for terrestrial organisms is available. Water may flow in rivers and streams to coastal *estuaries*, bodies of water with access to the open ocean as well as to these sources of fresh water. The movement of water from land to rivers, lakes, wetlands, and, ultimately, the ocean is *runoff*, and the area of land where runoff drains is a *watershed*.

The oceans are the main reservoir in the hydrologic cycle, holding 97 percent of all water on Earth. The fresh water we depend on for our survival accounts for less than 3 percent, and two-thirds of this small amount is tied up in glaciers, snowfields, and ice caps. Thus, considerably less than 1 percent of the planet's water is in a form that we can readily use—groundwater, surface fresh water, and rain from atmospheric water vapour.

Water moves to the atmosphere by evaporation in which liquid water is converted to a gaseous form. A second way that water vapour is transferred to the atmosphere is through transpiration, the release of water vapour through plant leaves during photosynthesis. As water evaporates or is transpired, the minerals and dissolved materials in the liquid form remain behind, and this distillation effectively purifies the water providing a constant supply of fresh water. Water is returned from the atmosphere through precipitation. Plants may take up this water, or it may flow as runoff to a receiving tributary, or it may percolate into the ground until it reaches the water table.

The hydrologic cycle FIGURE 5.10

In the hydrologic cycle, water moves from the ocean to the atmosphere, the land, and back to the ocean in a continuous process that supports life.

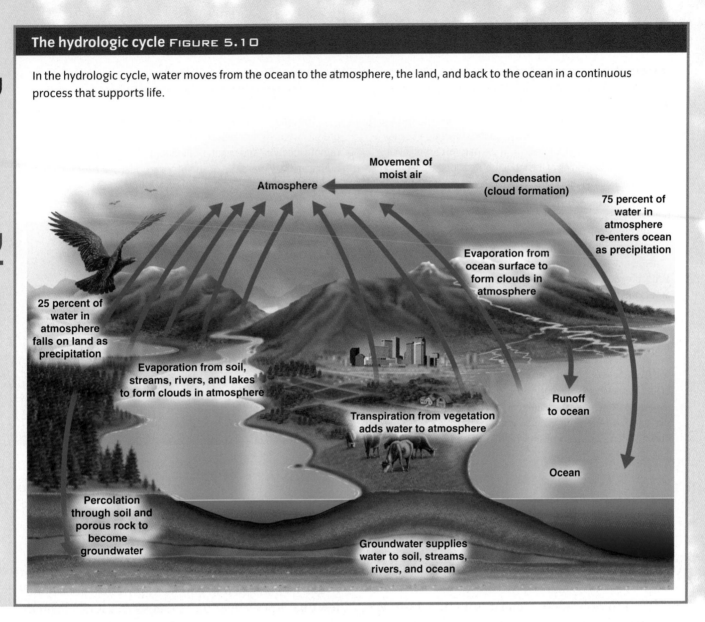

Movement of moist air

Atmosphere

Condensation (cloud formation)

75 percent of water in atmosphere re-enters ocean as precipitation

25 percent of water in atmosphere falls on land as precipitation

Evaporation from ocean surface to form clouds in atmosphere

Evaporation from soil, streams, rivers, and lakes to form clouds in atmosphere

Transpiration from vegetation adds water to atmosphere

Runoff to ocean

Ocean

Percolation through soil and porous rock to become groundwater

Groundwater supplies water to soil, streams, rivers, and ocean

Regardless of its physical form—solid, liquid, or vapour—or location, every molecule of water eventually moves through the hydrologic cycle.

THE NITROGEN CYCLE

Nitrogen is an essential part of biological molecules such as proteins and nucleic acids (for example, DNA) and defined as one of the six macronutrients. At first glance, a shortage of nitrogen for organisms appears impossible. The atmosphere, after all, is 78 percent nitrogen gas (N_2). But atmospheric nitrogen is so stable that it does not readily combine with other elements. Consequently, biologically useable forms of nitrogen such as ammonia and nitrate may be too low in concentration to promote high rates of plant growth in naturally undisturbed ecosystems.

There are five steps in the **nitrogen cycle**, in which nitrogen cycles between the abiotic environment and organisms: nitrogen fixation, nitrification, assimilation, ammonification, and denitrification (FIGURE 5.11).

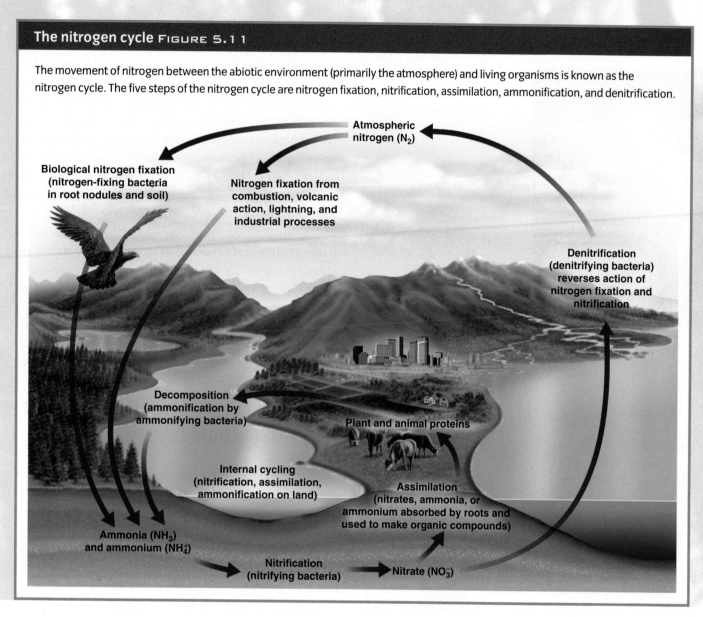

The nitrogen cycle FIGURE 5.11

The movement of nitrogen between the abiotic environment (primarily the atmosphere) and living organisms is known as the nitrogen cycle. The five steps of the nitrogen cycle are nitrogen fixation, nitrification, assimilation, ammonification, and denitrification.

Atmospheric nitrogen (N_2)

Biological nitrogen fixation (nitrogen-fixing bacteria in root nodules and soil)

Nitrogen fixation from combustion, volcanic action, lightning, and industrial processes

Denitrification (denitrifying bacteria) reverses action of nitrogen fixation and nitrification

Decomposition (ammonification by ammonifying bacteria)

Plant and animal proteins

Internal cycling (nitrification, assimilation, ammonification on land)

Assimilation (nitrates, ammonia, or ammonium absorbed by roots and used to make organic compounds)

Ammonia (NH_3) and ammonium (NH_4^+)

Nitrification (nitrifying bacteria)

Nitrate (NO_3^-)

Through these steps, nitrogen gas is transformed into ammonia and nitrates that form the biologically available sources to plants.

Nitrogen-fixing bacteria carry out *nitrogen fixation* in soil and aquatic environments. Nitrogen-fixing bacteria split atmospheric nitrogen and combine the resulting nitrogen atoms with hydrogen under anoxic (no oxygen) concentrations. Some nitrogen-fixing bacteria, *Rhizobium,* live inside swellings, or nodules, on the roots of legumes such as beans or peas and some woody plants (**FIGURE 5.12A** on page 144). In aquatic environments, photosynthetic bacteria called *cyanobacteria* perform most of the nitrogen fixation in specialized heterocyst cells (**FIGURE 5.12B**).

During *nitrification,* soil bacteria convert ammonia to nitrate. The process of nitrification furnishes these bacteria, called nitrifying bacteria, with energy. In *assimilation,* plants absorb ammonia or nitrate through their roots and convert the nitrogen into plant compounds such as proteins and nucleic acids. Animals assimilate nitrogen when they consume plants or other animals and convert the proteins into animal proteins.

The Cycling of Matter in Ecosystems   143

A Bacteria carry out nitrogen fixation in the nodules of a pea plant's roots.

B *Nostoc* is a cyanobacterium that fixes nitrogen. Nitrogen fixation occurs in specialized cells referred to as heterocysts.

Ammonification occurs when organisms produce nitrogen-containing waste products such as urine. These substances, plus the nitrogen compounds that occur in dead organisms, are decomposed, releasing the nitrogen into the abiotic environment as ammonia. The bacteria that perform this process are called ammonifying bacteria. Other bacteria perform *denitrification,* in which nitrate is converted back to nitrogen gas. Denitrifying bacteria prefer to live and grow in anoxic environments where there is little or no free oxygen. For example, they are found deep in the soil near the water table, an environment that is nearly oxygen-free.

Human activities have significantly influenced the nitrogen cycle. In the early 20th century, scientists discovered that there were industrial processes that could convert atmospheric nitrogen into ammonia, referred to as the Haber process. The ammonia could then be oxidized to create nitrates and nitrites. With this chemical advancement, humans could generate unlimited supplies of ammonia to meet demands to enhance crop production. There was no longer a dependency on nitrogen-fixing bacteria to generate these sources of available biologically useable sources of nitrogen. Indeed, the considerable increase in artificial fertilizer production has become the largest source of fixed nitrogen on Earth. It is projected that the amount of nitrogen deposition on Earth's surface will double within the next 25 years.

Runoff of nitrate-contaminated surface waters affects all biogeochemical processes as it ends up in watersheds, causing eutrophication (see Chapter 9). In coastal areas, excess nitrogen can cause red tides or dead zones. High levels of nitrates in groundwater affect drinking water and contribute to the occurrence of methemoglobinemia, or blue baby syndrome. Nitrogen oxides contribute to atmospheric problems such as smog, acid rain, and formation of greenhouse gases.

THE PHOSPHORUS CYCLE

Unlike the biogeochemical cycles just discussed, the **phosphorus cycle** does not have a significant atmospheric component and therefore has a patchy distribution with areas of high concentrations (for example, guano, or bird manure), formed by generations of seabirds congregating along coastal regions) and those with little phosphorus available. Phosphorus moves from land into living organisms, then from one organism to another, and finally back to the lithosphere (**FIGURE 5.13**). It moves slowly from one reservoir to another, and when it is trapped in the lithosphere it may remain there for an extended period.

Phosphorus is an important macronutrient used in biological molecules such as nucleic acids and ATP, a compound that is important in energy transfer reactions in cells. Consequently, the availability of this nutrient will strongly influence populations of plants. In excess it contributes to rapid growth of plants, while in low concentrations it may restrict or limit primary producers and other trophic levels. Like carbon and nitrogen, phosphorus moves through the food web as one organism consumes another.

Phosphorus moves through aquatic communities in much the same way it does through terrestrial

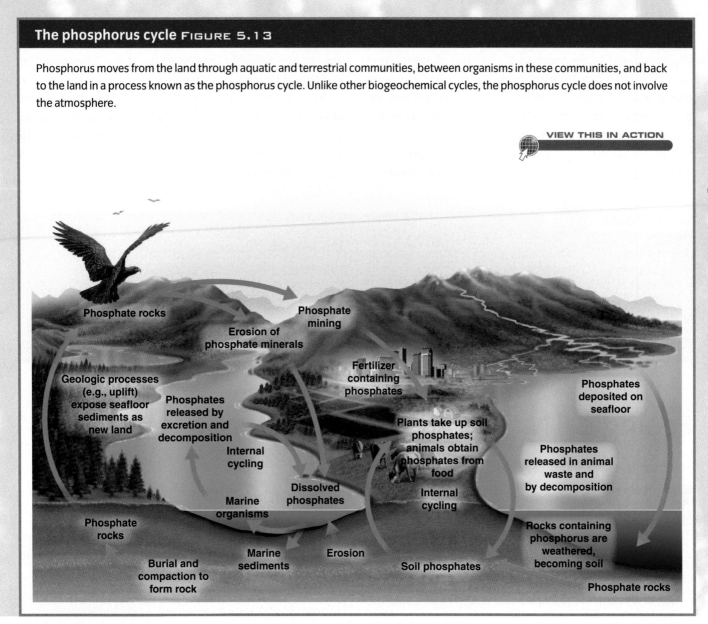

The phosphorus cycle FIGURE 5.13

Phosphorus moves from the land through aquatic and terrestrial communities, between organisms in these communities, and back to the land in a process known as the phosphorus cycle. Unlike other biogeochemical cycles, the phosphorus cycle does not involve the atmosphere.

VIEW THIS IN ACTION

Process Diagram

communities. Dissolved phosphorus enters aquatic communities as algae, and plants absorb and assimilate it; plankton and larger organisms obtain phosphorus when they consume the algae and plants. In turn, a variety of fishes and mollusks eat plankton. Ultimately, decomposers release inorganic phosphorus into the water, where it is available for aquatic producers to use again.

Phosphate can be lost from biological cycles. Some phosphate is carried from the land by streams and rivers to the ocean where it can be deposited on the seafloor and remain for millions of years.

A small portion of the phosphate in the aquatic food web finds its way back to the land. A few fishes and aquatic invertebrates are eaten by seabirds, which may defecate on land where they roost. Seabird guano contains large amounts of phosphate and nitrate. Once on land, these minerals may be absorbed by the roots of plants. The phosphate contained in guano may enter terrestrial food webs in this way, although the amounts involved are small.

In order for life to exist, chemical elements must be available at the right times and in the right amounts and concentrations. If this is not the case, a chemical element may become a limiting factor. A *limiting factor* is an environmental factor (such as a chemical element) that can prevent an organism's growth, distribution, or abundance. Some chemical elements can cycle quickly and are readily regenerated, while others can be tied up in immobile forms, returning slowly to be reused. Phosphorus is a limiting factor for algae growth and when present in small amounts will restrict the abundance and distribution

of algae. Higher phosphorus levels fuel algae populations to grow, and eventually they become dominant by consuming any available oxygen in the water. The reduction in dissolvable oxygen levels becomes an environmental resistance factor for other vegetation and aquatic life forms by limiting their growth, potentially leading to their death and decay. By increasing this one macronutrient in the water, there are many interconnected changes that occur and that affect the overall balance of the entire ecosystem.

CONCEPT CHECK **STOP**

HOW is photosynthesis involved in the carbon cycle?

HOW are living organisms involved in the hydrologic cycle?

What are the five steps of the nitrogen cycle? How are bacteria involved in the nitrogen cycle?

HOW does the phosphorus cycle differ from the other four cycles presented in this chapter?

Ecological Niches and Interactions

LEARNING OBJECTIVES

Describe the factors that contribute to an organism's ecological niche.

Distinguish between mutualism, commensalism, and parasitism.

Define *predation* and describe predator–prey relationships.

Define *competition* and distinguish between intraspecific and interspecific competition.

Y ou have seen that a diverse assortment of organisms inhabits each community and that these organisms obtain nourishment in a variety of ways. You have also considered energy flow and biogeochemical cycles. Now let's examine the way of life of a given species in its ecosystem. An ecological description of a species typically includes whether it is a producer, consumer, or decomposer. However, we need other details to provide a complete picture.

ECOLOGICAL NICHES

Every organism is thought to have its own role, or **ecological niche**, within the structure and function of an ecosystem. An ecological niche is difficult to define precisely because it takes into account all aspects of the organism's existence—all physical, chemical, and biological factors the organism needs to survive, remain healthy, and reproduce. Among other things, the niche includes the local environment in which an organism lives—its **habitat**. An organism's niche also encompasses what it eats, what organisms eat it, what organisms it competes with, and how the abiotic components of its environment, such as light, temperature, and moisture, interact with and influence it. A complete description of an organism's ecological niche involves numerous dimensions.

> **ecological niche** The totality of an organism's adaptations, its use of resources, and the lifestyle to which it is fitted.

An organism's ecological niche may be much broader potentially than it actually is in nature. Put differently, an organism is potentially capable of using much more of its environment's resources or of living in a wider assortment of habitats than it actually does. The potential, idealized ecological niche of an organism is its **fundamental niche**, but various factors, such as competition with other species, usually exclude it from part of its fundamental niche. The lifestyle an organism actually pursues and the resources it actually uses make up its **realized niche** (FIGURE 5.14).

When two species are similar, their ecological niches may appear to overlap. However, many ecologists think no two species indefinitely occupy the same niche in the same community. Resource partitioning is one way some species avoid or at least reduce niche overlap. **Resource partitioning** is the reduction in competition for environmental resources such as food among coexisting species as a result of the niche of each species differing from the niches of others in one or more ways. Evidence of resource partitioning in animals is well documented and includes studies in tropical forests of Central and South America that demonstrate little overlap in the diets of fruit-eating birds, primates, and bats that coexist in the same habitat. Although fruits are the primary food for several hundred bird, primate, and bat species, the wide variety of fruits available has allowed fruit eaters to specialize, thereby reducing competition. Resource partitioning also includes timing of feeding, location of feeding, nest sites, and other aspects of an organism's ecological niche (see the What a Scientist Sees feature on page 148).

ECOLOGICAL INTERACTIONS

No organism exists independently of other organisms. The producers, consumers, and decomposers of an ecosystem interact with one another in a variety of ways, and each forms associations with other organisms. Three main types of interactions occur among species in an ecosystem: symbiosis, predation, and competition.

The wildebeest's ecological niche FIGURE 5.14

Wildebeests live in herds on the open plains of east Africa and graze on the grasses. When the dry season begins, wildebeests living in East Africa's Serengeti Plain migrate more than 1200 kilometres in search of water and grass. Females usually bear one calf at a time. Adult wildebeests are swift runners. Lions and crocodiles prey on adult wildebeests. Hyenas, lions, cheetahs, leopards, and wild dogs prey on the calves.

Yellow-rumped warbler   Bay-breasted warbler   Cape May warbler   Black-throated green warbler   Blackburnian warbler

Resource Partitioning

Robert MacArthur's study of five American warbler species is a classic example of resource partitioning. Although it initially appeared that the niches of the species were nearly identical, MacArthur determined that individuals of each species spend most of their feeding time in different portions of spruces and other conifer trees. They also move in different directions through the canopy, consume different combinations of insects, and nest at slightly different times. The photo shows a male black-throated green warbler in a spruce tree.

Symbiosis In **symbiosis**, one species usually lives in or on another species. The partners in a symbiotic relationship may benefit, be unaffected, or be harmed by the relationship.

Symbiosis is the result of **co-evolution**, the interdependent evolution of two interacting species. Flowering plants and their animal pollinators are an excellent example of coevolution. Bees, beetles, hummingbirds, bats, and other animals transport pollen from one plant to another. During the millions of years over which these associations developed, flowering plants evolved several ways to attract animal pollinators. One of the rewards for the pollinator is food—nectar (a sugary solution) and pollen. Plants possess a variety of ways to get the pollinator's attention, most involving showy petals and scents.

While plants were acquiring specialized features to attract pollinators, animals coevolved specialized body parts and behaviours to aid pollination and obtain nectar and pollen as a reward. Coevolution is responsible for the hairy bodies of bumblebees, which catch and hold the sticky pollen for transport from one flower to another. Coevolution is also responsible for the long, curved beaks of certain Hawaiian birds that insert their beaks into tubular flowers to obtain nectar (**Figure 5.15** on page 150).

The thousands, or even millions, of symbiotic associations that result from coevolution fall into three categories: mutualism, commensalism, and parasitism (**Table 5.1**). One example of **mutualism**, an association in which both organisms benefit, is the interaction between acacia ants and the bull's horn acacia plant (**Figure 5.16** on page 150). The ants make hollow nests out of thorns at the base of the plant's leaves and gain special nutrients from the leaf tips. In return, the ants effectively protect the plant from

> **symbiosis** An intimate relationship or association between members of two or more species; includes mutualism, commensalism, and parasitism.

> **predation** The consumption of one species (the prey) by another (the predator).

invertebrate and vertebrate herbivores and clear away competing plants. Both ant and acacia depend on this association for survival.

Commensalism is a symbiotic relationship in which one species benefits and the other is neither harmed nor helped. One example of commensalism is the relationship between a tropical tree and its epiphytes, smaller plants, such as mosses, orchids, and ferns, that live attached to the bark of the tree's branches (**Figure 5.17** on page 150). An epiphyte anchors itself to a tree but typically doesn't obtain nutrients or water directly from the tree. Its location on the tree enables it to obtain adequate light, water (as rainfall dripping down the branches), and required nutrient minerals (which rainfall washes out of the tree's leaves). The epiphyte benefits from the association, whereas the tree is apparently unaffected.

Parasitism is a symbiotic relationship in which one species (the *parasite*) benefits at the expense of the other (the *host*). Parasitism is a successful lifestyle; more than 100 parasites live in or on the human species (**Figure 5.18** on page 150). A parasite, usually much smaller than its host, obtains nourishment from its host, but although a parasite may weaken its host, it rarely kills it quickly. (A parasite would have a difficult life if it kept killing off its hosts!) Some parasites, such as ticks, live outside the host's body; others, such as tapeworms, live within the host.

Predation **Predation** includes both animals eating other animals (for example, herbivore–carnivore interactions) and animals eating plants (producer–herbivore interactions). Predation has resulted in an "arms race," with the coevolution of predator strategies—more efficient ways to catch prey—and prey strategies—better ways to escape the predator.

Categories of symbiosis TABLE 5.1	Organism 1	Organism 2	Characteristic of relationship
Mutualism	Benefits	Benefits	Each organism depends on the other
Commensalism	Benefits	Not affected	Only one organism depends on the other
Parasitism	Benefits	Harmed	Host harmed, rarely killed; host usually much larger than parasite

Coevolution FIGURE 5.15
This Hawaiian honeycreeper uses its gracefully curved bill to sip nectar from the long, tubular flowers of the lobelia.

Mutualism FIGURE 5.16
Most common in Central America, the acacia ant gains shelter and nutrients from the acacia plant, in turn protecting the plant from predators. Photographed in Costa Rica.

Commensalism FIGURE 5.17
Epiphytes are small plants that attach to the branches and trunks of larger trees. Photographed in the Fiji Islands.

Parasitism FIGURE 5.18
Close-up of body lice feeding on a human arm. Each louse is about 3 mm long.

Predation FIGURE 5.19

Global Locator

A cheetah is in a full sprint as it pursues possible prey. Photographed in Masai Mara National Reserve, Kenya.

An efficient predator exerts a strong selective force on its prey, and over time the prey species may evolve some sort of countermeasure that reduces the probability of being captured. The countermeasure that the prey acquires in turn may act as a strong selective force on the predator.

The cheetah is the world's fastest animal and can sprint at 110 kilometres per hour for short distances (**FIGURE 5.19**) Orcas (formerly known as killer whales) hunt in packs and often herd salmon or tuna into a cove so that they are easier to catch. Any trait that increases hunting efficiency, such as the speed of a cheetah or the intelligence of an orca, favours the predators that pursue their prey. Ambush is another effective way to catch prey. The goldenrod spider is the same colour as the white or yellow flowers in which it hides. This camouflage prevents unwary insects that visit the flower for nectar from noticing the spider until it is too late.

Many potential animal prey, such as ground squirrels, run to their underground burrows to escape predators. Others have mechanical defences, such as the barbed quills of a porcupine or the shell of a pond turtle. Some animals live in groups—a herd of antelope, colony of honeybees, school of anchovies, or flock of pigeons—and this social behaviour decreases the likelihood of a predator catching one of them unaware because the group has so many eyes, ears, and noses watching, listening, and smelling for predators (**FIGURE 5.20A** on page 152).

Chemical defences are common among animal prey. The South American poison arrow frog has poison glands in its skin and bright warning colours that experienced predators avoid. Some animals blend into their surroundings and so hide from predators. Certain caterpillars resemble twigs so closely you would not guess that they are animals until they move (**FIGURE 5.20B**).

Plants also possess adaptations that protect them from being eaten. The presence of spines, thorns, tough leathery leaves, or even thick wax on leaves discourages foraging herbivores from grazing. Other plants produce an array of protective chemicals that are unpalatable or even toxic to herbivores. The nicotine found in tobacco is so effective at killing insects that it is an ingredient in many commercial insecticides.

A Richardson ground squirrels vocally communicate with each other with a basic vocabulary consisting of various but distinct squeals, chirps, whistles, teeth chatters, and alarm calls. If a squirrel spies a predator, it will alert the other squirrels, and all will scramble out of sight.

B Predators often overlook caterpillars that closely resemble small branches. Can you find the caterpillar? (*Hint:* The caterpillar is mimicking an adjacent twig.)

Avoiding predators FIGURE 5.20

Competition **Competition** occurs when two or more individuals attempt to use an essential common resource such as food, water, shelter, living space, or sunlight. Resources are often in limited supply in the environment, and their use by one individual decreases the amount available to others. For example, if a tree in a dense forest grows taller than surrounding trees, it absorbs more of the incoming sunlight (**FIGURE 5.21**). Less sunlight is available for nearby trees that the taller tree shades. Competition occurs among individuals within a population (*intraspecific competition*) and between different species (*interspecific competition*).

Competition is not always a straightforward, direct interaction. Consider a variety of flowering plants that live in a young pine forest and compete with conifers for resources such as soil moisture and soil nutrient minerals. Their relationship is more involved than simple competition. The flowers produce nectar that some insect species consume; these insects also prey on needle-eating insects, reducing the number of insects feeding on pines. It is therefore difficult to assess the overall effect of flowering plants on pines. If the flowering plants were removed from the community, would the pines grow faster because they were no longer competing for necessary resources? Or would the increased presence of needle-eating insects (caused by fewer omnivorous insects) inhibit pine growth?

> **competition** The interaction among organisms that vie for the same resources in an ecosystem (such as food or living space).

Competition FIGURE 5.21

Plants compete for light and growing space in a forest. Taller trees reduce the amount of sunlight available for shorter ones, as seen here in the Canadian boreal forest. (Photographed in the Northwest Territories)

Short-term experiments in which one competing plant species is removed from a forest community have in several instances demonstrated improved growth for the remaining species. However, few studies have tested the long-term effects on forest species of removing one competing species. These long-term effects may be subtle, indirect, and difficult to assess. They may reduce or negate the negative effects of competition for resources.

THE IMPORTANCE OF SPECIES TO ECOSYSTEMS

A species-based approach has commonly been used to determine if an ecosystem is functioning as expected, with indicator species selected as representatives of the well-being of the entire system. An *indicator species* is any biological species that defines a characteristic of the environment. Indicator species can be among the most sensitive species in a region, and sometimes act as an early warning to monitoring biologists. Indicator species may be chosen using different criteria, but three of the most often-used are the concepts of keystone species, umbrella species, or flagship species.

> **keystone species**
> A species that has a disproportionate effect on its environment relative to its biomass.

A **keystone species** is a species that has a disproportionate effect on its environment relative to its biomass. Such species affect many other organisms in an ecosystem and help to determine the types and numbers of various others species in a community. Such an organism plays a role in its ecosystem that is analogous to the role of a keystone in an arch. While the keystone is under the least pressure of any of the stones in an arch, the arch still collapses without it. Similarly, an ecosystem may experience a dramatic shift if a keystone species is removed, even though that species was a small part of the ecosystem by measures of biomass or productivity. Thus, a keystone species is a species that plays a critical role in maintaining the structure of an ecological community and whose impact on the community is greater than would be expected based on its relative abundance or total biomass.

Identifying and protecting keystone species are crucial goals of conservation biologists because if a keystone species disappears from an ecosystem, other organisms may become more common or more rare, or they may even disappear. The black-tailed prairie dog has been described as a keystone species in Grasslands National Park in Saskatchewan (**FIGURE 5.22** on page 154).

Although small, prairie dogs have a powerful influence on their environments and are therefore considered keystone species in their ecosystem.

Prairie grouse are considered an umbrella species across their North American range, meaning that if their habitat is maintained, their population will be as well. Preserving their habitat will also ensure that populations of other grassland species will be maintained. Here, a male sharp-tailed grouse displays on his "dancing ground" in an effort to attract females for mating activities.

The rodent exerts a profound influence on the structure of the ecosystem and the processes taking place within it. Animals may rely on the prairie dog as a source of food. Through their digging activity, prairie dogs enhance the habitat available for other species reliant on burrows. They also loosen soil that enhances the availability of nutrients, and they expose seeds that promote a greater diversity of grasses and forbs.

However, the burrowing nature of prairie dogs puts them into conflict with humans. Because of concern that cattle and livestock will be injured by the multitude of burrows, prairie dogs are poisoned, trapped, shot, flooded, and dynamited out of their homes. However, the resulting decline in the number of prairie dogs has caused a reduction in the number of owls, ferrets, snakes, foxes, and badgers on the prairie landscape. The region of the Grasslands National Park is one of the few places in Canada where the prairie dog still exists in the wild.

The concept of an **umbrella species** has been used to provide protection for other species using the same habitat (**FIGURE 5.23**). As the term implies, an umbrella species casts an "umbrella" over the other species by being more or equally sensitive to habitat changes. Animals identified as umbrella species typically have large home ranges that cover multiple habitat types. Therefore, protecting the umbrella species effectively protects many habitat types and the many species that depend on those habitats. Umbrella species can be used to help select the locations of potential reserves, find the minimum size of these conservation areas or reserves, and to determine the composition, structure and processes of ecosystems.

A **flagship species** is a species chosen to represent an environmental cause, such as an ecosystem in need of conservation. These species are chosen for their vulnerability, attractiveness, or distinctiveness to garner support and acknowledgement from the public at large. Thus, the concept of a flagship species holds that by giving publicity to a few key species, the support given to those species will successfully leverage conservation of entire ecosystems and all species contained therein.

With millions of species of concern, the identification of selected keystone species, umbrella species, or flagship species makes conservation decisions easier.

What are three aspects of the wildebeest's ecological niche?

What is one example of mutualism? Of parasitism?

What is one example of a predator–prey interaction? Of competition?

What is the difference between interspecific and intraspecific competition?

What is a keystone species? An umbrella species? A flagship species? Provide some examples.

Adaptive Ecosystem Management

LEARNING OBJECTIVES

Define *ecosystem management.*

Explain the major components of adaptive ecosystem management.

cosystem management is an evolving concept that is increasingly being used globally in recognition of the need for using ecological components in environmental decision making. So what is ecosystem management? If we consider the two terms independently, we know that an *ecosystem* as the biotic communities and their associated abiotic components that interact in a defined geographic area. The term *management* can be defined as meeting goals or objectives, and in the case of the environment, it may be characterized as meeting the goals or objectives of society with respect to the environment. Thus, when we put these terms together, **ecosystem management** can be defined as meeting the goals or objectives for the biotic communities and their associated abiotic components in a defined geographic area as established by society (**FIGURE 5.24**). This can indeed be a challenge!

ecosystem management An evolving concept that recognizes the need to use ecological components in environmental decision making.

One approach to ecosystem management, called *adaptive ecosystem management,* was developed by C. S. Holling and Carl J. Walters, ecologists from the University of British Columbia, in the 1970s. It is an approach to management that acknowledges uncertainty and the need for managers to learn from system

Ecosystem management FIGURE 5.24

Ecosystem management seeks to manage the goals and objectives for every organism using an environment. Here it would seek to preserve this stream for the fish living in it, the plants and animals that use it as a water source, and this flyfisher who uses it for recreation.

Ecosystem management can be quite a challenge. In order to achieve this monumental endeavour, an implementation strategy has been developed by the International Union for Conservation of Nature (IUCN) entitled the *ecosystem approach*. This strategy has been used successfully throughout the world, and with careful attention to detail, can be of benefit to a wide range of environmental decisions.

In their publication entitled *The Ecosystem Approach: Five Steps to Implementation*, the IUCN identifies a five-step approach to ecosystem management:

1. *The stakeholders and the ecosystem area:* Determining who the stakeholders are and defining the ecosystem area in question is the first step in the IUCN's ecosystem management approach. It is recognized that the definition of the geographic area under consideration is best done by consensus with the appropriate stakeholders participating in the decision making. This is generally the most time-consuming of the steps, but if done well, it builds considerable trust among the various interested parties.

2. *Ecosystem structure, function, and management:* This step requires the application of knowledge from a variety of sources, including "western" scientific knowledge, "local" knowledge from long-time residents, "traditional" knowledge as described through oral histories of Aboriginal groups, and others. Each type of knowledge should be considered, added to as required, and respected for its contribution to the overall management goal for the ecosystem.

3. *Economic issues:* The third step requires careful consideration of the key economic drivers that can be associated with different management choices. Wherever possible, negative economic deterrents should be avoided, and positive economic incentives should be promoted.

4. *Adaptive management over space:* Step 4 refers to the likely impact of an ecosystem on adjacent ecosystems. It takes into consideration other ecosystems, other stakeholders, other regions, and perhaps even other countries and is essential to building trust with a variety of different stakeholders. The use of adaptive resource management (ARM) approaches is particularly useful in achieving success.

5. *Adaptive management over time:* This step recognizes that unforeseen issues might affect the initial management goals and that development of alternative strategies might be needed to maximize success. This step also emphasizes the need for regular and science-based monitoring, once again best linked to the use of ARM. If a management plan can be revisited and revised, if necessary, on a regular basis, those involved in its creation and implementation are more likely to take ownership and trust the outcomes.

Chapter 13 presents a real-world example of the successful implementation of the ecosystem approach.

responses while they manage. Learning is acknowledged as a formal and integral objective of the management process, allowing managers to update and modify future action. Initially, adaptive management was applied to fisheries management, but it soon became more widely recognized and accepted in the 1990s and early 21st century with notable success in its application in the area of waterfowl harvest management in North America. Adaptive management is particularly useful in systems in which learning by experimentation is impractical.

Adaptive ecosystem management can be either passive or active. *Passive management* uses modelling to forecast information that may guide future management decisions. Conversely, *active management* includes changing management action altogether to test new hypotheses. For both active and passive adaptive management approaches, embracing risk and uncertainty to enhance learning and modifying action based on what has been learned are key components.

CONCEPT CHECK STOP

What is ecosystem management? Why is it important?

What are the key goals of adaptive ecosystem management?

ECOLOGIST

Ecologists study the relationships between living things and their environments, often gathering data directly from the site.

Imagine standing knee-deep in a fast-moving, frigid creek 20 metres from where it runs into a spectacular alpine lake. In front of you is a large fishing net strung between the creek's banks, and caught in it are five enormous bull trout. You are an ecologist and you've been here for two weeks gathering data on the endangered bull trout population.

Two decades ago, the province wanted to encourage sport fishing and tourism in the area, so it introduced rainbow and brown trout to the lake. These new species became direct competitors with the bull trout for food and habitat, sending the bull trout population into sharp decline. Years of study and work have been dedicated to reviving the bull trout population by removing the introduced fish. You are here to see if these measures are working.

As an ecologist, you spend a lot of time studying the population dynamics of the bull trout, and this time of year is always hectic for you. This is when the bull trout make a run up the creek to spawn, giving you your best opportunity to gather data. The fishing net contraption you have placed across the creek allows the trout to swim upstream to their spawning grounds but catches them before they return to the lake. Several times a day, you wade out to the nets to grab the fish that have been caught and bring them to your mobile station on the bank.

One at a time, you put the bull trout in a basin of water with a bit of anaesthetic that temporarily sedates the fish so you can work with each one for about 10 minutes. When the fish is sufficiently calm, you take it out of the basin and check for an identification chip implanted just under the skin. If the fish doesn't have a chip, it's probably a juvenile born last year, in which case you will implant a chip before putting it back in the water. Once you have identified the fish, you measure its length and weigh it on your portable scale. You then put the fish in another tank, where you will keep it until the anaesthetic's effects have worn off and the fish can be safely returned to the creek to continue on its way to the lake.

Once the spawn is over, you will compare the data from this year to years previous. The ID chip lets you track each fish individually so you can check if it is growing longer and gaining weight, indications of an abundant food supply. Also, the ID chip lets you measure recruitment rates by counting how many new juveniles are caught without chips, as well as death rates by counting how many fish from last year didn't return to the creek. These factors will allow you to evaluate the recovery of the lake's bull trout population.

After a couple of long weeks in the field, you will return to your office and begin analyzing all the data using statistical software to indicate the size and growth of the bull trout population and whether it is going to survive in the lake.

Ecologists study the relationships between living things and their environment. They study and monitor all kinds of aspects of natural and managed ecosystems, such as temperature and rainfall, competition for food and habitat, predation, disease, and human activities such as farming, hunting, and industry. Their work is used to answer questions of conservation, such as how many fish or deer can be harvested; questions of environmental protection, such as whether a species is in danger and what can be done about it; and questions of management and environmental stewardship, such as where parks and protected areas should be located.

For more information or to look up other environmental careers go to www.eco.ca.

HUMAN APPROPRIATION OF NET PRIMARY PRODUCTIVITY (HANPP)

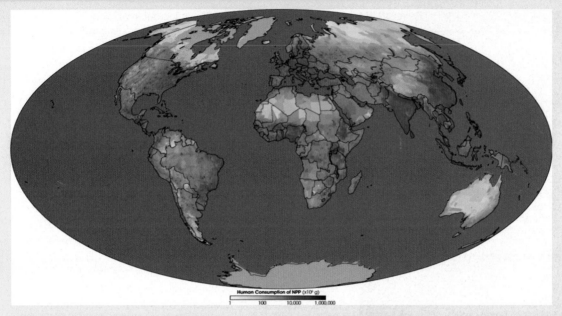

Human Consumption of NPP (x10⁶ g)
1   100   10,000   1,000,000

People everywhere depend on Earth's net primary production (NPP) for everything from food to clothing to fuel. This figure shows how much (millions of grams) of the Earth's NPP people in different parts of the world consume. High national consumption can be due to large populations with low per-person consumption levels (e.g., India), or to small populations with high per-person consumption levels (e.g., United States, European countries).

Humans have had a considerable impact on the structure and functioning of Earth's ecosystems through influences such as population growth, agricultural production, development, and infrastructure. Net primary production (NPP) can be considered the total biologically available energy resources available on Earth to provide the basic maintenance and growth requirements for life. **Human appropriation of net primary production (HANPP)** is an indicator of the area of land used by humans but also the amount of biomass consumed. It therefore provides some assessment of the availability of NPP to support the other species found in Earth's ecosystems. As such, it is an extremely valuable indicator of the "human domination of ecosystems" on a global scale.

The problem is not restricted to densely populated countries such as China and India, but can also be found in highly affluent countries such as the United States and Europe (see map). Many countries are consuming in excess of 100 percent of locally available NPP. Further, as human populations continue to grow exponentially, this will likely be accompanied by more land conversion and further demands on Earth's biomass, suggesting additional increases in HANPP in the future.

Increasing awareness of the implications of HANPP has given rise to policies and practices aimed at local and global efficiency in biomass use. One such program applied in Canada is the land use initiative known as Alternative Land Use Services (ALUS). ALUS is a creative incentive-based program designed by farmers in partnership with the Delta Waterfowl Foundation that recognizes the need for creative programs and policies that address global sustainability. ALUS, a volunteer-based program, provides farmers with a fee-for-service, in this case, rewards for the service of managing the landscape in ways that promote efficient use of biomass. Producers and Delta Waterfowl work together in a win-win situation to preserve waterfowl habitat so that there is no competition for land since the program benefits all involved.

SUMMARY

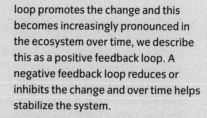

1 Defining Ecology and the Ecosystem Concept

1. **Ecology** is the study of the interaction among organisms and between organisms and their abiotic environment. An **ecosystem** is the biotic communities and their associated abiotic components that interact in a defined geographic area. All ecosystems undergo change as a result of **feedback loops**, which occur when a system's output serves as input to that same system. If the feedback loop promotes the change and this becomes increasingly pronounced in the ecosystem over time, we describe this as a positive feedback loop. A negative feedback loop reduces or inhibits the change and over time helps stabilize the system.

2. A **population** is a group of organisms of the same species that live together in the same area at the same time. A **community** is a natural association that consists of all the populations of different species that live and interact together within an area at the same time. A **landscape** is a region that includes several interacting ecosystems. The **biosphere** is the layer of Earth that contains all living organisms.

2 Energy Flow in Ecosystems

1. **Energy** is the capacity or ability to do work. According to the **first law of thermodynamics**, energy can't be created or destroyed, although it can change from one form to another. As a result of the **second law of thermodynamics**, when energy is converted from one form to another, some of it is degraded into heat, a less useable form that disperses into the environment.

2. A **producer** manufactures large organic molecules from simple inorganic substances. A **consumer** can't make its own food and uses the bodies of other organisms as a source of energy and bodybuilding materials. **Decomposers** are microorganisms that break down dead organic material and use the decomposition products to supply themselves with energy.

3. **Energy flow** is the passage of energy in a one-way direction through an ecosystem, from producers to consumers to decomposers.

3 The Cycling of Matter in Ecosystems

1. **Biogeochemical cycles** are the processes by which matter cycles from the living world to the nonliving, physical environment and back again. Carbon dioxide is the important gas of the carbon cycle; carbon enters the living world through photosynthesis and returns to the abiotic environment when organisms respire. The hydrologic cycle continuously renews the supply of water and involves an exchange of water between the land, the atmosphere, and organisms. There are five steps in the nitrogen cycle: nitrogen fixation, nitrification, ammonification, assimilation, and denitrification. The phosphorus cycle has no biologically important gaseous compounds; phosphorus erodes from rock and is absorbed by plant roots.

4 Ecological Niches and Interactions

1. An **ecological niche** is the totality of an organism's adaptations, its use of resources, and the lifestyle to which it is fitted. An organism's ecological niche includes its habitat, its distinctive lifestyle, and its role in the community.

2. **Symbiosis** is an intimate relationship or association between members of two or more species. Mutualism is a symbiotic relationship in which both species benefit. Commensalism is a symbiotic relationship in which one species benefits and the other species is neither harmed nor helped. Parasitism is a symbiotic relationship in which one species (the parasite) benefits at the expense of the other (the host).

3. **Predation** is the consumption of one species (the prey) by another (the predator). During coevolution between predator and prey, the predator evolves more efficient ways to catch prey (such as pursuit and ambush), and the prey evolves better ways to escape the predator (such as flight, association in groups, and camouflage).

4. **Competition** is the interaction among organisms that vie for the same resources in an ecosystem (such as food or living space). Competition occurs among individuals within a population (intraspecific competition) and between species (interspecific competition).

5 Adaptive Ecosystem Management

1. **Ecosystem management** can be defined as meeting the goals or objectives for the biotic communities and their associated abiotic components in a defined geographic area (the ecosystem) as established by society.

2. Adaptive ecosystem management is an approach that acknowledges uncertainty and the need for managers to learn from system responses while they manage. Embracing risk and uncertainty to enhance learning and modifying action based on what has been learned are key components of adaptive ecosystem management.

KEY TERMS

- **ecology** p. 128
- **population** p. 128
- **community** p. 128
- **ecosystem** p. 128
- **feedback loop** p. 129
- **landscape** p. 130
- **biosphere** p. 130
- **energy** p. 131

- **first law of thermodynamics** p. 131
- **photosynthesis** p. 132
- **second law of thermodynamics** p. 132
- **producers** p. 133
- **consumers** p. 134
- **decomposers** p. 134
- **energy flow** p. 135

- **biogeochemical cycles** p. 138
- **ecological niche** p. 146
- **symbiosis** p. 149
- **predation** p. 149
- **competition** p. 152
- **keystone species** p. 153
- **ecosystem management** p. 155

CRITICAL AND CREATIVE THINKING QUESTIONS

1. To function, ecosystems require an input of energy. Where does this energy come from?

2. After an organism uses energy, what happens to it?

3. What is a biogeochemical cycle? Why is the cycling of matter essential to the continuance of life?

4. Describe how organisms participate in each of these biogeochemical cycles: carbon, nitrogen, and phosphorus.

5. How are food chains important in biogeochemical cycles?

6. How are the many cichlid species in Lake Victoria an example of resource partitioning?

7. In both parasitism and predation, one organism benefits at the expense of another. What is the difference between the two relationships?

8. Some biologists think protecting keystone species would help preserve biological diversity in an ecosystem. Do you agree? Explain your answer.

9. How does the role of humans in the carbon cycle influence global climate change? How might your role in the carbon cycle compare with that of a young person on a remote South American farm that uses animal labour rather than machines?

10. Ecologists investigating interactions of two species at a study site first counted individuals of Species A and then removed all Species B individuals. Six months later, the ecologists again counted individuals of Species A. Viewing their results as graphed below, what is the likely ecological interaction between Species A and Species B? Explain your answer.

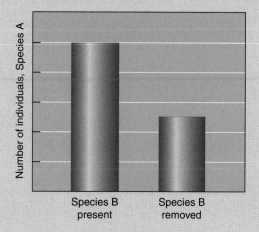

The figure below shows the components of a simple food chain. Use it to answer questions 11–13.

11. Identify the producers, consumers, and decomposers in the food chain. How many trophic levels are represented?

12. Describe or indicate the flow of food and energy within this system.

13. Which forms of energy are present within this chain?

What is happening in this picture ?

This caterpillar has inflated its thorax to make its body look like the head of a snake. Suggest a possible reason the caterpillar does this.

Note the two large spots. What do they resemble? Why would this animal have such conspicuous spots?

If a hungry bird saw this caterpillar, do you think it would have second thoughts before eating it? Why or why not?

Ecosystems in the Environment | 6

THE TALL GRASS PRAIRIE

The tall grass prairie, native to central North America, is a unique and complex ecosystem that includes a wide variety of wild plants, insects, birds, and wildlife. As the name suggests, this ecosystem is dominated by grasses that grow over 2 metres tall alongside forbs, shrubs, and colourful flowers including rare orchids and lilies, all thriving on some of the deepest, most productive soils in North America (see picture). The biodiversity of the tall grass prairie is among the highest in the northern hemisphere. In addition to being home to over 200 plant species, the tall grass prairie provides habitat for a variety of animals such as deer, moose, voles, frogs, songbirds, and butterflies, to name just a few.

In Canada, the tall grass prairies once stretched across 6000 square kilometres in the Red River Valley of Manitoba, and 1200 square kilometres in southern Ontario. Today, this grassland ecosystem is disappearing and is considered the rarest ecosystem in Canada. The productivity and richness of the soils found in the tall grass prairies attracted settlers who transformed these grasses to cultivated cereal and forage crops, urban areas, industrial sites, and forest cover, preserving the grasslands only in areas deemed unsuitable to develop or plow. These changes to the tall grass prairie ecosystem illustrate how misguided human activities can result in more harm than good.

One of the largest remaining blocks of tall grass prairie in Canada is found in the southeastern portion of Manitoba. Over the last two decades, several conservation organizations began securing land creating the Tall Grass Prairie Preserve. Some of these lands are now protected under provincial legislation. Similarly, The Ojibway Park, Tallgrass Prairie Heritage Park in Ontario protects the small remaining tall grass prairie found there. The small size of the remaining tall grass prairie limits any management efforts to re-create pre-settlement prairie conditions where large predators such as wolves and grizzly bears could be found. Today, management of the remaining tall grass prairie includes a combination of techniques such as prescribed burning, rotational grazing, and bio-controls.

NATIONAL GEOGRAPHIC

Classifying Terrestrial Ecosystems: Biomes

LEARNING OBJECTIVES

Define *biome* and discuss how biomes are related to climate.

Briefly describe the nine major terrestrial biomes, giving attention to the climate, soil, and characteristic organisms of each.

Earth has many different climates, which are based primarily on the long-term conditions of temperature and precipitation. These similar climatic conditions are repeated in different geographical areas and result in similar assemblages of biological organisms, known as **biomes**. Biomes are defined as having climatically and geographically similar conditions that result in certain communities of plants, animals, and soil organisms.

> **biome** A large, relatively distinct terrestrial region with similar climate, soil, plants, and animals regardless of where it occurs in the world.

Biomes are classified differently around the world by factors such as plant structures (such as trees, shrubs, and grasses), leaf types (such as broadleaf and needle-leaf), plant spacing (forest, woodland, savanna), and climate. Unlike ecozones, biomes are not defined by genetic, taxonomic, or historical similarities. Biomes are often identified with particular patterns of ecological succession and climax vegetation (quasi-equilibrium state of the local ecosystem). The most widely used systems of classifying biomes correspond to latitude (or temperature zoning) and humidity.

Biomes are therefore found in a number of geographic locations and are not restricted to any one continent. Each biome has a characteristic assemblage of plants and animals that have adapted and evolved to these climatic conditions. Because it is so large in area, a biome encompasses many interacting ecosystems (**FIGURE 6.1**). In terrestrial ecology, a

The world's terrestrial biomes FIGURE 6.1

Although sharp boundaries are shown in this highly simplified map, biomes actually grade together at their boundaries. Use the legend below to identify the locations of the different biomes.

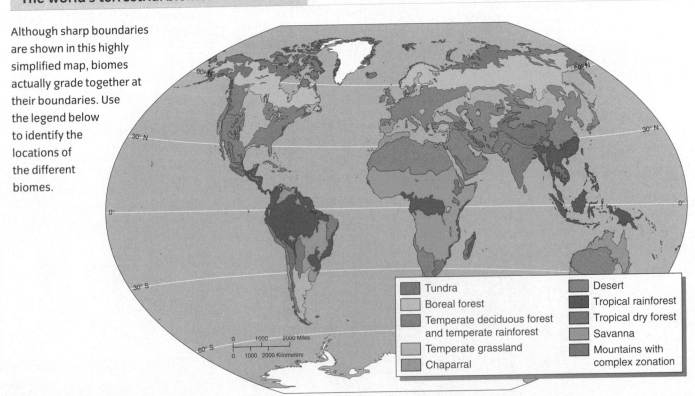

Tundra
Boreal forest
Temperate deciduous forest and temperate rainforest
Temperate grassland
Chaparral

Desert
Tropical rainforest
Tropical dry forest
Savanna
Mountains with complex zonation

biome is considered the next level of ecological organization above community, ecosystem, and landscape.

Near the poles, temperature is generally the overriding climate factor defining a biome, whereas in temperate and tropical regions, precipitation is more significant than temperature, as shown in **FIGURE 6.2**. Other *abiotic* factors to which certain biomes are sensitive include extreme temperatures as well as rapid temperature changes, fires, floods, droughts, and strong winds. Elevation also affects biomes: Changes in vegetation with increasing elevation resemble the changes in vegetation observed in going from warmer to colder climates (**FIGURE 6.3** on pages 166 and 167).

Visualizing

How climate shapes terrestrial biomes FIGURE 6.2

Two climate factors, temperature and precipitation, have a predominant effect on biome distribution.

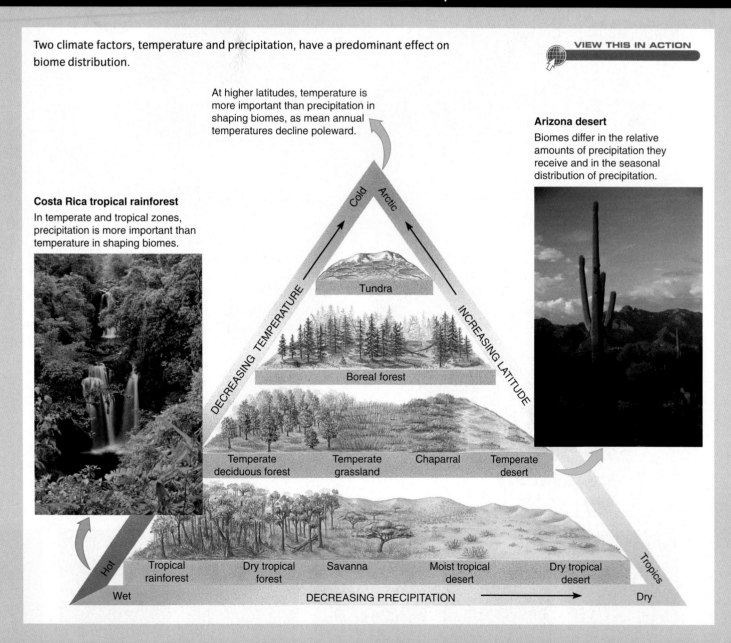

VIEW THIS IN ACTION

At higher latitudes, temperature is more important than precipitation in shaping biomes, as mean annual temperatures decline poleward.

Arizona desert
Biomes differ in the relative amounts of precipitation they receive and in the seasonal distribution of precipitation.

Costa Rica tropical rainforest
In temperate and tropical zones, precipitation is more important than temperature in shaping biomes.

DECREASING TEMPERATURE

Cold   Arctic

Tundra

Boreal forest

INCREASING LATITUDE

Temperate deciduous forest   Temperate grassland   Chaparral   Temperate desert

Hot

Tropical rainforest   Dry tropical forest   Savanna   Moist tropical desert   Dry tropical desert

Tropics

Wet   DECREASING PRECIPITATION →   Dry

The same type of vegetation can occur at many different locations.

GLOBAL LAND COVER COMPOSITION
Three characteristics underlie these categories: life form (woody, herbaceous, or bare); leaf type (needle or broad); and leaf duration (evergreen or deciduous).

EVERGREEN NEEDLELEAF FOREST
Tree height exceeds 5 m; more than 60% of this land is canopied by forest. Typical example: boreal (northern) region. On tree plantations, trees are logged for paper and building products.

EVERGREEN BROADLEAF FOREST
More than 60% of the land is covered by a forest canopy, with trees over 5 m. Such forests dominate in the tropics and are home to great biodiversity, when not cleared for mechanized farms, ranches, and tree plantations.

Land Cover
Forest
- Evergreen needleleaf
- Evergreen broadleaf
- Deciduous needleleaf
- Deciduous broadleaf
- Mixed forest
- Woodland

Grassland or shrubland
- Wooded grassland
- Closed shrubland
- Opened shrubland
- Grassland

- Cropland
- Barren (Desert and Polar Ice)
- Built-up

DECIDUOUS NEEDLE-LEAF FOREST
A forest canopy covers more than 60% of the land; tree height exceeds 5 m. This class is dominant only in Siberia, taking the form of larch forests.

DECIDUOUS BROADLEAF FOREST
More than 60% of the land is covered by a forest canopy; tree height exceeds 5 m. In temperate regions, much of this forest has been converted to cropland.

MIXED FOREST
Both needle and deciduous types of trees appear. Mixed forest is largely found between temperate deciduous and boreal evergreen forests.

WOODLAND
Land has herbaceous or woody understory; trees exceed 5 m and may be deciduous or evergreen. Highly degraded in long-settled human environments, such as West Africa.

WOODED GRASSLAND
Woody or herbaceous understories are punctuated by trees. Examples are African savanna as well as open boreal border land between trees and tundra.

CLOSED SHRUBLAN
Found where prolonge or dry seasons limit p growth. This cover is dominated by bushes shrubs not exceeding Tree canopy is less than 10%.

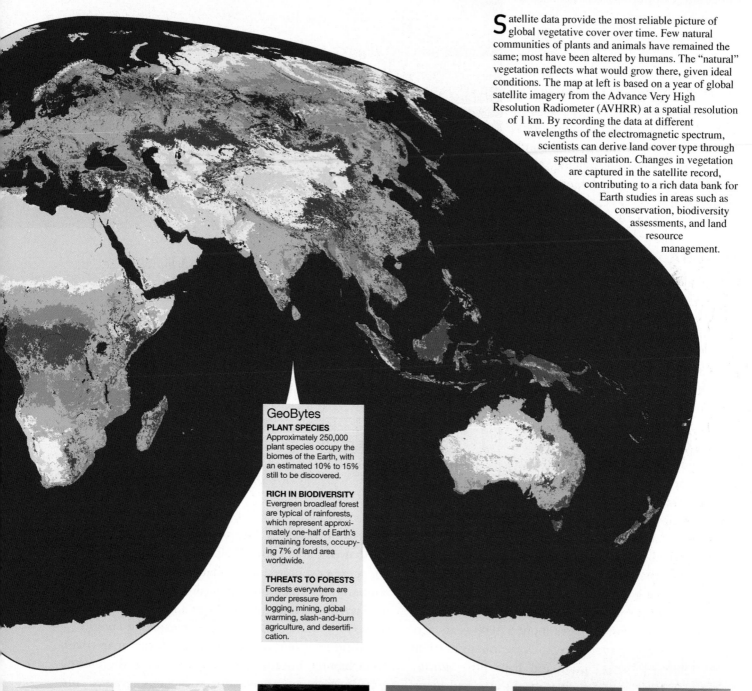

S atellite data provide the most reliable picture of global vegetative cover over time. Few natural communities of plants and animals have remained the same; most have been altered by humans. The "natural" vegetation reflects what would grow there, given ideal conditions. The map at left is based on a year of global satellite imagery from the Advance Very High Resolution Radiometer (AVHRR) at a spatial resolution of 1 km. By recording the data at different wavelengths of the electromagnetic spectrum, scientists can derive land cover type through spectral variation. Changes in vegetation are captured in the satellite record, contributing to a rich data bank for Earth studies in areas such as conservation, biodiversity assessments, and land resource management.

GeoBytes

PLANT SPECIES
Approximately 250,000 plant species occupy the biomes of the Earth, with an estimated 10% to 15% still to be discovered.

RICH IN BIODIVERSITY
Evergreen broadleaf forest are typical of rainforests, which represent approximately one-half of Earth's remaining forests, occupying 7% of land area worldwide.

THREATS TO FORESTS
Forests everywhere are under pressure from logging, mining, global warming, slash-and-burn agriculture, and desertification.

OPEN SHRUBLAND
Shrubs are dominant, not exceeding 2 m in height. They can be evergreen or deciduous. This type occurs in semi-arid or severely cold areas.

GRASSLAND
Occurring in a wide range of habitats, this landscape has continuous herbaceous cover. The American Plains and central Russia are the premier examples.

CROPLAND
Crop-producing fields constitute over 80% of the land. Temperate regions are home to large areas of mechanized farming; in the developing world, plots are small.

BARREN (DESERT)
The land never has more than 10% vegetated cover. True deserts, such as the Sahara, as well as areas succumbing to desertification are examples.

BUILT-UP
This land cover type was mapped using digital population data from around the world. It represents the most densely inhabited areas.

BARREN (POLAR ICE)
Permanent snow cover characterizes this class, the greatest examples of which are in the polar regions, as well as on high elevations in Alaska and the Himalayas.

Canada's terrestrial land base has been classified into biomes using the Canadian Committee on Ecological Land Classification (CCELC) System as developed by Natural Resources Canada. This system has generally been accepted as the standard for terrestrial landscapes in Canada and is used by government agencies, industry, and nongovernment organizations at various levels for management and conservation decision making. The classification system is largely based upon tree species.

TUNDRA

Tundra (or *arctic tundra*) occurs in the extreme northern latitudes where the snow melts seasonally (**FIGURE 6.4**). The southern hemisphere has no equivalent of the arctic tundra because it has no land in the corresponding latitudes. A similar ecosystem located in the higher elevations of mountains, above the tree line, is called *alpine tundra*.

Although the arctic tundra's growing season is short, the days are long. Above the Arctic Circle, the sun does not set at all for many days in midsummer, although the amount of light at midnight is one-tenth that at noon. There is little precipitation, and most of the yearly 10 to 25 centimetres of rain or snow falls during the summer months.

Most tundra soils formed when glaciers began retreating after the last ice age, about 17,000 years ago.

> **tundra** The treeless biome in the far north that consists of boggy plains covered by lichens and mosses; it has harsh, cold winters and extremely short summers.

These soils are usually nutrient poor and have little *detritus*, such as dead leaves and stems, animal droppings, or remains of organisms. Although the tundra's surface soil thaws during summer, beneath it lies a layer of *permafrost*, permanently frozen ground that varies in

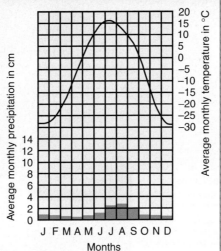

Arctic tundra FIGURE 6.4

Because of the tundra's short growing season and permafrost, only small, hardy plants grow in the northernmost biome that encircles the Arctic Ocean. Photographed in the Yukon Territory. The Ogilvie Mountains are in the background. The climate graph shows monthly temperatures and precipitation for Fort Yukon, Alaska.

depth and thickness. Permafrost impedes drainage, so the thawed upper zone of soil is usually waterlogged during summer. Limited precipitation, combined with low temperatures, flat topography (or surface features), and the layer of permafrost, produces a landscape of broad, shallow lakes and ponds, sluggish streams, and bogs.

Tundra supports relatively few species compared with other biomes, but the species that do occur there often exist in great numbers. Mosses, lichens, grasses, and grass-like sedges are the dominant plants. Stunted trees and shrubs grow only in sheltered locations. As a rule, tundra plants seldom grow taller than 30 centimetres.

Animals adapted to live year-round in the tundra include polar bears, lemmings, voles, weasels, arctic foxes, snowshoe hares, ptarmigan, snowy owls, and muskox.

In summer, caribou migrate north to the tundra to graze on sedges, grasses, and dwarf willow. Dozens of bird species also migrate north in summer to nest and feed on abundant insects. Mosquitoes, blackflies, and deerflies survive winter as eggs or pupae and appear in great numbers during summer weeks.

Tundra recovers slowly from even small disturbances. Oil and natural gas exploration and development and military use have caused damage to tundra likely to persist for hundreds of years. The tundra is expected to go through a more rapid change in temperatures due to the polar amplification associated with the climate change phenomenon. This is further discussed in Chapter 10.

BOREAL FOREST

Just south of the tundra is the **boreal forest**, or northern coniferous forest, which stretches across North America and Eurasia (**FIGURE 6.5**). There is no biome comparable to the boreal forest in the southern hemisphere. Winters are extremely cold and severe, although not as harsh as those in the tundra. Boreal forest receives little precipitation, perhaps 50 centimetres per year, and its soil is typically acidic and mineral poor, with a thick surface layer of partly decomposed pine and spruce needles. Where permafrost occurs, it is found deep under the surface. Boreal forest has numerous ponds and lakes dug by ice sheets during the last ice age.

Black and white spruces, balsam fir, eastern larch, and other conifers (cone-bearing evergreens) dominate the boreal forest. Conifers have many drought-resistant adaptations, such as needle-like leaves whose minimal surface area prevents water loss by evaporation. Such an adaptation helps conifers withstand the drought of the northern winter when roots cannot absorb water through the frozen ground. Being evergreen, conifers resume photosynthesis as soon as warmer temperatures return.

boreal forest

A region of coniferous forest (such as pine, spruce, and fir) in the northern hemisphere; located just south of the tundra. Also called taiga.

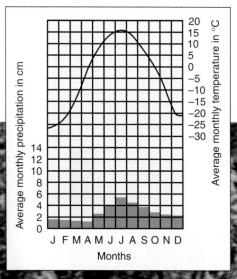

Boreal forest FIGURE 6.5

A herd of caribou runs past a boreal forest. These coniferous forests occur in cold regions of the northern hemisphere adjacent to the tundra. Photographed in the Northwest Territories. The climate graph shows monthly temperatures and precipitation for Fort Smith, Northwest Territories.

The animal life of the boreal forest consists of some larger species such as caribou, which migrate from the tundra for winter; wolves; brown and black bears; and moose. However, most boreal mammals are small to medium sized, including rodents, rabbits, and smaller predators such as lynx, sable, and mink. Birds are abundant in the summer but migrate to warmer climates for winter. Insects are plentiful.

Currently, boreal forest is the world's top source of industrial wood and wood fibre. Extensive logging, natural gas and oil exploration, hydroelectric development, and farming have contributed to loss of boreal forest.

TEMPERATE RAINFOREST (COAST FOREST)

A coniferous **temperate rainforest**, known as coast forest in the CCELC system, occurs on the northwest coast of North America. Similar vegetation exists in southeastern Australia and in southern South America. Annual precipitation in this biome is high,

temperate rainforest
A coniferous biome with cool weather, dense fog, and high precipitation.

from 200 to 380 centimetres, and condensation of water from dense coastal fogs augments the precipitation. The proximity of temperate rainforest to the coastline moderates its temperature so that the seasonal fluctuation is narrow; winters are mild, and summers are cool. Temperate rainforest has relatively nutrient-poor soil, though its organic content may be high. Cool temperatures slow the activity of bacterial and fungal decomposers. Thus, needles and large fallen branches and trunks accumulate on the ground as litter that takes many years to decay and release nutrient minerals to the soil.

The dominant vegetation in the North American temperate rainforest is large evergreen trees such as western hemlock, Douglas fir, western red cedar, Sitka spruce, and western arborvitae (**FIGURE 6.6**). Temperate rainforests are rich in epiphytes, smaller plants that grow on the trunks and branches of large trees. Epiphytes in this biome are mainly mosses, club mosses, lichens, and ferns, all of which also carpet the ground. Squirrels, wood rats, mule deer, elk, numerous bird species, and several species of amphibians and reptiles are common temperate rainforest animals.

Temperate rainforest FIGURE 6.6

This temperate biome has large amounts of precipitation. The climate graph shows monthly temperatures and precipitation for Nanaimo, British Columbia.

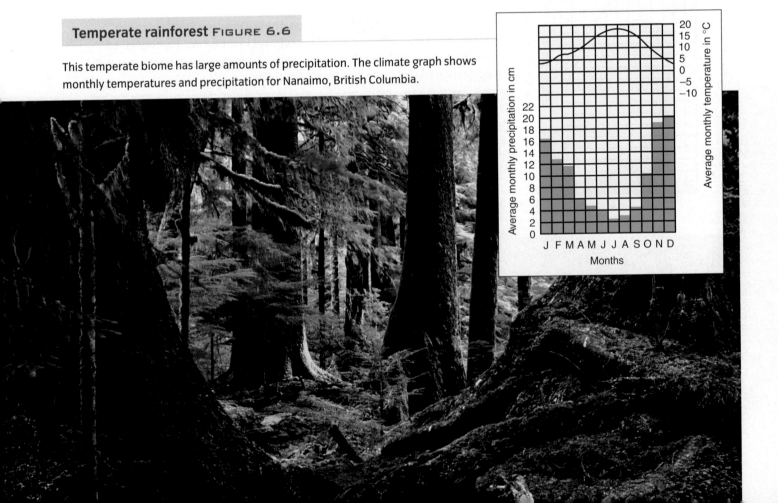

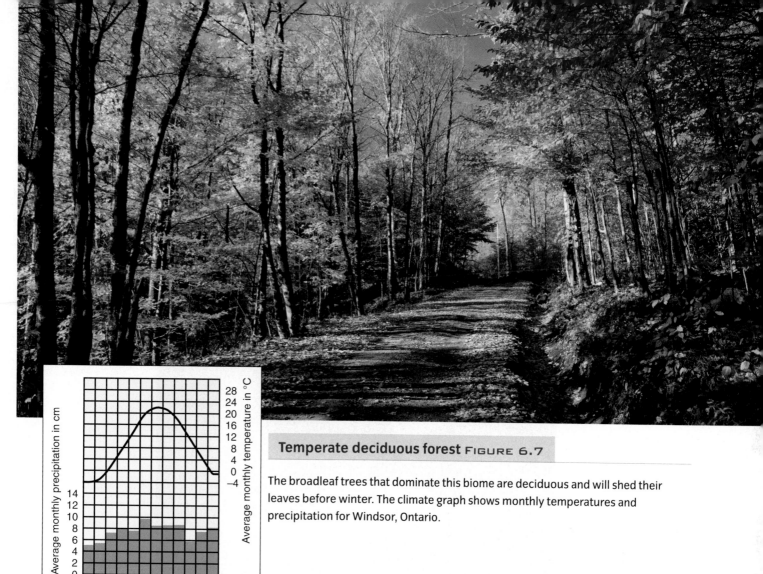

Temperate deciduous forest FIGURE 6.7

The broadleaf trees that dominate this biome are deciduous and will shed their leaves before winter. The climate graph shows monthly temperatures and precipitation for Windsor, Ontario.

TEMPERATE DECIDUOUS FOREST (GREAT LAKES ST. LAWRENCE FOREST)

> **temperate deciduous forest**
>
> A forest biome that occurs in temperate areas where annual precipitation ranges from about 75 to 126 centimetres.

Hot summers and cold winters characterize the **temperate deciduous forest**, or the Great Lakes St. Lawrence Forest in CCELC terms. Typically, the soil of a temperate deciduous forest consists of a topsoil rich in organic material and a deep, clay-rich lower layer. Broadleaf hardwood trees that lose their leaves seasonally, such as oak, hickory, and beech, dominate the temperate deciduous forests of southeastern Canada (**FIGURE 6.7**).

The trees of the temperate deciduous forest form a dense canopy that overlies saplings and shrubs.

Temperate deciduous forests originally contained a variety of large mammals, such as puma, wolves, and bison, which are now absent. Other more common animals include deer, bears, and many small mammals and birds.

In Europe and North America, logging and land clearing for farms, tree plantations, and cities destroyed much of the original temperate deciduous forest. Where it has regenerated, temperate deciduous forest is often in a semi-natural state that humans have modified for recreation, livestock foraging, timber harvest, and other uses. Many forest organisms have successfully re-established themselves in these returning forests.

The Carolinian forest, which extends from the Carolinas in the United States up to southwestern Ontario, is one of Canada's most endangered habitats, with close to 80 percent of the forest having been lost to agriculture and human development. Much of the forest that remains is isolated, fragmented, and threatened by further human encroachment. This is a problem when you consider that close to 40 percent of the species on the Canadian national endangered and threatened species list live in the Carolinian Forest.

The Acadian flycatcher and the hooded warbler (shown in the photo) are two examples of threatened and endangered songbirds that rely on the Canadian Carolinian forest for essential breeding habitat. The isolated and fragmented remnants of the forest are not able to support these species, as they prefer large and mature forests for nesting. The Acadian flycatcher, an endangered species in Canada, has only 35 to 50 nesting pairs,

while the hooded warbler, listed as threatened, has 150 to 210 nesting pairs found in each year. Recovery action, cooperative efforts among stakeholders, and conservational planning and action are necessary to protect the remaining forest.

In 1994, the Acadian flycatcher and the hooded warbler received their respective "endangered" and "threatened" listings under the Committee on the Status of Endangered Wildlife in Canada (COSEWIC), and in 1996 the Canadian Wildlife Service established the Acadian Flycatcher and Hooded Warbler Recovery Team, which developed a national recovery strategy and action plan to preserve these two species. An essential part of their strategy calls on the stakeholders of the Carolinian forest to protect, manage, restore, reconnect, and conserve remaining large woodlot areas that are essential for nesting and that act as critical habitat for a variety of wildlife species.

Hooded warbler

TEMPERATE GRASSLAND

Summers are hot, winters are cold, and rainfall is often uncertain in **temperate grassland.** Average annual precipitation ranges from 25 to 75 centimetres. Grassland soil has considerable organic material because the aboveground portions of many grasses die off each winter and contribute to the organic content of the soil, while the roots and rhizomes (underground stems) survive underground.

> ■ **temperate grassland**
> A grassland with hot summers, cold winters, and less rainfall than is found in the temperate deciduous forest biome.

Many grasses are sod formers—that is, their roots and rhizomes form a thick, continuous underground mat.

Moist temperate grasslands, also known as *tall grass prairies,* have largely disappeared from the Canadian prairies except in preserved locations in Manitoba. Trees grow sparsely except near rivers and streams, but grasses taller than a person grow in great profusion in the deep, rich soil. Periodic wildfires and grazing have helped to maintain grasses as the dominant vegetation in grasslands.

Temperate grassland FIGURE 6.8

This short grass prairie contains a mixture of grasses and other herbaceous flowering plants. The climate graph shows monthly temperatures and precipitation for Swift Current, Saskatchewan.

More than 90 percent of the North American grassland encountered by European settlers has been converted to farmland, and the remaining prairie is highly fragmented. Today, the tall grass prairie is considered North America's rarest biome. Tall grass prairie formerly supported large herds of grazing animals, such as bison and pronghorn elk. The principal predators were wolves, although in sparser, drier areas coyotes took their place. Smaller animals included prairie dogs and their predators (foxes, black-footed ferrets, and various birds of prey), grouse, reptiles such as snakes and lizards, and great numbers of insects. Refer to the chapter opening vignette for more on the tall grass prairies.

Short grass prairies are temperate grasslands that receive less precipitation than moist temperate grasslands but more precipitation than deserts. In Canada, short grass prairies are found primarily in Saskatchewan and Alberta and extend across the border into parts of Montana, Wyoming, South Dakota, and other midwestern states (FIGURE 6.8). Grasses that grow knee high or lower dominate short grass prairies. Plants grow less abundantly than in the moister grasslands, and bare soil is occasionally exposed.

Mixed grass prairies are positioned between the tall and short grass prairies and are dependent on an intermediate precipitation regime. There are patches of mixed grass prairies found in Manitoba, Saskatchewan, and Alberta as well as the northern United States. These grassland systems characteristically have two layers of grasses—one reaching about 30 centimetres above the ground, and the other about 120 centimetres. Both bunch and sod-forming grasses are present, as well as many forbs. Roots extend to depths of about 12 metres. The mixed grass prairie is an important faunal boundary. It is the home of many grassland animals, including prairie dogs, pronghorn elk, swift fox, and black-tailed jack rabbit. True bison habitat also occurs in the mixed prairie. The decline of the bison emphasizes that the prairie has been a constantly changing ecosystem.

tropical rainforest
A lush, species-rich forest biome that occurs where the climate is warm and moist throughout the year.

TROPICAL RAINFOREST

Tropical rainforest occurs where temperatures are warm throughout the year and precipitation occurs almost daily. Therefore, we do not find these biomes in the northern hemisphere. However, due to their extremely important contribution to biodiversity and global carbon uptake, they deserve recognition in our discussion.

Tropical rainforests are found in places close to the equator in Central and South America, Africa, and Southeast Asia. There, the annual precipitation is typically between 200 and 450 centimetres. Tropical rainforest commonly occurs in areas with ancient, highly weathered, mineral-poor soil. Little organic matter accumulates in such soils; because temperatures are high year-round, bacteria, fungi, and detritus-feeding ants and termites decompose organic litter quite rapidly. Roots quickly absorb nutrient minerals from the decomposing material.

Of all the biomes, the tropical rainforest is unexcelled in species richness and variety (**FIGURE 6.9**). No single species dominates this biome. The trees are typically evergreen flowering plants. A fully developed tropical rainforest has at least three distinct stories, or layers, of vegetation. The topmost story, or emergent layer,

consists of the crowns of very tall trees, some 50 metres or more in height, which are exposed to direct sunlight. The middle story, or canopy, which reaches a height of 30 to 40 metres, forms a continuous layer of leaves that lets in very little sunlight to support the smaller plants in the sparse understory. Only 2 to 3 percent of the light bathing the forest canopy reaches the forest understory. The vegetation of tropical rainforests is not dense at ground level except near stream banks or where a fallen tree has opened the canopy. Tropical rainforest trees support extensive communities of epiphytic plants such as ferns, mosses, orchids, and bromeliads.

Not counting bacteria and other soil-dwelling organisms, about 90 percent of tropical rainforest organisms are adapted to live in the canopy. Rainforests shelter the most abundant and varied insects, reptiles, and amphibians on Earth. The birds, often brilliantly coloured, are also varied. Some (parrots, for example) are specialized to consume fruit and others (hummingbirds and

Tropical rainforest FIGURE 6.9

A broad view of tropical rainforest vegetation along a riverbank in Borneo, Southeast Asia. Except at riverbanks, tropical rainforests have a closed canopy that admits little light to the rainforest floor. The climate graph shows monthly temperatures and precipitation for Belem, Brazil.

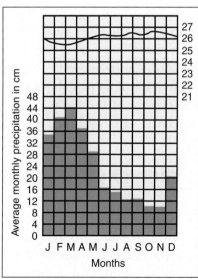

sunbirds, for example) to consume nectar. Most rainforest mammals, such as sloths and monkeys, are adapted to live only in the trees and rarely climb down to the ground, although some large, ground-dwelling mammals, including elephants, are also found in rainforests.

Human population growth and industrial expansion in tropical countries threaten the survival of tropical rainforests. (See Chapter 8 for more discussion of the ecological impacts of rainforest destruction.)

CHAPARRAL

Some hilly temperate environments have mild winters with abundant rainfall combined with very dry summers. Such *Mediterranean climates*, as they are called, occur not only in the area around the Mediterranean Sea, but also in the North American southwest, southwestern and southern Australia, central Chile, and southwestern South Africa. On the mountain slopes of southern California, this Mediterranean-type biome is known as **chaparral**

chaparral A biome with mild, moist winters and hot, dry summers; vegetation is typically small-leafed evergreen shrubs and small trees.

(FIGURE 6.10). Chaparral soil is thin and often not very fertile. Wildfires occur naturally in this environment and are particularly frequent in late summer and autumn.

Chaparral vegetation looks strikingly similar in different parts of the world, even though the individual species differ by location. A dense thicket of evergreen shrubs—often short, drought-resistant pine or scrub oak trees that grow one to three metres tall—usually dominates chaparral. These plant species have evolved adaptations that equip them to live where precipitation is seasonal. During the rainy winter season, the environment may be lush and green, and during the hot, dry summer, the plants lie dormant. The hard, small, leathery leaves of trees and shrubs resist water loss.

Many plants are also specifically fire adapted and grow best in the months following a fire. Such growth is possible because fire releases into the soil the nutrient minerals present in the above ground parts of the plants that burned. The seeds and underground parts of plants that survive fire make use of the newly available nutrient minerals and sprout vigorously during winter rains. (For more on the role

Chaparral FIGURE 6.10

Chaparral vegetation consists mainly of drought-resistant evergreen shrubs and small trees. Hot, dry summers and mild, rainy winters characterize the chaparral. Photographed on Mount Tamalpais in the Marin Hills, California. The climate graph shows monthly temperatures and precipitation for Culver City, California.

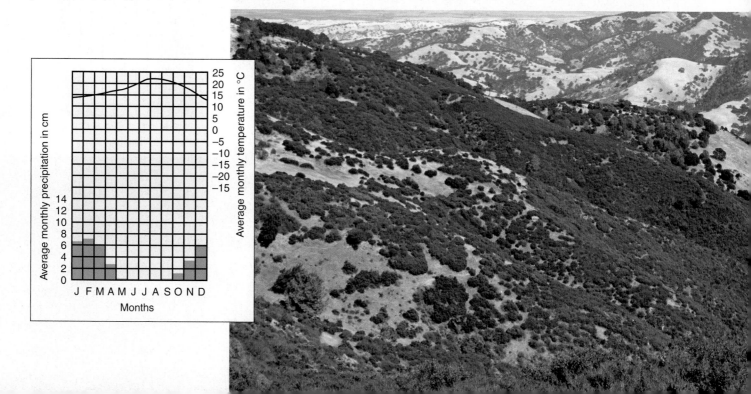

fire plays in nature and on how humans have disrupted this role, see the case study at the end of this chapter.) Mule deer, wood rats, chipmunks, lizards, and many species of birds are common animals of the chaparral.

SAVANNA

Savanna, a tropical grassland, is found in areas of low rainfall or, more commonly, intense seasonal rainfall with prolonged dry periods. Temperatures in tropical savannas vary little throughout the year. Precipitation is the overriding climate factor: annual precipitation is 85 to 150 centimetres. Savanna soil is somewhat low in essential nutrient minerals, in part because it is heavily leached during rainy periods—that is, nutrient minerals filter out of

savanna A tropical grassland with widely scattered trees or clumps of trees.

the topsoil. Although the African savanna is best known, savanna also occurs in South America, western India, and northern Australia.

Savanna has wide expanses of grasses interrupted by occasional trees like the acacia, which bristles with thorns to provide protection against herbivores. Both trees and grasses have fire-adapted features, such as extensive underground root systems, that let them survive seasonal droughts as well as periodic fires.

Spectacular herds of herbivores such as wildebeest, antelope, giraffe, zebra, and elephants occur in the African savanna (**FIGURE 6.11**). Large predators, such as lions and hyenas, kill and scavenge the herds. In areas of seasonally varying rainfall, the herds and their predators may migrate annually.

Savanna in many places is being converted into rangeland for cattle and other domesticated animals. The problem is particularly serious in Africa, where human populations are growing most rapidly.

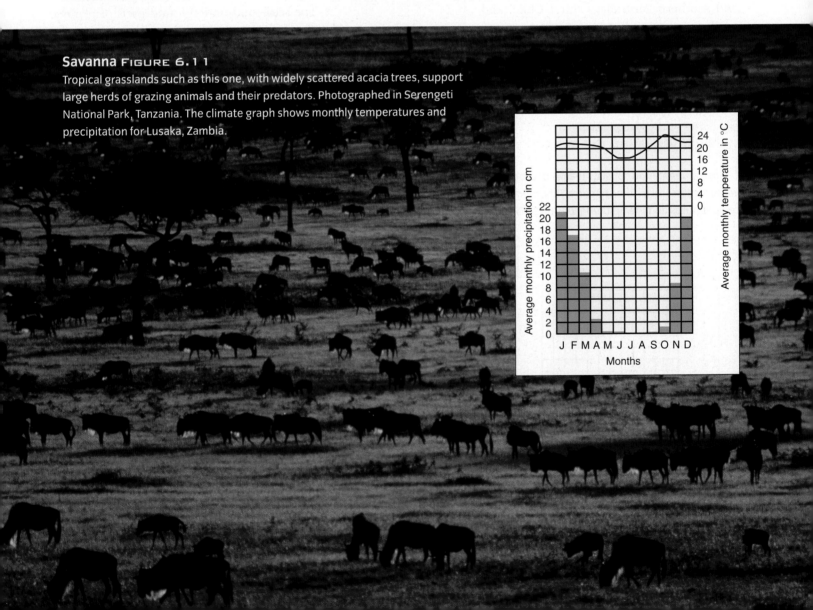

Savanna FIGURE 6.11

Tropical grasslands such as this one, with widely scattered acacia trees, support large herds of grazing animals and their predators. Photographed in Serengeti National Park, Tanzania. The climate graph shows monthly temperatures and precipitation for Lusaka, Zambia.

Desert FIGURE 6.12

This desert landscape includes tough-leaved yuccas and spine-covered prickly pear cacti. Inhabitants of deserts are strikingly adapted to the demands of their environment. Photographed in Big Bend National Park, Texas. The climate graph shows monthly temperatures and precipitation for Reno, Nevada.

DESERT

Desert consists of dry areas found in both temperate (cold deserts) and subtropical or tropical regions (warm deserts). The low water vapour content of the desert atmosphere results in daily temperature extremes of heat and cold, so that a major change in temperature occurs in each 24-hour period. Desert environments vary greatly depending on the amount of precipitation they receive, which is generally less than 25 centimetres per year. As a result of sparse vegetation, desert soil is low in organic material but is often high in mineral content, particularly salts. Plant cover is sparse in deserts, and much of the soil is exposed. Plants in North American deserts include cacti, yuccas, Joshua trees, and sagebrush (**FIGURE 6.12**). Desert plants are adapted to conserve water and as a result tend to have few, small, or no leaves. Cactus leaves are modified into spines, which discourage herbivores. Other desert plants shed their leaves for most of the year, growing only during the brief moist season.

Desert animals are typically small. During the heat of the day, they remain under cover or return to shelter

> **desert** A biome in which the lack of precipitation limits plant growth; deserts are found in both temperate and tropical regions.

periodically, emerging at night to forage or hunt. In addition to desert-adapted insects and arachnids such as tarantulas and scorpions, there are a few desert-adapted amphibians such as frogs and toads, and many reptiles such as the desert tortoise, Gila monster, and Mojave rattlesnake. Desert mammals in North America include rodents such as kangaroo rats, as well as mule deer and jackrabbits. Birds of prey, especially owls, live on the rodents and jackrabbits, and even the scorpions. During the driest months of the year, many desert animals tunnel underground, where they remain inactive.

In many areas, human development encroaches on deserts. Off-road vehicle use damages vegetation, and expansion of farms, cities, and residential areas places severe demands on limited groundwater.

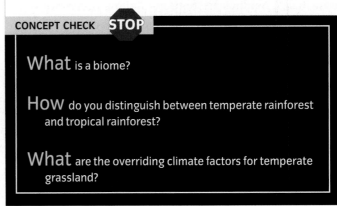

CONCEPT CHECK **STOP**

What is a biome?

How do you distinguish between temperate rainforest and tropical rainforest?

What are the overriding climate factors for temperate grassland?

Classifying Aquatic Ecosystems

The most fundamental division in aquatic ecology is probably between freshwater and saltwater environments. Salinity, which is the concentration of dissolved salts (such as sodium chloride) in a body of water, affects the kinds of organisms present in aquatic ecosystems, as does the amount of dissolved oxygen. Water greatly interferes with the penetration of light, so floating aquatic organisms that photosynthesize must remain near the water's surface, and vegetation anchored to lake floors or streambeds will grow only in relatively shallow water. In addition, low levels of essential nutrient minerals limit the number and distribution of organisms in certain aquatic environments.

Aquatic ecosystems contain three main ecological categories of organisms: free-floating plankton, strongly swimming nekton, and bottom-dwelling benthos. **Plankton** are usually small or microscopic organisms. They tend to drift or swim feebly, so, for the most part, they are carried about at the mercy of currents and waves. **Nekton** are larger, more strongly swimming organisms such as fishes, turtles, and whales. **Benthos** are bottom-dwelling organisms that fix themselves to one spot (sponges and oysters), burrow into the sand (worms and clams), or simply walk about on the bottom (crawfish and aquatic insect larvae).

MAJOR MARINE LIFE ZONES

The immense marine environment is subdivided into several zones (**FIGURE 6.13**):

- The intertidal zone
- The benthic (ocean floor) environment
- The two provinces—neritic and oceanic—of the pelagic (ocean water) environment

The *neritic province* is that part of the pelagic environment from the shore to where the water reaches a depth of 200 metres. It overlies the continental shelf. The *oceanic province* is that part of the pelagic environment where the water depth is greater than 200 metres, beyond the continental shelf.

The intertidal zone: Transition between land and ocean

Although high levels of light, nutrients, and oxygen make the **intertidal zone** a biologically productive habitat, it is a stressful one. On sandy intertidal beaches, inhabitants must contend with a constantly shifting environment that threatens to engulf them and gives them little protection against wave action.

Rocky shores provide fine anchorage for seaweeds and marine animals, but they are exposed to wave action when submerged during high tides and exposed to temperature changes and drying out when in contact with the air during low tides (**FIGURE 6.14**).

A rocky-shore inhabitant generally has some way of sealing in moisture, perhaps by closing its shell (if it has one), and a means of anchoring itself to the rocks. For example, mussels have tough, thread-like anchors secreted by a gland in the foot, and barnacles secrete a tightly bonding glue that hardens under water. Some organisms hide in burrows or under rocks or crevices at low tide. Some small crabs run about the splash line, following it up and down the beach.

intertidal zone The area of shoreline between low and high tides.

benthic environment The ocean floor, which extends from the intertidal zone to the deep ocean trenches.

The benthic environment

Most of the **benthic environment** consists of sediments (mainly sand and mud) where many bottom-dwelling animals, such as worms and clams, burrow. Bacteria are common in

Zonation in the ocean FIGURE 6.13

The intertidal zone, the benthic environment, and the pelagic environment make up the ocean. The pelagic environment consists of the neritic and oceanic provinces. (The slopes of the ocean floor aren't as steep as shown; they are exaggerated here to save space.)

Intertidal zone:

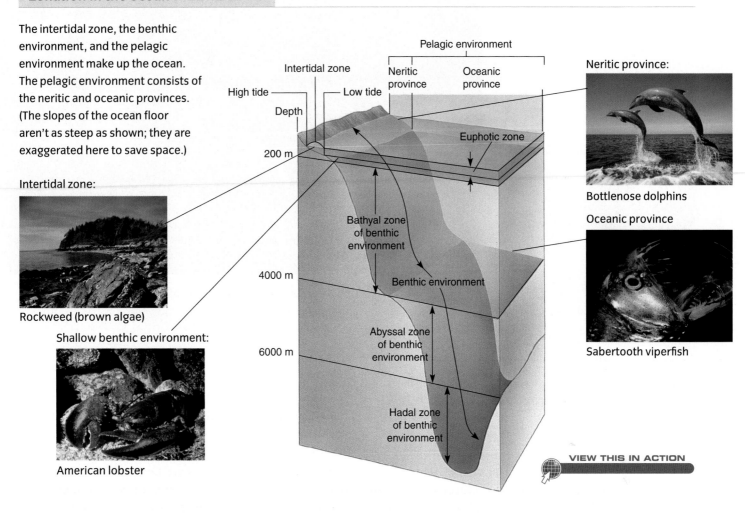

Rockweed (brown algae)

Shallow benthic environment:

American lobster

Pelagic environment

Intertidal zone

Neritic province | Oceanic province

High tide — Low tide

Depth

200 m

Euphotic zone

Bathyal zone of benthic environment

4000 m

Benthic environment

Abyssal zone of benthic environment

6000 m

Hadal zone of benthic environment

Neritic province:

Bottlenose dolphins

Oceanic province

Sabertooth viperfish

VIEW THIS IN ACTION

VIEW THIS IN ACTION

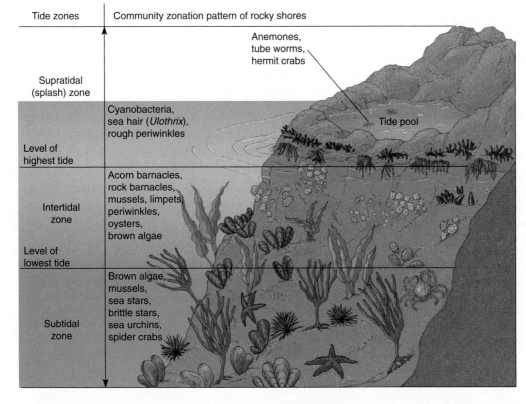

Tide zones	Community zonation pattern of rocky shores
Supratidal (splash) zone	
Level of highest tide	Cyanobacteria, sea hair (*Ulothrix*), rough periwinkles
Intertidal zone	Acorn barnacles, rock barnacles, mussels, limpets, periwinkles, oysters, brown algae
Level of lowest tide	
Subtidal zone	Brown algae, mussels, sea stars, brittle stars, sea urchins, spider crabs

Anemones, tube worms, hermit crabs

Tide pool

Zonation along a rocky shore FIGURE 6.14

Three zones are shown: the supratidal, or "splash" zone, which is never fully submerged; the intertidal zone, which is fully submerged at high tide; and the subtidal zone (part of the benthic environment), which is always submerged. Representative organisms are listed for each of these zones.

Coral reefs FIGURE 6.15

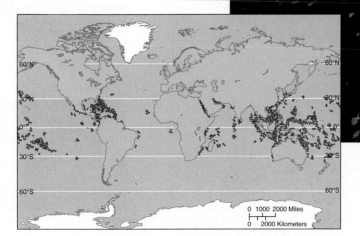

A This map shows the distribution of coral reefs around the world. There are more than 6000 of them worldwide.

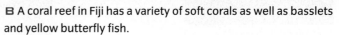

B A coral reef in Fiji has a variety of soft corals as well as basslets and yellow butterfly fish.

marine sediments, found even at depths more than 500 metres below the ocean floor. The deeper parts of the benthic environment are divided into three zones, from shallowest to deepest: the bathyal, abyssal, and hadal zones. The communities in the relatively shallow benthic zone that are particularly productive include coral reefs, sea grass beds, and kelp forests.

Corals are small, soft-bodied animals similar to jellyfish and sea anemones. Corals live in hard cups, or shells, of limestone (calcium carbonate) that they produce using the minerals dissolved in ocean water. When the coral animals die, the tiny cups remain and a new generation of coral animals grows on top of these. Over thousands of generations, a **coral reef** forms from the accumulated layers of limestone.

Coral reefs are found in warm (usually greater than 21° C), shallow sea water (**FIGURE 6.15A**). The living portions of coral reefs grow in shallow waters where light penetrates. The tiny coral animals require light for **zooxanthellae** (symbiotic algae) that live and photosynthesize in their tissues. In addition to obtaining food from the zooxanthellae that live inside them, coral animals capture food at night with stinging tentacles that paralyze plankton and small animals that drift nearby. The waters where coral reefs grow are often poor in nutrients, but other factors are favourable for high productivity, including the presence of zooxanthellae, appropriate temperatures, and year-round sunlight.

Coral reef ecosystems are the most diverse of all marine environments (**FIGURE 6.15B**). They contain hundreds of species of fishes and invertebrates, such as giant clams, snails, sea urchins, sea stars, sponges, flat worms, brittle stars, sea fans, shrimp, and spiny lobsters. The Great Barrier Reef occupies only 0.1 percent of the ocean's surface, but 8 percent of the world's fish species live there. The multitude of relationships and interactions that occur at coral reefs is comparable only to those of the tropical rainforest. As in the rainforest, competition is intense, particularly for light and space to grow.

Coral reefs are ecologically important because they both provide habitat for many kinds of marine organisms and protect coastlines from shoreline erosion. They provide humans with seafood, pharmaceuticals, and recreation and tourism dollars.

Sea grass bed FIGURE 6.16

Turtle grasses form underwater meadows that are ecologically important for shelter and food for many organisms.
Photographed in the Caribbean Sea, the Cayman Islands.

Sea grasses are flowering plants adapted to complete submersion in salty ocean water. They occur only in shallow water (to depths of 10 metres) where they receive enough light to photosynthesize efficiently. Extensive beds of sea grasses occur in quiet temperate, subtropical, and tropical waters. Eelgrass is the most widely distributed sea grass along the coast of North America; the world's largest eelgrass bed is in Izembek Lagoon on the Alaska Peninsula. The most common sea grasses in the Caribbean Sea are manatee grass and turtle grass (**FIGURE 6.16**). Sea grasses have a high primary productivity and are ecologically important: Their roots and rhizomes help stabilize sediments, reducing erosion, and they provide food and habitat for many marine organisms.

In temperate waters, ducks and geese eat sea grasses, and in tropical waters, manatees, green turtles, parrot fish, sturgeon fish, and sea urchins eat them. These herbivores consume only about 5 percent of sea grasses. The remaining 95 percent eventually enters the detritus food web and is decomposed when the sea grasses die. The decomposing bacteria are in turn consumed by animals such as mud shrimp, lugworms, and mullet (a type of fish).

Kelps, known to reach lengths of 60 metres, are the largest and most complex of all algae commonly called seaweeds (**FIGURE 6.17**). Kelps, which are brown algae, are common in cooler temperate marine waters of both the northern and southern hemispheres. They are especially abundant in relatively shallow waters (depths of about 25 metres) along rocky coastlines. Kelps are photosynthetic and are the primary food producers for the kelp "forest" ecosystem. Kelp forests provide habitats for many marine animals, such as tubeworms, sponges, sea cucumbers, clams, crabs, fishes, and sea otters.

Some animals eat kelp fronds, but kelps are mainly consumed in the detritus food web. Bacteria that decompose kelp provide food for sponges, tunicates, worms, clams, and snails. The diversity of life supported by kelp beds almost rivals that of coral reefs.

Kelp forest FIGURE 6.17

These underwater forests are ecologically important because they support many kinds of aquatic organisms. Photographed off the coast of California.

Otters in Alaskan waters

Sea otters play an important role in their environment. They feed on sea urchins, thereby preventing the urchins from eating kelp, which allows kelp forests to thrive. Now scientists have uncovered an alarming decline in sea otter populations in western Alaska's Aleutian Islands—a stunning 90 percent crash since 1990—that in turn poses wide-ranging threats to the coastal ecosystem there. The population of sea urchins in these areas is exploding, and kelp forests are being devastated. Strong evidence identifies orcas, or killer whales, as the culprits. Recently, orcas were observed for the first time preying on sea otters. Orcas generally feed on sea lions, seals, and fishes of all sizes. Sea otters, the smallest marine mammal species, are more like a snack to them than a desirable meal. So why are the orcas now choosing sea otters? Biologists suggest that it is because seal and sea lion populations have collapsed across the north Pacific.

In a scenario that is partly documented and partly speculative, the starting point of this disastrous chain of events is a drop in fish stocks, possibly caused by overfishing or climate change. With their food fish in decline, seal and sea lion populations have suffered and orcas have looked elsewhere for food. Such a change in the orcas' feeding behaviour transforms the food chain of kelp forests, putting orcas rather than otters at the top.

The neritic province: From the shore to 200 metres

Organisms that live in the pelagic environment's **neritic province** are all floaters or swimmers. The upper level of the pelagic environment is the **euphotic zone**, which extends from the surface to a maximum depth of 150 metres in the clearest open ocean water. Sufficient light penetrates the euphotic zone to support photosynthesis.

neritic province
The part of the pelagic environment that overlies the ocean floor from the shoreline to a depth of 200 metres.

Large numbers of phytoplankton (microscopic algae) produce food by photosynthesis and are the base of food webs. Zooplankton, including tiny crustaceans, jellyfish, comb jellies, and the larvae of barnacles, sea urchins, worms, and crabs, feed on phytoplankton. Zooplankton are in turn consumed by plankton-eating nekton (any marine organism that swims freely), such as herring, sardines, squid, baleen whales, and manta rays (**FIGURE 6.18**). These in turn become prey for carnivorous nekton such as sharks, tuna, porpoises, and toothed whales. Nekton are mostly confined to the shallower neritic waters (less than 60 metres deep) near their food.

Neritic province FIGURE 6.18

A manta ray moves slowly through the water, swallowing vast quantities of microscopic plankton as it swims. The wingspan of a mature manta ray can reach about 6 metres. Note the remoras that are hitching a ride.

The oceanic province: Most of the ocean

The **oceanic province** is the largest marine

> **oceanic province**
> The part of the pelagic environment that overlies the ocean floor at depths greater than 200 metres.

environment, representing about 75 percent of the ocean's water. Most of the oceanic province is loosely described as the "deep sea." (The average depth of the ocean is 4000 metres.) All but the surface waters of the oceanic province have cold temperatures, high pressure, and an absence of sunlight. These environmental conditions are uniform throughout the year.

Fishes of the deep waters of the oceanic province are strikingly adapted to darkness and scarcity of food (**FIGURE 6.19**). These drifting or slow swimming animals of the oceanic province often have reduced bone and muscle mass. Many have light-producing organs to locate one another for mating or food capture.

Most organisms of the deep waters of the oceanic province depend on **marine snow**, organic debris that drifts down into their habitat from the upper, lighted regions of the oceanic province. Organisms of this little-known realm are filter feeders, scavengers, and predators. Many are invertebrates, some of which attain great sizes. The giant squid measures up to 18 metres in length, including its tentacles.

FRESHWATER ECOSYSTEMS

Freshwater ecosystems include lakes and ponds (standing-water ecosystems), rivers and streams (flowing-water ecosystems), and marshes and swamps (freshwater wetlands). Specific abiotic conditions and characteristic organisms distinguish each freshwater ecosystem. Although freshwater ecosystems occupy only about 2 percent of Earth's surface, they play an important role in the hydrologic cycle: They help recycle precipitation that flows into the ocean as surface runoff. (See Chapter 5 for a detailed explanation of the hydrologic cycle.) Large bodies of fresh water help moderate daily and seasonal temperature fluctuations on nearby land regions, and freshwater habitats provide homes for many species.

Zonation characterizes **standing-water ecosystems**. A large lake has three zones: the littoral, limnetic, and profundal zones (see What a Scientist Sees on page 184). The *littoral zone* is a shallow-water area along the shore of a lake or pond. Emergent vegetation, such as cattails and bur reeds, as well as several deeper-dwelling aquatic plants and algae, live in the littoral zone. The *limnetic zone* is the open water beyond the littoral zone—that is, away from the shore. The limnetic zone extends down as far as sunlight

> **standing-water ecosystem**
> A body of fresh water surrounded by land and whose water does not flow; a lake or a pond.

Oceanic province FIGURE 6.19

Found at dark depths of 700 to 3000 metres, the spiky fanfin anglerfish attracts prey with its glowing lure. Its fin rays allow it to sense movement in the dark water. Photographed in Monterey Bay Canyon, California.

Zonation in a large lake

The zonation in this lake is not apparent to a visitor.

(Inset, below) A lake is a standing-water ecosystem surrounded by land. The littoral zone is the shallow-water area around the lake's edge. The limnetic zone is the open, sunlit water away from the shore. The profundal zone, under the limnetic zone, is below where light penetrates.

Limnetic zone

Littoral zone

Profundal zone

penetrates to permit photosynthesis. The main organisms of the limnetic zone are microscopic plankton. Larger fishes also spend most of their time in the limnetic zone, although they may visit the littoral zone to feed and reproduce. The deepest zone, the *profundal zone,* is beneath the limnetic zone of a large lake; smaller lakes and ponds typically lack a profundal zone. Because light does not penetrate effectively to this depth, plants and algae do not live there. Detritus drifts into the profundal zone from the littoral and limnetic zones; bacteria decompose this detritus. This marked zonation is accentuated by **thermal stratification**, in which the temperature changes sharply with depth.

Flowing-water ecosystems are highly variable. The surrounding environment changes greatly between a river's source and its mouth (**FIGURE 6.20**). Certain parts of the stream's course are shaded by forest, while other parts are exposed to direct sunlight. Groundwater may well up through sediments on the bottom in one particular area, making the water temperature cooler in summer or warmer in winter than adjacent parts of the stream or river. The kinds of organisms found in flowing water vary greatly from one stream to another, depending primarily on the strength of the current. In streams with fast currents, some inhabitants have adaptations such as suckers, with which they attach themselves to rocks to prevent being swept away. Some stream inhabitants have flattened bodies to slip under or between rocks. Other inhabitants such as fish are streamlined and muscular enough to swim in the current.

> **flowing-water ecosystem** A freshwater ecosystem such as a river or stream in which water flows in a current.

Features of a typical river FIGURE 6.20

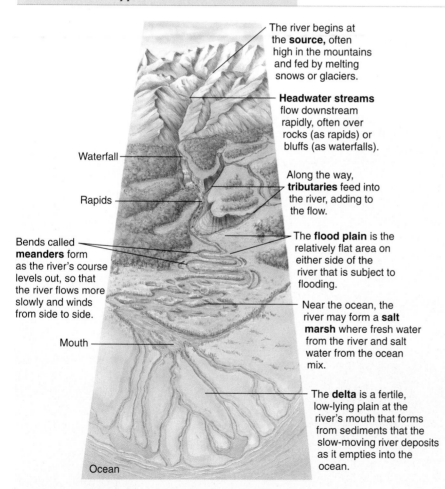

The river begins at the **source,** often high in the mountains and fed by melting snows or glaciers.

Headwater streams flow downstream rapidly, often over rocks (as rapids) or bluffs (as waterfalls).

Waterfall

Rapids

Along the way, **tributaries** feed into the river, adding to the flow.

The **flood plain** is the relatively flat area on either side of the river that is subject to flooding.

Bends called **meanders** form as the river's course levels out, so that the river flows more slowly and winds from side to side.

Near the ocean, the river may form a **salt marsh** where fresh water from the river and salt water from the ocean mix.

Mouth

The **delta** is a fertile, low-lying plain at the river's mouth that forms from sediments that the slow-moving river deposits as it empties into the ocean.

Ocean

A A river flows from its source to the ocean.

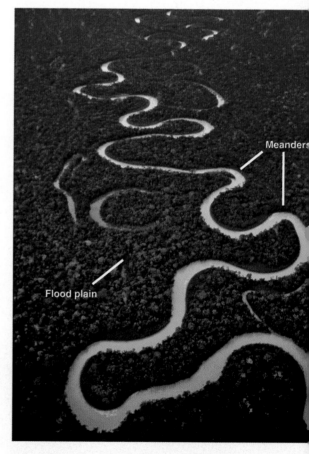

B Aerial view of meanders in the Tambopata River, Peru.

Meanders

Flood plain

FRESHWATER WETLANDS

Freshwater wetlands include **marshes**, dominated by grass-like plants, and **swamps**, dominated by woody trees or shrubs (**FIGURE 6.21**). Wetland soils are waterlogged for variable periods and are therefore anaerobic (without oxygen). They are rich in accumulated organic materials, partly because anaerobic conditions discourage decomposition.

Wetlands provide excellent wildlife habitat for migratory waterfowl and other bird species, as well as for beaver, otters, muskrats, and game fish. In addition to providing unique wildlife habitat, wetlands serve other important environmental functions, known as **ecosystem services**. When rivers flood their banks, wetlands are capable of holding or even absorbing the excess water, thereby helping to control flooding. The floodwater then drains slowly back into the rivers, providing a steady flow of water throughout the year. Wetlands also serve as groundwater recharging areas. One of their most important roles is to trap and hold pollutants in the flooded soil, thereby cleansing and purifying the water. Community developers are now recognizing the important services associated with wetlands and constructing modified engineered wetlands in residential developments as a way to reduce urban runoff of pollutants into receiving rivers and tributaries.

Freshwater wetland FIGURE 6.21

Freshwater wetlands and potholes once covered the Canadian prairies. As a result of land use changes, wetlands are under threat of human development.

BRACKISH ECOSYSTEMS: ESTUARIES

estuary A coastal body of water, partly surrounded by land, with access to the open ocean and a large supply of fresh water from a river.

Where the ocean meets the land, there may be one of several kinds of ecosystems: a rocky shore, a sandy beach, an intertidal mud flat, or a tidal **estuary**. Water levels in an estuary rise and fall with the tides; salinity fluctuates with tidal cycles, the time of year, and precipitation. Salinity also changes gradually within the estuary, from fresh water at the river entrance, to *brackish* (somewhat salty) water, to salty ocean water at the mouth of the estuary. Because estuaries undergo significant daily, seasonal, and annual variations in physical factors such as temperature, salinity, and depth of light penetration, estuarine organisms must have a high tolerance for changing conditions.

Temperate estuaries usually feature **salt marshes**, shallow wetlands in which salt-tolerant grasses grow (**Figure 6.22A**). Salt marshes perform many ecosystem services, including providing biological habitats, trapping sediment and pollution, supplying groundwater, and buffering storms by absorbing their energy, which prevents flood damage elsewhere.

Mangrove forests, the tropical equivalent of salt marshes, cover perhaps 70 percent of tropical coastlines (**Figure 6.22B**). Like salt marshes, mangrove forests provide valuable ecosystem services. Their interlacing roots are breeding grounds and nurseries for several commercially important fishes and shellfish, such as

A A restored salt marsh at the Chignecto National Wildlife Area in Nova Scotia.

Estuaries Figure 6.22

B A mangrove forest in the Caroline Islands, Micronesia. Mangrove roots grow into deeper water as well as into mudflats that are exposed at low tide. Many animals live among the complex root systems of mangrove forests.

CAROLINE IS.

Global Locator

crabs, shrimp, mullet, and spotted sea trout. Mangrove branches are nesting sites for many species of birds, such as pelicans, herons, egrets, and roseate spoonbills. Mangrove roots stabilize the submerged soil, thereby preventing coastal erosion and providing a barrier against the ocean during storms.

Both salt marsh and mangrove forest ecosystems have experienced significant losses due to coastal development. Salt marshes have been polluted and turned into dumping grounds; mangrove forests have been logged and used as aquaculture sites. However, there has been increasing effort to restore and conserve coastal salt marshes in Atlantic Canada. The Chignecto National Wildlife Area is designed to protect the unique ecological and aesthetic features of the area, and to maintain habitat diversity.

CONCEPT CHECK STOP

What are the four main life zones in the oceans? What are some of the characteristic features of biota living in these zones?

Which environmental factors shape flowing-water ecosystems? Which shape standing-water ecosystems?

What is a freshwater wetland? How does a freshwater wetland differ from an estuary?

What are some ecosystem services of salt marshes and mangrove swamps?

Population Responses to Changing Ecosystem Conditions: Evolution

LEARNING OBJECTIVES

Define *evolution*.

Explain how evolution by natural selection consists of four observations of the natural world.

Describe various types of evidence that support evolution.

cientists think all of Earth's remarkable variety of organisms descended from earlier species by a process known as **evolution**. The concept of evolution dates back to the time of Aristotle, but **Charles Darwin** (1809–1882), a 19th-century naturalist, proposed the mechanism of evolution that today's scientific community still accepts. As you will see, the environment plays a crucial role in Darwin's theory of evolution.

evolution
The cumulative genetic changes in populations that occur during successive generations.

It occurred to Darwin that in a population, inherited traits favourable to survival in a given environment tended to be preserved over successive generations, whereas unfavourable traits were eliminated. The result is *adaptation*, an evolutionary modification that improves a species' chance of survival and reproductive success in a given environment. Eventually the accumulation of adaptive modifications might result in a new species.

Darwin proposed the theory of evolution by natural selection in his monumental book *The Origin of Species by Means of Natural Selection*, which was published in 1859.

Since that time, scientists have accumulated an enormous body of observations and experiments that support Darwin's theory. Although biologists still do not agree completely on some aspects of the evolutionary process, the concept that evolution by natural selection has taken place and is still occurring is now well documented.

NATURAL SELECTION

Evolution occurs through the process of **natural selection**. As favourable traits increase in frequency in successive generations, and as unfavourable traits decrease or disappear, the collection of characteristics of a given population changes. Natural selection is the process by which successful traits are passed on to the next generation and unsuccessful ones are weeded out. It consists of four observations about the natural world:

1. *Overproduction*. Each species produces more offspring than will survive to maturity. Natural populations have the reproductive potential to increase their numbers continuously over time (**FIGURE 6.23**).

2. *Variation*. The individuals in a population exhibit variation. Each individual has a unique combination of traits, such as size, colour, and ability to tolerate harsh environments. Some traits improve the chances of an individual's survival and reproductive success, whereas others do not. It is important to remember that the variation necessary for evolution by natural selection must be inherited so that it can be passed to offspring.

3. *Limits on population growth, or a struggle for existence.* There is only so much food, water, light, growing space, and so on available to a population, and organisms compete with one another for the limited resources available to them. Because there are more individuals than the environment can support, not all of an organism's offspring will survive to reproductive age. Other limits on population growth include predators and diseases.

4. *Differential reproductive success*. Individuals that possess the most favourable combination of characteristics (those that make individuals better adapted to their environment) are more likely than others to survive, reproduce, and pass their traits to the next generation. Sexual reproduction is the key to natural selection: The best-adapted individuals are those that reproduce most successfully, whereas less-fit individuals die prematurely or produce fewer or inferior offspring. Over time, enough changes may accumulate in geographically separated populations (often with slightly different environments) to produce new species (**FIGURE 6.24** on page 190).

One premise on which Darwin based his theory of evolution by natural selection is that individuals transmit traits to the next generation. However, Darwin could not explain *how* this occurs or *why* individuals within a population vary. Beginning in the 1930s and 1940s, biologists combined the principles of genetics with Darwin's theory of natural selection. The resulting unified explanation of evolution is known as the **modern synthesis** (where *synthesis* refers to combining parts of previous theories). The modern synthesis explains Darwin's observation of variation among offspring in terms of **mutation**, or changes in DNA. Mutations provided the genetic variability on which natural selection acts during evolution.

A vast body of evidence supports evolution, most of which is beyond the scope of this text. This evidence includes observations from the fossil record, comparative anatomy, biogeography (the study of the geographic locations of organisms), and molecular biology (**FIGURE 6.25** on page 191). In addition, evolutionary hypotheses are tested experimentally.

> **natural selection**
> The tendency of better-adapted individuals—those with a combination of genetic traits best suited to environmental conditions—to survive and reproduce, increasing their proportion in the population.

Overproduction FIGURE 6.23

If each breeding pair of elephants were to produce six offspring that lived and reproduced, in 750 years a single pair of elephants would have given rise to more than 15 million elephants! Yet elephants have not overrun the planet. Photographed in Botswana.

Darwin's finches FIGURE 6.24

Charles Darwin was a ship's naturalist on a five-year voyage around the world. During an extended stay in the Galápagos Islands off the coast of Ecuador, he studied the plants and animals of each island, including 14 species of finches.

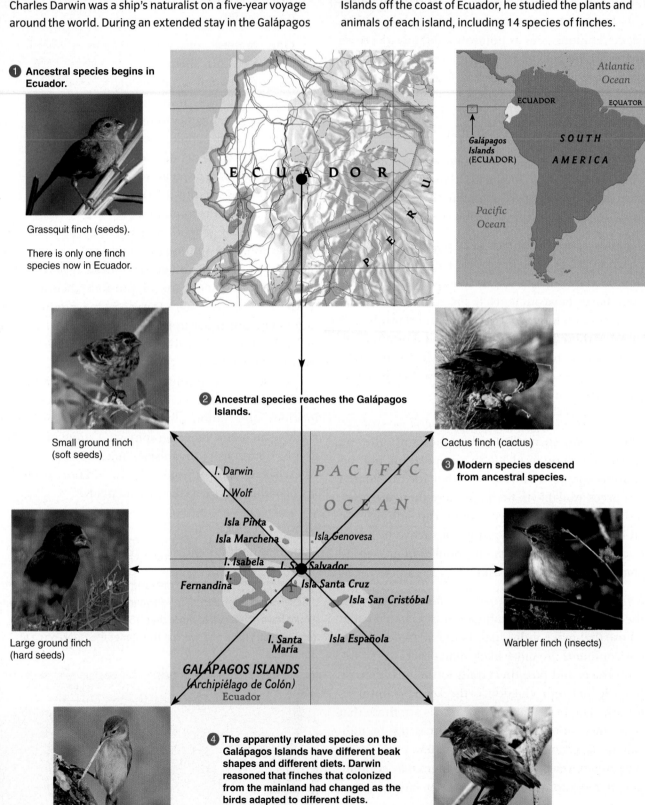

1 Ancestral species begins in Ecuador.

Grassquit finch (seeds).

There is only one finch species now in Ecuador.

2 Ancestral species reaches the Galápagos Islands.

Small ground finch (soft seeds)

Cactus finch (cactus)

3 Modern species descend from ancestral species.

Large ground finch (hard seeds)

Warbler finch (insects)

4 The apparently related species on the Galápagos Islands have different beak shapes and different diets. Darwin reasoned that finches that colonized from the mainland had changed as the birds adapted to different diets.

Woodpecker finch (insects)

Medium ground finch (moderate seeds)

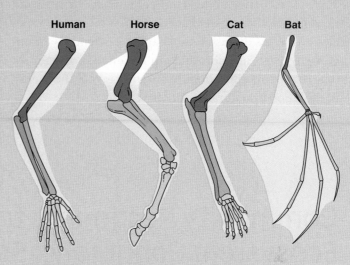

B Comparative anatomy Similarities among organisms demonstrate how they are related. These similarities among four vertebrate limbs illustrate that, while proportions of bones have changed in relation to each organism's way of life, the forelimbs have the same basic bone structure.

A The fossil record Fossils deposited in rock layers, which can be dated, show how organisms evolved over time. These fish fossils from Liaoning Province, China, date from 120 million years ago.

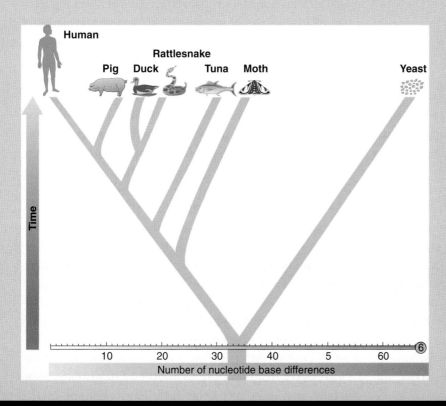

C Molecular biology The organisms pictured here all share a particular enzyme, but in the course of evolution, mutations have resulted in changes in the gene that codes for that enzyme. This diagram shows the nucleotide base differences in this gene among humans and other organisms. Note that organisms thought to be more closely related to humans have fewer differences than organisms that are more distantly related to humans.

On the basis of these kinds of evidence, virtually all biologists accept the principles of evolution by natural selection, although they don't agree on all the details. They try to better understand certain aspects of evolution, such as the role of chance and how quickly new species evolve. As discussed in Chapter 1, science is an ongoing process, and information obtained in the future may require modifications to certain parts of the theory of evolution by natural selection.

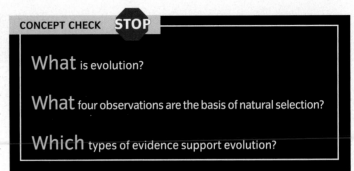

CONCEPT CHECK **STOP**

What is evolution?

What four observations are the basis of natural selection?

Which types of evidence support evolution?

Community Responses to Changing Ecosystem Conditions: Succession

LEARNING OBJECTIVES

Define *ecological succession.*

Distinguish between primary and secondary succession.

A community of organisms does not spring into existence full-blown. By means of **ecological succession**, a given community develops gradually through a sequence of species. Certain organisms colonize an area; over time others replace them, and eventually the replacements are themselves replaced by still other species.

 ecological succession The process of community development over time, which involves species in one stage being replaced by different species.

The actual mechanisms that underlie succession are not clear. In some cases, it may be that a resident species modifies the environment in some way, thereby making it more suitable for a later species to colonize. It is also possible that prior residents lived there in the first place because there is little competition from other species. Later, as more invasive species arrive, the original species are displaced.

Ecologists initially thought that succession inevitably led to a stable and persistent community, known as a *climax community*, such as a forest. But more recently, this traditional view has fallen out of favour. The apparent stability of a "climax" forest is probably the result of how long trees live relative to the human life span. It is now recognized that mature climax communities are not in a state of stable equilibrium but rather in a state of continual disturbance. Over time, a mature community changes in species composition and in the relative abundance of each species, despite the fact that it retains an overall uniform appearance.

Succession is usually described in terms of the changes in the plant species growing in a given area, although each stage of the succession may also have its own kinds of animals and other organisms. Ecological succession is measured on the scale of tens, hundreds, or thousands of years, not the millions of years involved in the evolutionary time scale.

Primary succession is the change in species composition over time in a previously uninhabited environment (**FIGURE 6.26**). No soil exists when primary succession begins. Bare rock surfaces, such as recently formed volcanic lava and rock scraped clean by glaciers, are examples of sites where primary succession may take place. Details vary from one site to another, but on bare rock, lichens are often the most important element in the

Primary succession on glacial moraine FIGURE 6.26

A After a glacier's retreat, lichens initially colonize the barren landscape, followed by mosses and small shrubs.

During the past 200 years, glaciers have retreated throughout Canada. Although these photos were not taken in the same area, they show some of the stages of primary succession on glacial moraine (rocks, gravel, and sand that a glacier deposits).

Process Diagram

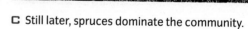

B At a later time, dwarf trees and shrubs colonize the area.

C Still later, spruces dominate the community.

pioneer community, which is the initial community that develops during primary succession. Lichens secrete acids that help break the rock apart, beginning the process of soil formation. Over time, mosses and drought-resistant ferns may replace the lichen community, followed in turn by tough grasses and herbs. Once soil accumulates, low shrubs may replace the grasses and herbs; over time, forest trees in several distinct stages would replace the shrubs. Primary succession on bare rock from a pioneer community to a forest community often occurs in this sequence: lichens → mosses → grasses → shrubs → trees.

Secondary succession is the change in species composition that takes place after some disturbance destroys the existing vegetation; soil is already present (**FIGURE 6.27** on page 194). Abandoned farmland or an open area caused by a forest fire are common examples of sites where secondary succession occurs. Biologists have studied secondary succession on abandoned farmland extensively. Although it takes more than 100 years for secondary succession to occur at a single site, a single researcher can study old-field succession in its entirety by observing different sites undergoing succession in

Secondary succession on an abandoned field FIGURE 6.27

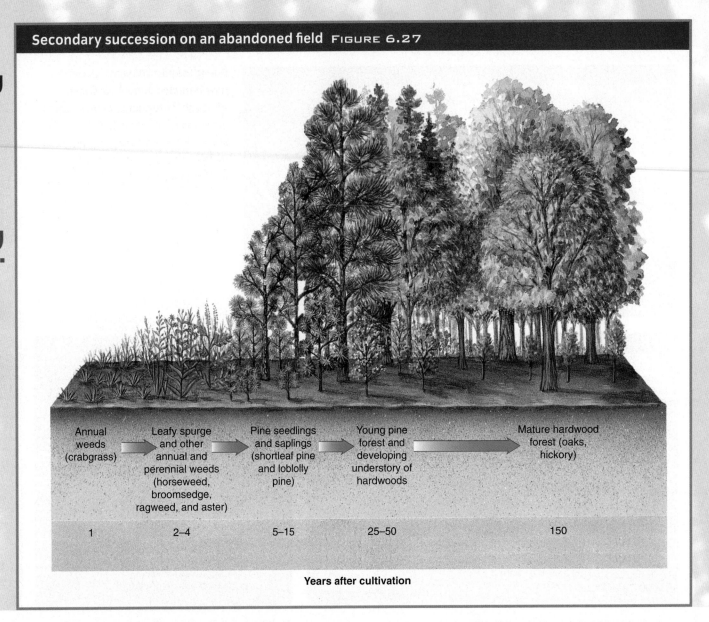

Annual weeds (crabgrass)	Leafy spurge and other annual and perennial weeds (horseweed, broomsedge, ragweed, and aster)	Pine seedlings and saplings (shortleaf pine and loblolly pine)	Young pine forest and developing understory of hardwoods	Mature hardwood forest (oaks, hickory)
1	2–4	5–15	25–50	150

Years after cultivation

the same general area. (The biologist may examine regional tax records to determine when each field was abandoned.) Secondary succession on abandoned farmland might proceed in this sequence: crabgrass → leafy spurge and other weeds → pine trees → hardwood trees.

Intermediate disturbance hypothesis

Disturbances, which occur in most habitats, can determine which species are able to live in a habitat and how common each species is within a habitat. When a habitat experiences a high level of disturbance, such as a forest experiencing a forest fire or a coral reef experiencing damage from a severe storm, the species that recover quickly will colonize the disturbed area and rapidly take over.

The *intermediate disturbance hypothesis* is a theory that attempts to predict changes in species abundance and diversity at changing disturbance levels. The intermediate disturbance hypothesis suggests that intermediate levels of disturbance will lead to the greatest level of species diversity. In a habitat that is rarely disturbed, the most competitive, dominant species will take over. In a habitat that frequently experiences a high level of disturbance, only the species that recover quickly will survive. As a result, an intermediate level of disturbance leads to a higher level of species diversity. Given this, ecosystem managers may begin to consider fire and other disturbances that are a part of the natural processes as helpful in rejuvenating the land and yielding greater overall biodiversity.

The capacity of an ecosystem to cope with disturbances without shifting into a qualitatively different state is known as *ecosystem resilience*. A resilient ecosystem has the ability to withstand disturbance and rebuild itself after a disturbance. Ecosystems with little resilience are vulnerable to the effects of a disturbance and may take a long time to recover, if at all. The new state of the ecosystem unable to recover may not be biologically or economically sound and the effects on the ecosystem may be irreversible.

CONCEPT CHECK STOP

What is ecological succession?

How does primary succession differ from secondary succession?

ECO CANADA | **CAREER FOCUS**

ENVIRONMENTAL MANAGER

Environmental managers work in both the private and public sectors to ensure that proposed projects have as little environmental impact as possible and meet all applicable legislative requirements.

Imagine you are standing at the front of a large boardroom introducing yourself to 15 environmental experts. You are an environmental manager and this is your new team. You have been hired by one of the country's largest oil and gas companies to create an environmental management strategy for a new refinery it proposes to build next year. Your job is to ensure that the proposed refinery complies with all provincial and federal environmental regulations, both in construction and operation. Your management strategy will act as a step-by-step guide for how compliance will be achieved and maintained.

As an environmental manager, you are not only an expert on environmental policy and legislation, but also a skilled manager of staff. You begin this project with research, coordinating each team member and assigning some to examine provincial and federal legislation, others to study similar refineries and their strategies for complying with regulations, and another group to prepare an environmental impact assessment. You also consult with project stakeholders, for example, area residents, government officials, and the company's executives. All this research and consultation will give you an idea of what must be done to ensure compliance, as well as identify any constraints that might affect your management strategy, for example, time or budget constraints. With this information, you can determine the scope of work required and the kinds of resources needed, including how many people and with what expertise.

From there, you put together your environmental strategy action plan, which will be presented to all stakeholders and will detail what the company wants to achieve, how it will do it, how long it will take, and how performance will be measured. Once all concerned parties approve the action plan, you can begin implementation. When fully realized, your environmental management strategy will ensure that the construction and operation of the refinery satisfy all environmental requirements and comply with regulations.

Environmental managers work in both the public and private sectors and are responsible for managing projects to lessen environmental impacts and ensure that all applicable legislative requirements are fulfilled. They are also involved in activities such as environmental awareness projects, sustainable development, fundraising, and public consultation programs. Often responsible for managing the work of others, environmental managers may also be involved in training personnel on environmental issues. To be a good environmental manager, you need a broad understanding of environmental issues combined with expertise and a lot of experience in project development and management.

For more information or to look up other environmental careers go to www.eco.ca.

BOREAL FOREST FIRES

The northern boreal ecoregion is the largest biome in the world, covering 12 million square kilometres and accounting for close to one-third of the planet's total forest area. In North America, the boreal ecoregion extends from Alaska to Newfoundland and from the edges of the Great Lakes in the south to the tundra line of the north. The boreal forest is characterized by spruces, firs, conifers and some deciduous tree species.

Fire is an essential disturbance factor in the boreal forest. The boreal ecoregion has evolved over the last 10,000 to 15,000 years with fire maintaining the forest's health and diversity. Fire destroys old, diseased trees and the associated pests, and re-starts plant succession, creating a mosaic of vegetation height and types that are necessary to support various wildlife species. Some of the boreal tree species, such as aspen and jack pine, require fire to stimulate their reproductive cycles. The fire takes the warm air up into the tops of the trees, which dries the cones and makes them split, allowing the seeds to be dispersed. Forest fires also clear the undergrowth of the forest floor, which creates an ideal environment for the seeds to germinate and grow.

In Canada, the number of fires and the area burned by fires can vary substantially from year to year. On average, there are 8000 reported wildfires in Canada each year that burn roughly 2.5 million hectares of land. Usually, only 3 percent of the fires in Canada reach a size larger than 200 hectares, such as the massive fire southwest of Princeton, British Columbia, in 2001, which burned out 2900 hectares (see photo A).

B Prescribed burning programs are widely used to reintroduce fire disturbance to certain areas and to mitigate the effects of wildfires.

A The 2001 forest fire near Princeton, BC, started in a travelling motor home. The fire took months to fully extinguish at a total cost of about $6 million.

Fire is often suppressed, however, since forest fires threaten people's safety and property and damage valuable timber. In many areas of Canada, *prescribed burning* is used to accomplish the goals of forest disturbance to increase forest diversity and health and reduce the amount of dry fuel in forested areas while lessening human fears about fire (see photo B). The Ecosystem Management by Emulating Natural Disturbance (EMEND) study illustrates efforts by Natural Resources Canada to investigate the effects of harvest and fire disturbances on the boreal forest. Similarly, the Boreal Fire Effects Model (BORFIRE) was developed by Natural Resources Canada to study the effects of future fire regimes on the boreal forest, which are, because of climate change, anticipated to be of higher intensity and severity, and the forest fire season to be of a longer duration. BORFIRE investigates fire disturbance impact on tree communities, specifically on tree mortality, tree recruitment, tree growth, biomass decomposition, and biomass consumption by fire.

For more information on BORFIRE and EMEND, visit Natural Resources Canada at www.nrcan-rncan.gc.ca.

SUMMARY

1 Classifying Terrestrial Ecosystems: Biomes

1. A **biome** is a large, relatively distinct terrestrial region with characteristic climate, soil, plants, and animals, regardless of where it occurs; a biome encompasses many interacting ecosystems. Near the poles, temperature is generally the overriding climate factor in determining biome distribution, whereas in temperate and tropical regions, precipitation is more significant.

2. **Tundra** is the treeless biome in the far north that consists of boggy plains covered by lichens and small plants such as mosses; it has harsh, very cold winters and extremely short summers. **Boreal forest** is a region of coniferous forest in the northern hemisphere, located just south of the tundra. **Temperate rainforest** is a coniferous biome with cool weather, dense fog, and high precipitation. **Temperate deciduous forest** is a forest biome that occurs in temperate areas where annual precipitation ranges from about 75 to 126 centimetres. **Tropical rainforest** is a lush, species-rich forest biome that occurs where the climate is warm and moist throughout the year. **Chaparral** is a biome with mild, moist winters and hot, dry summers; vegetation is typically small-leafed evergreen shrubs and small trees. **Temperate grassland** is grassland with hot summers, cold winters, and less rainfall than is found in the temperate deciduous forest biome. **Savanna** is tropical grassland with widely scattered trees or clumps of trees. **Desert** is a biome in which the lack of precipitation limits plant growth; deserts are found in both temperate and tropical regions.

2 Classifying Aquatic Ecosystems

1. The vast ocean is subdivided into major life zones. The biologically productive **intertidal zone** is the area of shoreline between low and high tides. The **benthic environment** is the ocean floor, which extends from the intertidal zone to the deep ocean trenches. Most of the benthic environment consists of sediments where many animals burrow. Common benthic habitats include sea grass beds, kelp forests, and coral reefs. The pelagic environment is divided into two provinces. The **neritic province** is the part of the pelagic environment from the shore to where the water reaches a depth of 200 metres. Organisms that live in the neritic province are all floaters or swimmers. The **oceanic province**, "the deep sea," is the part of the pelagic environment where the water depth is greater than 200 metres. The oceanic province is the largest marine environment, comprising about 75 percent of the ocean's water.

2. In aquatic ecosystems, important environmental factors include salinity, amount of dissolved oxygen, and availability of light for photosynthesis.

3. Freshwater ecosystems include standing-water, flowing-water, and freshwater wetlands. A **standing-water ecosystem** is a body of fresh water surrounded by land and whose water does not flow, such as a lake or pond. A **flowing-water ecosystem** is a freshwater ecosystem such as a river or stream in which the water flows in a current. **Freshwater wetlands** are marshes and swamps—lands that shallow fresh water covers for at least part of the year; wetlands have a characteristic soil and water-tolerant vegetation. An **estuary** is a coastal body of water, partly surrounded by land, with access to the open ocean and a large supply of fresh water from a river. Water in an estuary is brackish rather than truly fresh. Temperate estuaries usually contain salt marshes, whereas tropical estuaries are lined with mangrove forests.

3 Population Responses to Changing Ecosystem Conditions: Evolution

1. **Evolution** is the cumulative genetic changes in populations that occur during successive generations.

2. **Natural selection** is the tendency of better-adapted individuals—those with a combination of genetic traits best suited to environmental conditions—to survive and reproduce, increasing their proportion in the population. Natural selection is based on four observations established by Charles Darwin: (1) Each species produces more offspring than will survive to maturity. (2) The individuals in a population exhibit inheritable variation in their traits. (3) Organisms compete with one another for the resources needed to survive. (4) Individuals with the most favourable combination of traits are most likely to survive and reproduce, passing their genetic traits to the next generation.

3. Scientific evidence supporting evolution comes from the fossil record, comparative anatomy, biogeography, and molecular biology.

4 Community Responses to Changing Ecosystem Conditions: Succession

1. **Ecological succession** is the process of community development over time, which involves species in one stage being replaced by different species.

2. Primary succession is the change in species composition over time in an environment that was not previously inhabited by organisms; examples include bare rock surfaces, such as recently formed volcanic lava and rock scraped clean by glaciers. Secondary succession is the change in species composition that takes place after some disturbance destroys the existing vegetation; soil is already present. Examples include abandoned farmland and open areas caused by forest fires.

KEY TERMS

CRITICAL AND CREATIVE THINKING QUESTIONS

1. What two climate factors are most important in determining an area's characteristic biome?

2. What climate and soil factors produce each of the major terrestrial biomes?

3. In which biome do you live? Where would you place your biome in the figure below? How would that compare with your placement of the biome in northern Siberia or the biome dominating northern Africa and Saudi Arabia?

4. If your biome does not match the description given in this book, how do you explain the discrepancy?

5. Which biomes are best suited for agriculture? Explain why each of the biomes you did not specify is less suitable for agriculture.

6. What environmental factors are most important in determining the kinds of organisms found in aquatic environments?

7. Distinguish between freshwater wetlands and estuaries, and between flowing-water and standing-water ecosystems.

8. Identify which of the ocean life zones would be home to each of the following organisms: giant squid, kelp, tuna, and mussels. Explain your answers.

9. During the mating season, male giraffes slam their necks together in fighting bouts to determine which male is stronger and can therefore mate with females. Explain how the long necks of giraffes may have evolved, using Darwin's theory of evolution by natural selection.

10. Describe the stages in old-field succession.

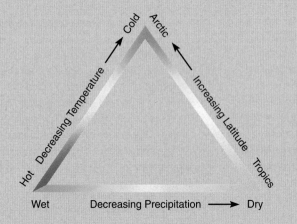

11. Although most salamanders have four legs, the aquatic salamander shown below resembles an eel. It lacks hind limbs and has very tiny forelimbs. Propose a hypothesis to explain how limbless salamanders evolved according to Darwin's theory of natural selection.

12. How could you test the hypothesis you proposed in question 11?

13. Which biome discussed in this chapter is depicted by the information in the graph below? Explain your answer.

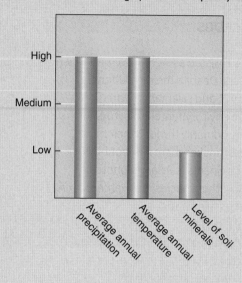

What is happening in this picture ?

■ This picture shows expensive homes built in the chaparral of the Santa Monica Mountains. Based on what you have learned in this chapter, what environmental problem might threaten these homes?

■ Sometimes people have removed the chaparral vegetation to prevent fires from damaging their homes. Where that has occurred, the roots no longer hold the soil in place. What could happen when the winter rains come?

Conservation of Biological Diversity

DISAPPEARING FROGS

Amphibians have existed as a group for more than 350 million years. Despite their evolutionary resilience, frogs, toads, and salamanders are remarkably sensitive environmental indicators. As adults, frogs breathe primarily through their moist, permeable skin, which makes them susceptible to environmental contaminants (see larger photo). Amphibians are bellwether species. *Bellwether species* provide early warnings of environmental damage that may affect other species.

Since the 1970s many of the world's frog populations have dwindled or disappeared. Initially, it was believed that environmental and habitat changes were solely responsible; however, we now know that the emergence of a deadly fungal disease that affects the frog's skin, called chytridiomycosis, has had devastating effects on frog populations. Nearly one-third of the world's amphibian species are threatened with extinction and 50 percent are in serious decline. We are witnessing the largest mass extinction since the disappearance of dinosaurs. It's hard to believe that this could happen within one human lifetime!

Abnormally large numbers of amphibian deformities have also been reported (see inset). Frog deformities have now been found on four continents. Scientists have demonstrated that several factors produce amphibian deformities, including parasites, which affect development in frog embryos; parasite infestations; and multiple environmental stressors, such as habitat loss, disease, and air and water pollution.

Amphibian Ark, a program of the Amphibian Conservation Action Plan, selects specific amphibian species to be raised and bred in captivity in zoos and aquariums until they can be safely reintroduced into their wild environments. The program hopes to save hundreds of amphibian species that would otherwise become extinct. Captive breeding of amphibians is difficult, however, and researchers are running out of the space they need in their race against time.

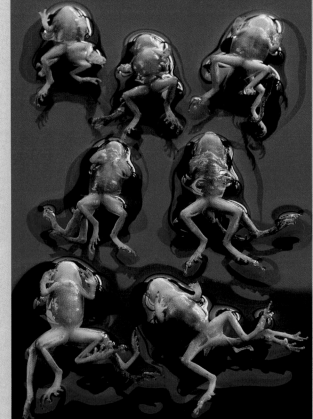

200

Defining and Measuring Biological Diversity

The variation among organisms is referred to as **biological diversity** or **biodiversity**. Although this description may appear to be simplistic, there have been numerous definitions of biological diversity put forward over the years, all of which are related but vary in the specific features they emphasize.

In 1980, biologists Elliott Norse and Roger McManus presented one of the first definitions of biological diversity as being "the amount of genetic variability within species and the number of species in a community of organisms." Since then many researchers and authorities have tried to fine-tune the definition, either for effective ecosystem management reasons, or for legislative reasons. In 1990, the U.S. Biodiversity Act proposed the following definition:

> *Biological diversity means the full range of variety and variability within and among living organisms and the ecological complexes in which they occur, and encompasses ecosystem or community diversity, species diversity, and genetic diversity.*

In the 1990s, the topic of biodiversity received a lot of attention. Many committees and forums were put together to try to piece together strategies to maintain the biological diversity of our planet. In 1991, the

biological diversity The variety of life in all forms, at all levels, and in all combinations in a defined area. It can be considered at the level of genes, species, populations, communities, ecosystems, and landscapes. The area is defined by humans, and can vary in size from something as large as the Earth to a country, province, region, municipality, or a puddle in your backyard.

Keystone Center in Colorado issued a report on biodiversity on federal lands. In it, they included the following definition:

> *In the simplest of terms, biological diversity is the variety of life and its processes; and it includes the variety of living organisms, the genetic differences among them, and the communities and ecosystems in which they occur.*

In 1992, the World Resources Institute, the World Conservation Union (now known as the International Union for Conservation of Nature) and the United Nations Environment Programme developed a global diversity strategy in which they put forward a more specific definition:

> *Biodiversity is the totality of genes, species, and ecosystems in a region . . . Biodiversity can be divided into three hierarchical categories—genes, species, and ecosystems—that describe quite different aspects of living systems and that scientists measure in different ways.*

In April 2005, Environment Canada updated its working definition of biodiversity as describing "the variety of plants, animals, and other lifeforms on Earth. It includes diversity within species, between species, and of ecosystems."

Despite how important biodiversity is for conservation efforts, it is clear that nailing down a definitive definition of biological diversity is difficult. There are obviously some similarities among all of these definitions, so why the challenge of finding just the right description? Reed Noss, a scientist at the University of Central Florida, summarized the problems with defining biological diversity and what needs to be taken away from all of these variations:

> *Biodiversity is not simply the number of genes, species, ecosystems, or any other group of things in a defined area . . . A definition of biodiversity that is altogether simple, comprehensive, and fully operational (i.e., responsive to*

real-life management and regulatory questions) is unlikely to be found. More useful than a definition, perhaps, would be a characterization of biodiversity that identifies the major components at several levels of organization.

What all of these different definitions tell us is that biodiversity is affected at all levels of biological organization, from genes to species to populations to communities to ecosystems to landscapes (**Figure 7.1**), and any attempts to use a definition of biological diversity in a conservation context needs to address all these levels to be successful.

The levels of biological diversity Figure 7.1

Traditionally, natural resource managers focused on specific species; however, management is now inclusive of all levels, in a hierarchal system, which is recognized as necessary to the conservation of overall biological diversity and ecosystem integrity.

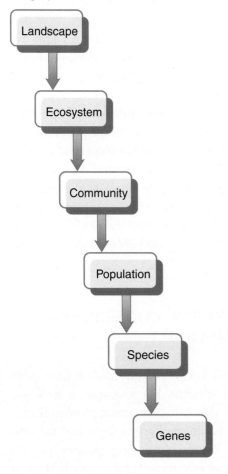

MEASURING BIOLOGICAL DIVERSITY AT THE SPECIES LEVEL

A **species** is a group of distinct organisms capable of interbreeding with one another in the wild but which do not interbreed with organisms outside of their group. We do not know exactly how many species exist. In fact, biologists now realize how little we know about Earth's diverse organisms. For instance, scientists estimate there may be as few as 5 million or as many as 100 million different species inhabiting Earth. To date, about 1.8 million species have been scientifically named and described, including about 270,000 plant species, 45,000 vertebrate animal species, and some 950,000 insect species. About 10,000 new species are identified each year.

Most often, biological diversity is described at the species level and consists of three different measurements: species richness, species evenness, and species dominance. The number of different species present in a community is **species richness**. But sometimes the same total number of species may be found in two communities, making it difficult to establish which is more biologically diverse by using the concept of species richness alone. Therefore, the relative number of species present in a community, or **species evenness**, is often used. A final important concept is that of **species dominance**, which is simply the most abundant species in a community.

species richness The number of different species in a community.

species evenness The relative number of species present in a community.

species dominance The most abundant species in a community.

Species richness varies greatly from one community to another. It is related to the abundance of potential ecological niches. A complex community, like the tropical rainforest or a coral reef, offers a greater variety of potential ecological niches than does a simple community like mountain chaparral (**Figure 7.2** on page 204).

Species richness is inversely related to the geographic isolation of a community. Isolated island communities are much less diverse than communities in similar environments found on continents for two reasons. Many species

A Chilean dry scrub (chaparral) has low species richness and supports few bird species.

CHILE

Global Locator

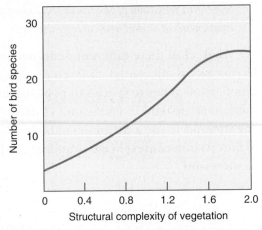

B The structural complexity of vegetation (*x*-axis) is based on its height and density. Note that species richness in birds increases as vegetation becomes more structurally complex.

have difficulty reaching and successfully colonizing the island, and locally extinct species are not readily replaced in isolated environments such as islands or mountaintops.

Species richness is also inversely related to environmental stress. Only those species capable of tolerating extreme environmental conditions can live in an environmentally stressed community, such as a polluted stream or a polar region exposed to a harsh climate. Species richness is also reduced when one species enjoys a decided dominance within a community, because it may appropriate a disproportionate share of resources, thus crowding out other species.

Species richness is usually greater at the edges of adjacent communities than in their centres. This is because an **ecotone**, a transitional zone where communities meet, contains all or most of the ecological niches of the adjacent communities as well as some niches unique to the ecotone. The change in species composition produced at ecotones is the **edge effect**.

Environmental history greatly affects species richness. Tropical rainforests are probably old, stable communities that have undergone few climate changes over time, which allowed a large number of species to evolve. In contrast, glaciers have repeatedly altered temperate

and arctic regions during Earth's history. An area recently vacated by glaciers will have a low species richness because few species will as yet have had a chance to enter it and become established.

Species richness is also unevenly distributed at global and local levels. Globally, species richness generally follows a latitudinal gradient, in that more species are present nearer to the equator. This is because there is more solar energy, heat, and humidity closer to the equator, which leads to more plant growth that supports larger numbers of animals. Locally, richness varies with habitat type, as more structurally diverse habitats harbour more ecological niches.

THE IMPORTANCE OF CONSERVING BIOLOGICAL DIVERSITY

Humans depend on the contributions of thousands of species for their survival. For example, insects are instrumental in several ecological and agricultural processes, including pollination of crops, weed control, and insect pest control. Bacteria and fungi provide us with foods, antibiotics and other medicines, and biological processes such as nitrogen fixation (see Chapter 5). However, relatively few species have been evaluated for their potential usefulness to humans. There are approximately 270,000 known plant species, but as many as 250,000 of them have yet to be assessed for industrial, medicinal, or agricultural potential. The same is true for most of the millions of microorganisms, fungi, and animals.

Ecosystem services and species richness

The living world functions much like a complex machine. Each ecosystem is composed of many parts that are organized and integrated to maintain the ecosystem's overall performance. The activities of all organisms are interrelated; we depend on one another and on the physical environment, often in subtle ways.

Ecosystems supply human societies with many environmental benefits, or **ecosystem goods and services** (**TABLE 7.1**). Forests are a source of lumber, but they also contribute to watershed dynamics by providing fresh water, moderating flooding severity and occurrences, and reducing soil erosion. Many flowering plant species depend on insects to transfer pollen for reproduction. Soil dwellers develop and maintain soil fertility for plants. Bacteria and fungi perform the crucial task of decomposition, which allows nutrients to cycle in the ecosystem. Conservationists maintain that ecosystems with greater species richness supply ecosystem services better than ecosystems with lower species richness.

> ■ **ecosystem goods and services**
> Environmental benefits, such as food, fibre, clean air, clean water, and fertile soil, provided by ecosystems.

The Millennium Ecosystem Assessment assessed the consequences of ecosystem change for human well-being. From 2001 to 2005, the assessment involved the work of more than 1360 experts worldwide. Their findings provide a state-of-the-art scientific appraisal of the condition and trends in the world's ecosystems and the services they provide, as well as the scientific basis for action to conserve and use them sustainably. For more information on this groundbreaking effort, see www.millenniumassessment.org.

For example, beavers play an important role in their natural ecosystem (**FIGURE 7.3** on page 206).

Ecosystem services TABLE 7.1

Ecosystem	Services provided
Forests	Purify air and water Produce and maintain soil Absorb carbon dioxide (carbon storage) Provide wildlife habitat Provide humans with wood and recreation
Freshwater systems (rivers, lakes, and groundwater)	Moderate water flow and mitigate floods Dilute and remove pollutants Provide wildlife habitat Provide humans with drinking and irrigation water, food, transportation corridors, electricity, and recreation
Wetlands, marshes, and potholes	Purify water Provide flood and drought protection Provide important habitat for endangered waterfowl
Grasslands	Purify air and water Produce and maintain soil Absorb carbon dioxide (carbon storage) Provide wildlife habitat Provide humans with livestock and recreation
Coasts	Provide a buffer against storms Dilute and remove pollutants Provide wildlife habitat Provide humans with food, harbours, transportation routes, and recreation

Through their damming activities, beavers alter water levels and produce a variety of moist and dry meadows, wet forests, marshes, bogs, streams, and open water that change the climate, nutrient flow, vegetation, wildlife, hydrology, and geomorphology of an entire watershed. Dead trees left by beavers or killed by beaver flooding become new homes for a plethora of animals and microbes. When beavers cut aspen, one of their favourite food and building materials, the cut parent tree sends up new suckers. The aspen suckers respond to the cutting of the parent tree by producing bitter alkaloids that beavers do not like, creating a balance between beaver cutting and aspen growth; however, the new aspen growth is perfect for moose and elk. Tree cutting also changes the tree canopy by opening up the understory to sunlight, which changes the ecological succession of the forest. Light-loving plants grow, creating new homes and food sources

Beavers in the environment FIGURE 7.3

Beavers play an important role in their natural ecosystem by creating a diversity of habitats, by controlling large amounts of energy and nutrient flow, and by increasing the overall biodiversity of an area.

Genetic diversity in corn FIGURE 7.4

The variation in corn kernels and ears is evidence of the genetic diversity in the species *Zea mays*.

for a variety of different species. The excess wood chips left behind from beaver cutting also become new shelter and food sources for insects, small mammals, and birds. Downed vegetation and debris picked up by water during flooding produces large amounts of nutrients that flow into the water where growing plants and animals trap them and then begin to cycle them. This increase in nutrients and overall biodiversity draws more fish, birds, and animal species into beaver ponds. As you can see, one little beaver can have a big impact on the ecosystem it lives in!

You might think that the loss of some species from an ecosystem might not endanger the rest of the organisms. However, if enough species are removed, the entire ecosystem will change. Species richness within an ecosystem provides the ecosystem with resilience, the ability to recover from environmental changes or disasters.

Genetic diversity Genetic diversity is the genetic variety *within* all populations of a species (FIGURE 7.4). The maintenance of a broad genetic base is critical for each species' long-term health and survival. Consider economically important crop plants.

During the 20th century, plant scientists developed genetically uniform, high-yielding varieties of important food crops like wheat. However, genetic uniformity resulted in increased susceptibility to pests and disease. By crossing the "super strains" with more genetically diverse relatives, disease and pest resistance can be reintroduced into such plants.

Genetic engineering, the incorporation of genes from one organism into a different species, makes it possible to use organisms' genetic resources on a wide scale. Genetic engineering has provided new vaccines, more productive farm animals, and disease-resistant agricultural plants.

Evolution has taken hundreds of millions of years to produce the genetic diversity found on our planet today. This diversity may hold solutions to today's problems and to future problems we have not begun to imagine. It would be unwise to allow such an important part of our heritage to disappear.

Medicinal, agricultural, and industrial importance of organisms the genetic resources of organisms are vitally important to the pharmaceutical industry, which incorporates hundreds of chemicals

Medicinal value of the rosy periwinkle
FIGURE 7.5

The rosy periwinkle produces chemicals effective against certain cancers. Drugs from the rosy periwinkle have increased the chance of surviving childhood leukemia from about 5 percent to more than 95 percent.

derived from plants and other organisms into its medicines (FIGURE 7.5). Many of the natural products taken directly from marine organisms are promising anticancer or antiviral drugs. The AIDS (acquired immune deficiency syndrome) drug AZT (azidothymidine), for example, is a synthetic derivative of a compound from a sponge. The 20 bestselling prescription drugs in the world are either natural products, natural products that are slightly modified chemically, or synthetic drugs whose chemical structures were obtained from organisms.

The agricultural importance of plants and animals is indisputable, because we must eat to survive. However, the number of different kinds of foods we eat is limited compared with the total number of edible species available in any given region. Many species are probably nutritionally superior to our common foods.

Modern industrial technology depends on a broad range of products from organisms. Plants supply oils and lubricants, perfumes and fragrances, dyes, paper, lumber, waxes, rubber and other elastic latexes, resins, poisons, cork, and fibres. Animals provide wool, silk, fur, leather, lubricants, waxes, and transportation, and they are important in medical research. The armadillo, for example, is used for research in Hansen's disease (leprosy) because it is one of only two species known to be susceptible to that disease (the other species is humans). Certain beetles produce steroids with birth-control potential, and fireflies produce a compound that may be useful in treating viral infections.

Aesthetic, ethical, and spiritual value of organisms

Organisms not only contribute to human survival and physical comfort, they provide recreation, inspiration, and spiritual solace. Our natural world is a thing of beauty largely because of the diversity of living forms found in it. Artists have attempted to capture this beauty in drawings, paintings, sculpture, and photography; and poets, writers, architects, and musicians have created works reflecting and celebrating the natural world.

Traditionally, many human cultures have viewed themselves as superior, subduing and exploiting other forms of life for human benefit. An alternative view is that organisms have intrinsic value and that as stewards of the life forms on Earth, humans should protect their existence (see Chapter 2 for a discussion of *environmental ethics*).

CONCEPT CHECK STOP

What is biological diversity?

Distinguish between species richness, species evenness, and species dominance.

What are two determinants of species richness? Give an example of each.

What are ecosystem goods and services? Describe some ecosystem goods and services that a wetland provides.

Extinction, Species at Risk, and Invasive Species

Extinction, the death of a life form, occurs when the last member of a species dies. Once a species is extinct, it will never reappear. Biological extinction is the fate of all species, much as death is the fate of all individuals. Biologists estimate that for every 2000 species that have ever lived, 1999 of them are extinct today.

> **extinction** The elimination of a species from Earth.

During the time in which organisms have occupied Earth, a continuous, low-level extinction of species, or **background extinction**, has occurred. Perhaps five or six times, a second kind of extinction, **mass extinction**, has occurred, in which a large number of species disappear during a relatively short period of geologic time.

The causes of past mass extinctions are not well understood, but biological and environmental factors were probably involved. Major climate change or a catastrophe such as a collision between a large asteroid or comet and Earth could have triggered the mass extinction.

Although extinction is a natural biological process, it is greatly accelerated by human activities (**FIGURE 7.6**). The burgeoning human population has spread into almost all areas of Earth. Whenever humans invade an area, the habitats of many organisms are disrupted or destroyed, which contributes to their extinction.

The International Union for the Conservation of Nature (IUCN) is the world's main authority on the conservation status of species. Since 1948 they have annually published the Red List of Threatened Species, widely considered to be the most objective and authoritative system for classifying species in terms of the risk of extinction. A series of regional red lists are produced by countries or organizations, which assess the risk of extinction to species within a political management unit.

The IUCN Red List is set upon precise criteria to evaluate the extinction risk of thousands of species and subspecies. These criteria are relevant to all species and all regions of the world. The aim is to convey the urgency of conservation issues to the public and to policy makers, as well as help the international community to try to reduce species extinction.

SPECIES AT RISK

Not all species are in imminent danger of becoming extinct, but not all species are flourishing, either. The latter are classified as species at risk in Canada and are covered under the Species at Risk Act, commonly referred to as SARA (which is covered in detail later in this chapter).

Under SARA, there are five categories into which species at risk can be classified. Extinct species are those that are no longer present anywhere on Earth. **Extirpated species** are those that are no longer present in the wild in Canada, but formerly existed in our country. **Endangered species** are those that are facing imminent extinction or extirpation. **Threatened species** show a high probability of becoming

> **extirpated species** A species that is no longer present in the wild in Canada.

> **endangered species** A species that is facing imminent extinction or extirpation.

> **threatened species** A species that has a high probability of becoming endangered.

Representative endangered or extinct species FIGURE 7.6

Currently, Earth's biological diversity is disappearing at an unprecedented rate. Conservation biologists estimate that species are now becoming extinct at a rate of 100 to 1000 times the natural rate of background extinctions.

special concern species A species that is particularly sensitive to human activities or natural events.

endangered. **Special concern species** are not yet in serious danger, but are of concern because of characteristics that make them particularly sensitive to human activities or natural events. For an up-to-date description of Canada's species at risk, check www. sararegistry.gc.ca

Endangered and threatened species represent a decline in biological diversity because as their numbers decrease, their genetic variability is severely diminished. Long-term survival and evolution depend on genetic diversity, so a decline in genetic diversity heightens the risk of

Extinction, Species at Risk, and Invasive Species   209

extinction for endangered and threatened species, compared with species that have greater genetic variability.

AREAS OF DECLINING BIOLOGICAL DIVERSITY

Declining biological diversity is a concern throughout Canada, but is most serious in the prairie region. About 99 percent of Canada's tall grass prairie has vanished, as has 90 percent of the mixed grass zone. These areas have been replaced by cultivated agriculture, which has established an agro-ecosystem across the Great Plains of North America. In the U.S., the states of Hawaii (where 63 percent of species are at risk) and California (where about 29 percent of species are at risk) are experiencing severe declining biological diversity. At least two-thirds of Hawaii's native forests are gone.

As serious as declining biological diversity is in North America, it is even more serious abroad, particularly in tropical rainforests, which are being destroyed faster than almost all other ecosystems; approximately 1 percent of tropical rainforests are being cleared or severely degraded each year. The forests are making way for human settlements, banana plantations, oil and mineral explorations, and other human activities.

Tropical rainforests are home to thousands or even millions of species, and they represent where most of the Earth's biodiversity resides. Scientists have warned that without changes to tropical forestry practices, within our lifetimes 50 to 75 percent of the world's species could go extinct. Many species in tropical rainforests are **endemic** (that is, they are not found anywhere else in the world), and the clearing of tropical rainforests contributes to their extinction.

> **endemic species** Organisms that are native to or confined to a particular region.

Perhaps the most unsettling outcome of tropical **deforestation** is its disruptive effect on evolution. In Earth's past, mass extinctions were followed over millions of years by the evolution of new species as replacements for those that died out. In the past, tropical rainforests may have supplied ancestral organisms from which other organisms evolved. Destroying tropical rainforests may be reducing nature's ability to replace its species.

EARTH'S BIODIVERSITY HOTSPOTS

In the 1980s ecologist Norman Myers of Oxford University coined the term **biodiversity hotspots**. In 2000, using plants as their criteria, Myers and ecologists at Conservation International identified 25 biological hotspots around the world (see What a Scientist Sees). Interestingly, these 25 hotspots for plants contain 29 percent of the world's endemic bird species, 27 percent of endemic mammal species, 38 percent of endemic reptile species, and 53 percent of endemic amphibian species. Fifteen of the 25 hotspots are tropical, and nine are mostly or solely islands. Many biologists recommend that conservation planners focus on preserving land in these hotspots to reduce the mass extinction of species currently underway.

> **biodiversity hotspots** Relatively small areas of land that contain an exceptional number of endemic species and that are at high risk from human activities.

HUMAN CAUSES OF SPECIES ENDANGERMENT

Scientists generally agree that the single greatest threat to biological diversity is loss of habitat. Pollution, overexploitation, and the spread of invasive species are also important.

Habitat destruction, fragmentation, and degradation
Most species are at risk today because of the destruction, fragmentation, or degradation of habitats by human activities. We demolish or alter habitats when we build roads, parking lots, bridges, and buildings; clear forests to grow crops or graze domestic animals; and log forests for timber. We drain marshes to build on aquatic habitats, thus converting them to terrestrial ones, and we flood terrestrial habitats when we build dams. Exploration for and mining of minerals, including fossil fuels, disrupts the land and destroy habitats. Habitats are altered by outdoor recreation, including using off-road vehicles, hiking off-trail, golfing, skiing, and camping. Because most organisms are utterly dependent on a particular type of environment,

Where Is Declining Biological Diversity the Greatest Problem?

 VIEW THIS IN ACTION

A If you've visited a zoo and seen a black-and-white ruffed lemur, you may know that lemurs are found in nature only on Madagascar and the small neighbouring Comoro Islands. However, most people do not realize that the ruffed lemur comes from a biological hotspot.

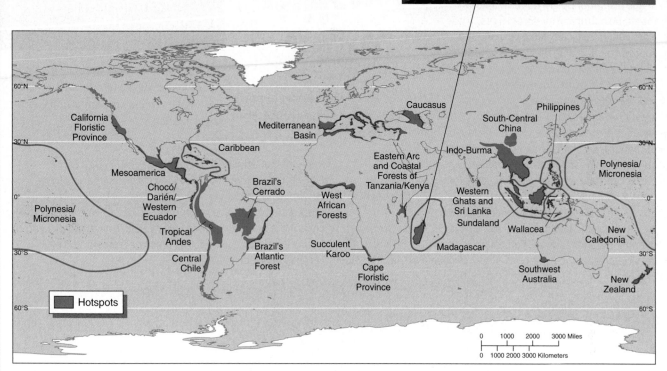

B Ecologists have identified the area of Madagascar and neighbouring islands as one of the world's 25 biological hotspots. These hotspots, which are rich in endemic species, are at great risk from human activities. All of Madagascar's 33 lemur species are currently in danger of extinction.

habitat destruction reduces their biological range and ability to survive.

As the human population has grown, the need for increased amounts of food has resulted in a huge conversion of natural lands into croplands and permanent pastures. According to the Food and Agriculture Organization of the United Nations, total agricultural lands (croplands and pasturelands) currently occupy 38 percent of Earth's land area. Agriculture also has a major impact on aquatic ecosystems because of the diversion of water for irrigation. Little habitat remains for many endangered species. The grizzly bear, for example, occupies about 2 percent of its original habitat. Human population growth and the extraction of resources have destroyed most of the grizzly's wilderness habitat.

Habitat fragmentation, the breakup of large areas of habitat into small, isolated patches (that is, islands), is a major threat to the long-term survival of many species. In ecological terms, *island* refers not only to any

Many species of migratory songbirds, favourites among North American bird lovers, are in decline, and North Americans' coffee habits may be playing a role. In the tropics, high-yield farms cultivating coffee in full sunlight, known as *sun plantations*, are rapidly replacing traditional *shade plantations.* Shade plantations grow coffee plants in the shade of tropical rainforest trees. These trees support a vast diversity of songbird species that winter in the tropics (one study counted 150 species in five hectares), as well as large numbers of other vertebrates and insects. In contrast, sun plantations provide poor bird habitat. Sun-grown varieties of coffee, treated with large inputs of chemical pesticides and fertilizers, outproduce the shade-grown varieties but lack the diverse products that come from the shade trees. About half of the region's shade plantations have been converted to sun plantations since the 1970s.

As a result, songbird populations have declined alarmingly during this period. Various conservation organizations and development agencies, such as Nature Canada, BirdLife

The SMBC allows certified shade-grown and organic coffee farmers to use this label on their product.

The prothonotary warbler spends its summers in eastern North America and its winters in Central and South America.

International, and the Smithsonian Migratory Bird Center (SMBC), have initiated programs to certify coffee as "shade-grown," which allow consumers the chance to support the preservation of tropical rainforest. Shade-grown coffee typically costs more than sun-grown coffee because it is hand-picked, involves more care in selecting only ripe beans, and is often certified organic.

land mass surrounded by water (**Figure 7.7A**), but also to any isolated habitat surrounded by an expanse of unsuitable territory. Accordingly, a small patch of forest surrounded by agricultural and suburban lands is considered an island (**Figure 7.7B**). Habitat destruction, fragmentation, and degradation are happening around the world, causing many species to become extinct and reducing the genetic diversity within many surviving species.

Africa provides a vivid example of the conflict over land use between growing human populations and other species. For example, African elephants require a lot of natural landscape in which to forage for the hundreds of kilograms of food that they consume daily. In Africa people are increasingly pushing into the elephants' territory to grow crops and graze farm animals. The elephants often trample or devour crops, ruining a year's growth of crops in a single night; they have even killed people. Farmers cannot shoot at or kill elephants because they are a protected species. (Before elephants were listed as a protected species, their numbers had declined precipitously because of overhunting by ivory hunters.) Researchers have found that elephants move out of areas when they become too crowded with people. Unfortunately, the wild areas to which elephants can move are steadily shrinking (**Figure 7.7C**).

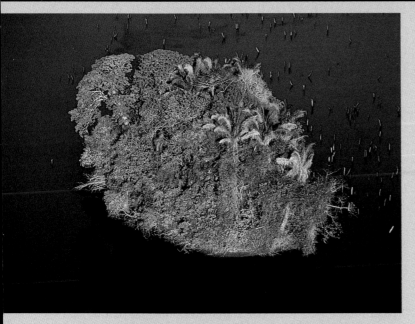

A Destruction of the world's wildlife habitats This tiny island, located in the Panama Canal, was once a hilltop in a forest that was flooded when the Panama Canal was constructed.

B Isolating wildlife habitats Roads and agricultural lands effectively isolate the scattered remnants, or "islands" of forest.

AFRICA
TANZANIA

Global Locator

C Crowded refuge Seeking sanctuary from human encroachment, an overpopulation of endangered African elephants in this Tanzanian crater refuge is exploiting its limited resources too quickly for the ecosystem to recover.

Pollution Humans produce acid precipitation, reduce the stratospheric ozone layer, and change the climate so that even wilderness habitats that are "completely" natural and undisturbed are degraded and unable to support the species diversity they once did. Acid precipitation is known to have contributed to the decline of large stands of forest trees and the biological death of many freshwater lakes. Because ozone in the upper atmosphere shields the ground from a large proportion of the sun's harmful ultraviolet (UV) radiation, ozone depletion in the upper atmosphere represents a threat to all terrestrial life. Climate warming, caused in part by an increase in atmospheric carbon dioxide released when fossil fuels are burned, is another threat. Such habitat modifications reduce the biological diversity of species with particularly narrow and rigid environmental requirements.

Other types of pollutants that affect organisms include industrial and agricultural chemicals, organic pollutants from sewage, acid mine drainage, and thermal pollution from heated industrial wastewater.

Overexploitation

Sometimes species become endangered or extinct as a result of deliberate efforts to eradicate or control their numbers. Ranchers, hunters, and government agents have in the past reduced populations of large predators such as the wolf and grizzly bear. Some animals are killed because their lifestyles cause problems for humans. The Carolina parakeet, a beautiful green, red, and yellow bird endemic to the southern United States, was exterminated as a pest by farmers because it ate fruit and grain crops. It was extinct by 1920.

Prairie dogs and pocket gophers were poisoned and trapped so extensively by ranchers and farmers between 1900 and 1960 that they disappeared from most of their original geographic range. As a result of sharply decreased numbers of prairie dogs, the black-footed ferret, the natural predator of these animals, became endangered. A successful captive-breeding program has allowed black-footed ferrets to be reintroduced into the wild and reproduce successfully, though some populations have been decimated by disease.

Unregulated hunting, or overhunting, was a factor contributing to the extinction of certain species in the past but is now strictly controlled in most countries. The passenger pigeon was one of the most common birds in North America in the early 1800s, but a century of overhunting resulted in its extinction in the early 1900s. Unregulated hunting was one of several factors that caused the near extinction of the American bison. Lessons have been learned from these past mistakes, and now modern wildlife management solutions, such as the North American Wildlife Conservation Model discussed in Chapter 2, are used as a necessary tool to ensure that biological diversity is conserved. It may seem peculiar to consider hunting as a method of conservation, but species have evolved with predation, and human predators are in many cases replacing others that have become less abundant. In these situations, hunting is used as a tool for managing ecosystems in a sustainable fashion.

Illegal commercial hunting, or **poaching**, endangers many larger animals, such as the tiger, cheetah, and snow leopard, whose beautiful furs are quite valuable. Rhinoceroses are slaughtered primarily for their horns, which are used for ceremonial dagger handles in the Middle East and for purported medicinal purposes in Asian medicine. Bears are killed for their gallbladders, used in Asian medicine to treat ailments from indigestion to heart ailments. Caimans (reptiles similar to crocodiles) are killed for their skins and made into shoes and handbags. Although these animals are legally protected, the demand for their products on the black market has caused them to be hunted illegally.

In West Africa, poaching has contributed to the decline in lowland gorilla and chimpanzee populations. The meat (called *bushmeat*) of these rare primates and other protected species, such as anteaters, elephants, and mandrill baboons, provides an important source of protein for indigenous people. Bushmeat is also sold to urban restaurants. This demand for a meat source increases the incidence of poaching.

Live organisms collected through **commercial harvest** end up in zoos, aquariums, biomedical research laboratories, circuses, and pet stores. Several million birds are commercially harvested each year for the pet trade, but unfortunately many of them die in transit, and many more die from improper treatment after they are in their owners' homes. At least 40 parrot species are now threatened or endangered, in part because of unregulated commercial trade (**FIGURE 7.8**).

Animals are not the only organisms threatened by excessive commercial harvest. Many unique and rare plants have been collected from nature to the point that

Illegal animal trade FIGURE 7.8

These green parrots, captured illegally in the Amazon rainforest, are held for sale at a market in Inquitos, Peru.

they are endangered. These include carnivorous plants, wildflowers, grasses, ferns, certain cacti, and orchids.

Invasive species

The introduction of a nonnative or foreign species into an ecosystem in which it did not evolve often upsets the balance among the organisms living in that area and interferes with the ecosystem's normal functioning. The foreign species may compete with native species for food or habitat or may prey on them. Foreign species whose introduction causes economic or environmental harm are called **invasive species** (**FIGURE 7.9**). Although invasive species may be introduced into new areas by natural means, humans are usually responsible for such introductions, either knowingly or unknowingly.

The introduction of invasive species to North America began many centuries ago. Prior

invasive species
Foreign species that spread rapidly in a new area if free of predators, parasites, or resource limitations that may have controlled their population in their native habitat.

Invasive species FIGURE 7.9

The Asian long-horned beetle is an example of an invasive species that has had devastating consequences to Canadian forests. Both the Canadian and U.S. governments have had to spend millions of dollars to control and attempt to eradicate the species. Maple trees in particular are a favourite host species for the beetle.

to European exploration and expansion during the 15th century, Eurasia and North America were relatively isolated from each other. However, with European expansion came the introduction of a staggering number of invasive plant and animal species. These biological invaders were on occasion introduced purposefully, but the majority of them were introduced accidentally by humans. The magnitude of invasion over the last 500 years is overwhelming. There are 2000 to 3000 invasive plant species in North America, which represent 10 to 25 percent of the total flora in Canada and the United States. The Hawaiian Islands provide an exceptional example of the pervasiveness of invasive species, where 40 percent of all bird species are invasive, 94 percent of all mammal species are invasive, and 100 percent of all reptile species are invasive. In many cases, human activity, such as habitat modification or hunting of natural competitors or predators, help the invading species thrive.

The introduction of exotic species often causes significant damage to native ecosystems, resulting in problems for agriculture, forestry, and other resources used by humans. These invasive species often spread at low population levels until they become established and then populations continue to increase in a logistical manner. This pattern of population growth makes invaders difficult to initially detect and remove, as populations are not noted until growth has accelerated and it is too late to stop the invasion. Canada, similar to all other countries in the world, is in the midst of an invasion where, next to habitat destruction, alien species are the leading cause of extinctions and pose a serious threat to overall biological diversity.

A country's vulnerability directly correlates to the amount of international trade it conducts, so as our economy grows, so does the number of invasive species. For example, cargo-carrying ocean vessels carry *ballast water* from their ports of origin to increase their stability in the ocean. When they reach their destinations, this water is discharged into local bays, rivers, or lakes. Ballast water may contain clams, mussels, worms, small fishes, and crabs, along with millions of microscopic aquatic organisms. These organisms, if they establish themselves, may threaten the area's aquatic environment and contribute to the extinction of native organisms.

One of North America's greatest biological threats— and its most costly aquatic invader—is the zebra mussel, a native of the Caspian Sea, probably introduced through ballast water flushed into the Great Lakes by a foreign

Zebra mussels FIGURE 7.10

A clump of zebra mussels, which have caused billions of dollars in damage in addition to displacing native clams and mussels.

ship in 1985 or 1986 (FIGURE 7.10). Zebra mussels are proficient invaders. They have few effective predators or parasites, an adult female can release 5 million eggs annually, and the larvae are free-swimming and easily dispersed in fresh water. Adult zebra mussels can survive out of water in moist conditions for up to two weeks, allowing them to be transported on boats from one body of water to another.

Zebra mussels have caused billions of dollars in damage in the Great Lakes. They plug cooling water systems of boat motors and of water intake systems of cottages, towns, cities, and industries; they reduce human recreation; and they exclude other mollusks and plants from the area. An estimated 140 different species of native North American shellfish may face extinction because of the zebra mussel. Zebra mussels also remove a large amount of phytoplankton from the water, which causes changes in the food chain and reduces the available plankton for invertebrates and fish. It is estimated that zebra mussels cost commercial fishermen on Lake Erie an average of $400 million in decreased catch per year. The U.S. Coast Guard estimates economic losses and control efforts associated with the zebra mussel cost the United States about $5 billion each year.

Invasive species in Canada are managed by the Canadian Food Inspection Agency (CFIA). In 2004 the Invasive Alien Species (IAS) plan was created with the purpose of minimizing the risk of invasive alien plant and animal species to the Canadian economy, environment, and society. Through prevention, early detection, rapid response, and management, the IAS plan hopes to mitigate the harmful effects of invasive alien species. Challenges such as the number of partners and stakeholders involved, the sheer number of invasive species, the large number of sectors and pathways, limited resources, and conflict interests from various stakeholders present numerous obstacles for the CFIA. In order to successfully reduce the presence of invasive alien species in Canada, there needs to be effective communication and cooperation between all parties.

CONCEPT CHECK STOP

What is background extinction? Mass extinction?

What is the difference between an extirpated species, a threatened species, and an endangered species?

How do human activities cause species to become endangered or extinct?

Conserving Earth's Biological Diversity

LEARNING OBJECTIVES

Define conservation and preservation and distinguish between them.

Briefly describe the goals of the Species at Risk Act.

Relate the purpose of the World Conservation Strategy.

Clearly biological diversity is important, but it is also declining. So what is being done to conserve biodiversity around the world? Conservation can take many forms and much research has been (and continues to be) done on conserving Earth's biological diversity. This section explores the legislation that has been introduced to protect Earth's biodiversity and some of the scientific methods that are being used to implement conservation efforts, such as landscape ecology and conservation biology.

CONSERVATION VERSUS PRESERVATION OF RESOURCES

Resources are any part of the natural environment used to promote the welfare of people or other species. Some examples of resources include air, water, soil, forests, minerals, and wildlife. **Conservation** is the sensible and careful management of natural resources. It can also be defined as ecosystem management within social and economic constraints, without depleting natural ecosystem diversity, and acknowledging the naturally dynamic character of

conservation The sensible and careful management of natural resources.

biological systems. Humans have practised conservation of natural resources for thousands of years. More then 3000 years ago, the Phoenicians terraced hilly farmland to prevent soil erosion. More than 2000 years ago, the Greeks practised crop rotation to maintain yields on farmlands, and the Romans practised irrigation. Other Europeans gradually adopted and further refined these and other conservation techniques (**FIGURE 7.11A** on page 218).

Conservation is different from preservation. Conservation involves sustainability—that is, using resources

Conservation and preservation FIGURE 7.11

A Plowing and planting fields in curves that conform to the natural contours of the land conserves soil by reducing erosion.

B Wapusk National Park near Churchill, Manitoba, preserves habitat and populations of polar bears.

without inflicting excessive environmental damage so that the resources are available not only for current needs but also for the needs of future generations.

> **preservation**
> Setting aside undisturbed areas, maintaining them in a pristine state, and protecting them from human activities.

Preservation is concerned with setting aside undisturbed areas, maintaining them in a pristine state, and protecting them from human activities that might alter their "natural" state. Preservation also involves protecting species or landscapes without reference to human requirements, or sometimes, to natural changes in living systems (**FIGURE 7.11B**).

Conservation did not become a popular movement until the early 20th century. At that time, expanding industrialization, coupled with enormous growth in the human population, began to increase pressure on the world's supply of natural resources. See Chapter 2 for more information on the history of conservation efforts and the environmental movement.

THE SPECIES AT RISK ACT

One of the most important ways of conserving Earth's biological diversity is through legislation. Many governments have introduced legislation to protect national biodiversity. In Canada, the **Species at Risk Act (SARA)** was passed in 2002 to provide legal authority to the federal government for ensuring the conservation of biological diversity at the national level. SARA was formulated to meet the following objectives:

- To prevent Canadian indigenous species, subspecies, and distinct populations from becoming extirpated or extinct
- To provide for recovery of endangered or threatened species
- To encourage management of other species to prevent them from becoming at risk

In order to achieve these objectives, the Act created the Committee on the Status of Endangered Wildlife in Canada (COSEWIC), an independent body of experts responsible for assessing and identifying to the Minister of Environment designations for different species at risk. The designations can fall into one of five categories, as discussd earlier: extinct species, extirpated species, endangered species, threatened species, and special concern species. For any species listed as extirpated, endangered, or threatened under SARA (FIGURE 7.12), it is illegal to kill, harm, harass, or capture the species. Possessing, collecting, buying or selling, or trading any part of a listed species, or causing damage or destruction to their habitat, is also illegal under SARA. These restrictions do not apply for wildlife listed as species of special concern.

Many environmentalists and scientists in Canada have protested against SARA, suggesting that the Act is too weak to protect species and associated habitat adequately. Many studies indicate that far too often politics and economics override science and COSEWIC recommendations, resulting in species being turned down

Visualizing

Examples of species protected under SARA FIGURE 7.12

The Species at Risk Act covers all types of natural life forms, from mosses and lichens to fish to terrestrial mammals. There are currently thousands of species listed as either extirpated, endangered, or threatened under SARA. What would Canada lose if any of these species were no longer around?

A A blue whale

B An American badger

C A burrowing owl

D A yucca moth

from achieving protected status. However, the Act does represent an important step toward protecting Canada's biological diversity. It also acknowledges government commitment to sustainable stewardship of nature and wildlife for future generations.

Along with Canadian federal legislation, each province has implemented additional legislation, regulations, and guidelines for species at risk within their respective jurisdictions. These provincial rules are in addition to the federal statute and are generally supplementary, thereby providing additional protection to individual species of concern.

INTERNATIONAL CONSERVATION POLICIES AND LAWS

In addition to the numerous conservation laws put in place by many of the world's nations, international efforts have been organized to protect global biological diversity as well. The **World Conservation Strategy**, a plan designed to conserve biological diversity worldwide, was formulated in 1980 by the World Conservation Union (now known as the International Union for Conservation of Nature, or IUCN), the World Wildlife Fund, and the UN Environment Programme. In addition to conserving biological diversity, the World Conservation Strategy seeks to preserve the vital ecosystem services on which all life depends for survival and to develop sustainable uses of organisms and the ecosystems they comprise.

The Convention on Biological Diversity produced by the 1992 Earth Summit requires that each signatory nation inventory its own biodiversity and develop a **national conservation strategy**, a detailed plan for managing and preserving the biological diversity in that specific country. Currently, 188 nations participate in the Convention on Biological Diversity.

The exploitation of endangered species is somewhat controlled through legislation. At the international level, 160 countries participate in the **Convention on International Trade in Endangered Species of Wild Flora and Fauna (CITES)**, which came into effect in 1975. Originally drawn up to protect endangered animals and plants considered valuable in the highly lucrative international wildlife trade, CITES bans hunting, capturing, and selling of endangered or threatened species and

Illegal trade in products made from endangered species FIGURE 7.13

A merchant in Myanmar deals in wildlife products.

regulates trade of organisms listed as potentially threatened (**FIGURE 7.13**). Unfortunately, enforcement of this treaty varies from country to country. Even where enforcement exists, the penalties are not severe. As a result, illegal trade continues in rare, commercially valuable species.

The goals of CITES often stir up controversy over issues such as who actually owns the world's wildlife and whether global conservation concerns take precedence over competing local interests. These conflicts often highlight socioeconomic differences between wealthy consumers of CITES products and poor people who trade the endangered organisms.

The case of the African elephant exemplifies these controversies. Listed as an endangered species since 1989 to halt the slaughter of elephants driven by the ivory trade, the species seems to have recovered in southern Africa. However, the African people living near the elephants want to cull the herd periodically and sell elephant meat, hides, and ivory for profit. In the late 1990s and early 2000s, CITES transferred elephant populations in Namibia, Botswana, and South Africa to a less restrictive listing to allow a one-time trade of legally obtained stockpiled ivory (from animals that died of natural causes). The money earned from the sale of ivory is funding elephant conservation programs and community development projects for people living near the elephants.

CONCEPT CHECK STOP

How does conservation differ from preservation?

What are the goals of the Species at Risk Act?

What is the World Conservation Strategy?

Landscape Ecology and Conservation Biology

LEARNING OBJECTIVES

Define the term *landscape ecology* and discuss its importance in biodiversity conservation.

Define *conservation biology* and compare in situ and ex situ conservation.

Describe restoration ecology.

LANDSCAPE ECOLOGY

Landscape ecology is the interdisciplinary study of the relationship between spatial patterns and ecological processes on a number of different landscape scales and organizational levels. Landscape ecology blends biological science, physical science, and analytical study with natural and social sciences to achieve a holistic landscape perspective. In basic terms, landscape ecology looks at the "big picture" or landscape level of how structure affects the abundance, behaviour, and function of the ecosystem as a whole (FIGURE 7.14 on page 222).

landscape ecology The study of the relationship between spatial patterns and ecological processes on a number of different landscape scales and organizational levels.

Often, landscape ecology focuses on large, heterogeneous areas of study and assesses human-caused landscape changes. For this reason, the concept of **scale**, or the distance on a map image and the corresponding distance on Earth, is central in landscape ecology. When looking at a landscape as a whole, researchers focus on **patches** of land (a patch is a relatively homogeneous area that differs from its surroundings). An area's **composition** refers to the number and the relative abundance of patch types that are represented on a landscape. For example, the amount of forest, wetland, and road density on a landscape would represent its composition. Landscape **structure** is determined by the composition, configuration, and proportion of various patches across a landscape, whereas landscape **function** refers to the interactions among elements within the landscape based on their life cycles.

The study of landscape ecology is an essential component of biodiversity conservation. It has been particularly important for research on climate change. Research in northern regions of Canada have examined landscape ecological processes, such as snow accumulation, melting, freeze–thaw action, percolation, soil moisture changes, and temperature regimes. This work studies ecosystem function across several spatial and temporal gradients over long periods of time, which gives researchers important data on the effects of climate change. Today, landscape ecology continues to develop with new innovative applications, advanced technologies such as remote sensing and geographic information system (GIS), and powerful quantitative models used to examine a changing environment.

Landscape ecology looks at the "big picture" FIGURE 7.14

A landscape ecologist would be interested in studying the dynamics of the various ecosystems found in the vicinity of this East Coast fishing village, including the marine ecosystem with its fish and other organisms in the ocean, the fishermen (human system) who depend on them for their livelihood, and the effects of the village itself on the surrounding terrestrial and marine ecosystems.

CONSERVATION BIOLOGY

conservation biology The scientific study of how humans impact organisms and of the development of ways to protect biological diversity.

Studies in the field of **conservation biology** cover everything from the processes that influence biological diversity to the protection and restoration of endangered species to the preservation of entire ecosystems and landscapes.

Conservation biology includes two problem-solving techniques to save organisms from extinction: in situ and ex situ conservation. **In situ conservation**, which includes the establishment of parks and reserves, concentrates on preserving biological diversity in nature. With increasing demands on land, in situ conservation cannot guarantee the preservation of all types of biological diversity. Sometimes only ex situ conservation can save a species. **Ex situ conservation** involves conserving biological diversity in human-controlled settings. The breeding of captive species in zoos (**FIGURE 7.15**) and the seed storage of genetically diverse plant crops are examples of ex situ conservation.

Protecting habitats Protecting animal and plant habitats—that is, conserving and managing the ecosystem as a whole—is the single best way to preserve biological diversity. Because human activities adversely affect the sustainability of many ecosystems, direct conservation management of protected areas is often required (**FIGURE 7.16**).

Currently, more than 3000 national parks, sanctuaries, refuges, forests, and other protected areas exist worldwide. Protected areas are not always effective in preserving biological diversity, however. Many existing protected areas are too small or too isolated from other protected areas to efficiently conserve species. In developing countries where biological diversity is greatest, there is little money or expertise to manage protected areas. Finally, many of the world's protected areas are in lightly populated mountain areas, tundra, and the driest deserts, places that often have spectacular scenery but relatively few species. In contrast, ecosystems in which biological diversity is greatest often receive little

Ex situ conservation FIGURE 7.15

Zoos serve a larger function than just providing an opportunity for people to see different species from around the world. In these human-controlled settings, endangered or threatened animal species can be bred to avoid the species' complete extinction. (Photographed at the Calgary Zoo)

Some challenges in conservation management FIGURE 7.16

VIEW THIS IN ACTION

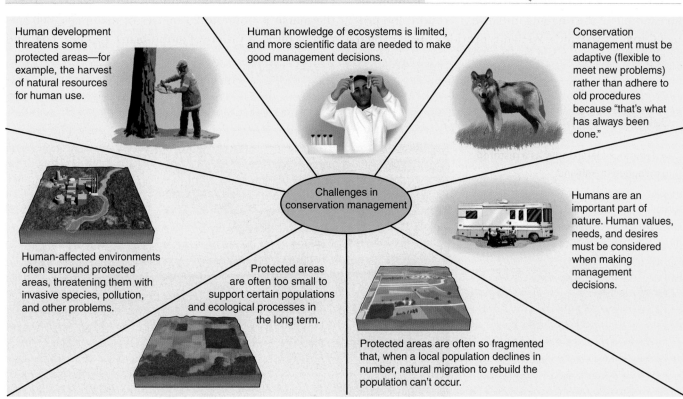

Human development threatens some protected areas—for example, the harvest of natural resources for human use.

Human knowledge of ecosystems is limited, and more scientific data are needed to make good management decisions.

Conservation management must be adaptive (flexible to meet new problems) rather than adhere to old procedures because "that's what has always been done."

Challenges in conservation management

Human-affected environments often surround protected areas, threatening them with invasive species, pollution, and other problems.

Protected areas are often too small to support certain populations and ecological processes in the long term.

Protected areas are often so fragmented that, when a local population declines in number, natural migration to rebuild the population can't occur.

Humans are an important part of nature. Human values, needs, and desires must be considered when making management decisions.

Landscape Ecology and Conservation Biology   223

protection. Protected areas are urgently needed in tropical rainforests, deserts, the tropical grasslands and savannas of Brazil and Australia, many islands and temperate river basins, and dry forests all over the world.

Restoring damaged or destroyed habitats

When preserving habitats is not possible, scientists can reclaim disturbed lands and convert them into areas with high biological diversity. In **restoration ecology**, ecological principles are used to

restoration ecology The study of the historical condition of a human-damaged ecosystem, with the goal of returning it as close as possible to its former state.

help return a degraded environment to a more functional and sustainable one. Restoration of disturbed lands creates biological habitats and provides additional benefits, such as the regeneration of soil damaged by agriculture or mining. The disadvantages of restoration include the expense and the time required to restore an area. Even so, restoration is an important aspect of in situ conservation, as restoration may reduce extinction.

In Canada, significant efforts have been made to restore numerous types of biological habitats. Oak Hammock Marsh, located near Stonewall, Manitoba, is an example of the ongoing initiatives to restore and protect Canada's wetlands. Wetlands provide critical habitat for many wildlife species and provide benefits for human society through their hydrological functions; these benefits include storage and eventual release of surface water, recharge of local and regional groundwater supplies, and erosion prevention, among others.

Oak Hammock Marsh is a 450-square-kilometre area that first existed as natural wetlands until it was drawn down for agriculture. In the 1950s the original wetlands had almost completely disappeared. The Province of Manitoba began to acquire much of the land in the area in the 1960s with the idea of restoring it to its natural wetland form. In the early 1970s, through a partnership between Ducks Unlimited Canada and the Province of Manitoba, the construction of dykes began, creating impoundments. Today, four cells and two compartments (a coot cell and a duck pond) now exist, all heavily managed (**FIGURE 7.17**). A diversion on Wavey Creek brings water into these cells, all gravity driven, and water control structures empty water out of cells into various creeks. This allows cell managers to create flood or mudflat conditions to keep the cells at various levels of vegetative growth, creating optimal conditions for many bird and waterfowl species. In 1972, with the creation of the initial perimeter dykes, the marsh saw an immediate increase in waterfowl populations using the area. Today the marsh is home to 25 species of mammals, 300 species of birds, a variety of amphibians, reptiles, fish, and

Oak Hammock Marsh FIGURE 7.17

Oak Hammock Marsh today as a restored and managed marshland.

NATURE CONSERVANCY OF CANADA (NCC)

The Nature Conservancy of Canada (NCC), a private nonprofit group, is the leading national land conservation organization in Canada. Through successful partnerships with governments, corporations, and landowners, NCC works to protect valuable ecosystems by securing property and committing to long-term stewardship. Over the last 45 years, NCC and their associated partners have conserved more than 2 million acres of land nationwide. Much of the land that NCC protects is ecologically valuable and in some cases is vulnerable and disappearing.

NCC strives to secure land through purchases, donations, and conservation agreements. The Ecogift Program, a federal program administered by Environment Canada, was created in 1995 to provide tax incentives to landowners wanting to protect their ecologically sensitive lands. NCC also receives land donated as capital properties and as assets that are sold, with the proceeds being used to purchase land with higher priority conservation needs. Other efforts include conservation agreements that are established between private landowners and NCC to permanently limit land use to protect its conservation value.

In March 2007, NCC and the Government of Canada signed a funding agreement to establish and manage the Natural Areas Conservation Fund. This fund is supported by $225 million from the Government of Canada and NCC's commitment to deliver a fund-matching conservation initiative called the Natural Areas Conservation Program. As a part of this program, NCC is able to sub-grant up to $15 million of the money received from the government to enable other qualified organizations in specific land acquisition projects that are consistent with NCC's goals. These initiatives allow NCC to secure valuable partnerships and increase support to land conservation projects across Canada.

For more information on NCC please visit their website at www.natureconservancy.ca

The Happy Valley Forest is one of the largest remaining intact deciduous forests on Canada's Oak Ridges Moraine (located north of Toronto). Over the past 30 years, the NCC has helped protect more than 1000 hectares of the Oak Ridges Moraine.

---

invertebrates, and a plethora of flora. During the migration season, Oak Hammock Marsh can see over 400,000 waterfowl each day.

Conserving species

Zoos, aquariums, botanical gardens, and other conservation organizations often play a critical role in saving individual species on the brink of extinction. Eggs or seeds may be collected from nature, or the few remaining wild animals may be captured and bred in research environments (FIGURE 7.18A on page 226). But attempting to save a species on the brink of extinction is expensive, and only a small proportion of endangered species can be saved.

Conservation organizations are an essential part of the effort to maintain biological diversity through species and habitat conservation. These groups help educate policy makers and the public about the importance of biological diversity. In certain instances, they galvanize public support for important biodiversity preservation efforts. They provide financial support for conservation projects, from basic research to the purchase of land that is a critical habitat for a particular species or group of species (FIGURE 7.18B).

Reintroducing endangered species to nature

The ultimate goal of the captive-breeding programs practised by zoos, aquariums, and other conservation organizations is to produce offspring in captivity and then release them into nature so wild populations are restored. However, only one of every

10 reintroductions using animals raised in captivity is successful. Before attempting a reintroduction, conservation biologists conduct a feasibility study. This includes determining (1) what factors originally caused the species to become extinct in nature, (2) whether these factors still exist, and (3) whether any suitable habitat still remains. Captive-breeding programs are sometimes unsuccessful because it is impossible to teach critical survival skills to animals raised in captivity.

Seed banks

More than 100 seed collections, called **seed banks** or *gene banks*, exist around the world and collectively hold more than 3 million samples at low temperatures (**FIGURE 7.18C**). They offer the advantage of storing a large amount of plant genetic material in a small space. Seeds stored in seed banks are safe from habitat destruction, climate warming, and general neglect. There have even been some instances of using seeds from seed banks to reintroduce to nature a plant species that had become extinct.

Seed banks have some disadvantages. Seeds of many types of plants, such as avocados and coconuts, cannot be stored because they do not tolerate being dried out, a necessary step in freezing the seeds. Seeds do not

Visualizing

Efforts to conserve species FIGURE 7.18

A Captive breeding Lucky (right) is the first whooping crane born to parents that were raised in captivity and then released, and the first in almost 70 years to be born in the wild in North America. Here Lucky stands with one of his parents at their central Florida home.

C Seeds from a seed bank Shown are seeds to be stored in a seed bank in Sussex, England.

B Minimum critical size of ecosystems project When a protected area is set aside, it is important to know what the minimum size of that area must be so it is not affected by encroaching species from surrounding areas. Shown are one-hectare and 10-hectare plots of a long-term study on the effects of habitat fragmentation on Amazon rainforest by the World Wildlife Fund and Brazil's National Institute for Amazon Research. Preliminary data indicate that the smaller forest fragments do not maintain their ecological integrity.

NATIONAL GEOGRAPHIC

remain alive indefinitely and must be germinated periodically so new seeds can be collected. Also, growing, harvesting, and returning seeds to storage is expensive. Perhaps the most important disadvantage of seed banks is that plants stored in this manner remain stagnant in an evolutionary sense. Thus, they may be less fit for survival when they are reintroduced into nature. Despite their shortcomings, seed banks are increasingly viewed as an important method of safeguarding seeds for future generations.

CONCEPT CHECK STOP

What is landscape ecology?

What is conservation biology? When is in situ conservation used? When is ex situ conservation used?

What are the goals of restoration ecology?

ECO CANADA | **CAREER FOCUS**

CONSERVATION BIOLOGIST

Conservation biologists focus on protecting biological diversity and minimizing the effects of human activities on the natural world as much as possible.

Imagine you are crouched on the edge of a small river, carefully observing a pack of wolves ambling along the opposite bank. You are a conservation biologist and you have been studying and gathering data on this particular group of wolves for the past few days. The River Pack, as it has been dubbed by your research team, is losing valuable habitat, and its population is declining quickly. You and your team of researchers are here to determine what can be done to protect habitat and save this pack.

As a conservation biologist, you must look at a number of issues to determine what this wolf pack needs to survive. For starters, you gather data on the life history of these wolves, including their average life span, reproductive and mortality rates, and average litter size. You also study their food and habitat requirements and look at how much of each is available and how this affects the pack's range. You will also study other wildlife in the area, specifically the wolves' predators and prey. For example, the River Pack hunts primarily deer, so you are interested in the area's deer population, including its size and reproductive rate. You will then examine the area's human population and the influences it has on local wildlife, particularly with respect to habitat loss. You will also research wolf studies from other areas to evaluate the success of different strategies that have been implemented to protect populations. As you determine the best way to preserve habitat and protect the River Pack population, you will balance all these factors.

Conservation biologists focus on how to protect and restore biodiversity, particularly understanding and minimizing human impacts on the natural world. They emphasize the use of planning and sustainable management practices to prevent species extinction and repair damage to ecosystems. Conservation biologists are involved in a number of projects that address issues such as managing threats to species and habitat, identifying and protecting endangered species, identifying sensitive areas for protection, and contributing to the recovery of threatened populations through breeding programs. They are also involved in building partnerships between government agencies, conservation groups, universities, and private landowners.

For more information or to look up other environmental careers go to www.eco.ca.

REINTRODUCING WOLVES TO YELLOWSTONE

Grey wolves once ranged across North America from northern Mexico to Greenland, but they were trapped, poisoned, snared, and hunted to extinction in most areas by 1960. Under the provisions of the U.S. Endangered Species Act, grey wolf populations in the northern Rocky Mountains were listed as endangered in 1974. Many scientists recommended reintroducing wolves as a way to restore their populations, but the controversial proposal was not acted on for more than two decades. Beginning in 1995, the U.S. Fish and Wildlife Service captured a small number of grey wolves in Canada and released them into Yellowstone National Park in Wyoming (see photograph). The population thrived and has increased to an estimated 300 individuals in the Yellowstone area.

The wolves in Yellowstone prey on elk, mule deer, moose, and bison. In some areas, intensive hunting by wolf packs has helped reduce the park's elk population, which was at an all-time high before the wolves returned. When elk are not managed properly, they overgraze their habitat, and thousands starve during hard winters. The reduction and redistribution of Yellowstone's elk population has relieved heavy grazing pressure on various plant species. As a result of a more lush and varied plant composition, herbivores such as beavers and snow hares have increased in number, which in turn supports small predators such as foxes, badgers, and martens.

Wolf packs have also severely reduced some coyote populations, allowing populations of the coyotes' prey, like ground squirrels, chipmunks, and pronghorns, to increase. Scavengers like ravens, magpies, bald eagles, wolverines, and bears benefit from dining on scraps from wolf kills.

The reintroduction of wolves did not occur without a fight. Ranchers and farmers who live in the area were against the reintroduction because their livelihood depends on livestock being safe from predators. To address this concern, ranchers are allowed to kill Yellowstone wolves that attack their cattle and sheep, and federal officers can remove any wolf that threatens humans or livestock. Also, ranchers are reimbursed the full market value for cattle, sheep, and other livestock lost to wolves.

SUMMARY

1 Defining and Measuring Biological Diversity

1. It is difficult to nail down a definitive definition of **biological diversity**, but in essence it is the variety of life in all forms, at all levels, and in all combinations in a defined area. The area is defined by humans, and can vary in size from something as large as the Earth to a country, province, region, municipality, or a puddle in your backyard.

2. **Species richness** is the number of different species in a community. **Species evenness** is the relative number of species in a community. **Species dominance** is the most abundant species in a community.

3. High species richness is associated with communities that are ecologically complex, not isolated, ecologically old and stable, and not subject to environmental stress. Species richness is also higher when no one species dominates the community.

4. Ecosystems with greater species richness are better able to supply **ecosystem goods and services**: environmental benefits such as clean air to breathe, clean water to drink, and fertile soil in which to grow crops.

2 Extinction, Species at Risk, and Invasive Species

1. **Extinction** is the elimination of a species from Earth. Background extinction, a continuous, low-level extinction of species, has occurred throughout the time in which organisms have occupied Earth. Mass extinction, in which many species disappear during a relatively short period of geologic time, has occurred at certain times in Earth's history.

2. An **extirpated species**, as defined by the Species at Risk Act, is a species that is no longer present in the wild in Canada. An **endangered species** faces threats that may cause it to become extinct or extirpated within a short period. A species is defined as **threatened** when it has a high probability of becoming endangered.

3. Humans cause species endangerment through habitat destruction, fragmentation, and degradation; pollution; the overexploitation of biological resources; and the spread of invasive species. **Endemic species** are organisms that are native to, or whose range is limited to, a specific place. **Biodiversity hotspots** are areas that contain particularly high numbers of endemic species. **Invasive species** are foreign species, usually introduced by humans, that spread rapidly in a new area where they are free of predators, parasites, or resource limitations that may have controlled their population in their native habitat.

3 Conserving Earth's Biological Diversity

1. **Conservation** is the sensible and careful management of natural resources, such as air, water, soil, forests, minerals, and wildlife. **Preservation** is concerned with setting aside undisturbed areas, maintaining them in a pristine state, and protecting them from human activities.

2. The goals of the Species at Risk Act are to prevent Canadian indigenous species, subspecies, and distinct populations from becoming extirpated or extinct, provide for recovery of endangered or threatened species, and encourage management of other species to prevent them from becoming at risk.

3. The World Conservation Strategy, formulated by the International Union for the Conservation of Nature, the World Wildlife Fund, and the UN Environment Programme, seeks to conserve biological diversity worldwide, to preserve vital ecosystem services, and to develop sustainable uses of organisms and their ecosystems.

4 Landscape Ecology and Conservation Biology

1. **Landscape ecology** is the interdisciplinary study of the relationship between spatial patterns and ecological processes on a number of different landscape scales and organizational levels. Landscape ecology looks at the "big picture" or landscape level of how structure affects the abundance, behaviour, and function of the ecosystem as a whole.

2. **Conservation biology** is the scientific study of how humans impact organisms and of the development of ways to protect biological diversity. Conservation biology involves two problem-solving tools: in situ conservation includes the establishment of parks and reserves, concentrating on preserving biological diversity in nature; ex situ conservation involves conserving biological diversity in human-controlled settings such as zoos and seed banks.

3. **Restoration ecology** is the study of the historical condition of a human-damaged ecosystem, with the goal of returning it as close as possible to its former state.

KEY TERMS

- **biological diversity** p. 202
- **species richness** p. 203
- **species evenness** p. 203
- **species dominance** p. 203
- **ecosystem goods and services** p. 205
- **extinction** p. 208

- **extirpated species** p. 208
- **endangered species** p. 208
- **threatened species** p. 208
- **special concern species** p. 209
- **endemic species** p. 210
- **biodiversity hotspots** p. 210

- **invasive species** p. 215
- **conservation** p. 217
- **preservation** p. 218
- **landscape ecology** p. 221
- **conservation biology** p. 222
- **restoration ecology** p. 224

CRITICAL AND CREATIVE THINKING QUESTIONS

1. Is biological diversity a renewable or nonrenewable resource? Why could it be seen both ways?

2. Relate the terms species richness, species evenness, and species dominance to the evaluation of an ecosystem that has high richness but low evenness to one that has low richness and high evenness.

3. Give at least five important ecosystem goods and services provided by living organisms.

4. If we preserve species solely on the basis of their potential economic value—such as a source of a novel drug—do they lose their "value" after we have capitalized on a newly discovered chemical? Why or why not?

5. What are the four main causes of species endangerment and extinction? Which cause do biologists consider most important?

6. What are invasive species?

7. Why are frogs and other amphibians considered bellwether species?

8. If you had the assets and authority to take any measure to protect and preserve biological diversity, but could take only one, what would it be?

9. The most recent version of the World Conservation Strategy includes stabilizing the human population. How would stabilizing the human population affect biological diversity?

10. In *A Sand County Almanac*, Aldo Leopold wrote, "To keep every cog and wheel is the first precaution of intelligent tinkering." How does his statement relate to this chapter?

11–13. The Nature Conservancy evaluated the extent of human-caused habitat disturbance in the world's various biomes. Use these data (graph below) to answer the following questions.

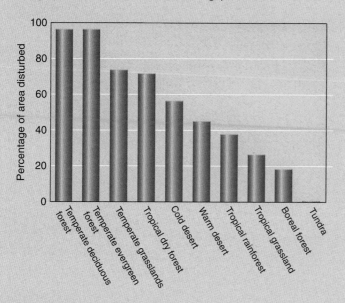

11. On which global region have humans had the greatest impact: polar, temperate, or tropical? Suggest why human impact has been greatest in this region.

12. What percentage of tropical rainforests has been disturbed by human activities? What percentage of temperate deciduous forests? Given these two values, why do you think people are more concerned about destruction in tropical rainforests than in temperate deciduous forests?

13. Which biome has had the lowest percentage of habitat disturbance? Suggest two possible reasons why human impacts on this biome may increase greatly in the future.

What is happening in this picture ?

At Zoo Atlanta in the United States, an actor playing an African park guard and a volunteer teach children about ivory poaching.

What are the goals of this effort?

Which conservation law bans hunting elephants for their ivory?

Do you think displaying elephants in zoos helps or hurts the species? Why?

Land Resources and Uses

WAPUSK NATIONAL PARK

Wapusk National Park, representing one of the wildest and most remote landscapes in Canada, was established in 1996. The park is located 45 kilometres south of Churchill, Manitoba, and is a part of the vast, low-lying Hudson Bay ecozone, the transitionary zone between the treeless tundra and the boreal forest. Due to its remote location, this is truly a wilderness park with no roads or visitor facilities to disturb the area.

Wapusk, which is the Cree word for "white bear," is an 11,476-square-kilometre park that protects, as its name would suggest, one of the world's largest known polar bear maternity denning areas (see photograph). Despite the continuous permafrost, this national park supports a wide variety of wildlife—close to 200 species of waterfowl and shorebirds and 44 species of mammals, including the Cape Churchill caribou herd. The national park supports a northern forest of white spruce, larch, and willow, and the tundra zone is made up of beach ridges, sedge meadows, peat land, and tundra ponds. The park's marine coastline consists of unique habitats found in the salt marshes, dunes, and beaches. There is also an extensive intertidal zone with up to 10 kilometres between high and low tides. These coastal areas provide habitat for hundreds of thousands of nesting, staging, and feeding birds, including rare species such as the Hudsonian godwit, Ross's and ivory gulls, Caspian tern, peregrine falcon, great grey owls, and sandhill cranes. Some current stressors in Wapusk National Park are climate change, long-range transport of contaminants, and overabundant snow geese populations, to name just a few.

Historically, Inuit, Dene, and Cree all lived in this region for more than 3000 years. In the 17th century, Metis and European traders arrived searching for the Northwest Passage to the orient. As part of the Wapusk National Park management plan, Parks Canada confirms the commitment to a greater inclusion of First Nations people in partnerships, economic tourism opportunities, and program development to tell the stories of the land and its people.

Global versus Canadian Land Resource Base and Uses

LEARNING OBJECTIVE

Summarize current land ownership in Canada.

The overall terrestrial land base of Earth is about 148,940,000 square kilometres. This amount of area is considerable indeed, and has resulted in the development of systems of ownership and management that vary from country to country and from ecosystem to ecosystem. These lands could include important resources such as minerals and fossil fuels, land that possesses historical or cultural significance, and land that provides critical biological habitat.

Canada contributes 9,984,670 square kilometres, or about 6 percent, of the world's total land area (**TABLE 8.1**). In Canada, ownership of land is most commonly divided into some form of private or Crown tenure. If privately held, rights of owners generally follow concepts of British common law where the owner holds a title to the land, which is protected by property laws.

However, only 11 percent of land in Canada is owned privately. The remaining 89 percent is government-owned or **Crown land**, which is land owned by either the federal or provincial governments. Under Section 92(a) of the British North America Act, the responsibility for management and administration of land in Canada is generally under the authority of the respective provincial

Land and freshwater area in Canada TABLE 8.1

	Total area	Land	Fresh water	% of total area
		square kilometres		
Canada	9,984,670	9,093,507	891,163	100.0
Newfoundland and Labrador	405,212	373,872	31,340	4.1
Prince Edward Island	5,660	5,660	0	0.1
Nova Scotia	55,284	53,338	1,946	0.6
New Brunswick	72,908	71,450	1,458	0.7
Quebec	1,542,056	1,365,128	176,928	15.4
Ontario	1,076,395	917,741	158,654	10.8
Manitoba	647,797	553,556	94,241	6.5
Saskatchewan	651,036	591,670	59,366	6.5
Alberta	661,848	642,317	19,531	6.6
British Columbia	944,735	925,186	19,549	9.5
Yukon	482,443	474,391	8,052	4.8
Northwest Territories	1,346,106	1,183,085	163,021	13.5
Nunavut	2,093,190	1,936,113	157,077	21.0

government of the area in question. Each provincial government in Canada has developed different approaches to the fulfillment of its responsibility. For example, in Manitoba, Crown land is managed by a number of regulations and various legislation. These laws allocate land for many different uses, including agriculture, forestry, parks, and wildlife protection. However, there are some exceptions to provincial authority for land. The Government of Canada is responsible for specific areas, such as national parks, national historic sites, and national marine conservation areas.

Government-owned land provides vital ecosystem services that benefit humans living far from public forests, grasslands, deserts, and wetlands. These services include wildlife habitat, flood and erosion control, groundwater recharge, and the breakdown of pollutants.

Undisturbed public land is an ecosystem that scientists use as a benchmark, or point of reference, to determine the impact of human activity. Geologists, zoologists, botanists, ecologists, and soil scientists are some of the scientists who use government-owned land for scientific inquiry. These areas provide perfect settings for educational experiences not only in science but also in history, because they can be used to demonstrate the condition of the land when humans originally settled here (FIGURE 8.1A).

Public or Crown land is also important for its recreational value, providing places for hiking, swimming, boating, rafting, sport hunting, and fishing (FIGURE 8.1B). Wild areas—forest-covered mountains, rolling prairies, barren deserts, and other undeveloped areas—are important to the human spirit. We can escape the tensions of the civilized world by retreating, even temporarily, to the solitude of natural areas.

In addition to privately owned land and Crown land, certain parcels of land are protected under the Constitution Act of 1982, acknowledging Aboriginal peoples' legal rights negotiated through treaties over a hundred years ago. Under the terms of treaties, Aboriginal peoples agreed to share the land in return for specific rights. Unfortunately, these treaties have not always been adhered to, which has led to significant Aboriginal land claims in recent decades. In addition, Aboriginal lands are generally still subject to government regulation and, as a result, there are fundamental disagreements between Aboriginal peoples and various levels of government as to what Aboriginal treaty rights include and to what degree these rights can be realized.

The many uses of Crown land FIGURE 8.1

A Public land provides many educational opportunities.

B National parks and other Crown land also provide all Canadians with opportunities to enjoy the vast natural beauty of our country. Photographed in Tombstone Territorial Park, Yukon.

As the country of Canada expanded, various uses of the land resource base began to become prevalent. Historically, these uses can be traced sequentially from agriculture to forestry to mining, with more recent development of parks and protected areas. For each land use, this chapter examines its extent, effect, and implications for the Canadian and global environment.

CONCEPT CHECK STOP

What percentage of land in Canada is privately owned?

Why are Crown lands so important?

Agricultural Land Uses

LEARNING OBJECTIVES

Contrast industrialized agriculture with subsistence agriculture and describe three kinds of subsistence agriculture.

Describe the environmental impacts of industrialized agriculture, including land degradation and habitat fragmentation.

Describe alternative ways to control pests.

Define desertification and explain its relationship to overgrazing.

AGRICULTURE IN CANADA

Agriculture is a business. It is responsible for producing what we eat and a lot of products we don't eat but use for other things, and it uses new technology and plain old common sense. Canada produces a wide range of agricultural products and is one of the largest agricultural producers in the world. Every year, Canada exports over $24 billion worth of agriculture and food products to 180 countries around the world.

The five largest agricultural production sectors in Canada are grains and oilseeds, livestock (red meats), dairy, horticulture, and poultry and eggs. In 2002, Canada exported $914.5 million in distilled spirits, beer, and wine and was the world's sixth largest fish and seafood exporter. In addition, Canada makes 85 percent of the world's maple syrup, producing close to 35,000 tonnes valued in 2002 at $164 million. Canada is also one of the top five producers of organic grains and oilseeds, with 5 percent of Canadian grain farms reported as organic.

And this is just a small sampling of the agricultural products Canada produces and exports!

Agriculture in Canada federally falls under the responsibility of the Department of Agriculture and Agri-Food, also referred to as Agriculture and Agri-Food Canada (AAFC). AAFC is responsible for policies relating to agriculture production, farming income, research and development, inspection, and the regulation of plants and animals. There are a number of organizations that AAFC is responsible for, including the Canadian Dairy Commission, the Canadian Food Inspection Agency, the Canadian Grain Commission, Farm Credit Canada, the National Farm Products Council, and the Prairie Farm Rehabilitation Administration, among others.

THE PRINCIPAL TYPES OF AGRICULTURE

Agriculture can be roughly divided into two types: industrialized and subsistence. Most farmers in highly developed countries and some in developing countries practise *high-input* or **industrialized agriculture**. This type of agriculture relies on large inputs of capital and energy (in the form of fossil fuels) to make and run machinery, irrigate crops, and produce agrochemicals like commercial inorganic fertilizers and pesticides (**FIGURE 8.2**) Industrialized agriculture produces high **yields** (the amount of food produced per unit of

industrialized agriculture Modern agricultural methods, which require a large capital input and less land and labour than traditional methods.

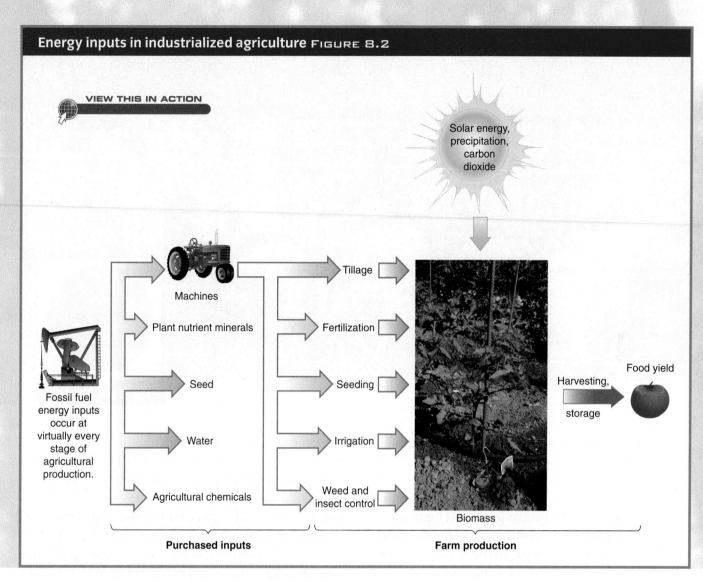

Energy inputs in industrialized agriculture FIGURE 8.2

VIEW THIS IN ACTION

Solar energy, precipitation, carbon dioxide

Machines

Plant nutrient minerals

Fossil fuel energy inputs occur at virtually every stage of agricultural production.

Seed

Water

Agricultural chemicals

Tillage

Fertilization

Seeding

Irrigation

Weed and insect control

Biomass

Harvesting, storage

Food yield

Purchased inputs

Farm production

land), which allows forests and other natural areas to remain wild instead of being converted to agricultural land. However, the productivity of industrialized agriculture comes with costs, such as soil degradation and increased pesticide resistance in agricultural pests; we discuss these and other problems later in the chapter.

Most farmers in developing countries practise **subsistence agriculture**, the production of enough food to feed oneself and one's family, with little left over to sell or reserve for hard times. Subsistence agriculture

subsistence agriculture
Traditional agricultural methods, which depend on labour and a large amount of land to produce enough food to feed oneself and one's family.

also requires a large input of energy, but from humans and draft animals rather than from fossil fuels.

Some types of subsistence agriculture require large tracts of land. **Shifting cultivation** is a form of subsistence agriculture in which short periods of cultivation are followed by longer periods of fallow (land left uncultivated), during which the land reverts to forest. Shifting cultivation supports relatively small populations. **Slash-and-burn agriculture** is a type of shifting cultivation that involves clearing small patches of tropical forest to plant crops. Because tropical soils lose their productivity quickly when they are cultivated, farmers using slash-and-burn agriculture must move from one area of forest to another every three years or so.

Agricultural Land Uses 237

Nomadic sheep herders in Kenya FIGURE 8.3

Global Locator

AFRICA
KENYA

Nomadic herding, in which livestock is supported by land too arid for successful crop growth, is a similarly land-intensive form of subsistence agriculture (FIGURE 8.3). Nomadic herders must continually move their livestock to find adequate food for them.

Intercropping is a form of intensive subsistence agriculture that involves growing a variety of plants simultaneously on the same field. When certain crops are grown together, they produce higher yields than when they are grown as **monocultures**. (A monoculture is the cultivation of only one type of plant over a large area.) **Polyculture** is a type of intercropping in which several kinds of plants that mature at different times are planted together. Where polyculture is practised in the tropics, fast- and slow-maturing crops are often planted together so that different crops can be harvested throughout the year.

CHALLENGES OF AGRICULTURE

Canada has more than 67.6 million hectares of **prime farmland**, land that has the soil type, growing conditions, and available water to produce food, forage, fibre, and oilseed crops. Of this land, over 530,000 hectares are now used to grow organic food, with another 48,000 hectares in transition to organic certification. But there is a decline in the area of prime farmland in Canada, as there is in most places in the world. Close to 94 percent of Canada's total land area is unsuitable for farming, and of that small percentage of land that will support agricultural endeavours, only 0.5 percent is designated as prime agricultural farmland where there is no significant limitation to farming activity. Pressures on prime farmland from urbanization, poor farming practices, and other nonagricultural activities mean that this already small percentage of viable farmland is shrinking even more and doing so at an alarming rate. Statistics Canada, for example, reported that between 1971 and 2001 over 14,000 square kilometres of Canada's prime agricultural land had been permanently lost to urban uses. Other challenges include coping with declining numbers of domesticated plant and animal varieties, improving crop and livestock yields, addressing environmental impacts, and controlling agricultural pests.

Loss of agricultural land There is considerable concern that much of our prime agricultural land is falling victim to urbanization and suburban sprawl by being converted to parking lots, housing developments,

Suburban spread onto agricultural land FIGURE 8.4

A housing development encroachs on farmland.

and shopping malls (**FIGURE 8.4**). In certain areas of Canada, loss of rural land is a significant problem.

Global decline in domesticated plant and animal varieties

A global trend is currently underway to replace the many local varieties of a particular crop or domesticated farm animal with just a few kinds. When farmers abandon traditional varieties in favour of more modern ones, which are bred for uniformity and maximum production, the traditional varieties frequently face extinction. This represents a great loss in genetic diversity, because each variety's characteristic combination of genes gives it distinctive nutritional value, size, colour, flavour, resistance to disease, and adaptability to different climates and soil types.

To preserve older, more diverse varieties of plants and animals, many countries, including Canada, are collecting **germplasm**: seeds, plants, and plant tissues of traditional crop varieties, and the sperm and eggs of traditional livestock breeds.

Increasing crop yields

Until the 1940s, agricultural yields of various countries, whether highly developed or developing, were generally equal. Advances by research scientists since then have dramatically increased food production in highly developed countries. Greater knowledge of plant nutrition has resulted in fertilizers that promote high yields. The use of pesticides to control insects, weeds, and disease-causing organisms has also improved crop yields.

The green revolution

By the middle of the 20th century, serious food shortages occurred in many developing countries coping with growing populations. During the 1960s, high-yield varieties of wheat and rice were developed and introduced to Asian and Latin American countries, giving these nations the chance to provide their people with adequate supplies of food (**FIGURE 8.5** on page 240). But the high-yield varieties required intensive industrial cultivation methods, including the use of commercial inorganic fertilizers, pesticides, and mechanized machinery, to realize their potential. These agricultural technologies were passed from highly developed nations to developing nations.

Using modern cultivation methods and the high-yield varieties of certain staple crops to produce more food per acre of cropland is known as the **green revolution**. Some of the success stories of the green revolution are remarkable. During the 1920s, Mexico produced less than 700 kilograms of wheat per hectare annually. During the green revolution years that began in 1965, Mexico's annual wheat production rose to more than 2400 kilograms per hectare. Indonesia, which formerly imported more rice than any other country in the world, today produces enough rice to feed its people and export some.

Development of high-yield rice varieties FIGURE 8.5

Traditional rice plants on the left are taller and do not yield as much grain (clusters at top of plant) as the more modern varieties shown in the middle and on the right. The rice plant in the middle was developed during the 1960s by crossing a high-yield, disease-resistant variety with a dwarf variety to prevent the grain-heavy plants from falling over. Improvements since the green revolution have been modest, as the rice variety developed during the 1990s (right) shows. Some researchers think rice and certain other genetically improved crops are near their physical limits of productivity.

Tall conventional plant   Improved high-yielding plant   Low-tillering ideotype (new plant type)

Critics of the green revolution argue that it has made developing countries dependent on imported technologies, such as agrochemicals and tractors, at the expense of traditional agriculture. The two most important problems associated with higher crop production are the high energy costs built into this type of agriculture and the environmental problems caused by the intensive use of commercial inorganic fertilizers and pesticides (discussed later in this chapter).

Increasing crop yields in the post–green revolution era

In 1999, the International Food Policy Research Institute projected that the world demand for rice, wheat, and corn would increase 40 percent between 2000 and 2020. Such a rise in demand would require a corresponding rise in grain production to feed the human population, as well as the livestock needed to satisfy the appetites of the increasing numbers of affluent people who can afford to buy meat.

This challenge cannot be met by increasing the amount of land under cultivation, as the best arable lands are already being cultivated. Projected freshwater shortages, rising costs of agricultural chemicals, and deteriorating soil quality caused by intensive agricultural techniques may further constrain productivity. As Figure 8.5 demonstrates, recent progress in coaxing more grain out of crops genetically improved during the green revolution has resulted in diminishing returns.

Grain yields have continued to rise since the 1960s, but in recent years the rates of increase have not been as great as they were previously.

Despite these problems, most plant geneticists think we can produce enough food in the 21st century to meet demand if countries spend more money in support of a concerted scientific effort to improve crops. Many scientists think genetic engineering is one of the keys to breeding more productive varieties. Additionally, modern agricultural methods, such as water-efficient irrigation, will have to be introduced to developing countries that do not currently have them if we are to continue increasing crop yields.

Environmental impacts

Industrialized agriculture has many environmental effects (**FIGURE 8.6A**). The agricultural use of fossil fuels and pesticides produces air pollution. Untreated animal wastes and agricultural chemicals such as fertilizers and pesticides cause water pollution, which reduces biological diversity, harms fisheries, and leads to outbreaks of nuisance species.

Industrialized agriculture has favoured the replacement of traditional family farms by large agribusiness conglomerates. In North America, most cattle, hogs, and poultry are now raised in feedlots and livestock factories (**FIGURE 8.6B**). The large concentrations of animals in livestock factories create many environmental problems, including air and water pollution.

A Environmental effects of industrialized agriculture

Individual agriculture practices may have multiple environmental effects.

Water issues	Air pollution	Land degradation	Loss of biological diversity
• Groundwater depletion from irrigation • Pollution from fertilizer and pesticide runoff • Sediment pollution from eroding soil particles • Pollution from animal wastes (livestock factories) • Enrichment of surface water from fertilizer runoff and livestock wastes	• Pesticide sprays • Soil particles from wind erosion • Odours from livestock factories • Greenhouse gases from combustion of fossil fuels • Other air pollutants from combustion of fossil fuels	• Soil erosion • Loss of soil fertility • Soil salinization • Soil pollution (pesticide residues) • Waterlogged soil from improper irrigation	• Habitat fragmentation (clearing land and draining wetlands) • Monocultures (lack of diversity in croplands) • Stressors from air and water pollution • Stressors from pesticides • Replacement of many traditional crop and livestock varieties with just a few

B Hog factory

The hogs remain indoors and are fed and watered by machine at timed intervals throughout the day. Although livestock factories are efficient and produce meat relatively inexpensively, they cause environmental issues such as problems related to sewage disposal.

Land degradation is a reduction in the potential productivity of land. Soil erosion, which is exacerbated by large-scale mechanized operations, causes a decline in soil fertility, and the eroded sediments damage water quality. Other types of degradation are compaction of soil by heavy farm machinery and waterlogging and salinization (salting) of soil from unsustainable agricultural practices. In Prince Edward Island, potato farming has caused nitrate contamination of soils.

Clearing grasslands and forests and draining wetlands to grow crops result in **habitat fragmentation** that reduces biological diversity. Many species are endangered or threatened as a result of habitat loss to agriculture. The most dramatic example of habitat loss in North America is tall grass prairie, more than 90 percent of which has been converted to agriculture.

Controlling agricultural pests

Any organism that interferes in some way with human welfare or activities is a **pest**. Some weeds, insects, rodents, bacteria, fungi, nematodes (microscopic worms), and other pest organisms compete with humans for food; other pests cause or spread disease. People try to control pests, usually by reducing the size of the pest population. **Pesticides** are the most common way of doing this, particularly in agriculture.

The ideal pesticide is a **narrow-spectrum pesticide** that kills only the intended organism and does not harm any other species. The perfect pesticide would readily break down, either by natural chemical decomposition or by biological organisms, into safe materials such as water, carbon dioxide, and oxygen. The ideal pesticide would stay exactly where it was put and would not move around in the environment. Unfortunately, no pesticide is perfect. Most pesticides are **broad-spectrum pesticides**. Some pesticides do not degrade readily, or they break down into compounds as dangerous as—if not

land degradation Natural or human-induced processes that decrease the future ability of the land to support crops or livestock.

habitat fragmentation The breakup of large areas of habitat into small, isolated patches.

pesticide Any toxic chemical used to kill pests.

biological control A method of pest control that involves the use of naturally occurring disease organisms, parasites, or predators to control pests.

integrated pest management (IPM) A combination of pest control methods that, if used in the proper order and at the proper times, keeps the size of a pest population low enough to prevent substantial economic loss.

more dangerous than—the original pesticide. And most pesticides move around the environment quite a bit.

Although pesticides have some benefits, such as controlling mosquito populations to reduce the incidence of malaria in many parts of the world and protecting crops, there are numerous disadvantages to using pesticides. Many insects, weeds, and disease-causing organisms have developed or are developing resistance to pesticides. Pesticide resistance forces farmers to apply progressively larger quantities of pesticides. Pesticide residues contaminate our food supply and reduce the number and diversity of beneficial microorganisms in the soil. Fishes and other aquatic organisms are sometimes killed by pesticide runoff into lakes, rivers, and estuaries. Further, pesticides kill more than just the pest being targeted. This means that beneficial insects and sometimes the pest's natural enemies are also being killed.

Given the many problems associated with them, pesticides are clearly not the final solution to pest control. Fortunately, pesticides are not the only weapons in our arsenal. Alternative ways to control pests include cultivation methods, **biological controls**, pheromones and hormones, reproductive controls, genetic controls, quarantine, and irradiation (**Table 8.2**).

Integrated pest management

Many pests are not controlled effectively with a single technique; a combination of control methods is often more effective. **Integrated pest management (IPM)** combines the use of a variety of biological, cultivation, and pesticide controls tailored to the conditions and crops of an individual farm (**Figure 8.7**). Biological and genetic controls, including genetically modified crops designed to resist pests, are used as much as possible, and conventional pesticides are used sparingly and only when other methods fail. When pesticides are required, the least toxic pesticides are applied in the lowest possible effective quantities. Thus, IPM

Alternative methods of controlling agricultural pests TABLE 8.2

Pest control method	How it works	Disadvantages
Cultivation methods	Interplanting mixes different plants, as by alternating rows; strip cutting alternates crop harvest by portion—remaining portions protect natural predators, parasites of pests	No appreciable disadvantages; more care must be taken in harvest
Biological controls	Naturally occurring predators, parasites, or disease organisms are used to reduce pest populations	Organism introduced for biological control can unexpectedly affect environment or other organisms
Pheromones and hormones	Sexual attractants (pheromones) lure pest species to traps; synthetic regulatory chemicals (hormones) disrupt pests' growth and development	Might affect beneficial species
Reproductive controls	Sterilizing some members of pest population reduces population size	Expensive; must be carried out continually
Genetic controls	Selective breeding develops pest-resistant crops	Plant pathogens evolve rapidly, adapting to disease-resistant host plant; plant breeders forced to constantly develop new strains
Quarantine	Governments restrict importation of foreign pests, diseases	Not foolproof; pests are accidentally introduced
Irradiating foods	Harvested foods are exposed to ionizing radiation that kills potentially harmful microorganisms	Consumers concerned about potential radioactivity (not a true risk); irradiation forms traces of potentially carcinogenic chemicals (free radicals)

Tools of integrated pest management (IPM)
FIGURE 8.7

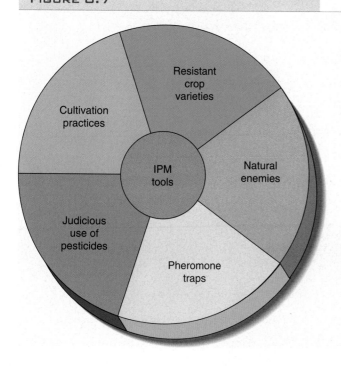

allows the farmer to control pests with a minimum of environmental disturbance and often at minimal cost.

To be effective, IPM requires a thorough knowledge of the life cycles and feeding habits of the pests as well as all their interactions with their hosts and other organisms. The timing of planting, cultivation, and treatments with biological controls is determined by carefully monitoring the concentration of pests. IPM is an important part of sustainable agriculture (discussed later in this chapter).

Problems with pesticides
Although pesticides have their benefits, they also have several problems. For one thing, the prolonged use of a particular pesticide can cause a pest population to develop **genetic resistance** to the pesticide. In the 50 years during which pesticides have been widely used, at least 520 species of insects and mites and at least 84 weed species have evolved genetic resistance to certain pesticides.

An organism exposed to a chemically stable pesticide that takes years to break down may accumulate high concentrations of the toxin, a phenomenon known as *bioaccumulation*. Organisms at higher levels on food webs tend to have greater concentrations of bioaccumulated pesticide stored in their bodies than those lower on food webs, through a process known as *biological magnification* (see Figure 4.10 on page 111).

RANGELANDS FOR LIVESTOCK PRODUCTION

Rangelands are grasslands, in both temperate and tropical climates, that serve as important areas of food production for humans by providing fodder for livestock such as cattle, sheep, and goats (**FIGURE 8.8**). Rangelands provide habitat for native plants and animals and support a broad range of economic benefits including biological research, development of nutraceuticals, and recreational activities such as hunting and eco-tourism. The predominant vegetation of rangelands includes grasses, forbs (small plants other than grasses), and shrubs.

rangeland Land that is not intensively managed and is used for grazing livestock.

Historically, North American grasslands evolved through a co-evolution of native plants and grazing regimes. Prior to settlement, bison, deer, elk, and pronghorn applied grazing pressures to grasslands. Grasslands were further shaped by the activities of ice movement, drought, flooding, topography, and climate. Bison established seasonal grazing behaviours that were based on forage availability, climate, their shelter needs, and fire patterns. Although bison applied considerable grazing pressure during portions of the year, they also moved on, creating long periods of rest for the grasslands to recover. Proper grazing patterns can stimulate plant growth,

Rangeland FIGURE 8.8

Rangeland is considered a renewable resource when its carrying capacity—the number of animals it can sustain without suffering deterioration—is not exceeded.

maintain optimal leaf area and enhance the nutritive value, accelerate nutrient cycling, and remove excess litter.

Rangeland degradation and desertification

Grasses, the predominant vegetation of rangelands, have a *fibrous root system,* in which many roots form a diffuse network in the soil to anchor the plant. Plants with fibrous roots hold the soil in place quite well, thereby reducing soil erosion. Grazing animals eat the leafy shoots of the grass, and the fibrous roots continue to develop, allowing the plants to recover and regrow to their original size.

Carefully managed grazing is beneficial for grasslands. Because rangeland vegetation is naturally adapted to grazing, when grazing animals remove mature vegetation, the activity stimulates rapid regrowth. At the same time, the hooves of grazing animals disturb the soil surface enough to allow rainfall to more effectively reach the root systems of grazing plants. Several studies report that moderate levels of grazing encourage greater plant diversity.

The **carrying capacity** of a rangeland is the maximum number of animals the natural vegetation can sustain over an indefinite period without deterioration of the ecosystem. When the carrying capacity of a rangeland is exceeded, **overgrazing** of grasses and other plants occur. When plants die, the ground is left barren, and the exposed soil is susceptible to erosion. Sometimes plants that do not naturally grow in a rangeland, but which can tolerate the depleted soil, invade an overgrazed area.

Most of the world's rangelands lie in semi-arid areas that have natural extended droughts. Under normal conditions, native grasses in these dry lands can survive a severe drought. But when an extended drought occurs in conjunction with overgrazing, once-fertile rangeland may be converted to desert as reduced grass cover allows winds to erode the soil. Even when the rains return, the degradation may be so extensive that the land cannot recover. Water erosion removes the little remaining topsoil and the sand left behind forms dunes.

This progressive degradation, which induces unproductive desert-like conditions on formerly productive rangeland (or tropical dry forest), is **desertification**

■ **overgrazing** When too many grazing animals consume the plants in a particular area, leaving the vegetation destroyed and unable to recover.

■ **desertification** Degradation of once-fertile rangeland or tropical dry forest into nonproductive desert.

(**Figure 8.9** on page 246). It reduces the agricultural productivity of economically valuable land, forces many organisms out, and threatens endangered species. Worldwide, desertification seems to be on the increase. The United Nations estimates that each year since the mid-1990s, 3560 square kilometres—an area about two-thirds the size of Prince Edward Island—has turned into desert.

Rangeland trends in Canada

Rangelands occupy 47 percent of the Earth's total land area. In Canada, there are only about 70 million acres that are suitable for livestock grazing compared with the United States, where 770 million acres, or close to one-third of the total land area, are rangelands. Much of the private rangeland in Canada is under increasing pressure from developers who want to subdivide the land into lots for houses and condominiums. To preserve the open land, conservation groups often pay ranchers for **conservation easements** that prevent future owners from developing the land.

Much of the public rangeland in Canada is managed as provincial Crown land, but Agriculture and Agri-Food Canada also plays an important role through their Agri-Environmental Services Branch (AESB). Currently, AESB manages 915,000 hectares of rangelands in Western Canada, of which 85 percent is native rangeland. The overall goal of management of these rangelands is to preserve the productivity and biodiversity of native grasslands by ensuring grazing practices maintain the land in "good" range condition, meaning 50 to 75 percent of the biomass of the original vegetation is maintained. If rangelands are in good condition, then they support the conservation of biodiversity, reduced soil degradation, and potential increase of carbon sequestration. The AESB conducts land biodiversity inventories and monitors and documents resource use on rangelands to measure range conditions, range condition trends, and species composition. Unfortunately, most of the native grasslands have been cultivated or reseeded, and it is estimated that less than half of the prairie rangelands are classified as being in good condition.

Desertification in the African Sahel region, Niger FIGURE 8.9

As the goats consume the remaining grass and shrubs, the dunes of the Sahara Desert will inevitably continue to encroach upon the pastureland. There is still much that scientists do not understand about desertification and the processes related to it, such as to what extent desertification results from natural fluctuations in climate versus population pressures and human activities.

Global Locator

Restoration of grasslands or rangelands is slow and costly, and more is needed. Rangeland management includes seeding in places where plant cover is sparse or absent, conducting controlled burns to suppress shrubby plants, constructing fences to allow rotational grazing, controlling invasive weeds, and protecting habitats of endangered species. Today, careful analysis of the historical livestock use of grasslands, the soil type and moisture conditions, and the species composition of the land is used to determine the carrying capacity of each pasture. Adjustments to stocking rates are made annually depending on pasture conditions. During periods of drought it is especially important to reduce stocking rates to avoid the rangeland from becoming overgrazed. Under proper management conditions, livestock can be used to mimic the historical grazing patterns of ungulates and improve the overall health and function of the grassland.

The AESB provides three essential service programs to encourage preservation of the rangelands while also helping farmers earn a living: the Community Pasture Program, the Cover Crop Protection Program (CCPP), and the Prairie Shelter Belt Program. The Community Pasture Program was created in the 1930s to help reclaim badly eroded lands and has, to date, successfully restored 145,000 hectares of poor-quality land back to grass cover. The program works with land owners and uses careful management of grazing cattle as the primary

Grasslands National Park, representing the Prairie Grassland natural region of Canada, is located in southwestern Saskatchewan and is one of the country's newest national parks. The park will eventually cover 900 square kilometres along the Canada–U.S. border.

The area is characterized by unique landscapes and a harsh climate and is home to many specially adapted plants and animals. The park and its surrounding area provide a home to the country's only black-tailed prairie dog species along with other endangered species including the pronghorn, antelope, sage grouse, burrowing owl, ferruginous hawk, prairie rattlesnake, and eastern short-horned lizard. In December 2005, plains bison were reintroduced to Grasslands National Park after 120 years

of absence (see photograph). Prior to European settlement the area was home to millions of free-ranging bison. By the 1880s, however, the large bison herds were nearly gone.

The large open range of the grasslands in this area later attracted ranchers with large cattle operations. These ranchers soon realized the difficulties of living in such a demanding environment. Those who persevered combined their ranching activities with farming, creating the country communities that surround the park today.

For more information on Grasslands National Park, please visit the Parks Canada website at www.pc.gc.ca.

tool to maintain a healthy, diverse, and functioning prairie landscape. The CCPP was initiated in 2006 to provide financial assistance to Canadian agricultural producers who were unable to seed commercial crops because of flooding. This federal assistance supports farmers and protects flood-damaged soil from being seeded again immediately. The Prairie Shelter Belt Program promotes the environmental and economic benefits of integrating trees with agricultural systems by providing prairie farmers with research findings and tree seedlings for planting. Through these programs, efforts are being made to provide land managers with solutions to the challenges of applying effective stewardship of Canada's rangelands.

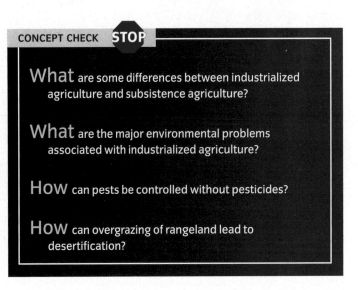

CONCEPT CHECK STOP

What are some differences between industrialized agriculture and subsistence agriculture?

What are the major environmental problems associated with industrialized agriculture?

How can pests be controlled without pesticides?

How can overgrazing of rangeland lead to desertification?

Forests

Forests, important ecosystems that provide many goods and services to support human society, occupy less than one-third of Earth's total land area. In Canada, forests account for about 45 percent of the country's land base and represent about 10 percent of global forest resources (**FIGURE 8.10**). The boreal forest biome in Canada is the largest area in which forestry is practised, but 45 percent of the forests cut in Canada are in British Columbia.

In terms of ownership, 94 percent of the land base on which forestry is practised in Canada is Crown land, with 71 percent of that area in provincial ownership.

A forest ecosystem can be described as a plant community dominated by trees and other woody plants, but that also contains shrubs, forbs, and grasses. Forests are generally distinguished from one another in terms of dominant species, species density, age of trees, or soil types. Forest ecosystems are usually divided into vertical layers: the canopy, shrub, and forest floor. And in all cases today, forest ecosystems must include all other organisms, including humans, and their respective needs.

Timber harvested from forests is used for fuel, construction materials, and paper products. Forests supply nuts, mushrooms, fruits, and medicines. Forests provide employment for millions of people worldwide and offer

Forest regions of Canada FIGURE 8.10

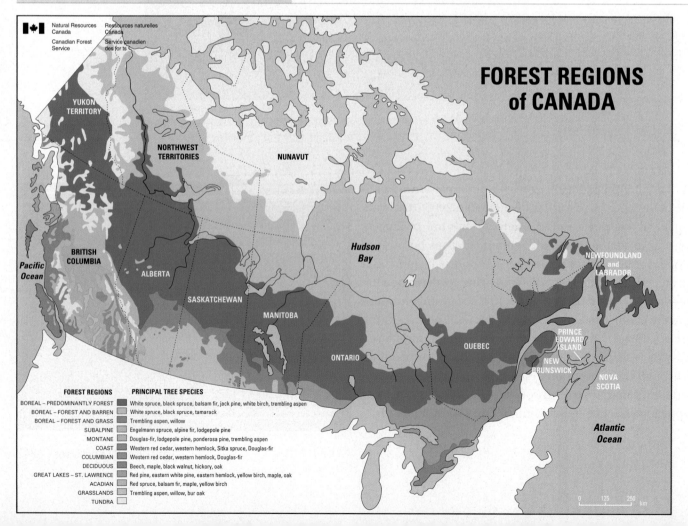

recreation and spiritual sustenance to an increasingly crowded world.

Forests also provide a variety of beneficial ecosystem services, such as influencing climate conditions. If you walk into a forest on a hot summer day, you will notice that the air is cooler and moister than it is outside the forest. This is the result of a biological cooling process called *transpiration,* in which water from the soil is absorbed by roots, transported through plants, and then evaporated from their leaves and stems. Transpiration provides moisture for clouds, eventually resulting in precipitation (**FIGURE 8.11**). Thus, forests help maintain local and regional precipitation.

Forests play an essential role in regulating global biogeochemical cycles like those for carbon and nitrogen. Photosynthesis by trees removes large quantities of heat-trapping carbon dioxide from the atmosphere and fixes it into carbon compounds while releasing the oxygen back into the atmosphere. Forests thus act as carbon "sinks," which may help mitigate global warming and which produce the oxygen that almost all organisms require for cellular respiration.

Tree roots hold vast tracts of soil in place, reducing erosion and mudslides. Forests protect watersheds because they absorb, hold, and slowly release water; this moderation of water flow provides a more regulated flow of water downstream, even during dry periods, and helps control floods and droughts. Forest soils remove impurities from water, improving its quality. In addition, forests provide a variety of essential habitats for many organisms, such as mammals, reptiles, amphibians, fishes, insects, lichens and fungi, mosses, ferns, conifers, and numerous kinds of flowering plants.

Role of forests in the hydrologic cycle FIGURE 8.11

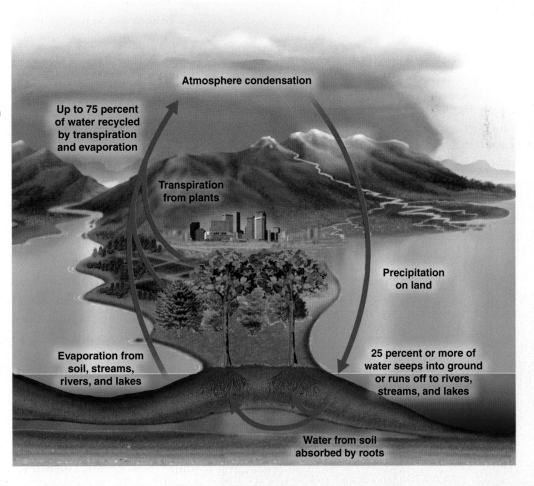

Forests return most of the water that falls as precipitation to the atmosphere by transpiration. When an area is deforested, almost all precipitation is lost as runoff.

Atmosphere condensation

Up to 75 percent of water recycled by transpiration and evaporation

Transpiration from plants

Precipitation on land

Evaporation from soil, streams, rivers, and lakes

25 percent or more of water seeps into ground or runs off to rivers, streams, and lakes

Water from soil absorbed by roots

Process Diagram

FOREST MANAGEMENT

Concerns about the effects of logging on biodiversity and sustainability have strongly influenced forest management decisions, policy debates, and the development of silvicultural or revegetation systems. As a result, foresters are faced with the challenge of creating forestry management plans that meet operational objectives yet also consider the biological, physical, and ecological impacts of their actions.

Commercial logging operations have long been controversial. Industry supporters suggest that logging is a viable method of disturbance that actually leads to regeneration of forest vegetative species (**FIGURE 8.12**), that it is an economically advantageous use of a sustainable resource, and that it supports community lifestyles. In contrast, critics maintain that the forest exists to satisfy a variety of other needs such as providing wildlife habitat, enabling recreational activities, and contributing an aesthetic value.

Forest regeneration after harvest
FIGURE 8.12

Most logging operations in Canada attempt to emulate natural disturbances (such as forest fires) during harvest. Although they cannot exactly duplicate natural disturbances, these practices do help maintain the diverse ecosystems present in forests.

The debate turns on whether logging, due to greater severity of disturbance and alteration of the forest age structure, impacts the ecosystem and wildlife more profoundly than do natural disturbances. To find out, researchers have studied the effects of logging on a variety of wildlife species. Many wildlife species take advantage of newly logged areas. Owls, for example, move into the fragmented habitat to search for displaced voles and field mice. Large ungulates such as moose and deer move into newly logged areas and feed on spread branches and seedlings. Predator populations, such as grey wolves, also benefit from a greater abundance of prey in the area. However, logging is not good for all species. Declines in woodland caribou populations have been attributed to human disturbance near caribou range. Researchers suggest that management of woodland caribou should incorporate buffers around habitat and ensure long-term monitoring of range occupancy. Further research is needed to determine the long-term effects logging has on wildlife.

In addition to its effects on wildlife, commercial logging has significant effects on various plant species. In some parts of North America, specific varieties of commercially important trees are planted, and those trees not as commercially desirable are thinned out or removed. *Traditional forest management* often results in low-diversity forests. In the southeastern United States, many tree plantations of young pine grown for timber and paper production are all the same age and are planted in rows a fixed distance apart (**FIGURE 8.13**). These "forests" are essentially monocultures—areas uniformly covered by one crop, like a field of corn. Herbicides are sprayed to kill shrubs and herbaceous plants between the rows. One of the disadvantages of monocultures is that they are more prone to damage from insect pests and disease-causing microorganisms. Also, because managed forests contain few kinds of food, they can't support the variety of organisms typically found in natural forests.

In recognition of the many ecosystem services performed by natural forests, a newer method of forest management, known as ecologically sustainable forest management, or simply **sustainable forestry**, is evolving.

> ■ **sustainable forestry** The use and management of forest ecosystems in an environmentally balanced and enduring way.

Tree plantation FIGURE 8.13

This intensively managed pine plantation in the southeastern United States is a monoculture, with trees of uniform size and age. Such plantations supplement harvesting of trees in wild forests to provide the United States with the timber it requires.

Sustainable forestry can be described as the art and science of controlling the establishment, growth, composition, health, and quality of forests and woodlands to meet the diverse needs and values of landowners and society on a sustainable basis. This requires that sustainable development considerations—economic, social, and environmental—are considered in the forest operation. Sustainable forestry maintains a mix of forest trees by age and species, rather than imposing a monoculture. This broader approach seeks to conserve forests for the long-term commercial harvest of timber and non-timber forest products. Sustainable forestry also attempts to sustain biological diversity by providing habitats for a variety of species, prevent soil erosion and improve soil conditions, and preserve watersheds that produce clean water. Effective sustainable forest management involves cooperation among environmentalists, loggers, farmers, indigenous people, and local, provincial, and federal governments.

When loggers use sustainable forestry principles, they set aside unlogged areas and **wildlife corridors** as sanctuaries for organisms. The purpose of wildlife corridors is to provide animals with escape routes, should they need them, and to allow animals to migrate so they can interbreed. (Small, isolated, inbred populations may have a higher risk of extinction.) Wildlife corridors may also allow large animals to maintain large territories. Recent research on wildlife corridors

> **wildlife corridor**
> A protected zone that connects isolated unlogged or undeveloped areas.

in fragmented landscapes suggests that they help certain wildlife populations persist. Additional research is needed to resolve the questions of effectiveness of wildlife corridors for all endangered species.

Methods for ecologically sustainable forest management are under development. Such practices vary from one forest ecosystem to another, in response to different environmental, cultural, and economic conditions. In Mexico, for example, many sustainable forestry projects involve communities that are economically dependent on forests. Because trees have such long life spans, scientists and forest managers of the future will judge the results of today's efforts.

Increasingly, forest companies are being required to ensure that their operations meet particular forest certification standards, which are checked by independent auditors. There are a variety of forest certification standards. The most basic is an inventory of operations through the International Standards of Operation (ISO). There are also three main certification standards: those from the Sustainable Forestry Initiative (SFI), the Canadian Standards Association (CSA), and the Forest Stewardship Council (FSC).

The FSC certification standard is the most difficult to achieve and requires forestry operations to meet stringent standards. The primary goal of FSC is to promote environmental responsibility. FSC standards, which include ensuring at least 20 percent of forests remain old growth and leaving large core blocks of forest that can contain no more than 5 percent new growth, ensure protection of biodiversity, the ethical treatment of indigenous communities, employee benefits, proper onsite cleanup and waste management, and modification for protection of species at risk. Becoming FSC certified is a lengthy, exhaustive, and expensive process for a forestry operation. The result, however, is recognition of the company's commitment to environmental responsibility. For more on FSC certified wood, see the EnviroDiscovery box.

Harvesting trees According to the Food and Agricultural Organization (FAO), 3.5 million cubic metres of wood (for fuelwood, timber, and other products) were harvested in 2005 (the latest available data).

EnviroDiscovery ECOLOGICALLY CERTIFIED WOOD

The Forest Stewardship Council ecologically certifies "green" wood.

Green timber has proven profitable for its national distributors, who have seen sales increases of 20 to 30 percent. Often, the consumer pays no additional premium, or only slightly more, for ecologically certified wood, which has become so popular that demand threatens to exceed supply.

Many homebuilders and homeowners are interested in "green" wood for flooring and other building materials *(see photograph).* Such wood is ecologically certified by a legitimate third party such as the Forest Stewardship Council (FSC) to have come from a forest managed with environmentally sound and socially responsible practices. Although these areas remain a small percentage of world forests, by early 2005 the FSC had certified as well managed more than 48 million hectares in 61 countries. Certification is based on sustainability of timber resources, socioeconomic benefits provided to local people, and forest ecosystem health, which includes such considerations as preservation of wildlife habitat and watershed stability.

Green forestry has its detractors. Traditional forestry organizations are sceptical about the reliability of FSC investigations and the economic viability of this type of forestry. Trade experts caution that government efforts to specify the purchase of certified timber could violate global free-trade agreements. Still, green timber is gaining market share, pleasing business owners and consumers alike, and offering the promise of better conservation practices in managed forests.

For more information on FSC Canada, please visit their website at www.fsccanada.org.

The five countries with the greatest tree harvests are the United States, Canada, Russia, Brazil, and China; these countries currently produce more than half the world's timber. About 50 percent of harvested wood is burned directly as fuelwood or used to make charcoal. Most fuelwood and charcoal are used in developing countries. Highly developed countries consume more than three-fourths of the remaining 50 percent for paper and wood products.

Loggers harvest trees in several ways—selective cutting, shelterwood cutting, seed tree cutting, and clear-cutting (see the What a Scientist Sees feature on page 254). **Selective cutting**, in which mature trees are cut individually or in small clusters while the rest of the forest remains intact, allows the forest to regenerate naturally.

The removal of all mature trees in an area over an extended period is **shelterwood cutting**. In the first year, undesirable tree species and dead or diseased trees are removed. Subsequent harvests occur at intervals of several years, allowing time for remaining trees to grow.

In **seed tree cutting**, almost all trees are harvested from an area; a scattering of desirable trees is left behind to provide seeds for the regeneration of the forest.

Clear-cutting is harvesting timber by removing all trees from an area, then either allowing the area to reseed and regenerate itself naturally or planting the area with one or more specific varieties of trees. Timber companies prefer clear-cutting because it is the most cost-effective way to harvest trees. However, clear-cutting over wide areas is ecologically unsound. It destroys biological habitats and increases soil erosion, particularly on sloping land, sometimes degrading land so much that reforestation doesn't take place. Clear-cut areas at lower elevations are usually regenerated successfully, whereas those at high elevations are difficult to regenerate. Obviously, the recreational benefits of forests are lost when clear-cutting occurs.

■ **clear-cutting** A logging practice in which all the trees in a stand of forest are cut, leaving just the stumps.

DEFORESTATION

The most serious problem facing the world's forests is **deforestation**. According to the

■ **deforestation**
The temporary or permanent clearance of large expanses of forest for agriculture or other uses.

latest estimates by the FAO, world forests shrank by about 36 million hectares each year between 2000 and 2005. This estimate represents a total loss of about 1 percent of global forested area. Causes of this decades-long trend of deforestation include fires caused by drought and land-clearing practices, expansion of agriculture, construction of roads, tree harvests, insects, disease, and mining.

Most of the world's deforestation is currently taking place in Africa and South America, according to the FAO. Africa lost about 3.2 percent of its forested area from 2000 to 2005, and South America lost about 2.5 percent. Central America and the Caribbean nations lost about 3.9 percent of their forests during this period. There was a very small loss of forested area in North America, whereas Europe and Asia actually gained forested areas, either by natural regrowth or by increasing forest plantations.

Results of deforestation Deforestation results in decreased soil fertility, as the essential mineral nutrients found in most forest soils leach away rapidly without trees to absorb them. Uncontrolled soil erosion, particularly on steep deforested slopes, affects the production of hydroelectric power as silt builds up behind dams. Increased sedimentation of waterways caused by soil erosion harms downstream fisheries. In drier areas, deforestation contributes to the formation of deserts. Regulation of water flow is disrupted when a forest is removed, so that the affected region experiences alternating periods of flood and drought.

Deforestation contributes to the extinction of many species. (See Chapter 7 for a discussion of the importance of tropical forests as repositories of biological diversity.) Many tropical species, in particular, have limited ranges within a forest, so they are especially vulnerable to habitat modification and destruction. Migratory species, including birds and butterflies, also suffer from deforestation.

Deforestation is thought to induce regional and global climate change. Trees release substantial amounts of moisture into the air; about 97 percent of the water that roots absorb from the soil evaporates

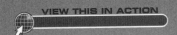

Harvesting Trees

A Aerial view of a large patch of clear-cut forest in British Columbia. Clear-cutting is the most common but most controversial type of logging. The obvious line is a road built to haul away the logs.

B (1 to 3) As a forest scientist looks at the clear-cut forest, he or she may think about the various kinds of tree harvesting that are less environmentally destructive than clear-cutting (4).

(1) In selective cutting, the older, mature trees are selectively harvested from time to time and the forest regenerates itself naturally.

(2) In shelterwood cutting, less desirable and dead trees are harvested. As younger trees mature, they produce seedlings, which continue to grow as the now-mature trees are harvested.

(3) Seed tree cutting involves the removal of all but a few trees, which are allowed to remain, providing seeds for natural regeneration.

(4) In clear-cutting, all trees are removed from a particular site. Clear-cut areas may be reseeded or allowed to regenerate naturally.

What a Scientist Sees

directly into the atmosphere, then falls back to Earth in the hydrologic cycle. When a large forested area is removed, local rainfall may decline, droughts may become more common in that region, and temperatures may rise slightly. Deforestation may also contribute to an increase in global temperature by releasing carbon originally stored in the trees into the atmosphere as carbon dioxide, which enables the air to retain heat. Researchers estimate that when an old-growth forest is harvested, it takes about 200 years for the replacement forest to accumulate the equivalent amount of carbon stored in the original trees.

Boreal forests Extensive deforestation in boreal forests due to logging began in the late 1980s. Boreal forests occur throughout Canada, Alaska, Scandinavia, and northern Russia, and are dominated by coniferous evergreen trees such as spruce, pine, fir, cedar, and hemlock. Boreal forests are an important source of industrial wood and wood fibre. The annual loss of boreal forests is estimated to encompass an area twice as large as the rainforests of Brazil.

At 1.3 billion acres, the Canadian boreal forest is one of the largest intact forest and wetland ecosystems remaining on Earth. It is a major source of North America's fresh water and home to the some of the planet's largest populations of wolves, grizzly bears, and woodland caribou. Its vast lakes and rivers offer up fish in abundance and its trees and wetlands provide nesting grounds for billions of songbirds and waterfowl. Hundreds of First Nations communities also depend on the boreal forest ecosystem for fish and wildlife. The boreal forest is under increasing pressure from logging, mining, and oil and gas operations, and only 10 percent has been protected to date, far less than what is scientifically recognized as necessary to sustain the ecosystem over time.

About 1 million hectares of forest in Canada—currently the world's biggest timber exporter—are logged annually, and most of Canada's forests are subject to logging contracts. On the basis of current harvest quotas, logging is unsustainable in Canada (**FIGURE 8.14**). Extensive tracts of Siberian forests in Russia are also harvested, although estimates are unavailable. Alaska's boreal forests are at risk because the

Logging in Canada's boreal forest FIGURE 8.14

About 80 percent of Canada's forest products are exported to the United States.

U.S. government may increase logging on public lands in the future.

Tropical forests and deforestation There are two types of tropical forests: tropical rainforests and tropical dry forests. *Tropical rainforests* prevail in warm areas that receive 200 centimetres or more of precipitation annually. Tropical rainforests are found in Central and South America, Africa, and Southeast Asia, but almost half of them are in just three countries: Brazil, Democratic Republic of the Congo, and Indonesia (**FIGURE 8.15A** on page 256). *Tropical dry forests* occur in other tropical areas where annual precipitation is less but is still enough to support trees. India, Kenya, Zimbabwe, Egypt, and Brazil are a few of the countries that have tropical dry forests.

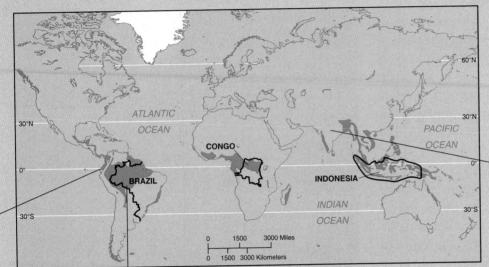

A Distribution of tropical rainforests Rainforests (green areas) in Central and South America, Africa, and Southeast Asia. Much of the remaining forested area is highly fragmented. The three countries with the largest area of tropical rainforest are outlined.

B Human settlements along a road in Brazil's tropical rainforest This satellite photograph shows numerous smaller roads extending perpendicularly from the main roads. As farmers settle along the roads, they clear out more and more forest (dark green) for croplands and pastures (tans and pinks).

D Deforestation for fuelwood Women gather firewood in the Ranthambore National Park buffer zone in India. Note the branches trimmed off the trees in the background. About 94 percent of wood removed from Indian forests is burned as fuel.

C Deforestation due to mining Forest was removed for this oil exploration drilling site in Ecuador.

Most of the remaining undisturbed tropical forests, which lie in the Amazon and Congo river basins of South America and Africa, respectively, are being cleared and burned at a rate unprecedented in human history. Tropical forests are also being destroyed at an extremely rapid rate in southern Asia, Indonesia, Central America, and the Philippines.

Why are tropical rainforests disappearing? Several studies show a strong statistical correlation between population growth and deforestation. More people need more food, and so forests are cleared for agricultural expansion.

However, tropical deforestation can't be attributed simply to population pressures because it is also affected by a variety of interacting economic, social, and governmental factors that vary from place to place. Government policies sometimes provide incentives that favour the removal of forests. The Brazilian government opened the Amazonian frontier, beginning in the late 1950s, by constructing the Belém-Brasilia Highway, which cut through the Amazon Basin. Such roads open the forest for settlement (**FIGURE 8.15B**). An example of economic conditions encouraging deforestation is the farmer who converts forest to pasture so that he can maintain a larger herd of cattle.

Keeping in mind that the origins of tropical deforestation are complex, three agents—subsistence agriculture, commercial logging, and cattle ranching—are considered its most immediate causes. Subsistence agriculture, in which a family produces just enough food to feed itself, accounts for perhaps 60 percent of tropical deforestation. Subsistence farmers carry out *slash-and-burn agriculture* (discussed earlier in this chapter). Subsistence farmers often follow loggers' access roads until they find a suitable spot. They cut down trees and allow them to dry; then they burn the area and plant crops immediately after burning. The yield from the first crop is often quite high because the nutrients that were in the burned trees are now available in the soil. Soil productivity subsequently declines at a rapid rate, and subsequent crops are poor. In a short time, the people farming the land must move to a new part of the forest and repeat the process. Cattle ranchers often claim the abandoned land for grazing because land not rich enough to support crops can still support livestock.

Slash-and-burn agriculture done on a small scale, with plenty of forest to shift around in so that there are periods of 20 to 100 years between cycles, is sustainable. The forest regrows rapidly after a few years of farming. But when millions of people try to obtain a living in this way, the land is not allowed to lie uncultivated long enough to recover.

Another 20 percent of tropical deforestation is the result of commercial logging. Vast tracts of tropical rainforests are harvested for export abroad. Most tropical countries allow commercial logging to proceed at a much faster rate than is sustainable. Unmanaged logging does not contribute to economic development; rather, it depletes a valuable natural resource faster than it can regenerate for sustainable use.

Approximately 12 percent of tropical deforestation provides open rangeland for cattle. Other causes of tropical rainforest destruction include the development of hydroelectric power, which inundates large areas of forest, and mining, particularly when ore smelters burn charcoal produced from rainforest trees (**FIGURE 8.15C**).

Tropical dry forests are also being destroyed at a rapid rate, primarily for fuelwood (**FIGURE 8.15D**). Often the wood cut for fuel is converted to charcoal, which is then used to power steel, brick, and cement factories. Charcoal production is extremely wasteful: 3.6 tonnes of wood produce only enough charcoal to fuel an average-sized iron smelter for five minutes.

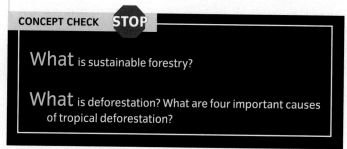

CONCEPT CHECK STOP

What is sustainable forestry?

What is deforestation? What are four important causes of tropical deforestation?

Uses of Minerals and Soils

ECONOMIC GEOLOGY: USEFUL MINERALS

Earth's outermost layer, the crust, contains many kinds of minerals that are of economic importance. In this section we focus on the economic and environmental impacts of extracting and using mineral resources. We then consider soil, the part of the crust where biological and physical processes meet.

Minerals are such an integral part of our lives that we often take them for granted. Steel, an essential building material, is a blend of iron and other metals. Beverage cans, aircraft, automobiles, and buildings all contain aluminum. Copper, which readily conducts electricity, is used for electrical and communications wiring. The concrete used in buildings and roads is made from sand and gravel, as well as cement, which contains crushed limestone. Sulphur, a component of sulphuric acid, is an indispensable industrial mineral. It is used to make plastics and fertilizers and to refine oil. Other important minerals include platinum, mercury, manganese, and titanium.

minerals Elements or compounds of elements that occur naturally in Earth's crust.

Earth's minerals are elements or (usually) compounds of elements and have precise chemical compositions. **Sulphides** are mineral compounds in which certain elements are combined chemically with sulphur, and **oxides** are mineral compounds in which elements are combined chemically with oxygen. Minerals are metallic or nonmetallic. **Metals** are minerals such as iron, aluminum, and copper, which are malleable, lustrous, and good conductors of heat and electricity. **Nonmetallic minerals**, such as sand, stone, salt, and phosphates, lack these characteristics.

Rocks are naturally formed mixtures of minerals that have varied chemical compositions. **Ore** is rock that contains a large enough concentration of a particular mineral to be profitably mined and extracted. *High-grade ores* contain relatively large amounts of particular minerals, whereas *low-grade ores* contain lesser amounts. Although some minerals are abundant, all are nonrenewable resources that are not replenished by natural processes on a human time scale.

Minerals: An economic perspective
At one time, most of the highly developed nations had abundant mineral deposits that enabled them to industrialize. In the process of industrialization, these countries largely depleted their domestic reserves of minerals so that they must increasingly turn to developing countries.

Canada's vast natural resources position it as one of the largest mining countries in the world, producing more than 60 different minerals and metals (**FIGURE 8.16**). Canada is also one of the world's leading exporters of minerals and mineral products. In 2006, Canada's minerals and mining products accounted for over 17 percent of Canada's total domestic exports, and mining and mineral processing contributed over $40 billion to Canada's gross domestic product. The value of Canadian mineral production totalled over $77 billion in 2006, with nickel as the top metallic mineral produced with shipments totalling $6 billion. Canada continues to be the third-largest producer of primary aluminum and diamonds. By 2010 Canada was responsible for approximately 20 percent of the world's total diamond production.

Mining and mineral production contribute extensively to employment within Canada. There were 369,000 jobs in mining and mineral production in 2006, and the mining sector is said to provide some of Canada's best jobs, as weekly average earnings are more than $1000, compared with the Canadian economy average of just under $750.

The Minerals and Metals Sector (MMS) of Natural Resources Canada encourages investment in Canada's

The Diavik Diamond Mine, located in the Northwest Territories, opened in 2003 and produces 8 million carats of diamonds every year.

minerals and metals, protects and maximizes economic benefits to Canada from Canadian mineral operations abroad, and helps developing countries maximize the benefits from their mineral supplies. The MMS promotes sustainable development, social responsibility, and the development of "mining-friendly" legislation protecting Canadian interests abroad.

Despite the economic advantages, mining and mineral production face some key management issues in Canada, such as water management, acid rock drainage, metal leaching, salt accumulation, release of contaminants from orphaned or abandoned mine sites, and wastewater treatment.

How minerals are extracted and processed

The process of making mineral deposits available for human consumption occurs in several steps. First, a particular mineral deposit is located. Geologic knowledge of Earth's crust and how minerals are formed is used to estimate locations of possible mineral deposits. Once these sites are identified, geologists drill or tunnel for mineral samples and analyze their composition. Second, mining extracts the mineral from the ground. Third, the mineral is processed, or refined, by concentrating it and removing impurities. Finally, the purified mineral is used to make a product.

surface mining
The extraction of mineral and energy resources near Earth's surface by first removing the soil, subsoil, and overlying rock strata.

subsurface mining
The extraction of mineral and energy resources from deep underground deposits.

Extracting minerals The depth of a particular mineral deposit determines whether surface or subsurface mining will be used. In **surface mining**, minerals are extracted near the surface. Surface mining is more common because it is less expensive than subsurface mining. However, even surface mineral deposits occur in rock layers beneath Earth's surface, so the overlying soil and rock layers, called overburden, must first be removed, along with the vegetation growing in the soil. Then giant power shovels scoop out the minerals.

There are two kinds of surface mining, open-pit surface mining and strip mining. Iron, copper, stone, and gravel are usually extracted by **open-pit surface mining**, in which a giant hole, called a quarry, is dug in the ground to extract the minerals (**FIGURE 8.17A** on page 260). In **strip mining**, a trench is dug to extract the minerals (**FIGURE 8.17B**). Then a new trench is dug parallel to the old one, and the overburden from the new trench is put into the old one, creating a hill of loose rock called a spoil bank.

Subsurface mining extracts minerals too deep in the ground to be removed by surface mining. It disturbs the land less than surface mining, but it is more expensive and more hazardous for miners. There is always a risk of death or injury from explosions or collapsing walls, and prolonged breathing of dust in subsurface mines can result in lung disease.

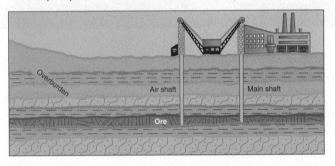

A An open-pit surface mine in Fort McMurray, Alberta.

B Strip mining removes overburden along narrow strips to reach the ore beneath.

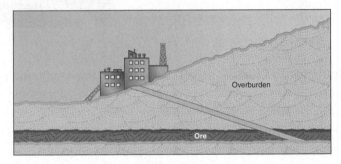

C In a shaft mine, a hole is dug straight through the overburden to the ore, which is removed up through the shaft in buckets.

D In a slope mine, an entry to the ore is dug at an angle so that the ore can be hauled out in carts.

Subsurface mining may be done with underground shaft mines or slope mines. A **shaft mine**, often used for mining coal, is a direct vertical shaft to the vein of ore (**FIGURE 8.17C**). The ore is broken up underground and then hoisted through the shaft to the surface in buckets. A **slope mine** has a slanting passage that makes it possible to haul the broken ore out of the mine in cars rather than to hoist it up in buckets (**FIGURE 8.17D**). Sump pumps keep a subsurface mine dry, and a second shaft is usually installed for ventilation.

Processing minerals Processing minerals often involves **smelting**. Purified copper, tin, lead, iron, manganese, cobalt, or nickel smelting is done in a blast furnace. The iron ore reacts with coke (modified coal) to form molten iron and carbon dioxide. The limestone reacts with impurities in the ore to form a molten mixture called **slag**. There is a vent near the top of the iron smelter for exhaust gases. If air pollution control devices are not installed, many dangerous gases are emitted during smelting.

ENVIRONMENTAL IMPLICATIONS OF MINERAL USE

The extraction, processing, and disposal of minerals clearly harm the environment. Mining disturbs and damages the land, and the processing and disposal of minerals pollute the air, soil, and water. Although pollution can be controlled and damaged lands can be restored, these remedies are costly. Historically, the environmental cost of extracting, processing, and disposing of minerals has not been incorporated into the actual price of mineral products to consumers.

Most highly developed countries have regulatory mechanisms in place to minimize environmental damage from mineral consumption, and many developing nations are in the process of putting them in place. Such mechanisms include policies to prevent or reduce pollution, restore mining sites, and exclude certain recreational and wilderness sites from mineral development.

Mining and the environment Mining, particularly surface mining, disturbs large areas of land. Coal mines occupy large areas of Eastern Canada. Because mining destroys existing vegetation, this land is particularly prone to erosion, with wind erosion causing air pollution and water erosion polluting nearby waterways and damaging aquatic habitats.

Open-pit mining of gold and other minerals uses huge quantities of water. As miners dig deeper, they eventually hit the water table and pump out the water to keep the pit dry. Farmers and ranchers in open-pit mining areas are concerned about depletion of the groundwater they need for irrigation. Environmentalists and others would like the mining operations to reinject the water into the ground after pumping it out.

Mining has contaminated thousands of kilometres of streams and rivers in North America. Rocks rich in minerals often contain high concentrations of heavy metals such as arsenic and lead. Rainwater seeping through the sulphide minerals in mine waste produces sulphuric acid, which dissolves the heavy metals and other toxic substances in the spoil banks. These acids, called **acid mine drainage**, are highly toxic and are washed into soil and water—including groundwater—by precipitation runoff (FIGURE 8.18). When such acids and toxic compounds make their way into nearby lakes and streams, particularly through "toxic pulses" of thunderstorms or spring snowmelt, they adversely affect aquatic life.

Environmental impacts of refining minerals

Approximately 80 percent of mined ore consists of

Copper ore tailings FIGURE 8.19

Tailings dumped in mountainous heaps cause air, soil, and water pollution and have serious effects on land use. In highly developed countries, responsible mining companies are developing methods to dispose of tailings in environmentally sustainable ways. However, in developing countries, where there are few environmental regulations, disposal of tailings is a far more serious problem.

impurities that become wastes after processing. These wastes, called **tailings**, are usually left in giant piles on the ground or in ponds near the processing plants (FIGURE 8.19). The tailings contain toxic materials such as cyanide, mercury, and sulphuric acid. Left exposed, they contaminate the air, soil, and water.

Acid mine drainage
FIGURE 8.18

Acid runoff that contains sulphuric acid has a characteristic orange-red colour, like that shown here. This runoff, which seeps into the groundwater or is washed away into nearby water bodies, is often also contaminated with lead, arsenic, cadmium, silver, and zinc.

Gold is a precious metal used primarily for jewellery and as a medium of exchange in many countries. Annual global gold production in 2006 was about 2500 tonnes, up from 1200 tonnes in 1980. Worldwide demand for gold is increasing, and the environment is suffering from the increased mining. The waste from mining and processing ore is enormous: six tonnes of wastes are produced to yield enough gold to make just two wedding rings. One technology, *cyanide heap leaching*, allows profitable mining when minuscule amounts of gold are present, but this process produces up to 6 million kilograms of waste for every kilogram of gold produced. The world's largest gold mine, located in Indonesia but owned by a U.S. company, dumps more than 100,000 tonnes of cyanide-contaminated waste into the local river each day. The highly toxic cyanide threatens waterfowl and fishes, as well as underground drinking water supplies.

Small-scale miners use other extraction techniques with destructive side effects: soil erosion, production of silt that clogs streams and threatens aquatic organisms, and contamination from mercury used to extract the gold. The environmental hazards of gold mining do not end when the gold is carried away: if not disposed of properly, mining wastes cause long-term problems such as acid mine drainage and heavy-metal contamination.

Smelting plants may emit large quantities of air pollutants during mineral processing, particularly sulphur. Unless expensive pollution control devices are added to smelters, the sulphur escapes into the atmosphere, where it forms sulphuric acid. (The environmental implications of the resulting acid precipitation are discussed in Chapter 10.) Pollution control devices for smelters are the same as the devices used for the burning of sulphur-containing coal—scrubbers and electrostatic precipitators.

Contaminants in ores include the heavy metals lead, cadmium, arsenic, and zinc. These toxic elements may pollute the atmosphere during the smelting process and cause harm to humans. In addition to airborne pollutants, smelters emit hazardous liquid and solid wastes that can pollute the soil and water.

One of the most significant environmental impacts of mineral production is the large amount of energy required to mine and refine minerals, particularly if they are being refined from low-grade ore. Most of this energy is obtained by burning fossil fuels, which depletes nonrenewable energy reserves and produces carbon dioxide and other air pollutants.

Restoration of mining lands

When a mine is no longer profitable to operate, the land can be reclaimed or restored to a semi-natural condition (**FIGURE 8.20**).

Reclamation prevents further degradation and erosion of the land, eliminates or neutralizes local sources of toxic pollutants, and makes the land productive for purposes other than mining. Restoring land degraded by mining—called **derelict land**—involves filling in and grading the area to the shape of its natural contours and then planting vegetation to hold the soil in place.

Restoration of mining lands FIGURE 8.20

Most mining operations attempt to restore the land they work on. Here, a former mine site has been replanted with trees. Photographed in Fort McMurray, Alberta.

Large reclamation efforts have been underway for decades in Sudbury, Ontario, which is considered one of the most ecologically disturbed regions in Canada. Sudbury was seriously damaged through the events of logging and mining and became known as a barren wasteland. In the early 1970s, the Sudbury community decided to restore the area that had been devastated by acid rain and resource extraction. Regrettably, the damaging activities of the past left a wasteland of toxic soils contaminated from metal deposition and acidic fumes. Many volunteers from the community, with the help of scientists from universities, provincial and federal politicians, and large corporations, worked together to lime, fertilize, and then seed the area. Over the last three decades, 10 million trees have been planted over nearly 35,000 acres.

SOIL PROPERTIES AND PROCESSES

The relatively thin surface layer of Earth's crust is **soil**, which consists of mineral and organic matter modified by the natural actions of agents such as weather, wind, water, and organisms. It is easy to take soil for granted. We walk on and over it throughout our lives but rarely stop to think about how important it is to our survival. Vast numbers and kinds of organisms, mainly microorganisms, inhabit soil and depend on it for shelter, food, and water. Plants anchor themselves in soil, and from it they receive essential minerals and water. Terrestrial plants could not survive without soil, and because we depend on plants for our food, humans could not exist without soil either.

> **soil** The uppermost layer of Earth's crust, which supports terrestrial plants, animals, and microorganisms.

Soil formation and composition

Soil is formed from *parent material*, rock that is slowly broken down, or fragmented, into smaller and smaller particles by biological, chemical, and physical **weathering processes**. It takes a long time, sometimes thousands of years, for rock to disintegrate into finer and finer mineral particles. Time is also required for organic material to accumulate in the soil. Soil formation is a continuous process that involves interactions between Earth's solid crust and the biosphere. The weathering of parent material beneath already-formed soil continues to add new soil.

Topography, a region's surface features (such as the presence or absence of mountains and valleys), is also involved in soil formation. Steep slopes often have little or no soil on them because soil and rock are continually transported down the slopes by gravity. Runoff from precipitation tends to amplify erosion on steep slopes. Moderate slopes and valleys, on the other hand, may encourage the formation of deep soils.

Soil is composed of four distinct parts: mineral particles, organic matter, water, and air. The mineral portion, which comes from parent material, is the main component of soil. It provides anchorage and essential nutrient minerals for plants, as well as pore space for water and air. Litter (dead leaves and branches on the soil's surface), animal dung, and the remains of plants, animals, and microorganisms constitute the organic portion of soil. Organisms such as bacteria and fungi gradually decompose this material.

The black or dark brown organic material that remains after extended decomposition is called **humus** (**Figure 8.21**). Humus, which is a mix of many organic compounds, binds to nutrient mineral ions and holds water.

Many soils are organized into distinctive horizontal layers called **soil horizons**. A **soil profile** is a vertical section from

> **soil horizons** The horizontal layers into which many soils are organized, from the surface to the underlying parent.

Soil rich in humus FIGURE 8.21

Humus is partially decomposed organic material, primarily from plant and animal remains. Soil rich in humus has a loose, somewhat spongy structure with several properties, such as increased water-holding capacity, that are beneficial for plants and other organisms living in it.

Soil Profile

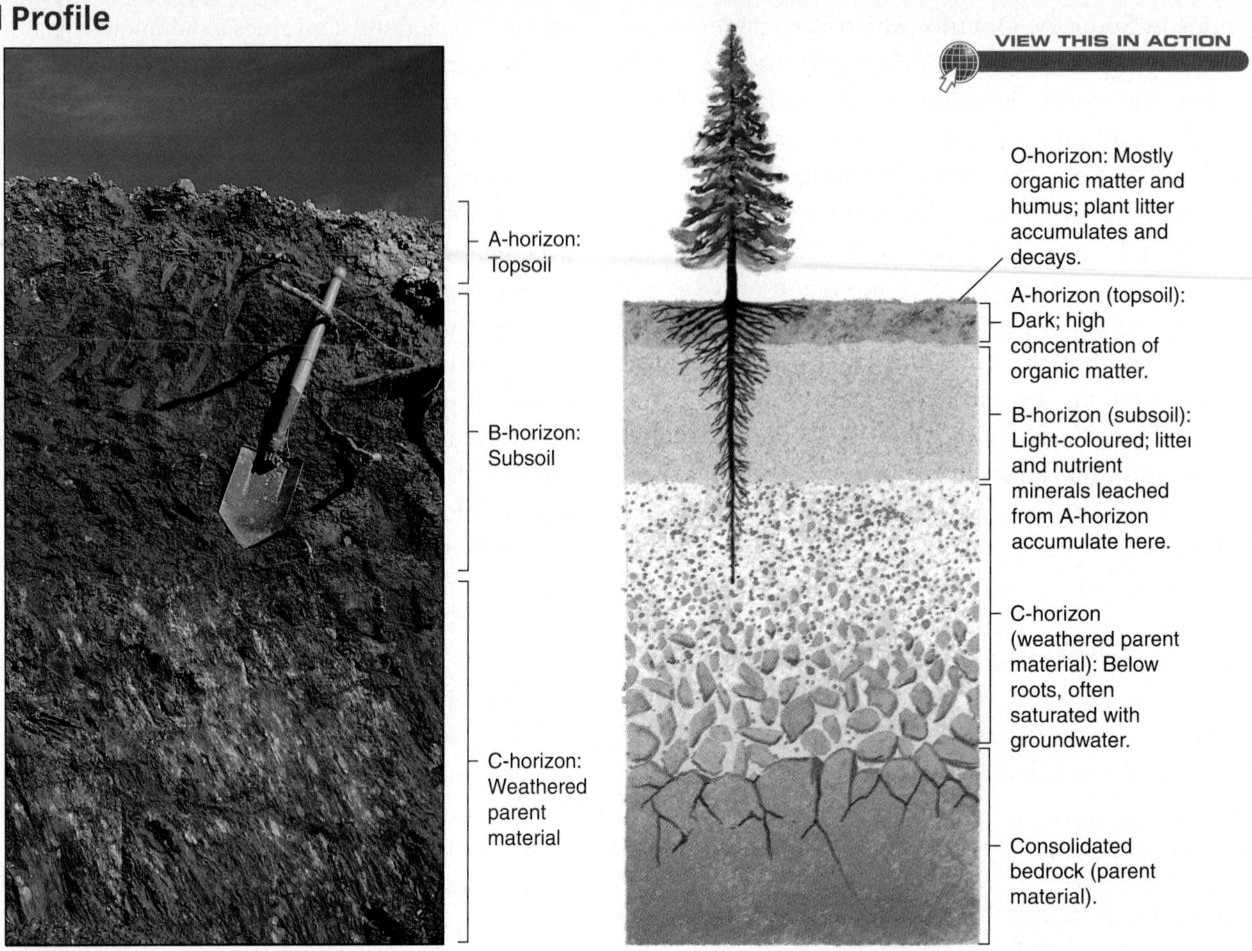

VIEW THIS IN ACTION

A-horizon: Topsoil

B-horizon: Subsoil

C-horizon: Weathered parent material

O-horizon: Mostly organic matter and humus; plant litter accumulates and decays.

A-horizon (topsoil): Dark; high concentration of organic matter.

B-horizon (subsoil): Light-coloured; litter and nutrient minerals leached from A-horizon accumulate here.

C-horizon (weathered parent material): Below roots, often saturated with groundwater.

Consolidated bedrock (parent material).

A This soil has no O-horizon because it is used for agriculture; the surface litter that would normally compose the O-horizon was plowed into the A-horizon. The shovel gives an idea of the relative depths of each horizon.

B A "typical" soil profile, as it appears to the trained eye of a soil scientist. Each horizon has its own chemical and physical properties.

surface to parent material, showing the soil horizons (see What a Scientist Sees). The topsoil (or A-horizon) is somewhat nutrient poor because of the leaching of many nutrients into deeper soil layers. **Leaching** is the removal of dissolved materials from the soil by water percolating downward.

Soil organisms
Soil organisms, which are usually hidden underground, are remarkably numerous. Organisms that colonize the soil ecosystem include plant roots, insects such as termites and ants, earthworms, moles, snakes, and groundhogs (**FIGURE 8.22**). Most numerous in soil are bacteria, which number in the hundreds of millions per gram of soil. Other microorganisms that are abundant in soil ecosystems include fungi, algae, microscopic worms such as nematodes, and protozoa.

In a balanced ecosystem, the relationship between soil and the organisms that live in and on it ensure soil fertility. Soil organisms provide several essential ecosystem services, such as maintaining soil fertility, preventing soil erosion, breaking down toxic materials, and cleansing water.

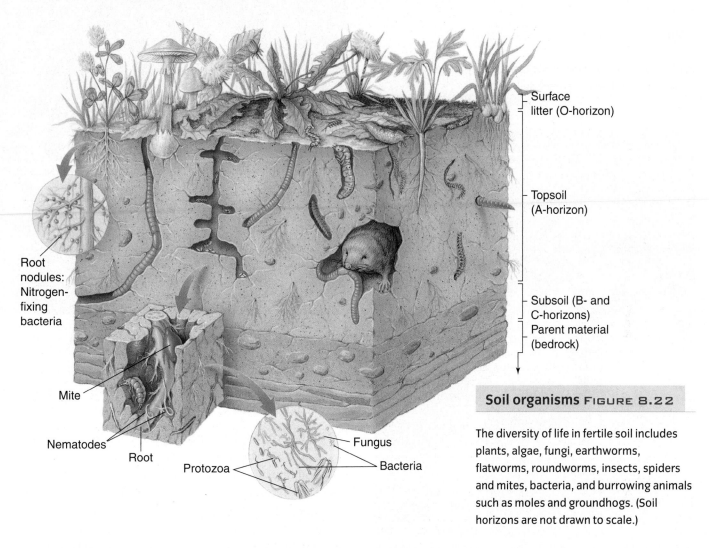

Surface
litter (O-horizon)

Topsoil
(A-horizon)

Subsoil (B- and
C-horizons)
Parent material
(bedrock)

Root
nodules:
Nitrogen-
fixing
bacteria

Mite

Nematodes

Root

Protozoa

Fungus

Bacteria

Soil organisms FIGURE 8.22

The diversity of life in fertile soil includes
plants, algae, fungi, earthworms,
flatworms, roundworms, insects, spiders
and mites, bacteria, and burrowing animals
such as moles and groundhogs. (Soil
horizons are not drawn to scale.)

Essential nutrient minerals such as nitrogen and phosphorus are cycled from the soil to organisms and back to the soil again. Decomposition, another ecosystem service, is part of **nutrient cycling**. Bacteria and fungi decompose plant and animal detritus and wastes, transforming large organic molecules into small inorganic molecules, including carbon dioxide, water, and nutrient minerals; the nutrient minerals are released into the soil to be reused (FIGURE 8.23 on page 266; also see Chapter 5). Nonliving processes are also involved in nutrient cycling: The weathering of the parent material replaces some nutrient minerals lost through erosion or agricultural practices.

SOIL PROBLEMS

Soil is as important as air and water for human survival. Yet humans disrupt soil systems that would be balanced in nature. We have had a harmful impact on soil resources worldwide, particularly by intensifying agricultural use. These human activities often cause or exacerbate soil problems such as erosion, mineral depletion, and soil pollution, all of which occur worldwide. Such activities do not promote **sustainable soil use**. Soil used in a sustainable way renews itself by natural processes year after year.

Soil erosion Water, wind, ice, and other agents promote **soil erosion**, a natural process often accelerated by human activities. Water and wind are

sustainable soil use The wise use of soil resources without a reduction in the amount or fertility of soil so it is productive for future generations.

soil erosion The wearing away or removal of soil from the land.

Nutrient cycling FIGURE 8.23

In a balanced ecosystem, nutrient minerals cycle from the soil to organisms and then back to the soil.

VIEW THIS IN ACTION

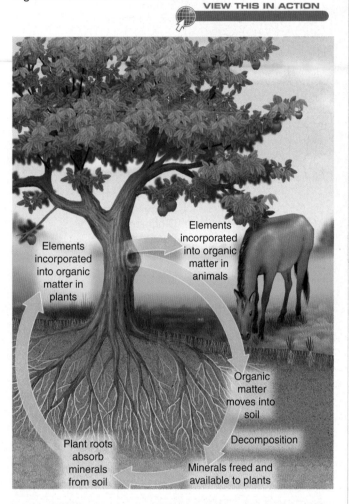

Elements incorporated into organic matter in animals

Elements incorporated into organic matter in plants

Organic matter moves into soil

Decomposition

Plant roots absorb minerals from soil

Minerals freed and available to plants

particularly effective in moving soil from one place to another. Rainfall loosens soil particles, which are transported by moving water (**FIGURE 8.24**). Wind loosens soil and blows it away, particularly if the soil is barren and dry. Erosion reduces the amount of soil in an area and therefore limits the growth of plants.

Humans often accelerate soil erosion with poor soil management. Poor agricultural practices are partly to blame, as are the removal of natural plant communities during road and building construction and unsound logging practices such as clear-cutting. Soil erosion has an impact on other natural resources as well. Sediment that gets into streams, rivers, and lakes affects water quality and fish habitats (see Chapter 9). If the sediments contain pesticide and fertilizer residues, they further pollute the water.

Soil erosion caused by water FIGURE 8.24

The branching gullies shown here are the most serious form of erosion and will continue to grow unless checked by some type of erosion control.

Sufficient plant cover limits soil erosion. Leaves and stems cushion the impact of rainfall, and roots help to hold soil in place. Although soil erosion is a natural process, the abundant plant cover in many natural ecosystems makes it negligible.

Soil pollution   **Soil pollution** is any physical or chemical change in soil that adversely affects the health of plants and other organisms living in or on the soil. Soil pollution is important not only in its own right but because so many soil pollutants tend to also pollute surface water, groundwater, and the atmosphere. For example, selenium, an extremely toxic natural element found in many western soils, leaches off irrigated farmlands and poisons nearby lakes, ponds, and rivers. This has caused death and deformity in thousands of migratory birds and other organisms. Most soil pollutants originate as agricultural chemicals such as fertilizers and pesticides. Other soil pollutants include salts, petroleum products, and heavy metals.

Irrigation of agricultural fields often results in their becoming increasingly saline, an occurrence known as **salinization** (**FIGURE 8.25**). In time, salt concentrations in soil can rise to such a high level that plants are poisoned or their roots become dehydrated.

Salinized soil FIGURE 8.25

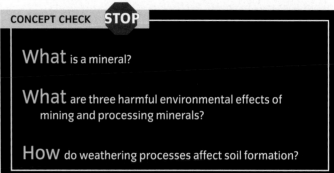

Irrigation water contains small amounts of dissolved salts. Over time, the salt accumulates in the soil. This irrigated soil has become too salty for plants to tolerate.

CONCEPT CHECK **STOP**

What is a mineral?

What are three harmful environmental effects of mining and processing minerals?

How do weathering processes affect soil formation?

Parks and Protected Areas

LEARNING OBJECTIVES

Describe the role of the federal and provincial governments in managing Canada's parks and protected areas.

State the primary objective of Parks Canada.

Explain what conservation efforts are underway in Canada's national and provincial parks.

Many acres of federal and provincial land in Canada are set aside as either national or provincial park property or as wilderness areas. Both types of land were established to encourage the protection of the natural environment, yet both experience conflicts associated with how best to use and manage these protected areas.

NATIONAL PARKS

National parks and protected areas in Canada are managed federally by **Parks Canada**. In 1911, Parks Canada (then known as the Dominion Parks Branch) was established under the Department of the Interior and was the world's first national park service. During the 19th century, Canada's parks were established and maintained for monumentalism (preserving parks as a tribute to the scenic grandeur nature has to offer). However, in the 20th century the focus of national parks turned to human recreation. By the end of the 20th century, there were increasing concerns for the environment, and the focus in Canada's national parks shifted once again toward science and ecotourism.

Canada's national parks and park reserves, of which there were 42 in 2010, range in size from under nine square kilometres in the St. Lawrence Islands National Park to the nearly 45,000 square kilometres of Wood Buffalo National Park. The total area in Canada covered by national parks is about 225,000 square kilometres, approximately 2 percent of Canada's total land mass (**FIGURE 8.26** on page 268). A relatively new addition to the Canadian park system are the National Marine Conservation Areas (NMCAs). Currently, Canada has three NMCAs, two located in Ontario (Fathom Five and Lake Superior) and one in Quebec (Saguenay-St. Lawrence).

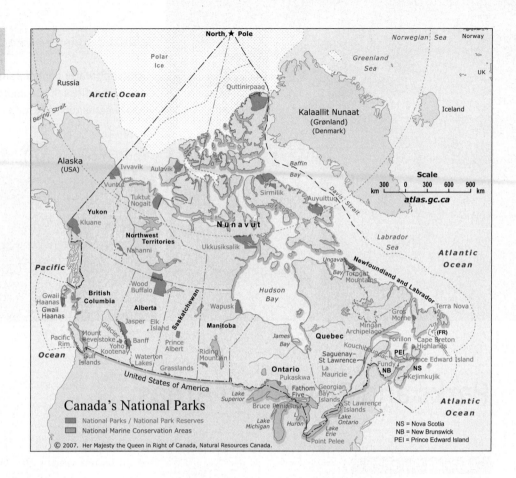

Canada's National Parks

National Parks / National Park Reserves
National Marine Conservation Areas

NS = Nova Scotia
NB = New Brunswick
PEI = Prince Edward Island

© 2007. Her Majesty the Queen in Right of Canada, Natural Resources Canada.

One of the primary goals of Parks Canada is to protect the natural landscapes that occur in Canada's 39 natural regions, located in every province and territory. Canada's national parks represent diverse landscapes that include mountains, plains, tundra, forests, lakes, glaciers, and more (**FIGURE 8.27**). Banff National Park, located on the western edge of Alberta, is Canada's oldest national park and was established in 1885, 20 years before Alberta became a province. Banff National Park is part of the Canadian Rocky Mountain Parks World Heritage Site, an area protecting over 20,000 square kilometres of the Rocky Mountains.

Canada's national parks and their resources are protected by federal legislation, the Canada National Parks Act, which prohibits activities such as mining, forestry, agriculture, and hunting. Under the Act, a mandate of Parks Canada is to ensure that the national parks are managed primarily to protect and restore their ecological integrity, meaning native plant and animal species are able to live, reproduce, and maintain long-term viability. This is an ecosystem management approach,

managing the system as a whole to include ecosystem biodiversity, preserve the overall structure and function of the ecosystem, and promote an understanding of the regional evolutionary and historic context that has shaped the ecosystem. The National Marine Conservation Areas are managed differently—for sustainable use (**FIGURE 8.28** on page 270). However, they still seek the protection of the area's overall ecological integrity.

To further achieve its mandate of ecological integrity, Parks Canada adopted a natural region system, the National Parks System plan, in 1970 to design and guide park expansion activities. In 1989, the endangered spaces campaign was launched by the Canadian Parks and Wilderness Society (CPAWS) and World Wildlife Fund Canada to encourage the development of the National Parks System, which is now over 60 percent complete. Feasibility studies were also initiated in 2005 to look at the further establishment of national parks in four areas: Wolf Lake in Yukon, South Okanagan–Lower Similkameen in British Columbia, Manitoba Lowlands in northwestern areas of Lake Winnipeg, and Mealy Mountains in Labrador.

Canada's national parks, which are found in every province and territory, protect the majestic landscapes of this country.

B Forillon National Park, located in Quebec at the farthest reach of the Gaspé Peninsula, is "where the land meets the sea."

A Banff National Park is a UNESCO World Heritage Site and was Canada's first national park (1885).

C Fundy National Park, located in southern New Brunswick, sees tidal fluctuation off the Bay of Fundy that is the highest in the world.

D Kluane National Park and Reserve is a UNESCO World Heritage Site and is home to Mount Logan, Canada's highest peak (5959 m).

NATIONAL
GEOGRAPHIC

National Marine Conservation Areas

FIGURE 8.28

National Marine Conservation Areas, like the Saguenay-St. Lawrence Marine Park shown here, are managed for sustainable use. They include the seabed, the water above it, and any species that live therein.

PROVINCIAL PARKS

Provincial parks are administered by agencies that are usually found within the government departments that manage either natural resources, tourism, or culture. For example, Ontario Parks is a special agency of the Ministry of Natural Resources, whereas in British Columbia the parks agency belongs to the Ministry of the Environment. Most provincial agencies have developed a park system that guides decisions. A management plan is usually prepared for each park to facilitate decisions about protecting the park's environment.

Park managers face many challenges such as fires, wildlife imbalances, diseases of animals and plants, and human impacts. They must also consider park visitors, which means managing and maintaining transportation routes, access, recreation facilities, education and information services, and safety features.

Each province has legislation to regulate the protection of the provincial parks. Many have argued that provincial parks legislation is not suitable to ensure protection of the environment and resource sustainability. In September 2007, a new, stronger piece of legislation came into effect in Ontario that replaced the province's old parks Act. This new legislation was created using extensive consultation, and its first priority, unlike any other provincial parks legislation, is ecological integrity.

CONSERVATION IN CANADIAN PARKS

European explorers considered natural areas as an unlimited resource to exploit. They appreciated prairies as valuable agricultural land and forests as immediate sources of lumber and eventual farmland. This outlook was practical as long as there was more land than people needed. But as the population increased and the amount of available land decreased, it was necessary to consider land as a limited resource. Increasingly, the emphasis has shifted from exploitation to preservation or conservation of the remaining natural areas.

Remember from Chapter 7 that there is a difference between *preservation* and *conservation*. Preservation is concerned with setting aside undisturbed areas, maintaining them in a pristine state, and protecting them from human activities. Conservation, on the other hand, is the sensible and careful management of natural resources, such as air, water, soil, forests, minerals, and wildlife. In Canada, national parks generally attempt to meet preservation objectives, whereas provincial parks tend to focus more on conservation.

Parks and protected areas in Canada must be managed in a manner that sustains them and respects their intrinsic values. These "special places" contribute to broader sustainable development and conservation strategies by:

- Maintaining ecological integrity and biodiversity of natural areas
- Preserving the commemorative integrity of historic places
- Promoting a conservation ethic, citizenship values based on respect for the environment and heritage, and ecosystem and cultural resource management
- Generally demonstrating conservation principles and approaches set out in various relevant United Nations reports.

The 12 percent challenge About 20 years ago the World Wildlife Fund issued the "12 percent challenge," where they attempted to conserve a target representation of at least 12 percent of every ecosystem in the world. (See FIGURE 8.29 on pages 272 and 273 for a map of the world's protected lands.) In Canada, the responsibility for achieving this target most often fell upon the provincial parks. In British Columbia, for example, the provincial government instituted a protected areas strategy (PAS) to achieve this 12 percent target and went about setting land aside for conservation.

Although this 12 percent level of protection reflects an overall commitment to conserve the Earth's ecosystems, many governments and conservation organizations interpreted this to be a 12 percent protection of the region's land area and did not consider the biologically important criteria of species richness areas. The vast majority of parks established to meet this initiative were less than 100 hectares in size, making them useful in protecting important populations of plants, but less useful in protecting wildlife. Studies indicate that medium and large mammals will become extinct over time in these smaller parks and will only survive in the largest parks and in connected park complexes. In addition, many provinces took for granted that land outside these protected areas would remain as wilderness. However, logging and other human activities threaten the land outside of these parks with becoming suboptimal habitats and may even act as sinks for moving animals. The 12 percent target sounded good politically, but scientifically it did not call for specific consideration of the long-term persistence of some provincial species and is not considered to be biologically defensible.

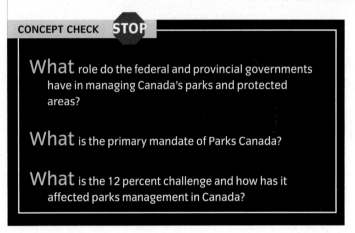

CONCEPT CHECK STOP

What role do the federal and provincial governments have in managing Canada's parks and protected areas?

What is the primary mandate of Parks Canada?

What is the 12 percent challenge and how has it affected parks management in Canada?

Example Approaches to Sustainable Land Resource Management

LEARNING OBJECTIVES

Define *sustainable agriculture* and contrast it with industrialized agriculture.

Identify the potential benefits and problems of genetic engineering.

Identify and summarize the major soil conservation methods.

Increasingly, attempts are being made to ensure that human uses of land resources are not resulting in the degradation of that same resource base. Some examples of land usage issues and possible resolutions have been provided throughout this chapter, but this section cements the idea of sustainability more strongly with respect to agricultural uses of land resources and soil conservation. These examples are rooted in the paradigm of sustainable development (see Figure 1.9 on page 18), and attention to sustainable development requires that not only environmental considerations be addressed in all land resource uses, but also social and economic issues.

SUSTAINABLE AGRICULTURE

Food production poses an environmental quandary. The green revolution and industrialized agriculture have unquestionably met the food requirements of most of the human population, even as that population has more

Protected ecosystems around the world FIGURE 8.29

GeoBytes

LARGEST NATIONAL PARK
North East Greenland National Park, Greenland, 972,000 sq km

LARGEST MARINE PARK
Northwestern Hawaiian Islands Marine National Monument, U.S., 360,000 sq km

LARGEST TROPICAL FOREST PARK
Tumucumaque National Park in the Brazilian Amazon, 24,135 sq km

BIODIVERSITY HOTSPOTS
Conservation International identifies world regions suffering from a severe loss of biodiversity.

WORLD HERITAGE SITES
The United Nations Educational, Scientific, and Cultural Organization (UNESCO) recognizes natural and cultural sites of "universal value."

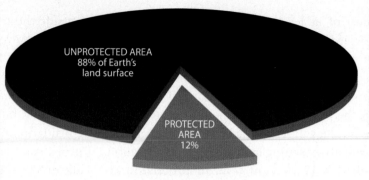

UNPROTECTED AREA
88% of Earth's land surface

PROTECTED AREA
12%

Protected areas worldwide represent 12 percent of the Earth's land surface, according to the U.N. Environmental Programme World Conservation Monitoring Centre. Only 0.5 percent of the marine environment is within protected areas—an amount considered inadequate by conservationists because of the increasing threats of overfishing and coral reef loss worldwide.

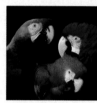

HAWAII VOLCANOES NATIONAL PARK, HAWAII
The park includes two of the world's most active volcanoes, Kilauea and Mauna Loa. The landscape shows the results of 70 million years of volcanism, including calderas, lava flows, and black sand beaches. Lava spreads out to build the island, and sea water vaporizes as lava hits the ocean at 1,149°C. The national park, created in 1916, covers 10 percent of the island of Hawaii and is a refuge for endangered species like the hawksbill turtle and Hawaiian goose. It was made a World Heritage site in 1987.

GALÁPAGOS NATIONAL PARK, ECUADOR
Galápago means "tortoise" in Spanish, and at one time 250,000 giant tortoises roamed the islands. Today about 15,000 remain, and three of the original 14 subspecies are extinct—and the Pinta Island tortoise may be extinct soon. In 1959, Ecuador made the volcanic Galápagos Islands a national park, protecting the giant tortoises and other endemic species. The archipelago became a World Heritage site in 1978, and a marine reserve surrounding the islands was added in 2001.

WESTERN UNITED STATES
An intricate public lands pattern—including national forests, wilderness areas, wildlife refuges, and national parks such as Arches (above)—embraces nearly half the surface area of 11 western states. Ten out of 19 World Heritage sites in the United States are found here. It was in the West that the modern national park movement was born in the 19th century, with the establishment of Yellowstone and Yosemite National Parks.

MADIDI NATIONAL PARK, BOLIVIA
Macaws may outnumber humans in Madidi, Bolivia's second largest national park, established in 1995. A complex community of plants, animals, and native Indian groups share this 18,900-sq-km reserve, part of the Tropical Andes biodiversity hotspot. Indigenous communities benefit from eco-tourism.

AMAZON BASIN, BRAZIL
Indigenous peoples help manage reserves in Brazil that are linked with Jaú National Park. The park and reserves are part of the Central Amazon Conservation Complex, a World Heritage site covering more than 60,000 sq km. It is the largest protected area in the Amazon Basin and one of the most biologically rich regions on the planet.

ARCTIC REGIONS
Polar bears find safe havens in Canadian parks, such as on Ellesmere Island, and in Greenland's huge protected area—Earth's largest—that preserves the island's frigid northeast. In 1996 countries with Arctic lands adopted the Circumpolar Protected Areas Network Strategy and Action Plan to help conserve ecosystems. Today 15 percent of Arctic land area is protected.

PROTECTED AREAS WORLDWIDE
What are protected areas? Most people agree that such territories are dedicated to protecting and maintaining biodiversity and are often managed through legal means. Yellowstone National Park, established in 1872, is often cited as the start of the modern era of protected areas. From a mere handful in 1900, the number of protected areas worldwide now exceeds 104,790, covering more than 20 million sq km. North America claims the most protected land of any region (in terms of total area), amounting to almost 18 percent of its territory. South Asia, at about 7 percent, has the least amount of land under some form of protection. Not all protected areas are created or managed equally, and management categories developed by the International Union for Conservation of Nature (IUCN) range from strict nature reserve to areas for sustainable use. Management effectiveness varies widely and can be affected by such factors as conservation budgets, and political stability. Throughout the world—but especially in tropical areas—protected areas are threatened by illegal hunting, overfishing, pollution, and the removal of native vegetation. Countries and international organizations no longer choose between conservation and development; rather the goal for societies is to balance the two for equitable and sustainable resource use.

For millennia, lands have been set aside as sacred ground or as hunting reserves for the powerful. Today, great swaths are protected for recreation, habitat conservation, biodiversity preservation, and resource management. Some groups may oppose protected spaces because they want access to resources now. Yet local inhabitants and governments are beginning to see the benefits of conservation efforts and sustainable use for human health and future generations.

Wildest biomes
- Wildlands
- Ice or snow cover

◀ **WILDEST AREAS**
Although generally far from cities, the world's remaining wild places play a vital role in a healthy global ecosystem. The boreal forests of Canada and Russia, for instance, help cleanse the air we breathe by absorbing carbon dioxide and providing oxygen. With the human population increasing by an estimated 1 billion over the next 15 years, many wild places could fall within reach of the plow or under a cloud of smog.

SAREKS NATIONAL PARK, SWEDEN
This remote 1970-sq-km park, established in 1909 to protect the alpine landscape, is a favourite of backcountry hikers. It boasts some 200 mountains more than 1800 m, high, narrow valleys, and about 100 glaciers. Sareks forms part of the Laponian Area World Heritage site and has been a home to the Saami (or Lapp) people since prehistoric times.

AFRICAN RESERVES
Some 120,000 elephants roam Chobe National Park in northern Botswana. Africa has more than 7500 national parks, wildlife reserves, and other protected areas, covering about 9 percent of the continent. Protected areas are under enormous pressure from expanding populations, civil unrest and war, and environmental disasters.

WOLONG NATURE RESERVE, CHINA
Giant pandas freely chomp bamboo in this 2000-sq-km reserve in Sichuan Province, near the city of Chengdu. Misty bamboo forests host a number of endangered species, but the critically endangered giant panda—among the rarest mammals in the world—is the most famous resident. Only about 1600 giant pandas exist in the wild.

KAMCHATKA, RUSSIA
Crater lakes, ash-capped cones, and diverse plant and animal species mark the Kamchatka Peninsula—a World Heritage site—located between the icy Bering Sea and Sea of Okhotsk. The active volcanoes and glaciers form a dynamic landscape of great beauty, known as "The Land of Fire and Ice." Kamchatka's remoteness and rugged landscape help fauna flourish, producing record numbers of salmon species and half of the Steller's sea-eagles on Earth.

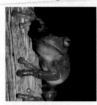

GUNUNG PALUNG NATIONAL PARK, INDONESIA
A tree frog's perch could be precarious in this 900-sq-km park on the island of Borneo, in the heart of the Sundaland biodiversity hotspot. The biggest threat to trees and animals in the park and region is illegal logging. Gunung Palung contains a wider range of habitats than any other protected area on Borneo, from mangroves to lowland and cloud forests. A number of endangered species, such as orangutans and sun bears, depend on the dense forests.

AUSTRALIA & NEW ZEALAND
Uluru, a red sandstone monolith (formerly known as Ayers Rock), and the vast Great Barrier Reef, one of the largest marine parks in the world, are outstanding examples of Australia's protected areas—which make up more than 10 percent of the country's area and conserve a diverse range of unique ecosystems. About a third of New Zealand is protected, and it is a biodiversity hotspot because of threats to flightless native birds, such as the kakapo and kiwi. Cats, stoats, and other predators, introduced to New Zealand by settlers, kill thousands of birds each year.

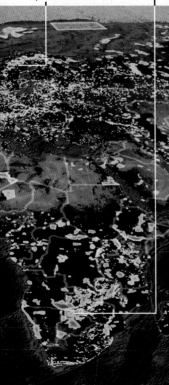

ANTARCTICA
The Antarctic Treaty, signed in 1959, regulates the continent and the marine environment (south of 60° south latitude) as a "natural reserve devoted to peace and science." The treaty is recognized as one of the most successful international agreements. The annual Antarctic Treaty Consultative Meeting brings national representatives together to discuss topics such as scientific cooperation and environmental protection.

than doubled since 1960. But we pay for these food gains with serious environmental problems, and we do not know if industrialized agriculture is sustainable for more than a few decades. To compound the issue, we must continue to increase food production to feed the growing human population—but the resulting damage to the environment may lessen our chances of continuing those food increases into the future.

Fortunately, the dilemma is not hopeless. Farming practices and techniques exist that ensure a sustainable output at yields comparable to industrialized agriculture. Farmers who practise industrialized agriculture can adopt these alternative agricultural methods, which cost less and are less damaging to the environment. Advances are also being made in sustainable subsistence agriculture.

In **sustainable agriculture** (also called alternative or low-input agriculture), certain modern agricultural techniques are carefully combined with traditional farming methods. Sustainable agriculture relies on beneficial biological processes and environmentally friendly chemicals that disintegrate quickly and do not persist as residues in the environment. The sustainable farm consists of field crops, trees that bear fruits and nuts, small herds of livestock, and even tracts of forest (**FIGURE 8.30**).

Such diversification protects the farmer against unexpected changes in the marketplace. The breeding of disease-resistant crop plants and the maintenance of animal

> **sustainable agriculture**
> Agricultural methods that maintain soil productivity and a healthy ecological balance while having minimal long-term impacts.

health rather than the continual use of antibiotics to prevent disease are important parts of sustainable agriculture. Biological diversity is maintained as a way to minimize pest problems. Water and energy conservation are also practised in sustainable agriculture.

An important goal of sustainable agriculture is to preserve the quality of agricultural soil. Crop rotation, conservation tillage, and contour plowing help control erosion and maintain soil fertility. Sloping hills converted to mixed-grass pastures erode less than hills planted with field crops, thereby conserving the soil and supporting livestock.

Animal manure added to soil not only cuts costs but decreases the need for high levels of commercial inorganic fertilizers. Legumes like soybeans, clover, and alfalfa convert atmospheric nitrogen into a form that plants can use in a process called biological nitrogen fixation, which lessens the need for nitrogen fertilizers.

Sustainable agriculture is not a single program but a series of programs adapted for specific soils, climates, and farming requirements. Some sustainable farmers—those who practise **organic agriculture**—use no pesticides or mineral fertilizers; others use a system of *integrated pest management (IPM)*, which incorporates the limited use of pesticides with pest-controlling biological and cultivation practices. (IPM was discussed earlier in this chapter.)

In growing recognition of the environmental problems associated with industrialized agriculture, more

Some goals of sustainable agriculture FIGURE 8.30

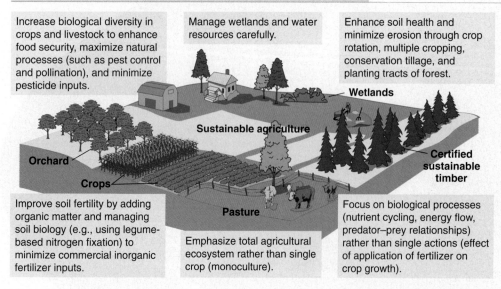

Increase biological diversity in crops and livestock to enhance food security, maximize natural processes (such as pest control and pollination), and minimize pesticide inputs.

Manage wetlands and water resources carefully.

Enhance soil health and minimize erosion through crop rotation, multiple cropping, conservation tillage, and planting tracts of forest.

Improve soil fertility by adding organic matter and managing soil biology (e.g., using legume-based nitrogen fixation) to minimize commercial inorganic fertilizer inputs.

Emphasize total agricultural ecosystem rather than single crop (monoculture).

Focus on biological processes (nutrient cycling, energy flow, predator–prey relationships) rather than single actions (effect of application of fertilizer on crop growth).

Wetlands · Orchard · Crops · Sustainable agriculture · Certified sustainable timber · Pasture

and more mainstream farmers are trying some methods of sustainable agriculture. These methods cause fewer environmental problems to the agricultural ecosystem, or **agroecosystem**, than industrialized agriculture. This trend away from using intensive techniques that produce high yields and toward methods that focus on long-term sustainability of the soil is sometimes referred to as the second green revolution.

GENETIC ENGINEERING: A SOLUTION OR A PROBLEM?

Genetic engineering is a controversial technology that has begun to revolutionize medicine and has the potential to improve agriculture as well. The agricultural goals

> **genetic engineering**
> The manipulation of genes (for example, taking a specific gene from one species and placing it into an unrelated species) to produce a particular trait.

of genetic engineering are not new. Using traditional breeding methods, farmers and scientists have developed desirable characteristics in crop plants and agricultural animals for centuries. It takes time, 15 years or more, to develop such genetically improved organisms using traditional breeding methods. Genetic engineering has the potential to accomplish the same goal in a fraction of that time.

Genetic engineering differs from traditional breeding methods in that desirable genes from any organism can be used, not just those from the species of the plant or animal being improved. If a gene for disease resistance found in soybeans would be beneficial to tomatoes, for example, the genetic engineer can splice the soybean gene into the tomato plant's DNA (**FIGURE 8.31**).

Genetic engineering FIGURE 8.31

This example of genetic engineering uses a plasmid, a small circular molecule of DNA (genetic material) found in many bacteria. (1) The plasmid of the bacterium *Agrobacterium* introduces desirable genes from another organism into a plant. After the foreign DNA is spliced into the plasmid (2), the plasmid is inserted into *Agrobacterium* (3), which then infects plant cells in culture (4). The foreign gene is inserted in the plant's chromosome (5), and genetically modified plants are then produced from the cultured plant cells (6, 7, 8).

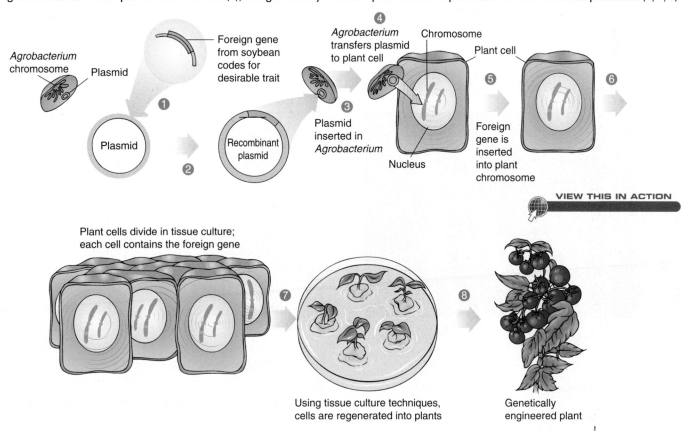

VIEW THIS IN ACTION

Traditional breeding methods could not do this, because soybeans and tomatoes belong to separate groups of plants and do not interbreed.

Genetic engineering has the potential to produce more nutritious food plants that contain all the essential amino acids (which no current food crop does) or that would be rich in necessary vitamins. Crop plants resistant to insect pests, viral diseases, drought, heat, cold, herbicides, or salty or acidic soils are also being developed.

The first **genetically modified (GM)** crops were approved for commercial planting in the United States in the early 1990s. The United States is the world's top producer of GM crops (**FIGURE 8.32A**). Since 2000, however, GM production in developing nations has increased faster than in highly developed countries (**FIGURE 8.32B**).

Genetic engineering has been used to develop more productive farm animals, including quicker-growing hogs and fishes. Perhaps the greatest potential contribution of animal genetic engineering is the production of vaccines against disease organisms that harm agricultural animals. For example, genetically engineered vaccines have been developed to protect cattle against the deadly viral disease rinderpest, which is economically devastating in parts of Asia and Africa.

Concerns about genetically modified foods During the late 1990s and early 2000s, opposition to genetically engineered crops increased in many countries in Europe and Africa. In 1999 the European Union placed a five-year moratorium on virtually all approvals of GM crops. One concern is that the inserted genes could spread in an uncontrolled manner from GM crops to weeds or wild relatives of crop plants and possibly harm natural ecosystems in the process. Scientists recognize this concern as legitimate and must take special precautions to avoid this possibility. Critics also worry that some consumers might develop food allergies to GM foods, although scientists routinely screen new GM crops for allergenicity.

The scientific consensus is that the risks associated with consuming food derived from GM varieties are the same as those associated with consuming food derived from varieties produced by traditional breeding techniques. A growing body of evidence, summarized in the Food and Agricultural Organization's *State of Food and Agriculture 2003–2004*, concludes that current GM crop plants are as safe for human consumption as crops grown by conventional or organic agriculture. However, more research on the environmental impact of GM crops is required. To that end, strict guidelines exist in areas of genetic engineering research in which there are unanswered questions about possible effects on the environment.

World production of GM crops FIGURE 8.32

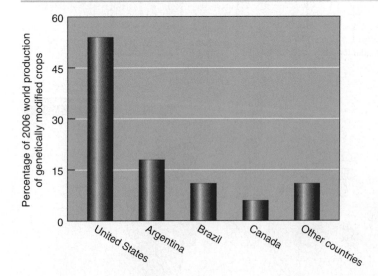

A The world's top producers of GM crops.

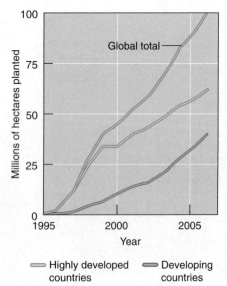

B The production of GM crops has increased significantly.

Monsanto Canada, a subsidiary of the American-based multinational agricultural and biotechnology company, offers Canadian farmers a full line of canola, corn, and soybean seed products. Monsanto uses biotechnology to make seeds grow easier, constantly looking for ways to maximize the potential of the seed both in terms of yield and technology to protect the seed better. Monsanto is the leading producer of genetically engineered seed and invests close to $1.5 million a day in research and development. In addition to seed products, Monsanto sells a wide range of herbicides and is the leading producer of the herbicide glyphosate commonly known as Roundup.

Monsanto has been controversial around the world and has been a primary target by environmental activists and those against the globalization movement. The company has been responsible for making some of the most toxic products ever sold, including polychlorinated biphenyls (PCBs), bovine growth hormone, and Agent Orange. In 1944, Monsanto and 15 other companies began to manufacture DDT, the use of which was banned in 1972 because of the serious environmental effects of the insecticide. Roundup has also been a source of controversy. Several studies have suggested that Roundup leads to the first stages of cancer (although a review of the toxicity of Roundup concluded that at present the herbicide does not pose a health risk to humans). Monsanto has also been identified by the U.S. Environmental Protection Agency as being partially responsible for 56 contaminated sites in the United States. Between 1965 and 1972, Monsanto paid contractors to illegally dump thousands of tonnes of highly toxic waste in landfills. They have been sued and have settled on multiple occasions for damaging the health of their employees or the residents living in close proximity to their production sites through pollution and poisoning.

Despite these controversies, Monsanto pledges their commitment to integrity, transparency, and quality. You be the judge! For more information on Monsanto visit their website at www.monsanto.ca. In addition, look for *The World According to Monsanto*, an in-depth documentary that looks at Monsanto's domination of the agricultural industry.

SOIL CONSERVATION AND REGENERATION

Only 11 percent of the world's soil is suitable for agriculture (**FIGURE 8.33** on page 278). We therefore need to protect the soil we use for agriculture. Although agriculture may cause or accelerate soil degradation, good soil conservation practices promote sustainable soil use. Conservation tillage, crop rotation, contour plowing, strip cropping, terracing, and shelterbelts minimize erosion and mineral depletion of the soil. Badly eroded and depleted land can be restored, but restoration is costly and time-consuming.

Conservation tillage and crop rotation

Conventional methods of tillage, or working the land, include spring plowing, in which the soil is cut and turned in preparation for planting seeds. Although conventional tillage prepares the land for crops, it greatly increases the likelihood of soil erosion. Conventionally tilled

conservation tillage A method of cultivation in which residues from previous crops are left in the soil, partially covering it and helping to hold it in place until the newly planted seeds are established.

crop rotation The planting of a series of different crops in the same field over a period of years.

fields contain less organic material and generally hold less water than undisturbed soil.

Conservation tillage is one of the fastest-growing trends in North American agriculture (**FIGURE 8.34** on page 279). In addition to reducing soil erosion, conservation tillage increases the organic material in the soil, which improves the soil's water-holding capacity. Decomposing organic matter releases nutrient minerals more gradually than when conventional tillage methods are employed. However, use of conservation tillage requires new equipment, new techniques, and greater use of herbicides to control weeds. Research in developing alternative methods of weed control for use with conservation tillage is under way.

Farmers who practise effective soil conservation measures often use a combination of conservation tillage and **crop rotation**. When the same crop is grown over and over in one place, pests for that crop accumulate to destructive levels, and the essential nutrient minerals for that crop are depleted in greater amounts. This makes the soil more prone to erosion, and it makes the crops less productive as well. Crop rotation is effective in decreasing insect damage and disease, reducing soil erosion, and maintaining soil fertility.

A typical crop rotation would be corn → soybeans → oats → alfalfa. Soybeans and alfalfa, both members of the legume family, increase soil fertility through their association with bacteria that fix atmospheric nitrogen into the soil. Thus, planting soybeans and alfalfa as part of crop rotation produces higher yields of the grain crops that are also part of the rotation.

Contour plowing, strip cropping, and terracing
Hilly terrain must be cultivated with care because it is more prone to soil erosion than flat land. Contour plowing, strip cropping, and terracing help control

■ **contour plowing**
Plowing that matches the natural contour of the land.

erosion of farmland with variable topography. In **contour plowing**, furrows run around hills rather than in straight rows. Strip cropping, a special type of contour plowing, produces alternating strips of different crops along natural contours (Figure 7.11A on page 218). For example, alternating a row crop such as corn with a closely sown crop such as wheat reduces soil erosion. Even more effective control of soil erosion is achieved when strip cropping is done in conjunction with conservation tillage.

Farming is undesirable on steep slopes, but if it must be done, **terracing** produces level areas and thereby reduces soil erosion from gravity or water runoff (**FIGURE 8.35**). Nutrient minerals and soil are retained on the horizontal platforms instead of being washed away.

Soil reclamation
Badly eroded land can be reclaimed by (1) preventing further erosion and

Visualizing

Soil conservation FIGURE 8.33

Soil and agriculture

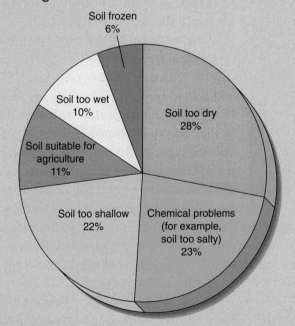

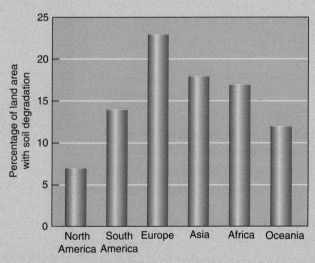

A Only 11 percent of the world's total land area has soil that is naturally suitable for agriculture. Soil that is too dry can be irrigated, whereas soil that is too wet can be drained.

B Degree of soil degradation (eroded, desertified, or salty soil) by continent. North America has the least degraded soils, and Europe, Africa, and Asia have the most degraded soils.

Conservation tillage FIGURE 8.34

Decaying residues from the previous year's crop (rye) surround young soybean plants in a farmer's field in Iowa. Conservation tillage reduces soil erosion as much as 70 percent because plant residues from the previous season's crops are left in the soil.

Terracing FIGURE 8.35

Global Locator

Terraces are small earthen embankments placed across a steep hillside or mountain. Terracing hilly or mountainous areas, such as these rice terraces on Bali, Indonesia, curbs water flow and reduces the amount of soil erosion. Farmers must maintain these terraced fields.

(2) restoring soil fertility. To prevent further erosion, the bare ground is seeded with plants; they eventually grow to cover the soil, stabilizing it and holding it in place.

The plants start to improve the quality of the soil almost immediately, as dead material decays into humus. The humus holds nutrient minerals in place, releasing them a little at a time; it also improves the water-holding capacity of the soil. One of the best ways to reduce the effects of wind erosion on soil is to plant **shelterbelts** that reduce the impact of wind (**FIGURE 8.36**).

shelterbelt
A row of trees planted as a windbreak to reduce soil erosion of agricultural land.

Restoration of soil fertility is a slow process. The land cannot be farmed or grazed until the soil has completely recovered. But the restriction of land use for an indefinite period may be difficult to accomplish. How can a government tell landowners they cannot use their own land? How can land use be restricted when people's livelihoods, and maybe even their lives, depend on it?

Soil conservation policies
In Canada, soil conservation policies are generally lagging behind efforts being made elsewhere, particularly in the United States. The U.S. Conservation Reserve Program (CRP) is a voluntary subsidy program that pays U.S. farmers to stop producing crops on highly erodible farmland.

Shelterbelts surrounding kiwi orchards FIGURE 8.36

Trees protect the delicate fruits from the wind and reduce wind erosion of farmland soil. Photographed on the North Island, New Zealand.

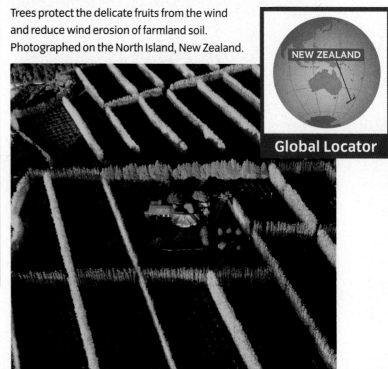

Global Locator

It requires planting native grasses or trees on such land and then "retiring" it from further use for 10 to 15 years. The CRP has benefitted the environment. Annual loss of soil on CRP lands planted with grasses or trees has been reduced by more than 90 percent. Because the vegetation is not disturbed once it is established, it provides biological habitat. Small and large mammals, birds of prey, and ground-nesting birds such as ducks have increased in number and kind on CRP lands. The reduction in soil erosion has improved water quality and enhanced fish populations in surrounding rivers and streams.

An important Canadian land-use initiative is Alternative Land Use Services (ALUS). ALUS is a unique incentive-based program designed by farmers in partnership with Delta Waterfowl that provides farmers with a fee-for-service, in this case, rewards for the service of managing the landscape to benefit ducks. It is a win-win program that addresses conservation of waterfowl and environmental enhancement of agricultural lands. Because the environment is a top election concern in Canada, ALUS is gaining momentum across the country. Prince Edward Island officially adopted ALUS in 2008 as part of their agricultural policy. ALUS promotes farmers and ranchers as conversationalists and communicates the benefits that agricultural producers provide to Canadians.

CONCEPT CHECK STOP

What is sustainable agriculture?

What are one pro and one con of genetically engineered crops?

How do conservation tillage, contour plowing, and shelterbelts contribute to soil conservation?

ECO CANADA **CAREER FOCUS**

REMEDIATION SPECIALIST

Environmental remediation involves removing contamination from soil, groundwater, and other natural sites. Remediation specialists put together plans to achieve those outcomes.

Imagine you are standing in an old farmyard, surrounded by empty, run-down buildings, dead trees, and patches of bare soil. You are a remediation specialist and this is the site of your next project. Forty years ago, the farm's owner retired and sold the land to a nearby chemical company that began using it as an illegal dump and burn site. When the company went out of business, the province investigated the site and discovered dangerous levels of soil and water pollution. Since then, the province has declared the farm an abandoned hazardous waste site and hired your firm to clean it up.

As a remediation specialist, you focus on cleaning up toxic sites and removing contaminants from soil and groundwater. For this project, you will work with a team of environmental geologists, hydrologists, and toxicologists to develop a remedial action plan for the site. You start by investigating which methods and technologies will work well for this particular site and how they can be incorporated into the plan.

Next, you develop interim remediation measures, including removing the contamination source. You will excavate more than 500 tonnes of impacted soil for off-site disposal at a hazardous waste landfill. To verify you have dug deep enough and far enough, you will test samples from the side walls and floor of the excavation for remaining contaminants before the pit is filled.

You will also develop a program to assess groundwater quality in the area using existing monitoring wells to collect samples for laboratory analysis. Once the interim cleanup is complete, you'll implement long-term measures for revegetation and water quality monitoring. It may take years, but your remedial action plan will direct cleanup and return this site to its former state.

Environmental remediation is the treatment and removal of contamination from soil, groundwater, and other media. Remediation specialists design and implement remedial action plans to clean up sites affected by substances such as automotive fuels, pesticides, and heavy metals. They select the most appropriate remedial techniques or processes for each project while adhering to federal and provincial environmental regulations. Remediation specialists may also work in conjunction with site assessors to determine the extent and type of contamination present at disturbed sites.

For more information or to look up other environmental careers go to www.eco.ca.

CASE STUDY

INDUSTRIAL ECOSYSTEMS

Traditional industries operate in a linear fashion: natural resources → products → wastes dumped back into the environment. However, natural resources are finite, and the environment's capacity to absorb waste is limited. The field of **industrial ecology** seeks to address these issues by using resources efficiently, regarding "wastes" as potential products, and creating **industrial ecosystems** that in many ways mimic natural ecosystems.

One pioneering industrial ecosystem in Kalundborg, Denmark (**FIGURE A**), consists of an electric power plant, an oil refinery, a pharmaceutical plant, a wallboard factory, a sulphuric acid producer, a cement manufacturer, a fish farm, greenhouses, and area homes and farms. These entities are linked in ways that resemble a food web in a natural ecosystem (**FIGURE B**).

The Kalundborg Industrial Ecosystem

A Kalundborg, Denmark

B Kalundborg's industrial ecosystem has significantly reduced resource consumption and waste production.

The coal-fired electric power plant originally cooled its waste steam and released it into the local fjord. The steam is now supplied to the oil refinery and pharmaceutical plant, and the surplus heat warms greenhouses, the fish farm, and area homes. Surplus natural gas from the oil refinery is sold to the power plant and the wallboard factory. Before selling the natural gas, the oil refinery removes excess sulphur from it (as required by law) and sells it to the sulphuric acid producer.

To meet environmental regulations, the power plant installed pollution control equipment to remove sulphur from its coal smoke. This sulphur, in the form of calcium sulphate, is sold to the wallboard plant and used as a gypsum substitute. The fly ash produced by the power plant goes to the cement manufacturer for use in road building.

To fertilize their fields, local farmers use sludge from the fish farm and a high-nutrient sludge generated at the pharmaceutical plant. Most pharmaceutical companies discard this sludge because it contains living microorganisms, but the Kalundborg plant heats the sludge to kill the microorganisms and convert a waste material into a commodity.

It took a decade to develop this entire industrial ecosystem. Although initiated for economic reasons, the industrial ecosystem has distinct environmental benefits, from energy conservation to a reduction of pollution.

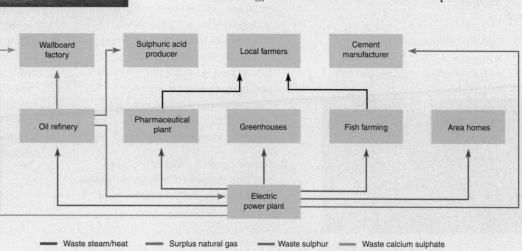

SUMMARY

1 Global versus Canadian Land Resource Base and Uses

1. Canada's total land surface represents about 6 percent of the world's land base. However, only 11 percent of Canada's land is privately owned. Crown land, which is land owned by either federal or provincial governments, is vitally important.

It provides numerous ecosystem services, opportunities for scientific research and education, and important recreational spots for Canadians and visitors.

2 Agricultural Land Uses

1. **Industrialized agriculture** uses modern methods requiring large capital input and less land and labour than traditional methods. **Subsistence agriculture** uses traditional methods dependent on labour and a large amount of land to produce enough food to feed a family. There are three types of subsistence agriculture. In slash-and-burn agriculture, small patches of tropical forests are cleared to plant crops. Nomadic herding, carried out on arid land, requires herders to move livestock continually to find food for them. Intercropping involves growing a variety of plants simultaneously on the same field.

2. Environmental problems caused by industrialized agriculture include air pollution from the use of fossil fuels and pesticides, water pollution from untreated animal wastes and agricultural chemicals, pesticide-contaminated foods and soils, and increased resistance of pests to pesticides. **Land degradation** decreases the future ability of the land to support crops or livestock. Clearing grasslands and forests and draining wetlands to grow crops have resulted in **habitat fragmentation,** the breakup of large areas of habitat into small, isolated patches.

3. A **pesticide** is any toxic chemical used to kill pests. Although there are benefits to pesticides, there are numerous disadvantages. Safer alternates to pesticides include **biological controls**, which use naturally occurring disease organisms, parasites, or predators to control pests. Pheromones, natural substances produced by animals to stimulate a response in other members of the same species, are used to attract and trap pest species. **Integrated pest management** is a combination of pest control methods that, if used in the proper order and at the proper times, keeps a pest population low enough to prevent substantial economic loss.

4. **Overgrazing** is the destruction of vegetation caused by too many grazing animals consuming the plants in a particular area, leaving them unable to recover. Overgrazing accelerates land degradation, which decreases the future ability of the land to support crops or livestock. **Desertification** is the degradation of once-fertile rangeland or tropical dry forest into nonproductive desert.

3 Forests

1. **Sustainable forestry** is the use and management of forest ecosystems in an environmentally balanced and enduring way. Sustainable forestry maintains a mix of forest trees, by age and species, rather than a monoculture, in which only one type of plant is cultivated over a large area. Adopting sustainable forestry principles requires setting aside sanctuaries and **wildlife corridors,** protected zones that connect isolated unlogged or undeveloped areas.

2. **Deforestation** is the temporary or permanent clearance of large expanses of forest for agriculture or other uses.

Clear-cutting is a logging practice in which all the trees in a stand of forest are cut, leaving just the stumps; clear-cutting over a wide area is ecologically unsound. The major causes of tropical deforestation are subsistence farming, commercial logging, and cattle ranching, all accelerated by growing human populations. Increased needs for fuelwood drives deforestation of tropical dry forests.

4 Uses of Minerals and Soils

1. A **mineral** is a metallic or nonmetallic element or compound of elements that occurs naturally in Earth's crust. Highly developed nations consume a disproportionate share of the world's minerals, but as developing countries become industrialized, their need for minerals increases.

2. Minerals are extracted through **surface** or **subsurface mining**. Surface mining destroys vegetation across large areas, increasing erosion. Open-pit mining uses huge quantities of water. Mining also affects water quality. Acid mine drainage is pollution caused when dissolved toxic materials wash from mines into nearby lakes and streams.

3. **Soil** is the uppermost layer of Earth's crust and supports terrestrial plants, animals, and microorganisms. Soil is formed from parent material—rock that is slowly fragmented into small particles by biological, chemical, and physical weathering processes. Soil is composed of mineral particles, organic matter, water, and air. **Soil horizons** are the horizontal layers into which many

soils are organized, from the surface to the underlying parent material.

5 Parks and Protected Areas

1. The federal government is responsible for Canada's national parks, of which there are 42. The provincial governments are responsible for managing their own provincial parks. Although both types of parks have a goal of protecting the natural landscape, national parks tend to focus more on *preservation*, and provincial parks more on *conservation*. Both types of parks offer many scientific, educational, and recreational opportunities.

2. Parks Canada is the arm of the federal government responsible for managing and maintaining Canada's national parks. Its primary mandate is to protect and restore the ecological integrity of each park, meaning that native plant and animal species are able to live, reproduce, and maintain long-term viability.

3. The 12 percent challenge was issued over 20 years ago by the World Wildlife Fund, who tried to encourage countries to implement strategies that would protect a target representation of at least 12 percent

of every ecosystem in the world. In Canada, it has been implemented primarily through provincial parks, which have set aside protected areas of land.

6 Example Approaches to Sustainable Land Resource Management

1. **Sustainable agriculture** uses methods that maintain soil productivity and a healthy ecological balance while minimizing long-term impacts. Unlike industrialized agriculture, sustainable agriculture relies on beneficial biological processes and environmentally friendly chemicals.

2. **Genetic engineering** is the manipulation of genes to produce a particular trait. Genetic engineering produces more productive livestock varieties, more nutritious crops, or crop plants resistant to pests, diseases, or drought. Concerns about genetic engineering include its potential to produce harmful organisms and to trigger food allergies.

3. Good soil conservation practices promote sustainable soil use. In **conservation tillage**, residues from previous crops partially cover the soil to help hold it in place until newly planted seeds are established. **Crop rotation**, the planting of different crops in a field over a period of years, decreases the insect damage, disease, and mineral depletion that occur when one crop is grown continuously. **Contour plowing**, which matches the natural contour of the land, helps control erosion of land with variable topography. Strip cropping produces alternating strips of different crops along natural contours. Terracing reduces soil erosion on steep slopes. A **shelterbelt** is a row of trees planted as a windbreak to reduce soil erosion.

KEY TERMS

- **industrialized agriculture** p. 236
- **subsistence agriculture** p. 237
- **land degradation** p. 242
- **habitat fragmentation** p. 242
- **pesticide** p. 242
- **biological control** p. 242
- **integrated pest management (IPM)** p. 242
- **rangeland** p. 244
- **overgrazing** p. 245

- **desertification** p. 245
- **sustainable forestry** p. 250
- **wildlife corridor** p. 251
- **clear-cutting** p. 253
- **deforestation** p. 253
- **minerals** p. 258
- **surface mining** p. 259
- **subsurface mining** p. 259
- **soil** p. 263
- **soil horizons** p. 263

- **sustainable soil use** p. 265
- **soil erosion** p. 265
- **sustainable agriculture** p. 274
- **genetic engineering** p. 275
- **conservation tillage** p. 277
- **crop rotation** p. 277
- **contour plowing** p. 278
- **shelterbelt** p. 279

CRITICAL AND CREATIVE THINKING QUESTIONS

1. Should private landowners have control over what they wish to do to their land? How would you as a landowner handle land use decisions that may affect the public? Present arguments for both sides of the issue.

2. Name two environmental problems associated with industrialized agriculture, and give at least three examples of ways that industrialized agriculture could be made more sustainable.

3. Distinguish between shifting cultivation and slash-and-burn agriculture.

4. What was the green revolution? Describe a few of its successes and shortcomings.

5. Is a major goal of integrated pest management (IPM) to eradicate the pest species? Explain your answer.

6. Distinguish between rangeland degradation and desertification. Why is moderate grazing beneficial to rangelands, yet overgrazing leads to erosion?

7. Why is deforestation a serious global environmental problem?

8. What are the environmental effects of clear-cutting on steep mountain slopes? On tropical rainforest land?

9. Give at least five ecosystem services provided by forests and other nonurban lands.

10. Suppose a valley contains a small city surrounded by agricultural land. The land is encircled by a mountain wilderness. Explain why the preservation of the mountain ecosystem would support both urban and agricultural land in the valley.

11. How many minerals have you come in contact with today? Which were metals; which were nonmetals?

12. What is the difference between surface and subsurface mining? Open-pit and strip mines? Shaft and slope mines? When is each most likely to be used?

13–14. The figure below represents the flow of minerals in a low-waste society.

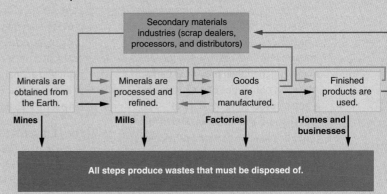

13. Which colours of arrows represent sustainable manufacturing, consumer reuse, and consumer recycling, respectively?

14. Would the flow of minerals be more or less complicated in a high-waste society? How would the wastes generated differ from those of the low-waste society represented here?

15. What are the roles of weathering, organisms, and topography in soil formation?

16. Certain pests that cause plant disease reside in the plant residues left on the ground with conservation tillage. Given that these organisms are often specific for the plants they attack, how can we control them?

17. How does the industrial ecosystem at Kalundborg, Denmark, resemble a natural ecosystem? What are some environmental benefits of this industrial ecosystem?

What is happening in this picture ❓

■ Special equipment carries out "hydroseeding," applying a mixture of grass seed, water, and fertilizers. What is the purpose of this effort?

■ Why do you think hydroseeding was chosen here rather than applying grass seed using equipment that rolls across the ground?

■ Hydroseeding is often used after human activities have disturbed large patches of ground; suggest one or two likely examples.

Aquatic Resources, Uses, and Management

9

DEPLETING BLUEFIN TUNA STOCKS

Stocks of the giant, or Atlantic, bluefin tuna *(Thunnus thynnus),* a fish highly prized for sushi, are classified as depleted in the Mediterranean Sea by the Food and Agriculture Organization of the United Nations. Once harvested sustainably through traditional trapping, bluefins in the Mediterranean are now fished—often illegally—at approximately four times the sustainable rate. Spotter aircraft locate fish stocks and alert huge fishing fleets, whose ships (see inset) cast purse seines to net schools. Captured bluefins are fattened in offshore pens (see large photo) before being butchered for market. The enormous economic value of the bluefin, which as an adult weighs an average of 250 kilograms, places it at great risk. Only recently have Mediterranean nations begun implementing conservation measures to protect the species.

Overfishing, the harvesting of fishes faster than they can reproduce, is not limited to the Mediterranean. Worldwide, about 30 percent of fish species have been overfished as the world demand for fish has grown and harvesting methods have become more sophisticated. In late 2006, a team of ecologists and economists estimated that if current trends in overfishing and ocean pollution aren't curbed, populations of virtually all harvested seafood species could collapse by 2048.

The depletion of Mediterranean bluefin tuna stocks teaches us that as our technologies (such as more efficient fishing techniques) advance, so do the impacts that we have on our environment. As a result, we must regulate these impacts (e.g., the overharvesting of fishes) more closely than ever.

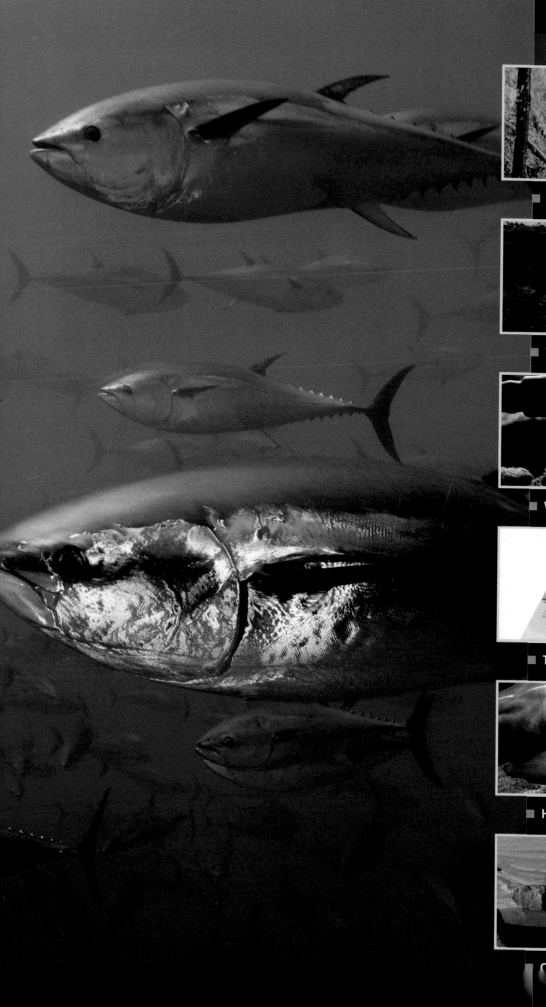

The Importance of Water

Young brick workers in India bathe with water from an irrigation pipe FIGURE 9.1

Global Locator

INDIA

Water resources are as vital to life as air. Fresh water is used by humans for consumption, domestic needs, industrial and manufacturing demands, energy and mining production, and agriculture. Yet fresh water is one of the rarest resources on Earth. As the human population continues to grow, the demands we place on the world's freshwater supply will increase. It is estimated that over the next two decades, human use of water will increase by 40 percent. The World Water Council warns that by 2020 we will need 17 percent more water than is available if we are going to be able to feed the world. The United Nations Development Programme (UNDP) indicates that 1 billion people, roughly one in every six people in this world, have no access to clean and safe water, and that over 2 billion, or one in three, lack access to safe sanitation.

All life forms, from unicellular bacteria to multicellular plants and animals, contain water. Humans are composed of approximately 60 percent water by body weight. We depend on water for our survival as well as for our convenience: we drink it, cook with it, wash with it (FIGURE 9.1), travel on it, and use an enormous amount of it for agriculture, manufacturing, mining, energy production, and waste disposal.

Although Earth has plenty of water, about 97 percent of it is saline and not consumable by most terrestrial organisms. Further, of the 2.5 percent of Earth's fresh water, less than 1 percent (or 0.01 percent of all water on Earth) is in a useable form for human consumption and withdrawal. Water is constantly recycling through Earth's hydrologic cycle. Due to population growth and

pollution, less and less is available per person per year, and global climate change adds new uncertainties to the equation. Efficiency, conservation, and technology can help ensure that the water you absorb today will still be useable and clean hundreds of years from now.

THE HYDROLOGIC CYCLE AND OUR SUPPLY OF FRESH WATER

Water is constantly recycling through Earth's **hydrologic cycle**, from the ocean to the atmosphere through a process of **evaporation**, and to the land and back to the ocean through **precipitation** in the form of rain, snow, sleet, hail, and so on (see FIGURE 9.2; also see Figure 5.10 on page 142). Transpiration also contributes

The hydrologic cycle FIGURE 9.2

Water in the form of rain, snow, sleet, and hail continuously moves from the atmosphere to land and oceans, while evaporation transports water vapour from land and oceans into the atmosphere. Transpiration represents a major pathway by which plants and forests direct water vapour to the atmosphere through loss during photosynthesis.

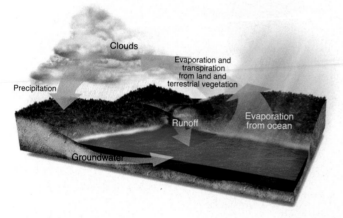

Canada's watersheds FIGURE 9.3

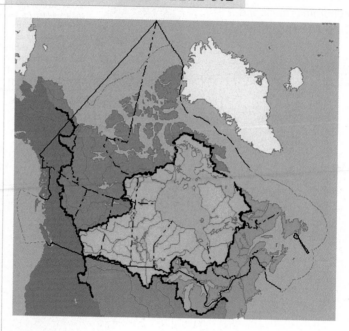

to the cycle as plants take up water from roots in the soil and release it to the atmosphere through pores on the undersides of their leaves during photosynthesis. The result is a balance of the water resources in the ocean, on land, and in the atmosphere. Through the continual process of the hydrologic cycle, there is a renewal of fresh water to landscapes, which is essential to support life.

Surface water is water found in streams, rivers, lakes, ponds, reservoirs, and **wetlands** (areas of land covered with water for at least part of the year). The **runoff** of precipitation from the land replenishes surface waters and is considered a renewable, although finite, resource. **Drainage basins**, or **watersheds**, represent the area where all surface water drains into the same river, lake, or ocean. The water starts out as tiny droplets that are funnelled into streams and eventually combine to form rivers and lakes. Canada's watersheds include the Atlantic Ocean, Hudson Bay, the Arctic Ocean, the Pacific Ocean, and the Gulf of Mexico (**FIGURE 9.3**). **FIGURE 9.4** on pages 290 and 291 shows Earth's primary watersheds.

surface water
Precipitation that remains on the surface of the land and does not seep down through the soil.

runoff
The movement of fresh water from precipitation and snowmelt to rivers, lakes, wetlands, and the ocean.

groundwater
The supply of fresh water under Earth's surface that is stored in underground aquifers.

aquifers
Underground reservoirs of permeable rock from which groundwater can be extracted.

Water collects underground to form reservoirs of **groundwater**. This originated as surface precipitation that slowly seeped into the tiny spaces in the ground between loose sand, gravel, and between bedrock cracks. Eventually, an impenetrable rock layer prevents further seepage into the ground. Groundwater collects in large reservoirs referred to as **aquifers**. These are often used as sources of fresh water for human use. Groundwater may reside in these underground aquifers for hundreds of years but eventually will be discharged into rivers, wetlands, springs, or the ocean. Thus, surface and groundwater are closely interconnected through the hydrologic cycle.

There are two types of aquifers found. In **unconfined aquifers**, groundwater seeps from the surface through an unsaturated zone until it reaches the water table. At this point, the ground is saturated and the groundwater collects in reservoirs. The height of the water table is the same as the unconfined aquifer. In contrast, a **confined aquifer** is sandwiched between two impermeable layers or confining beds

Earth's primary watersheds FIGURE 9.4

Earth's bodies of fresh water can cover enormous areas and cross many political boundaries.

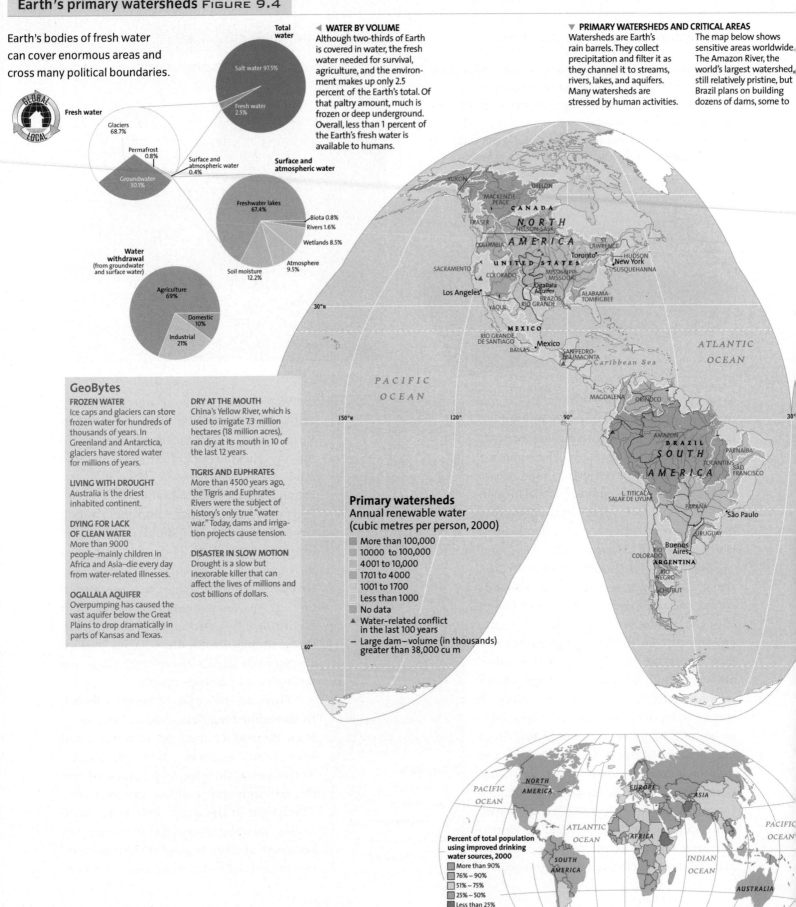

Total water

Salt water 97.5%

Fresh water 2.5%

Fresh water

Glaciers 68.7%

Permafrost 0.8%

Surface and atmospheric water 0.4%

Groundwater 30.1%

Surface and atmospheric water

Freshwater lakes 67.4%

Biota 0.8%

Rivers 1.6%

Wetlands 8.5%

Atmosphere 9.5%

Soil moisture 12.2%

Water withdrawal
(from groundwater and surface water)

Agriculture 69%

Domestic 10%

Industrial 21%

◄ **WATER BY VOLUME**
Although two-thirds of Earth is covered in water, the fresh water needed for survival, agriculture, and the environment makes up only 2.5 percent of the Earth's total. Of that paltry amount, much is frozen or deep underground. Overall, less than 1 percent of the Earth's fresh water is available to humans.

▼ **PRIMARY WATERSHEDS AND CRITICAL AREAS**
Watersheds are Earth's rain barrels. They collect precipitation and filter it as they channel it to streams, rivers, lakes, and aquifers. Many watersheds are stressed by human activities.

The map below shows sensitive areas worldwide. The Amazon River, the world's largest watershed, still relatively pristine, but Brazil plans on building dozens of dams, some to

GeoBytes

FROZEN WATER
Ice caps and glaciers can store frozen water for hundreds of thousands of years. In Greenland and Antarctica, glaciers have stored water for millions of years.

LIVING WITH DROUGHT
Australia is the driest inhabited continent.

DYING FOR LACK OF CLEAN WATER
More than 9000 people—mainly children in Africa and Asia—die every day from water-related illnesses.

OGALLALA AQUIFER
Overpumping has caused the vast aquifer below the Great Plains to drop dramatically in parts of Kansas and Texas.

DRY AT THE MOUTH
China's Yellow River, which is used to irrigate 7.3 million hectares (18 million acres), ran dry at its mouth in 10 of the last 12 years.

TIGRIS AND EUPHRATES
More than 4500 years ago, the Tigris and Euphrates Rivers were the subject of history's only true "water war." Today, dams and irrigation projects cause tension.

DISASTER IN SLOW MOTION
Drought is a slow but inexorable killer that can affect the lives of millions and cost billions of dollars.

Primary watersheds
Annual renewable water
(cubic metres per person, 2000)

- More than 100,000
- 10000 to 100,000
- 4001 to 10,000
- 1701 to 4000
- 1001 to 1700
- Less than 1000
- No data
- ▲ Water-related conflict in the last 100 years
- — Large dam—volume (in thousands) greater than 38,000 cu m

Percent of total population using improved drinking water sources, 2000
- More than 90%
- 76% – 90%
- 51% – 75%
- 25% – 50%
- Less than 25%
- No data available

ver aluminum smelters.
frica and Asia, lack
ccess to water and
ter-related diseases are
main problems. In
ope and the Middle East,
ruse, pollution, and

disagreement over
diverting water are the
major challenges. Hope
rests in better planning
and community-scale
projects.

It's as vital to life as air. Yet fresh water is one of the rarest resources on Earth. Only 2.5 percent of Earth's water is fresh, and of that the useable portion for humans is less than 1 percent of all fresh water, or 0.01 percent of all water on Earth. Water is constantly recycling through Earth's hydrologic cycle. But population growth and pollution are combining to make less and less available per person per year, while global climate change adds new uncertainty.

Efficiency, conservation, and technology can help ensure that the water you absorb today will still be useable and clean hundreds of years from now.

ACCESS TO FRESH WATER

ess to clean fresh water
ritical for human health.
in many regions, potable
ter is becoming scarce
ause of heavy demands
pollution. Especially

worrisome is the poisoning of aquifers—a primary source of water for nearly one-third of the world—by sewage, pesticides, and heavy metals.

▼ GLOBAL IRRIGATED AREAS AND WATER WITHDRAWALS

Since 1970, global water withdrawals have correlated with the rise in irrigated area. Some 70% of with-

drawals are for agriculture, mostly for irrigation that helps produce 40 percent of the world's food.

Freshwater withdrawal as a percentage of total water utilization, 2000

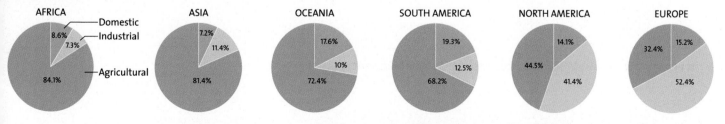

AFRICA		
Domestic	8.6%	
Industrial	7.3%	
Agricultural	84.1%	

ASIA: 7.2%, 11.4%, 81.4%

OCEANIA: 17.6%, 10%, 72.4%

SOUTH AMERICA: 19.3%, 12.5%, 68.2%

NORTH AMERICA: 14.1%, 44.5%, 41.4%

EUROPE: 15.2%, 32.4%, 52.4%

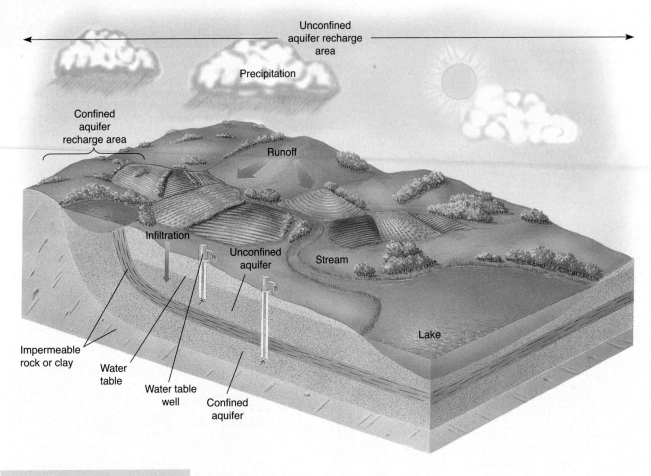

Groundwater FIGURE 9.5

Surface water infiltrates into deeper soil and porous rock layers until it reaches impermeable rock or clay. The boundary between saturated and unsaturated soils defines the depth of the water table. An unconfined aquifer has groundwater recharged by surface water directly above it. In a confined aquifer, groundwater is stored between two impermeable layers.

VIEW THIS IN ACTION

of rock that restrict groundwater movement and force this water to be under high pressure (**FIGURE 9.5**). Aquifers are maintained through recharge zones. In unconfined aquifers the unsaturated portion above the water table represents opportunity for water infiltrations that recharges the aquifer. In confined aquifers, recharge occurs at points where the aquifer contacts surface water bodies. Artificial recharge is done by injecting water into wells or by flooding the surface.

Most groundwater is considered a nonrenewable resource because it has taken hundreds or even thousands of years to accumulate and usually only a small portion of it is replaced each year by seepage of surface water.

PROPERTIES OF WATER

Water is composed of molecules of H_2O, each consisting of two atoms of hydrogen and one atom of oxygen. Water molecules are **polar**—that is, one end of the molecule has a positive electrical charge and the other end has a negative charge (**FIGURE 9.6**). The negative (oxygen)

Chemical properties of water FIGURE 9.6

A Water molecules are polar, with positively and negatively charged areas.

B The polarity causes hydrogen bonds (represented by dashed lines) to form between the positive areas of one water molecule and the negative areas of others. Each water molecule forms up to four hydrogen bonds with other water molecules.

end of one water molecule is attracted to the positive (hydrogen) end of another water molecule, forming a **hydrogen bond** between the two molecules. Hydrogen bonds are the basis for many of water's physical properties, including its high melting/freezing point (0° C) and high boiling point (100° C). Because most of Earth has a temperature between 0° C and 100° C, most water exists in the liquid form organisms need. As well, water absorbs a great deal of solar heat without substantial increases in temperature. This high heat capacity allows the ocean to have a moderating influence on climate, particularly along coastal areas.

Another important quality of water is that it seems to be lighter when it is in the form of ice and able to float on the more dense liquid form below. This is a unique feature of water related to the hydrogen bonding between molecules and is essential for life to exist in aquatic systems. For most substances, as energy is removed and the temperature decreased to near freezing conditions, the volume of space occupied by the vibrating atoms and between molecules is reduced and the compound becomes more dense and heavier. However, when water loses heat and begins to cool, the greatest density is achieved in its liquid state. With colder

temperatures, there is first an increase in density until the maximum is reached at 4° C. With even further decreases in temperature, there is now a decrease in the density of the compound and it apparently becomes lighter. This translates to ice cubes floating in a glass of water or to ice "floating" on the top of a lake.

The reason for this phenomenon relates to the increase in volume as water is transformed from a liquid to a solid. If you freeze a bottle of water, you will likely realize that the ice expands to occupy additional space (without any physical change in the amount of water), which may have caused the bottle to rupture. By increasing the volume, ice can become less dense (remember, density = mass/volume). So the ice that has formed on the top of the lake is lighter than the water below. If this were not the case, ice would sink and a new layer would form at the lake surface. This process would continue until the water body formed a giant ice cube. Aquatic organisms would freeze and would not be able to survive winter in these water bodies. As ice begins to melt and warm in the spring, the density difference lessens. The water column mixes and redistributes nutrients and sediments from the bottom toward the top of the lake and vice versa.

And finally, water is a *solvent,* meaning that it can dissolve many materials. In nature, water is never completely pure because it contains dissolved gases from the atmosphere (e.g., carbon dioxide dissolves in water to form carbonic acid) and dissolved mineral salts from the land. This feature is a major drawback in that it contributes to many environmental issues of pollution, for example, nutrient enrichment or eutrophication (discussed later in this chapter).

CONCEPT CHECK STOP

What is surface water? What is groundwater?

HOW do hydrogen bonds form between adjacent water molecules?

What are the unique properties of water?

Water Issues

WATER RESOURCE PROBLEMS

Water resource problems fall into three categories: too much water, too little water, and poor quality water. Flooding occurs when a river's discharge cannot be contained within its normal channel. Today's floods are more disastrous in terms of property loss than those of the past because humans often remove water-absorbing plant cover from the soil and construct buildings on flood plains. (A **flood plain** is the area bordering a river channel that has the potential to flood.)

These activities increase the likelihood of both floods and flood damage.

When a natural area—that is, an area undisturbed by humans—is inundated with heavy precipitation, the plant-protected soil absorbs much of the excess water. What the soil cannot absorb runs off into the river, which may then spill over its banks onto the flood plain. Because rivers meander, the flow is slowed, and the swollen waters rarely cause significant damage to the surrounding area. (See Figure 6.20 on page 185 for a diagram of a typical river, including its flood plain.)

When an area is developed for human use, construction projects replace much of this protective plant cover. Buildings and paved roads don't absorb water, so runoff, usually in the form of storm sewer runoff, is significantly greater in developed areas (**FIGURE 9.7**).

As the natural flow of water is altered through urbanization, there is an increasing need for municipalities to divert water away from homes through the construction of sewers and culverts. As described later in this chapter, approaches in diverting water include combined sewers, separate sewers, retention ponds,

---

How development changes the natural flow of water FIGURE 9.7

Shown is the fate of precipitation before (**A**) and after (**B**) the development of an urbanized environment dominated by buildings and pavement. Urbanization reduces evaporation and groundwater infiltration but increases amounts of surface runoff.

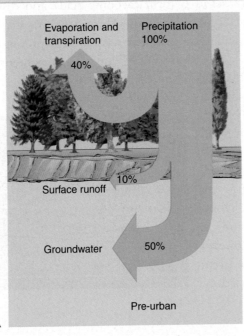

Evaporation and transpiration
Precipitation 100%
40%
Surface runoff 10%
Groundwater 50%
Pre-urban

A

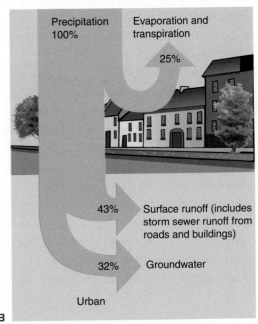

Precipitation 100%
Evaporation and transpiration
25%
43% Surface runoff (includes storm sewer runoff from roads and buildings)
32% Groundwater
Urban

B

and engineered wetlands (or naturalized stormwater retention ponds). People who build homes or businesses on the flood plain of a river will most likely experience flooding at some point (**FIGURE 9.8**).

Types of water use

Humans exploit water in a variety of ways. Left in place in the stream, river, and lake, **instream uses** include hydroelectric power generation, shipping and transportation, and many recreational activities. Often we remove water from the source through an intake, use it in an activity, and ultimately return or discharge all or a portion of it to the source water body. These uses are referred to as **water withdrawals**. Household and industrial uses, thermal and nuclear power generation, mining, manufacturing, and municipal demands are

> **water withdrawal**
> The removal of water from a source such as a lake or river where a portion of this water is returned to the source and is available to be used again.

> **water consumption** The removal of water for human use without any return of water to its original source.

common examples of withdrawals. The portion of water that is never returned to the source is used in **water consumption**. Too much consumption leads to insufficient volume downstream.

Agricultural crop irrigation is the largest consumptive use of water (**FIGURE 9.9** on page 296). It accounts for about 71 percent of the world's total water consumption. According to UNESCO, in 2000, about 57 percent of the world's freshwater withdrawal and 70 percent of its consumption took place in Asia, where the world's major irrigated lands are located. In 2006, the gross water use in Canada amounted to 60,527 million cubic metres. Of this, 30 percent was consumed in industrial and agricultural processes, while the rest was discharged back to receiving waters.

Visualizing

Flooding in Canada FIGURE 9.8

Flooding has been a major issue for municipalities and homesteads along the Red River basin in Manitoba. The Red River Floodway was built to divert water around the city of Winnipeg and to reduce the potential of residential flooding. The upstream community of Ste. Agathe instead built a dyke around the community. Unfortunately during the Flood of the Century in 1997, it experienced a breach in one of the walls that permitted floodwaters to enter the town and destroy many of the homes in the community (A). In 1996, a flood in Saguenay, Quebec (B), killed 10 people, forced 12,000 residents from their homes, and was estimated to have caused $700 million in damages.

A Ste. Agathe, Manitoba, 1997

B Saguenay, Quebec, 1996

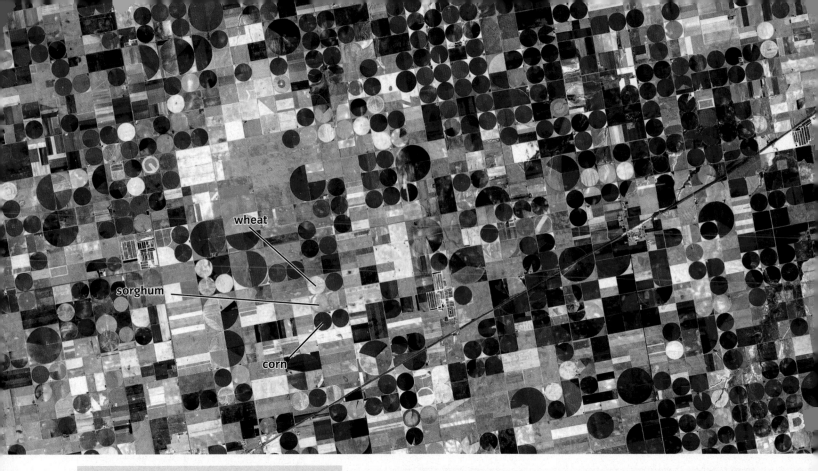

wheat

sorghum

corn

Agricultural use of water FIGURE 9.9

These fields use central-pivot irrigation, a method that minimizes evaporative water loss and gives the fields a distinctive circular shape. Each circle is the result of a long irrigation pipe that extends along the radius from the circle's centre to its edge and slowly rotates, spraying the crops. This satellite photo was taken in June; wheat fields are bright yellow, corn fields (dark green) are growing vigorously, and the sorghum crop (light green) is just starting to appear.

Aquifer depletion

Aquifer depletion from excessive removal of groundwater lowers the water table and prolonged consumption drains the aquifer dry, effectively eliminating it as a water resource. In Canada and worldwide, aquifers supply drinking water for roughly 25 percent of the population. Reports from 2006 indicated that two-thirds of these demands in Canada are from rural areas. The remaining demands are from livestock watering, irrigation, aquaculture (fish farms), and mineral and fossil fuel extraction. It is important to guarantee a sustainable yield from aquifers to ensure ecosystem stability and a freshwater supply for future generations.

Aquifer depletion is a more serious problem in the United States than in Canada.

aquifer depletion The removal of groundwater faster than it can be recharged by precipitation or melting snow.

salt water intrusion The movement of sea water into a freshwater aquifer near the coast when the water table drops due to withdrawal rates that exceed groundwater recharge.

In the United States, groundwater is being depleted and used at four times its replacement rate. One reason for this is that the withdrawal rate exceeds the natural rate of recharge. In extreme cases, land subsidence (or sinking) and compression reduce the storage capacity of the aquifer. **Salt water intrusion** can also occur along coastal areas when the water table drops and sea water begins to infiltrate the unsaturated porous land. Well water in such areas eventually becomes too salty for human consumption or other freshwater uses.

Overdrawing surface waters

Removing too much fresh water from a river or lake can have disastrous consequences in local ecosystems. Humans can remove

The Fraser basin FIGURE 9.10

The Fraser basin has undergone significant changes because of human alterations of landscapes that have caused changes to the hydrology and water availability in the region.

perhaps 30 percent of a river's flow without greatly affecting the natural environment. However, in some places considerably more is withdrawn for human use.

When surface waters are overdrawn, wetlands dry up and estuaries become saltier as less fresh water discharges into the ecosystem. Wetlands and estuaries, which serve as breeding grounds for many species of birds and other animals, also play a vital role in the hydrologic cycle. When these resources are depleted, the ensuing water shortages and reduced productivity have economic as well as ecological ramifications.

Climate change further complicates the water shortage issue. In Canada, the Fraser basin, an area in British Columbia drained by the Fraser River and its 13 main watersheds, has seen profound changes in the hydrology and water availability of the region (FIGURE 9.10). Rivers are peaking earlier in spring as the result of milder winters, increased winter melt, and winter river flow. Since less snowpack accumulates during winter, the amount of water stored in snow is reduced and less is available for withdrawal during summer months. Water depths in rivers and creeks are lower as are rates of groundwater recharge and the volume of the groundwater stored.

The reduction in water quantity in the area is aggravated by a reduction in water quality and the increasing human demand for water resulting from increasing temperatures and population growth. In Kamloops, the average resident consumes 800 litres of water per day,

compared with the average person in a developing country who uses less than 10 litres per day. Increasing temperatures means people are using more energy and water for daily activities such as cooling their homes and watering their lawns. Climate change will also impact the agricultural production in the area due to reduced irrigation capacity.

Attempts to mitigate and adapt to the effects of climate change and water use in the Fraser basin, such as by promoting alternative forms of transportation, increasing building efficiency, and using alternative sources of energy, are underway. Some adaptive measures include building infrastructure to protect communities against rising sea levels, storing rainwater, developing upstream storage infrastructure, exploiting groundwater, protecting wetlands and riparian habitat, and creating corridors to allow plant and animal migration routes. Community leaders in the Fraser basin recognize the need for water demand management, including public education initiatives promoting water conservation, promoting higher efficiency appliances and irrigation systems, initiating water meters and user charge systems, introducing incentives for green building projects, and using water recycling technologies.

Salinization of irrigated soil
Although irrigation improves the agricultural productivity of arid and semi-arid lands, it often causes salt to accumulate in the soil,

a phenomenon called **salinization**. Irrigation water contains small amounts of dissolved salts. Normally, as a result of precipitation runoff, rivers carry salt away. Irrigation water, however, normally soaks into the soil and does not run off the land into rivers. The continued application of such water, season after season, year after year, leads to the gradual accumulation of mineral salts in the soil (**Figure 9.11**). Given enough time, the concentration of these salts can rise to such a high level that plants are poisoned or their roots become dehydrated because water remains in the soils rather than being taken up by plants. In some instances, salinization renders soil unfit for crop production.

Agriculture and Agri-Food Canada has reported that nearly 12 percent of agricultural prairie land was rated as having a moderate, high, or very high risk of salinization in 2001, representing a significant improvement from the assessment conducted in 1981. Changes in agricultural practices (such as the introduction of permanent cover and extending crop rotation) were identified as the most important factors contributing to the improvements seen in soil health.

salinization The gradual accumulation of salt in soil, often as a result of improper irrigation methods.

water pollution A physical or chemical change in water that adversely affects the health of humans and other organisms.

WATER POLLUTION

Water pollution is a global problem that varies in magnitude and type of pollutant from one region to another. In many locations, particularly in developing countries, the main water pollution issue is providing individuals with disease-free drinking water. Soil management is linked closely with water pollution problems as are activities associated with the construction of roadways and other buildings. Potential impacts of poor soil management through these activities include sediment loading, nutrient additions, pesticide pollution, and pathogen contamination, which are discussed in greater detail later in this chapter.

Types of water pollution Water pollutants are divided into eight categories: sediment pollution, sewage and waste water, fecal coliform and other disease-causing agents, eutrophication, organic compounds, inorganic chemicals, radioactive substances, and thermal pollution. Causes and examples of each of these types of water pollution are summarized in **TABLE 9.1** (on the next page).

Salinized soil FIGURE 9.11

Irrigation water contains small amounts of dissolved salts. Over time, the salt accumulates in the soil. This irrigated soil has become too salty for plants to tolerate.

Types of water pollution TABLE 9.1

Type of pollution	Source	Examples	Effects
Sediment pollution	Erosion of agricultural lands, forest soils exposed by logging, degraded stream banks, overgrazed rangelands, strip mines, construction	Clay, silt, sand, and gravel, suspended in water and eventually settle out and smother benthic habitat	Reduces light penetration, limiting photosynthesis and disrupting food chain; clogs gills and feeding structures of aquatic animals; carries and deposits disease-causing agents and toxic chemicals
Sewage and waste water	Waste water from sanitary and stormwater sewers	Human wastes, soaps, detergents	Releases fecal coliforms that cause gastroenteritis and other public health concerns; increases biochemical oxygen demand (BOD); contributes to eutrophication,
Fecal coliform and other disease-causing agents	Wastes of infected individuals	Bacteria, viruses, protozoa, parasitic worms	Spread infectious diseases (cholera, dysentery, typhoid, infectious hepatitis, giardiasis, etc.)
Eutrophication	Human and animal wastes, plant residues, fertilizer runoff from agricultural and residential land	Nitrogen and phosphorus	Stimulates growth of excess plants and algae, causing nutrient enrichment and a high BOD; suspected of causing red tides and blooms of toxic algae
Organic compounds	Landfills, agricultural runoff, industrial wastes	Synthetic chemicals: pesticides, cleaning solvents, industrial chemicals, plastics	Contaminate groundwater and surface water; threaten drinking water supply; found in some bottled water; some are suspected endocrine disrupters
Inorganic chemicals	Industries, mines, irrigation runoff, oil drilling, urban runoff from storm sewers	Acids, salts, heavy metals such as lead, mercury, and arsenic	Contaminate groundwater and surface water; threaten drinking water supply; found in some bottled water; don't easily degrade or break down
Radioactive substances	Nuclear power plants, nuclear weapons industry, medical and scientific research facilities	Unstable isotopes of radioactive minerals such as uranium and thorium	Contaminate groundwater and surface water; threaten drinking water supply
Thermal pollution	Industrial runoff	Heated water produced during industrial processes, then released into waterways	Depletes water of oxygen and reduces amount of oxygen that water can hold; reduced oxygen threatens fish

We will explore how sediment, nutrient enrichment (phosphorus and nitrogen from various sources), and sewage discharges impact water quality as measured by fecal coliform populations, dissolved oxygen concentrations, biochemical oxygen demand, oxygen sags, and eutrophication. We will also consider how these sources of pollutants can influence the diversity and abundance of aquatic life downstream of the discharge site.

Sediment pollution Water and wind have the potential to cause major soil erosion events that transport silt, clay, and sand into waterways such as creaks, streams, rivers, and lakes, and cause a problem known

■ **sediment pollution** A type of water pollution caused by the transportation of silt, clay, and sand particles into waterways.

as **sediment pollution**. This pollution not only alters many of the physical features of the receiving water body such as the rocky bottom or light availability, but will also change the concentrations of dissolved substances such as nutrients and pesticides (**FIGURE 9.12**). Sediment deposition changes water flow and reduces water depth, which impacts not only the ecosystem but also recreation and navigation.

Water that is polluted with high sediment loads becomes cloudy with a high degree of murkiness or turbidity because of the entrained particulates. The turbid water impedes sunlight penetration and thus restricts plant photosynthesis and overall primary production of the water body. The sediments

The effects of sediment pollution FIGURE 9.12

The effects of sediment pollution can be significant. **A** indicates the undisturbed situation, which promotes a biologically diverse ecosystem, whereas **B** shows that sediments will reduce available light, bury productive habitat, and introduce chemicals that impact water quality.

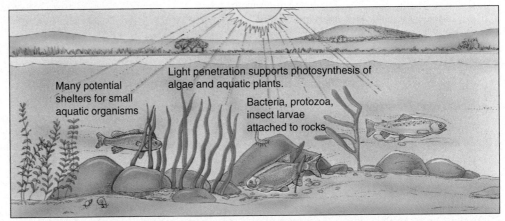

Many potential shelters for small aquatic organisms

Light penetration supports photosynthesis of algae and aquatic plants.

Bacteria, protozoa, insect larvae attached to rocks

A Stream ecosystem with low level of sediment

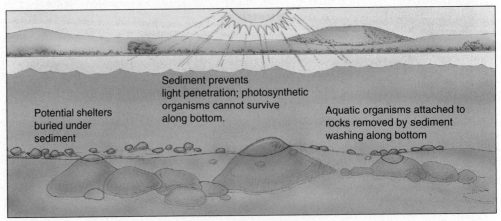

Potential shelters buried under sediment

Sediment prevents light penetration; photosynthetic organisms cannot survive along bottom.

Aquatic organisms attached to rocks removed by sediment washing along bottom

B Same stream with high level of sediment

disrupt the dynamics of the food chain, destroy critical fish spawning habitat, and suffocate benthic species such as freshwater mussels and clams that struggle to filter feed effectively and are buried by the sediment once it settles to the bottom. Sediment pollution further impairs water quality because many chemicals like pesticides, metals, and minerals (especially phosphorus) that are absorbed on the surface of particulates are carried directly to aquatic systems.

Waste water and sewage The discharge of **sewage** into water causes several pollution problems that can be harmful to ecosystems and human health. First, untreated sewage carries bacteria, viruses, and other disease-causing pathogens that contribute to gastroenteritis and other health problems (see Chapter 4). Water samples are often taken to monitor fecal or total coliform and results are typically reported as the number of colony-forming units per 100 millilitres of water. The acceptable limits vary depending on the intended use. For example, colony counts exceeding 200 per 100 mL result in beach closures and restrictions on recreational activities. In contrast, no detectable colonies are permitted in safe drinking water as defined by the Guidelines for Canadian Drinking Water Quality.

Another problem that results from the discharge of untreated sewage is the enrichment of the receiving river

sewage Waste water from drains or sewers (from toilets, washing machines, and showers); includes human wastes, soaps, and detergents.

biochemical oxygen demand (BOD) The amount of oxygen that microorganisms need in respiration to decompose biological wastes into carbon dioxide, water, and minerals.

with organic carbon. Populations of microorganisms grow quickly and consume dissolved oxygen during decomposition. Scientists monitor this as **biochemical oxygen demand (BOD)**, which measures the rate of dissolved oxygen uptake from water samples incubated at room temperature over a period of days. The higher the BOD detected, the greater the organic matter concentration and hence a higher rate of microbial activity.

Measurements taken at increasing distances from the discharge point highlight the changes observed in the river

Effect of sewage on dissolved oxygen, biochemical oxygen demand, and the aquatic community FIGURE 9.13

Note the initial oxygen depletion (blue line) and increasing BOD (red line) close to the sewage discharge. BOD increases rapidly while dissolved oxygen is consumed by decomposers. The stream gradually recovers as the sewage is diluted and degraded. The septic zone represents the downstream stretch where dissolved oxygen concentrations are too low to support fish populations.

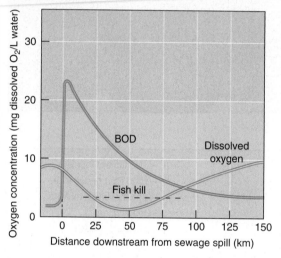

(**FIGURE 9.13**). A short distance downstream from the sewage discharge pipe, the microbial population grows rapidly and the BOD spikes. An oxygen sag is characteristic because of the extremely high microbial decomposition activities. Over time and distance from the discharge point, less organic carbon is available and microbial activities decline. Concentrations of dissolved oxygen recover to pre-discharge conditions.

Sewage also contains other pollutants that further degrade water. Inorganic nutrients (nitrogen and phosphorus) lead to eutrophication, which is described below; toxic metals such as mercury, lead, cadmium, chromium, and arsenic as well as pharmaceutical products have acute and chronic toxicity influences.

Eutrophication: A nutrient enrichment problem
Nitrogen (ammonia and nitrate) and phosphorus (orthophosphate) are macronutrients for plants and consequently restrict or stimulate population growth depending on their availability. Lakes, estuaries, and

A and **B** The average person looking at these two photographs would notice the dramatic differences between them but wouldn't understand the environmental conditions responsible for the differences. **A** shows an oligotrophic lake, whereas **B** shows a eutrophic lake.

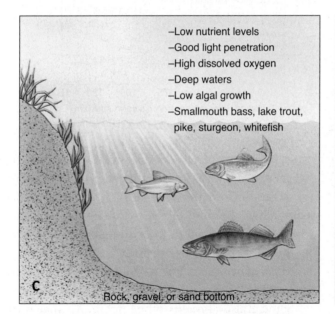

- Low nutrient levels
- Good light penetration
- High dissolved oxygen
- Deep waters
- Low algal growth
- Smallmouth bass, lake trout, pike, sturgeon, whitefish

Rock, gravel, or sand bottom

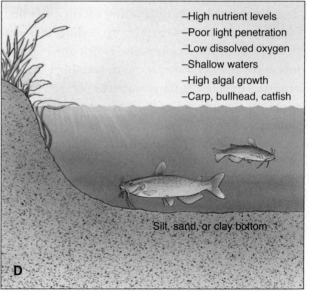

- High nutrient levels
- Poor light penetration
- Low dissolved oxygen
- Shallow waters
- High algal growth
- Carp, bullhead, catfish

Silt, sand, or clay bottom

C and **D** Aquatic ecologists understand the characteristics of oligotrophic and eutrophic lakes. An oligotropic lake has a low level of inorganic plant and algal nutrients (**C**), whereas a eutrophic lake has a high level of these nutrients (**D**).

slow-flowing streams that have minimal levels of these nutrients are considered most desirable and often pristine or oligotrophic water bodies. An **oligotrophic** lake is clear and supports only small populations of aquatic organisms because of the restricted or limiting concentrations of nitrogen and phosphorus available (see What a Scientist Sees, parts **A** and **C**). Often you can see light penetrate deep in the body of water and the bottom substrate of these lakes is typically rocky. These lakes support a high diversity of coldwater fish species such as smallmouth bass, lake trout, sturgeon, and whitefish that enjoy healthy concentration of dissolved oxygen.

In sharp contrast to an oligotrophic lake is one that has become eutrophied. **Eutrophication** occurs through the addition of nutrients that stimulate the prolific growth of plants along the shoreline and/or free floating in the water column. Light penetration is restricted in eutrophic lakes because of the suspension of algal cells that absorb light. These lakes typically have fewer species of fish with dominance by those better able to tolerate these conditions (see parts B and D of What a Scientist Sees). The rocky layer at the bottom of the lake is now covered by a muddy substrate.

Over time, oligotrophic water bodies tend to become more eutrophic through natural inputs of organic matter through leaf litter, sediment erosion, dying aquatic vegetation, and other organisms. However, human activities accelerate this process and what once took hundreds to thousands of years to transition can now happen within decades. We highlight the human-induced process as **cultural eutrophication** to distinguish it from a natural process. Cultural eutrophication results from enrichment of aquatic ecosystems by nutrients found predominantly in detergents and soaps, fertilizer runoff, and agricultural and municipal sewage.

Great attention has been directed to the dominance of cyanobacteria (blue-green algae) in eutrophic water bodies. Research conducted at the Experimental Lakes Area in northwestern Ontario identified a link between phosphorus enrichment and the occurrence of these algae. Recall from Chapter 5 that phosphorus is typically a limiting nutrient and often undetectable in unpolluted oligotrophic waters. When introduced into a water body, the blue-green algae have a competitive advantage in many situations as they can rapidly uptake phosphorus as

Experimental Lakes eutrophication experiment on Lake 226 FIGURE 9.14

Eutrophication studies were conducted in the Canadian Shield area of northwestern Ontario at the Experimental Lakes Area (ELA). The two basins of this lake were separated by a plastic curtain to examine the dynamics of eutrophication. The lower basin received additions of carbon, nitrogen, and phosphorus; the upper basin received carbon and nitrogen only. The blue-green colour observed in the lower basin is from cyanobacteria growth resulting from the addition of phosphorus.

well as obtain ammonia through nitrogen fixation of atmospheric nitrogen gas. This is especially important in those conditions where lakes receive disproportionately higher amounts of phosphorus. In these circumstances, other plant species struggle and give way to a rich, dense, gelatinous floating mat of blue-green algae (FIGURE 9.14). As the biomass of blue-green algae rise, concentrations of excretions and neurotoxins increase that may reach toxic levels over the growing season. This is the state of Lake Winnipeg, which is examined in the Case Study found at the end of this chapter.

Eutrophication also influences dissolved oxygen concentrations and BOD. During the daytime in summer, plants and algae grow vigorously to produce dissolved oxygen through photosynthesis. However, once darkness takes over in the early and late evening, photosynthesis haults but respiration continues. The large population of respiring plants and algae may lead to a reduction in the dissolved oxygen concentration, with the lowest concentration in early morning before sunrise. These eutrophic lakes therefore exhibit diurnal (morning and night) fluctuations in dissolved oxygen with peaks represented during high rates of photosynthesis and lows when photosynthesis has stopped and respiration continues. In contrast, oligotrophic lakes maintain a constant supply of dissolved oxygen that is sustained through atmospheric processes.

In the eutrophic lake, BOD increases when the high biomass of plants and algae die at the end of the summer season. As the decomposing mats sink to the bottom of the water column, microorganisms undergo high rates of respiration, consuming the dissolved oxygen in a similar process observed with the discharge of untreated sewage described above.

Winter fish kill has also been observed in Canadian lakes polluted by organic loading from sewage or eutrophication. Under unusually harsh winter conditions, a thick layer of ice forms and may be covered by a layer of snow, effectively sealing off the lake from gas diffusion processes with the atmosphere and impeding sunlight penetration. With this barrier, there is insufficient oxygen dissolved in the water column to support fish over the entire winter season and, consequently, they suffocate under these progressively anoxic conditions. These dead fish wash up along the shores and river banks in the spring, signifying the lack of available oxygen in the past winter period. Researchers are concerned that increasing the organic content in rivers and lakes through pollution increases the threat of more frequent fish kill events. Further, at these low dissolved oxygen levels, anaerobic (without oxygen) microorganisms produce compounds with unpleasant odours, further deteriorating water quality.

MAKING A DIFFERENCE

DAVID SCHINDLER

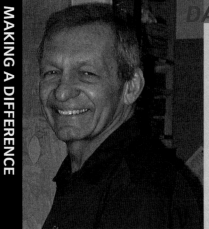

David Schindler, a Killam Memorial Professor of Ecology at the University of Alberta, is an interdisciplinary scholar teaching a wide range of courses, including limnology, philosophy, sociology, public policy, and the politics of science. Dr. Schindler, author of over 300 scientific publications, is widely recognized for his work on lake chemistry, and his research results are often used to create ecologically sound management policies.

Between 1968 and 1989, Schindler created and directed the Experimental Lakes Area (ELA) program near Kenora, Ontario. He is considered a master of experimental design, a reputation achieved largely through his insights in creating the ELA program, a multidisciplinary natural ecosystem research program that is now recognized as one of the most successful long-term studies of freshwater systems in the world. His research includes the impacts of eutrophication, acid rain, radioactive elements, and climate change. The ELA research results, which identify phosphate in detergents as a major lake pollutant, have led to dramatic changes in federal policy, manufacturing of products, and consumer buying habitats. His findings on the effects of acid rain resulted in changes in air quality legislation in Canada, the United States, and Europe. Recently, Schindler's research includes investigating fisheries management in mountain lakes, biomagnifications of organochlorines in food chains, effects of climate change and ultraviolet radiation in lakes, and global carbon and nitrogen budgets. He currently warns that the exploitation of Alberta's oil sands is devastating the Athabasca River.

Schindler, an officer of the Order of Canada, has received numerous international and national awards including the $1 million Gerhard Herzberg Canada Gold Medal for Science and Engineering in 2001 and the Douglas H. Pimlott Award for Conservation from the Canadian Nature Federation. He is a Fellow of the Royal Society of Canada, and in 2008, he was awarded the Alberta Order of Excellence. He has received 10 honorary doctorates from Canadian and American universities and his contributions to science and policy are internationally recognized.

Sources of water pollution

Water pollutants come from both natural sources and human activities. Natural sources of pollution such as mercury and arsenic tend to be local concerns, but human-generated pollution is generally more widespread.

The sources of water pollution are classified into two types: point source pollution and nonpoint source pollution. **Point source pollution** is discharged into the environment through pipes, sewers, or ditches from specific sites such as factories or sewage treatment plants (FIGURE 9.15).

Pollutants that enter bodies of water over large areas rather than at a single point cause **nonpoint source pollution**, also called *polluted runoff*. Nonpoint source pollution occurs when precipitation moves over and through the soil, picking up and carrying away pollutants that are eventually deposited in lakes, rivers, wetlands, groundwater, estuaries, and the ocean. Nonpoint source pollution includes agricultural runoff (such as sediment loading caused by soil erosion, nutrient enrichment, pathogens, pesticides, livestock wastes, and salt from irrigation), mining wastes (such as acid mine drainage), municipal wastes (such as fecal coliforms, eutrophication), construction sediments, and soil erosion (from fields, logging operations, and eroding stream banks). Although nonpoint sources cover more than one site and can be hard to identify, their combined effect can be significant and cover a substantial area.

Although sewage is the main pollutant produced by cities and towns, municipal water pollution also has a nonpoint source: urban runoff that carries a variety of contaminants (FIGURE 9.16 on page 306). (See the Improving Water Quality section for a further description of water treatment strategies.)

Agricultural activities, however, are the leading source of water quality impairment of surface waters globally. Agricultural practices produce several types of pollutants that contribute to nonpoint source pollution. Sediment pollution changes the quality of light and buries habitat, and fertilizer runoff causes nutrient enrichment. Animal wastes and plant residues in waterways produce high BOD, high levels of suspended solids, and nutrient enrichment. Highly toxic chemical pesticides may leach into the soil and from there into water or may find their way into waterways by adhering

point source pollution Water pollution that can be traced to a specific spot.

nonpoint source pollution Pollutants that enter bodies of water over large areas rather than being concentrated at a single point of entry.

Point source pollution FIGURE 9.15

Industrial runoff pours into a waste pit in the Amazon River basin. Natural gas burns from an adjacent pipe.

These pollutants may be carried from storm drains on streets to streams and rivers.

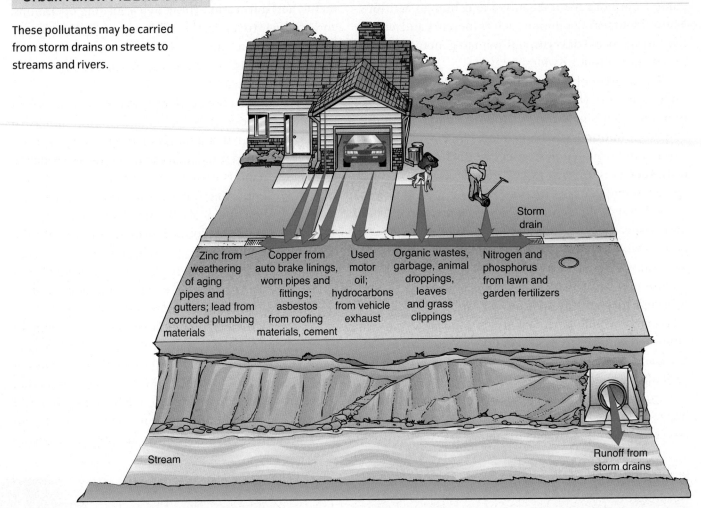

Zinc from weathering of aging pipes and gutters; lead from corroded plumbing materials

Copper from auto brake linings, worn pipes and fittings; asbestos from roofing materials, cement

Used motor oil; hydrocarbons from vehicle exhaust

Organic wastes, garbage, animal droppings, leaves and grass clippings

Nitrogen and phosphorus from lawn and garden fertilizers

Storm drain

Stream

Runoff from storm drains

to sediment particles. Soil erosion from fields and range-lands causes sediment pollution in waterways. To address runoff from animal wastes, government regulations are becoming more stringent, requiring livestock operations to develop plans for managing manure to prevent it from becoming polluted runoff.

Upland and riparian buffer zone management A key component to healthy rivers, streams, and lakes is a robust and well-developed riparian area comprising the vegetation along banks, shores, and the uplands farther away from these waterways (**FIGURE 9.17**). These buffer zones effectively act as natural "filters" for both surface and groundwater. Features of healthy buffer zones include high amounts of vegetation that serve to slow the movement of water runoff so that sediments begin to settle out of the surface flow, and the water and

pollutants can infiltrate into the ground where the runoff can be taken up by plants or decomposed by the rich soil community. Examples of these buffer zones include grassed waterways, strips of vegetation, and borders along cropped fields, as well as the trees, shrubs, and grasses that hug the shoreline along river and stream banks and lakes.

Important ecosystem services from a healthy riparian zone or buffer strip include the following:

- Filters nonpoint source pollutants (for example, agricultural pollutants, industrial sediment)

- Reduces water runoff and thereby increase sedimentation and retention of sediments and pollutants

- Acts as physical holding areas where vegetation reduces runoff by improving infiltration, enhancing evapotranspiration, and intercepting rainwater,

A Grassed waterways are a component of riparian areas that act as buffers to trap sediment and nutrients moving into the waterway from surrounding agricultural lands. The vegetation also stabilizes the banks and shores from the erosive action of the waterway itself.

B Rehabilitating eroding shorelines and degraded riparian areas by re-establishing natural vegetation ecosystems reduces runoff and soil erosion and contributes to improved water quality.

decreasing volume and velocity of runoff movement toward the water

- Creates biologically active zones where, depending on the vegetative community, riparian areas support the plant/microbial uptake of nitrates and denitrification (less nitrates reach the water source)

- Generates a cooling effect on water temperatures adjacent to riparian zones as the result of shading of water runoff and the water system, which has been shown to benefit some fish species

- Controls erosion and stabilizes wetland edges

- Protects fish habitat and spawning areas

- Effectively removes nutrients, nitrogen, phosphorus, waterborne pathogens, and pesticides from surface water (depending on vegetation cover of the riparian zone)

While these healthy riparian and buffer areas are critical in protecting water resources, they tend to suffer significant degradation from livestock grazing and cropping practices. Agriculture and Agri-Food Canada promotes best management strategies like exclusion fencing, concentrated stream access areas, grazing management plans, and remote (off-stream) watering as ways that producers can enhance their riparian areas while also promoting higher water quality for their surrounding watershed.

A number of government departments at different levels have legislative responsibilities for water management and water quality. Although there are various programs that relate to the protection and restoration of riparian areas, they are often underfunded and usually uncoordinated. Nonprofit organizations such as Ducks Unlimited Canada work hard to create science-based water management plans for agricultural producers and private land owners. The Department of Fisheries and Oceans and Agriculture and Agri-Food Canada are two departments that have created successful programs to protect and restore riparian areas (see Figure 9.17B). Together, along with a variety of partners, they have created a program to reduce soil erosion and bank instability by planting deep-rooted plants and trees, restoring riparian area vegetation, and restabilizing water banks.

Groundwater pollution

Groundwater contamination has received increasing attention in recent years. In unconfined aquifers common pollutants, such as pesticides, fertilizers, and organic compounds, seep

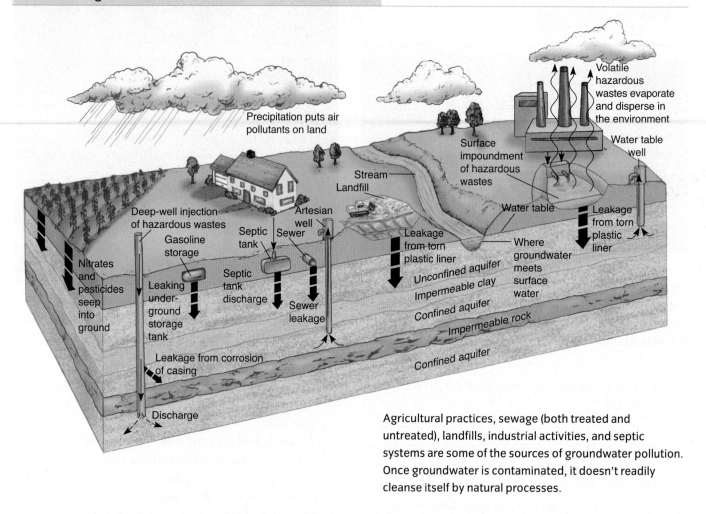

Precipitation puts air pollutants on land

Volatile hazardous wastes evaporate and disperse in the environment

Water table well

Surface impoundment of hazardous wastes

Stream
Landfill

Deep-well injection of hazardous wastes

Artesian well
Septic tank   Sewer

Gasoline storage

Water table

Leakage from torn plastic liner

Nitrates and pesticides seep into ground

Septic tank discharge

Leaking underground storage tank

Leakage from torn plastic liner

Where groundwater meets surface water

Unconfined aquifer
Impermeable clay
Confined aquifer

Sewer leakage

Impermeable rock

Leakage from corrosion of casing

Confined aquifer

Discharge

Agricultural practices, sewage (both treated and untreated), landfills, industrial activities, and septic systems are some of the sources of groundwater pollution. Once groundwater is contaminated, it doesn't readily cleanse itself by natural processes.

into groundwater from municipal sanitary landfills, underground storage tanks, backyards, golf courses, and intensively cultivated agricultural lands (FIGURE 9.18). While there is a tendency to believe that confined aquifers are protected from infiltration through the impermeable bedrock above the reservoir, these can be particularly challenging to restore should contaminants enter the aquifer.

Currently, most of the groundwater supplies in North America are of good quality and do not violate standards established to protect human health. However, areas that do experience local groundwater contamination face a challenge. Cleanup of polluted groundwater is costly, takes years, and in some cases is not technically feasible.

GLOBAL WATER ISSUES

As the world's population continues to grow and climate change advances, global water problems are expected to increase in severity to the point that by 2030, the United Nations warns, almost half of the world's population, and between 75 million and 250 million people living in Africa, will have water scarcity problems. This could displace between 24 million and 700 million people who will be forced to live as environmental refugees. The U.N. indicates this will have serious implications on human health as almost 80 percent of diseases in developing countries are associated with water. Local famines are expected to increase with greater frequency as water shortages spread across the globe.

In arid and semi-arid regions there is already evidence of dwindling sources of fresh water to support humans. In India, 17 percent of the world's population has access to only 4 percent of the world's fresh water, leaving approximately 8000 villages with no local water supply. Water supplies are precarious in much of China too, again owing to population pressures: water table levels are dropping, wells have gone dry, and much of the water in the Yellow River is diverted for irrigation, depriving downstream areas of water. Mexico is facing the most serious water shortages of any country in the western hemisphere. The main aquifer supplying Mexico City is dropping rapidly, and the water table is falling fast in Guanajuato, an agricultural state.

Sharing water resources among countries

In the 1950s, the then Soviet Union began diverting water that feeds into the Aral Sea to irrigate nearby desert areas. Since 1960 the Aral Sea has declined more than 50 percent in area (**FIGURE 9.19**). Its total volume is down 80 percent, and much of its biological diversity has disappeared. Millions of people living in the Aral Sea's watershed have developed serious health problems, probably because of storms lifting toxic salts from the receding shoreline into the air.

Following the breakup of the Soviet Union in 1991, responsibility for saving the Aral Sea shifted to the five Asian countries that share the Aral basin—Uzbekistan, Kazakhstan, Kyrgyzstan, Turkmenistan, and Tajikistan. Despite recent cooperative restoration efforts made by these nations and backed by the World Bank and the United Nations Environment Programme, the Aral Sea will probably never return to its former size and economic importance.

Like the Aral Sea, many of Earth's other watersheds cross political boundaries and face management issues associated with their shared use. In fact, three-fourths of the world's 200 or so major watersheds are shared between nations. International cooperation is therefore required to manage international rivers. The heavily populated drainage basin for the Rhine River in Europe crosses five countries—Switzerland, Germany, France, Luxembourg, and the Netherlands. All five nations recognize that international cooperation is essential to conserve and protect the supply and quality of the Rhine River. Their efforts have paid off: The main sources of pollution have been eliminated, and water in the Rhine River today is almost as pure as drinking water; long-absent fishes have returned; and projects are underway to restore riverbanks, control flooding, and clean up remaining pollutants.

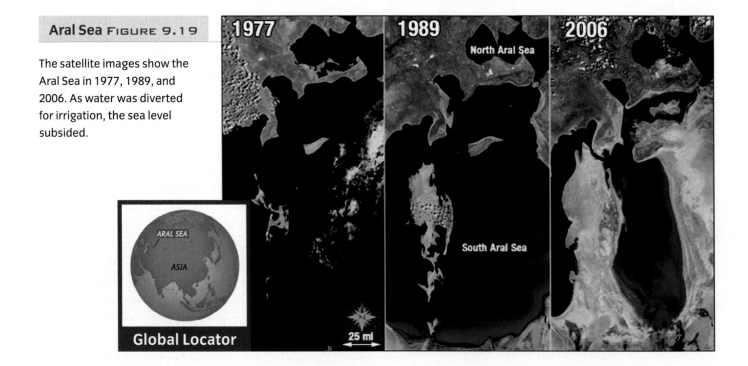

Aral Sea FIGURE 9.19

The satellite images show the Aral Sea in 1977, 1989, and 2006. As water was diverted for irrigation, the sea level subsided.

CONCEPT CHECK **STOP**

What problems are associated with overdrawing surface water? With aquifer depletion?

What is biochemical oxygen demand? How is BOD related to sewage? How is BOD related to eutrophication?

HOW does point source pollution differ from nonpoint source pollution? What are some examples of each?

Water Management

LEARNING OBJECTIVES

Define *sustainable water use.*

Give examples of water conservation in agriculture, industry, and individual homes and buildings.

Describe how most drinking water is purified in North America.

T he main goal of water management is to provide a sustainable supply of high quality water. **Sustainable water use** means careful human use of water resources so water is available for future generations and for existing nonhuman needs.

Water supplies are obtained by building dams, diverting water, and/or desalination (removing salt from sea water or salty groundwater). Conservation, which includes reusing water, recycling water, and improving water-use efficiency, augments water supplies and is an important aspect of sustainable water use. Political and economic policies are also important in managing water sustainably: When water is inexpensive, it tends to be wasted.

sustainable water use The wise use of water resources, without harming the essential functioning of the hydrologic cycle or the ecosystems on which present and future humans depend.

DAMS AND RESERVOIRS: MANAGING THE COLUMBIA RIVER

Dams generate electricity and ensure a year-round supply of water in areas with seasonal precipitation or snowmelt, but many people think their costs outweigh their benefits. The Columbia River, the fourth-largest river in North America, illustrates the impact of dams on natural fish communities. It is the largest hydroelectric power–producing river in North America with more than 100 dams, 19 of which are major generators of hydroelectric power (**FIGURE 9.20**). It stretches from British Columbia through Washington and Oregon before emptying into the Pacific Ocean. The Columbia and its tributaries are home to numerous fish, which migrate between freshwater streams and the Pacific Ocean. The river is used by numerous species of salmon, which are a vital part of the river's ecology and the local economy and have been for thousands of years.

The numerous dams along the river adversely affect these fish populations. The salmon population in the Columbia River system is only a fraction of what it was before the watershed was developed. The many dams that impede salmon migrations are widely considered the most significant factor in salmon decline.

Various efforts to assist migrating salmon have not proved particularly successful (**FIGURE 9.21**). Conservationists and biologists support using a natural approach of releasing water to flush young salmon downstream; they also support adopting a controversial

Grand Coulee Dam on the Columbia River
FIGURE 9.20

Shown are the dam and part of its reservoir, Lake Roosevelt. Dams provide electricity generation, flood control, and water recreation opportunities, but they disrupt or destroy natural river habitats and are expensive to build.

Fish ladder FIGURE 9.21

This ladder is located at the Bonneville Dam along the Oregon side of the Columbia River. Fish ladders help migratory fishes bypass dams in their migration upstream.

proposal to tear down several dams on the lower Snake River, a tributary of the Columbia River. Farmers and the hydroelectric companies strongly oppose these plans, which they fear would threaten their water supplies.

WATER CONSERVATION

Today there is more competition than ever among water users with different priorities, and water conservation measures are necessary to guarantee sufficient water supplies.

The United States and Canada are the top two countries in the world when it comes to water consumption per day per person. Canadians use an average of close to 200 litres of water per day per person, which means that major water conservation efforts need to be made in our homes and communities. There are many ways that each of us can conserve water in our daily lives by wasting less water, using water more efficiently, and not misusing water for the wrong job. This leads to the three common rules (or 3Rs) of water conservation: *reduce, retrofit,* and *repair.* TABLE 9.2 on page 312 outlines some of the easy changes you can make in your own home to reduce your water consumption.

Similar to the ecological footprint, researchers have devised a **water footprint** that can be used to examine the consumption and withdrawals of water at the global and national levels. Developed nations have the highest water footprint (FIGURE 9.22 on page 312). A trend that raises concern is the increasing demand for fresh water by developing nations producing agricultural goods intended for export to other nations. This is further complicated when these exporting nations lack the governance and mechanisms to create wise water management and conservation.

Reducing agricultural water consumption

Irrigation generally makes inefficient use of water. Traditional irrigation methods involve flooding the land or diverting water to fields through open channels. Plants absorb about 40 percent of the water that flood irrigation applies to the soil; the rest of the water usually evaporates into the atmosphere or seeps into the ground.

One of the most important innovations in agricultural water conservation is **microirrigation**,

> **microirrigation**
> A type of irrigation that conserves water by piping it to crops through sealed systems.

Reducing your household's water consumption TABLE 9.2

In the kitchen…	• Repair leaky faucets
	• Operate the dishwasher on energy-saving cycles and use it only for full loads
	• Wash dishes in a partially filled sink
	• Don't thaw food under running water
In the bathroom…	• Repair leaky faucets
	• Install water-efficient tap aerators, especially in the shower
	• Turn the water off while brushing your teeth
	• Reduce the length and frequency of each shower
	• Only partially fill the bathtub
	• Don't flush the toilet needlessly or use it as a garbage can
	• Install low-flow toilets (4- or 6-litre instead of 18-litre)
In the laundry room…	• Only wash full loads
	• Use shorter washing cycles
	• Wash in warm water instead of hot water (or better yet, use cold water and save on electricity too!)
	• Adjust the water level in your machine to the least amount necessary
Outside…	• Use rain basin collecting devices to reuse rainwater for watering grass and plants, washing your car, etc.
	• Use a pail and sponge to wash your car
	• Place a trigger nozzle on your hose
	• Cover the swimming pool to reduce the number of times more water needs to be added
	• Put your sprinklers on timers and use a heavy drop setting instead of a spray setting
	• Water only during cool times of the day and only the grass (not the pavement!)
	• Plant drought-resistant plants
	• Leave grass at least 6 centimetres long

For more information on what you can do in your neighbourhood and in your home to conserve water please visit Environment Canada—
Water Conservation Strategies at www.ec.gc.ca/water/en/info/pubs/speak/e_slides.htm. You can start by keeping a water log to see how much
water you use in your home and compare with others. See where you can make changes that will add some water-saving benefits and reduce
your water footprint.

Average national water footprint per capita FIGURE 9.22

What is your water footprint? Find out at www.waterfootprint.org/?page=cal/WaterFootprintCalculator

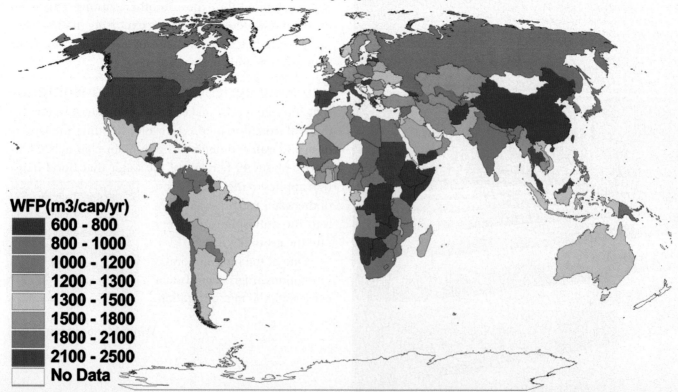

WFP(m3/cap/yr)
- 600 - 800
- 800 - 1000
- 1000 - 1200
- 1200 - 1300
- 1300 - 1500
- 1500 - 1800
- 1800 - 2100
- 2100 - 2500
- No Data

also called **drip** or **trickle irrigation**, in which pipes with tiny holes bored in them convey water directly to individual plants (FIGURE 9.23A on page 314). Microirrigation substantially reduces the water needed to irrigate crops—usually by 40 percent to 60 percent compared with traditional irrigation—and also reduces the amount of salt that irrigation water leaves in the soil.

Other measures that could save irrigation water include using lasers to level fields, which allows a more even water distribution, and making greater use of recycled waste water. A drawback of such techniques is their cost, which makes them unaffordable for most farmers in highly developed countries, let alone subsistence farmers in developing nations.

Reducing water consumption in industry

Electric power generators and many industrial processes require water. In the United States, five major industries—chemical products, paper and pulp, petroleum and coal, primary metals, and food processing—consume almost 90 percent of industrial water.

Stricter pollution-control laws provide some incentive for industries to conserve water. Industries usually recapture, purify, and reuse water to reduce their water use and their water treatment costs (FIGURE 9.23B). The U.S. Steel Corporation plant in Granite City, Illinois, for example, recycles approximately two-thirds of the water it uses daily. The Ghirardelli Chocolate Company in San Leandro, California, installed a recycling system to cool large tanks of its chocolate. The potential for industries to conserve water by recycling is enormous.

Reducing municipal water consumption

Like industries, regions and cities—and the households within them—recycle or reuse water to reduce consumption (FIGURE 9.23C). For example, homes and other buildings can be modified to collect and store grey water. *Grey water* is water that has been already used in sinks, showers, washing machines, and dishwashers. Grey water is recycled to flush toilets, wash cars, or sprinkle lawns. In contrast to water *recycling*, wastewater *reuse* occurs when water is collected and treated before being redistributed. The reclaimed water is generally used for irrigation.

In response to severe water shortages, California is exploring different ways to produce sources of fresh water. There are a number of proposals for large **desalination** plants that will draw water through intake pipes from the ocean and convert the sea water into drinking water. Although this may sound like a great plan because the Pacific Ocean is a vast expanse of available water, the process could harm fish and larvae, produce high levels of greenhouse gases, and place additional demands on sources of energy. As well, the outflow water back into the ocean has a higher salt content that will alter the salinity and thereby affect the aquatic biotic community.

Another California effort to reduce surface water demands is the promotion of "toilet to tap," whereby waste water is treated through a series of processes to generate high-quality drinking sources. Orange County Water District has designed a $481 million plant that uses microfiltration, reverse osmosis, ultraviolet light, and hydrogen peroxide disinfection to convert sewage water into drinking water. Treated water is pumped into the ground where natural processes further purify the water. Underground engineered aquifers store the water until needed. Currently, the plant is treating 280 million litres of sewage water each day and is meeting the water needs of over 500,000 people. This "groundwater replenishment system" is more economical than desalination but faces the negative public reaction to the idea of drinking former sewage water.

Cities also decrease water consumption by providing consumer education, requiring water-saving household fixtures, developing economic incentives to save water, and repairing leaky water supply systems. Also, increasing the price of water to reflect its true cost promotes water conservation.

IMPROVING WATER QUALITY

Water quality is improved by removing contaminants from the water supply before and after it is used. Technology assists in both processes.

Purification of drinking water

Most Canadian municipal water supplies are treated before the water is used so it is safe to drink. Untreated drinking water goes through a series of steps in purification before it is distributed to a customer's tap. In the first step, suspended particulates free floating in the raw water

A Microirrigation

Close-up of a drip irrigation on pipe system about to release a drop of water directly over a seedling, eliminating much of the waste associated with traditional methods of irrigation. Photographed in the Negev Desert, Israel.

B Industrial water conservation

The wastewater treatment facilities at the General Motors Orion Assembly Plant in Michigan play a role in the company's efforts to conserve water. The plant reduced water use by 14.7 percent between 2002 and 2004, a savings of almost 4 billion litres of water annually.

C Conserving water at home

In your bathroom, kitchen, and laundry room, you can take many steps to limit water use. Also, to reuse water, individual homes and buildings can be modified to collect and store "grey water," water already used in sinks, showers, washing machines, and dishwashers. This grey water is used when clean water is not required—for example, in flushing toilets, washing the car, and sprinkling the lawn.

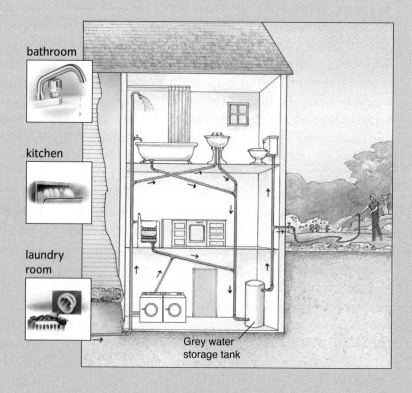

Floc floating at the surface of a basin
FIGURE 9.24

undergo a chemical reaction with aluminum sulphate or other flocculent that reduces the negative charges of the particulates and reduces the tendency of these to repel one another. Through a process of constant mixing, the particulates collide and begin to clump together, reducing the concentration of unwanted silts, bacteria, colour-causing particles, and viruses that would otherwise not settle out of the water for days, months, or even years. Through this step the clarity of the water is significantly improved, along with its taste and odour.

After the reaction has been provided sufficient time and large particulates or flocs are produced (FIGURE 9.24), the water moves into the next stage of treatment in which sedimentation is encouraged through a series of settling basins. Over several hours, water in the settling basins flows slowly and turbulence is minimized. This promotes the settling of previously suspended particulates to the bottom of the basin as a result of gravitational forces. The layer of particulates that form over an extended period are periodically removed, dewatered, and disposed of to landfill.

From the sedimentation basins, water flows into the filter beds where the remaining particulates or unwanted flocs are removed. The most common filtration method is a rapid filtration process in which layers of material of varying size remove solids from the water. The first layer of rapid filtration design is typically anthracite coal, and the second layer is sand. Water and flocs flow through the filter and the suspended particulates are trapped.

Rapid filtration is not commonly used in developing nations. Instead, drinking water passes through a slow filtration treatment process of biological purification. In this approach, a biofilm of living microorganisms is encouraged to develop on a layer of sand. This biofilm not only prevents the flow of particulates by trapping them in the mucilaginous matrix but also reduces the organic load by respiration and decomposition activities among the layer of bacteria and fungi. Since the slow filtration process requires significantly more space and longer contact times than the rapid filtration method, it is not commonly used in larger municipalities.

Following filtration, water may be held in storage tanks for a period of time. In the final stages of treatment, water is stabilized and disinfected on the way to the distribution pumps. As a stabilizing step, polyphosphate is commonly added to the water to prevent pipe corrosion that could be very costly should it be permitted to occur. In many municipalities, water is fluoridated during this process. Fluoridation of water is a major advancement in the effort to reduce tooth decay. Fluoride is added to the water as it leaves the storage tank and in this way the fluoride dosage can be modified and controlled to achieve the desired concentration.

Water undergoes a final process of disinfection. A common approach in disinfection has been the use of sodium hypochlorite (a strong bleach). This is added to the finished water and generally the chlorine remains in the water until it reaches the customer's tap. There have been some problems detected with chlorination. For example, resistant pathogens such as *Giardia* and *Cryptosporidium* can remain viable after treatment and threaten public health. Further, residual dissolved trihalomethanes that may form as a result of chlorine disinfection have been detected at sufficient concentrations to make them a health concern because of their carcinogenic qualities.

Increasingly, treatment plants are receiving upgrades to incorporate new technologies such as ozone or ultraviolet (UV) light disinfection. UV light is very effective at inactivating pathogens and resistant dormant cysts. It does require water to have a low level of dissolved organics so the UV light can pass through it without being absorbed by these compounds. One disadvantage of this technology is that water may need to be treated again prior to distribution as there are no residual disinfectant properties that remain.

Municipal sewage treatment Ideally, waste water should undergo several treatment stages prior to being discharged into receiving water bodies such as rivers, lakes, or the ocean. This treatment process is intended to improve water quality in the major areas of sediment loading, fecal coliform loading, high organic loading and associated biochemical oxygen demand, and nutrient removal. The intention is to reduce the risk of sediment pollution, microbial decomposition and oxygen sags, and to ensure a healthy environment to promote high aquatic diversity.

While many cities have developed the infrastructure to at least pre-treat waste water to some level prior to discharge, five cities in Canada remain without necessary treatment plants (Victoria, Saint John, Halifax, St. John's, and Dawson City) and continue to dump raw and untreated sewage directly into rivers, lakes, and oceans. In contrast, other cities, such as Calgary, have achieved the highest level of treatment, including using advanced treatment to remove pollutants such as nitrogen and phosphorus, pathogens, and

■ **primary treatment**

Treatment of waste water that involves removing suspended and floating particles by mechanical processes.

■ **secondary treatment**

Biological treatment of waste water to decompose suspended organic material; secondary treatment reduces the water's biochemical oxygen demand.

heavy metals, applying UV disinfection techniques, and ensuring regular toxicity testing.

There are three treatment stages associated with wastewater purification. Sanitary waste water enters the treatment facility where it undergoes an initial stage of **primary treatment** that removes suspended and floating particles such as sand and silt, by mechanical screens and gravitational settling (**FIGURE 9.25**, left side). The solid material that settles out at this stage is called **primary sludge**. The liquid phase is transferred to the next vessel where **secondary treatment** begins. In this process, rates of decomposition are enhanced by increasing the chamber temperature, injecting additional aerobic bacteria from sludge, and bubbling the sewage with oxygen (**FIGURE 9.25**, right side). After

Primary and secondary sewage treatment FIGURE 9.25

Process Diagram

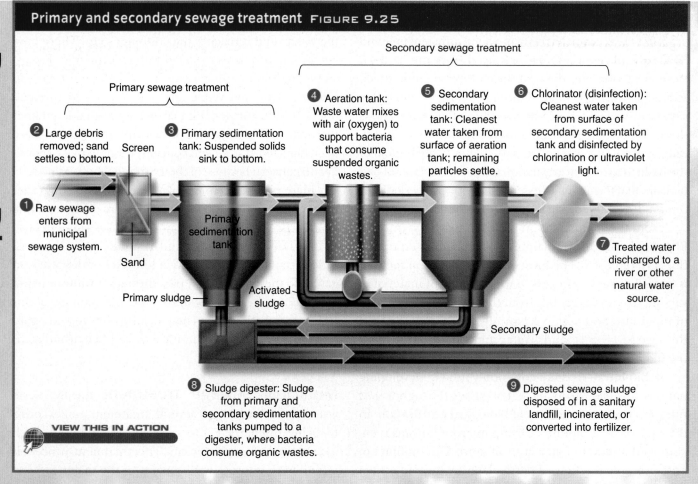

VIEW THIS IN ACTION

several hours of processing to reduce the organic load, the particles and microorganisms are allowed to settle out from the **effluent**, forming **secondary sludge**, a slimy mixture of bacteria-laden solids. Waste water that has undergone primary and secondary treatment has significantly lower organic wastes and therefore will not degrade receiving rivers with elevated BOD.

Even after primary and secondary treatments, waste water still contains pollutants, such as dissolved minerals, heavy metals, viruses, and organic compounds. Advanced wastewater treatment methods, or **tertiary treatment**, include a variety of biological, chemical, and physical processes. Tertiary treatment reduces phosphorus and nitrogen loading that contributes to eutrophication.

Disposal of primary and secondary sludge is a major problem associated with wastewater treatment. Sludge typically undergoes a process of digestion to reduce the organic load as well as decrease the number of disease-causing pathogens. Once this is completed, sludge may be handled by application to soil as fertilizer, incinerated, or disposed of in a sanitary landfill.

Stormwater management

Land drainage is an important issue for municipalities as there must be a system in place to ensure that precipitation will not flow through or collect in any unwanted places where it could cause damage. Paving streets, driveways, sidewalks, and walkways impairs the natural flow of water into soils, which then forces water to travel along the surface. Consequently, cities must establish a network of stormwater collection sewers and transport runoff to a receiving body of water such as a river. Stormwater represents a real challenge since amounts can be variable depending on weather conditions, and the runoff contains sediments and soil, pet waste, salt, pesticides, fertilizer, oil and grease, litter, and other pollutants. Consequently, the water quality of this stormwater runoff may be worse than that of sewage.

Winnipeg provides a good example of how some municipalities have managed stormwater runoff. In some older districts of Winnipeg, combined sewers were first used to collect and combine sanitary sewage with stormwater. This was then directed to the sewage treatment plant to remove pollutants prior to discharge to the river (**FIGURE 9.26**). While the idea that all collected stormwater could receive treatment was appealing, it was soon realized that heavy periods of precipitation overburdened the treatment facility and the only solution

Stormwater runoff in Winnipeg FIGURE 9.26

In some older districts of Winnipeg, there is a combined sewer system in place where both sanitary sewage and stormwater runoff are treated at the same wastewater plant. Most new housing areas have separate systems for household sewage and stormwater because of the problems associated with heavy rainfall overburdening the wastewater treatment plant.

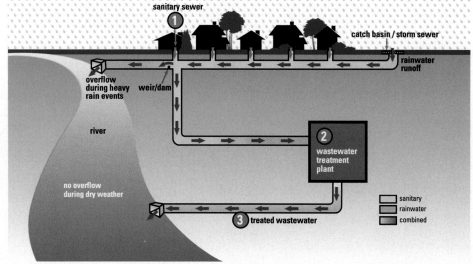

Copyright 2010, City of Winnipeg

was to divert excess waste water to the river without treatment. This meant that, on occasion, raw sewage would also be diverted to the river, leading to spikes in fecal coliform counts and higher BOD for several days at downstream locations. Winnipeg has since opted to build separate sewers in newer suburbs so that sanitary sewage is collected and sent to the sewage treatment facility while stormwater is collected and directed to the river or a conventional **stormwater retention pond**.

The goal in moving stormwater to retention ponds is to slow the rate of discharge to the receiving river and provide the opportunity to modify the quality of the water, much like a natural buffer provides in agricultural applications. Sediments have the opportunity to settle to the bottom of the pond along with some of the dissolved nutrients and pesticides. Fecal coliform counts gradually decline and microbial decomposition of organic matter occurs prior to discharge in the river, reducing the risk of an oxygen sag.

However, retention ponds are often plagued by blue-green algal blooms that produce toxins that present health issues to surrounding residents. There can be high rates of BOD when these blooms begin to die and decomposing microorganisms begin to proliferate. Odour issues develop over summer months and in the fall, and the ponds attract migrating geese in huge numbers that contribute large amounts of fecal matter to the water body and generally become a nuisance to the community. Management of these ponds may mean herbicide application to reduce the plant productivity and impair algal growth, mechanical removal of plant biomass, and/or aeration to increase concentrations of dissolved oxygen.

A more recent design element in stormwater retention ponds is the creation of engineered wetlands. These naturalized stormwater retention ponds provide critical ecosystem services that improve both water quality and aesthetic features for the community. These are becoming the norm in urban stormwater management. Wetlands trap incoming nutrients and sediments and reduce fecal coliform counts within only a few days. Because wetlands have a fringe of macrophytes (rooted plants that grow out from the water), these plants will typically remove nutrients before they flow into the water column, and competing algal populations are limited in growth because of the lack of available nutrients. Furthermore, since the rooted vegetation will not be discharged to the river, very little of the incoming pollutants can make their way downstream and into the river.

In Winnipeg, Native Plant Solutions (a subsidiary of Ducks Unlimited) has implemented these ponds in several communities and has monitored their performance over the past several years (**FIGURE 9.27**). The results have been resoundingly good and have reinforced the importance of riparian buffers to water quality. Data point to significant improvements over all other stormwater management approaches. In some years, between

An engineered stormwater retention pond FIGURE 9.27
This stormwater retention pond was created by Natural Plant Solutions in a suburb of Winnipeg to manage stormwater pollution.

95 percent and 98 percent reductions were found in water quality parameters—namely in reduced concentrations of phosphorus detected in water samples and extremely low counts of cyanobacteria. Further, these wetlands tend to be less appealing to migrating geese because of physical features that impair the geese's abilities to detect predators.

The use of wetlands to purify water has extended beyond stormwater management to include sewage treatment. Several communities have adopted wetlands to treat waste water and replace sewage treatment facilities (FIGURE 9.28). Beginning in 1978, the small town of Arcata, California, restored and constructed a series of freshwater wetlands in a former industrial area and then routed the town's waste water through these wetlands. The marshes absorb and assimilate contaminants normally removed through more expensive treatment methods. The highly productive marsh ecosystem—the Arcata Marsh and Wildlife Sanctuary—also provides wildlife habitat for many organisms and opportunities for human recreation.

CONTROLLING WATER POLLUTION

Many governments have passed legislation to control water pollution. In Canada, the Constitution Act, 1867, gives the federal and each provincial government the jurisdiction to pass laws concerning water, the environment, and public health. Traditionally, the provincial governments have regulated water management; however, the federal government does take an active role in certain water-related issues. This constitutional division of power means that the federal and provincial governments share jurisdiction over water and water-related issues. The federal government primarily focuses its responsibility for water management on fisheries and navigation and on water issues that lie on or across international borders.

The Canadian Environmental Quality Guidelines are published by the federal government and provide environmental quality guidelines for air, water, soil, sediment, and tissue. As well, the Canadian Council of Ministers of the Environment (CCME) Water Quality Guidelines, also published by the federal government, address water quality initiatives, including developing guidelines for agricultural water uses, developing sediment guidelines to protect aquatic life, developing tissue residue guidelines, and examining other water quality issues. The federal government also plays a role in relation to water quality with respect to toxic substances.

The federal government has proposed a new Canada Health Protection Act that would give the federal Minister of Health a role in regulating drinking water. The Guidelines for Canadian Drinking Water Quality are published by Health Canada on behalf of the Federal-Provincial-Territorial Committee on Drinking Water.

These guidelines are based on the most current scientific research related to health effects, aesthetic effects (taste, odour), and operational considerations (for example, a substance may interfere with or impair a treatment process or technology). They ensure a uniform set of standards for drinking water. The new Canada Health Protection Act would extend regulations to apply to the production of bottled water and water served on passenger conveyances.

Drinking water for First Nations peoples is the joint responsibility of First Nations band councils and the federal government. Together they have the responsibility to design, construct, maintain, and operate the water facilities to First Nations reserves and northern communities in accordance with federal or provincial standards (whichever is more stringent). Indian and Northern Affairs Canada (INAC) provides funding to First Nations peoples to assist them in providing clean water services to their communities. INAC also monitors the design, construction, and maintenance of this infrastructure. The water must meet the Guidelines for Canadian Drinking Water Quality.

The Canada Water Act, enacted in 1970 and administered by Environment Canada, contains requirements to govern water quality, and an advisory committee was established to assist with the implementation of the Act. The Act authorizes federal–provincial planning of subcommittees, programs, and agreements with respect to water resource management. It also regulates discharges of wastes into defined water quality management areas and establishes federal water quality management programs for interjurisdictional waters. The Act requires the Minister of the Environment to annually report to Parliament on all operations under the Act.

The Fisheries Act, enacted in 1868, is administered by the Department of Fisheries and Oceans and is aimed at protecting fish and fish habitat. The Act contains strong provisions related to water pollution, and therefore the protection of water quality. It prohibits the harmful alteration, disruption, or destruction of fish habitat and the deposition of "deleterious substances" into or near waters that are frequented by fish. It also imposes penalties for the unlawful deposition of harmful substances. The penalties under the Act include fines up to $1 million, jail time, profit-stripping, licence suspension, and restoration orders.

Preventing water pollution at home

Although individuals cause little water pollution, the collective effect of municipal water pollution, even in a small neighbourhood, can be quite large. There are many things you can do to protect surface waters and groundwater from water pollution (see TABLE 9.3).

Preventing water pollution TABLE 9.3

Location	What you can do
Bathroom	Never throw unwanted medicines down the toilet.
Kitchen	Use the smallest effective amount of toxic household chemicals such as oven cleaners, mothballs, drain cleaners, and paint thinners. Substitute less hazardous chemicals wherever possible. Dispose of unwanted hazardous household chemicals at hazardous waste collection centres.
Driveway/Car	Never pour used motor oil or antifreeze down storm drains or on the ground. Recycle these chemicals at service stations or local hazardous waste collection centres.
	Clean up spilled oil, brake fluid, and antifreeze, and sweep sidewalks and driveways instead of hosing them off. Dispose of dirt properly; don't sweep it into gutters or storm drains.
	Wash your car on a surface that will absorb the water (grass or gravel).
	Drive less. Air pollution emissions from automobiles eventually get into surface water and groundwater. Toxic metals and oil by-products deposited on roads by vehicles are washed into surface waters by precipitation.

TABLE 9.3 (CONTINUED)

Location	What you can do
Lawn and garden	Pick up pet waste and dispose of it in the garbage or toilet. If left on ground, it eventually washes into waterways, where it can contaminate shellfish and enrich water.
	Replace some lawn areas with trees, shrubs, and ground covers, which absorb up to 14 times more precipitation and require little or no fertilizer. To reduce erosion, use compost or mulch to cover bare ground in your gardens.
	Use fertilizer sparingly; excess fertilizer leaches into groundwater or waterways. Never apply fertilizer near surface water.
	Place compost or mulch on your garden to prevent soil from washing away.
	Avoid mowing 3 to 5 metres from the edge of a water body (leave a buffer zone)
	Make sure that gutters and downspouts drain onto water-absorbing grass or gravelled areas instead of onto paved surfaces.
Landscapes	Notify your local government if you see high sediment loads entering streets off a construction site.

CONCEPT CHECK STOP

What is sustainable water use?

How can individuals conserve and manage water resources?

How is most drinking water purified in North America?

The Global Ocean

LEARNING OBJECTIVES

Discuss the roles of winds and the Coriolis effect in producing global water flow patterns, including gyres.

The ocean is a vast wilderness, much of it unknown. It teems with life—from warm-blooded mammals such as whales to soft-bodied invertebrates such as jellyfish. The ocean is essential to the hydrologic cycle that provides us with water. It affects cycles of matter on land, influences our climate and weather (discussed in Chapter 10), and provides foods that enable millions of people to survive. The ocean dominates Earth, and its condition determines the future of life on our planet. If the ocean dies, then we do as well.

The global ocean is a huge body of salt water that surrounds the continents and covers almost three-fourths (71 percent) of Earth's surface. It is a single, continuous body of water, but geographers divide it into four sections separated by the continents: the Pacific, Atlantic, Indian, and Arctic Oceans. The Pacific is the largest: It covers one-third of Earth's surface and contains more than half of Earth's water. (Refer to Figure 9.4 on pages 285 and 286 for a visual representation of the size of the global oceans.)

Oceans derive their salinity from the dissolved minerals that once originated on land but were transported to the oceans over millions of years by erosional processes. Sodium and chloride represent the largest mineral content of oceans followed by sulphate, magnesium, calcium, potassium, and bicarbonate. Overall, these minerals contribute to a salinity of between 33 and 37 parts per thousand, but this depends on processes that include evaporation and runoff from freshwater sources.

The persistent prevailing winds blowing over the ocean produce currents, which are mass movements of surface ocean water (FIGURE 9.29). The prevailing winds generate **gyres**, circular ocean currents. In the North Atlantic Ocean, the tropical trade winds tend to blow toward the west, whereas the westerlies in the midlatitudes blow toward the east. This helps establish a clockwise gyre in the North Atlantic. As described in Chapter 10, these currents are influential to global climate and energy transport around the Earth.

> **gyres** Large, circular ocean current systems that often encompass an entire ocean basin.

The **Coriolis effect** influences the paths of surface, or shallow, ocean currents just as it does the winds (see Chapter 10, specifically Figure 10.8 on page 346). Earth's rotation from west to east causes surface ocean currents to swerve to the right in the northern hemisphere, helping establish the circular, clockwise pattern of water currents. In the southern hemisphere, ocean currents swerve to the left, thereby moving in a circular, counterclockwise pattern.

OCEAN–ATMOSPHERE INTERACTION

The ocean and the atmosphere are strongly linked, with wind from the atmosphere affecting the ocean currents and heat from the ocean affecting atmospheric circulation. One of the best examples of the interaction

Visualizing

Surface ocean currents FIGURE 9.29

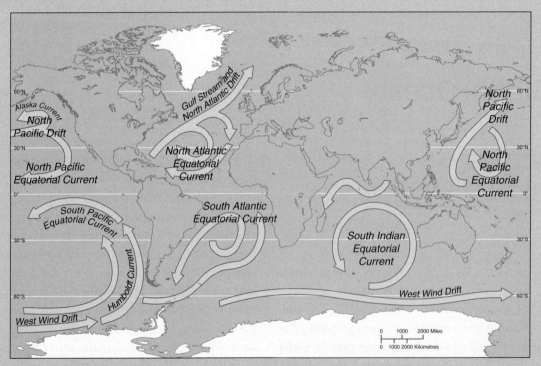

Winds largely cause the basic pattern of ocean currents. The main ocean currents flow clockwise in the northern hemisphere and counterclockwise in the southern hemisphere, partly because of the Coriolis effect.

between ocean and atmosphere is the **El Niño-Southern Oscillation (ENSO)** event, which is responsible for much of Earth's interannual (from one year to the next) climate variability. As a result of ENSO, some areas are drier, some wetter, some cooler, and some warmer than usual. Normally, westward-blowing trade winds restrict the warmest waters to the western Pacific near Australia (**FIGURE 9.30A**). Every three to seven years, however, the trade winds weaken, and the warm mass of water expands eastward to South America, increasing surface temperatures in the usually cooler east Pacific (**FIGURE 9.30B**). Ocean currents, which normally flow westward in this area, slow down, stop altogether, or even reverse and go eastward. The name for this phenomenon, El Niño (in Spanish, "the boy child"), refers to the Christ child, because the warming usually reaches the fishing grounds off Peru just before Christmas. Most ENSOs last between one and two years.

ENSO devastates the fisheries off South America. Normally, the colder, nutrient-rich deep water is about 40 metres below the surface and **upwells** (comes to the surface) along the coast, partly in response to strong trade winds (**FIGURE 9.31A** on page 324). During an ENSO event, however, the deep water is about 150 metres below the surface, and the warmer surface temperatures and weak trade winds prevent upwelling (**FIGURE 9.31B**). The lack of nutrients in the water results in a severe decrease in the populations of anchovies and many other fishes. During the 1982–1983 El Niño, one of the worst ever recorded, the anchovy population decreased 99 percent. Other species, such as shrimp and scallops, thrive during an ENSO event.

ENSO alters global air currents, directing unusual, sometimes dangerous, weather to areas far from the tropical Pacific where it originates. By one estimate,

ENSO FIGURE 9.30

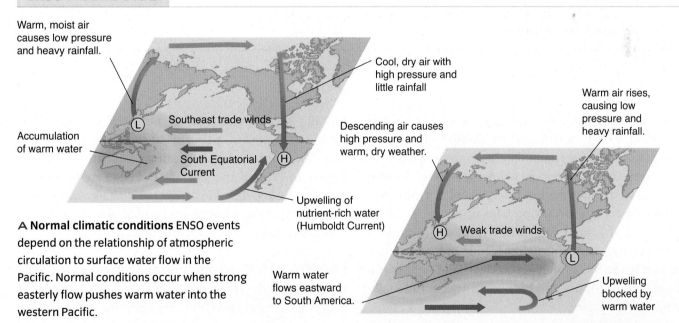

Warm, moist air causes low pressure and heavy rainfall.

Accumulation of warm water

Southeast trade winds

South Equatorial Current

Cool, dry air with high pressure and little rainfall

Descending air causes high pressure and warm, dry weather.

Upwelling of nutrient-rich water (Humboldt Current)

Warm air rises, causing low pressure and heavy rainfall.

Weak trade winds

Warm water flows eastward to South America.

Upwelling blocked by warm water

A Normal climatic conditions ENSO events depend on the relationship of atmospheric circulation to surface water flow in the Pacific. Normal conditions occur when strong easterly flow pushes warm water into the western Pacific.

B ENSO conditions An ENSO event occurs when easterly flow weakens, allowing warm water to collect along the South American coast. Note the relationship between precipitation and the location of pressure systems. During an ENSO event, northern areas of the contiguous United States are typically warmer during winter, whereas southern areas are cooler and wetter.

The Global Ocean   323

Process Diagram

Strong trade winds

Warm water moves away from coast.

Cool, nutrient-rich water upwells to surface.

Weak trade winds

Warm water stays along coast.

Cool, nutrient-rich water remains offshore at great depths.

A Coastal upwelling, where deeper waters come to the surface, occurs in the Pacific Ocean along the South American coast. Upwelling provides nutrients for microscopic algae, which in turn support a complex food web.

B Coastal upwelling weakens considerably during years with El Niño-Southern Oscillation (ENSO) events, temporarily reducing fish populations.

the 1997–1998 ENSO, the strongest on record, caused more than 20,000 deaths and $33 billion in property damage worldwide. It resulted in heavy snows in parts of the western United States; ice storms in eastern Canada; torrential rains that flooded Peru, Ecuador, California, Arizona, and western Europe; and droughts in Texas, Australia, and Indonesia. An ENSO-caused drought—the worst in 50 years—particularly hurt Indonesia. Fires that had been deliberately set to clear land for agriculture got out of control and burned an area in Indonesia the size of New Jersey.

Climate scientists observe and monitor sea surface temperatures and winds to better understand and predict the timing and severity of ENSO events. The **TAO/TRITON array** consists of 70 moored buoys in the tropical Pacific Ocean. These instruments collect oceanic and weather data during normal conditions and

El Niño events. The data are transmitted to scientists onshore by satellite. Scientists at the National Oceanic and Atmospheric Administration's Climate Prediction Center forecasted the 1997–1998 ENSO six months in advance, using data from TAO/TRITON. Such forecasts give governments time to prepare for the extreme weather changes associated with ENSO.

La Niña El Niño isn't the only periodic ocean temperature event to affect the tropical Pacific Ocean. **La Niña** (in Spanish, "the girl child") occurs when the surface water temperature in the eastern Pacific Ocean becomes unusually cool and westbound trade winds become unusually strong. During the spring of 1998, the surface water of the eastern Pacific cooled 6.7 °C in just 20 days. La Niña often occurs after an El Niño event and is considered part of the natural oscillation of ocean temperature.

Like ENSO, La Niña affects weather patterns around the world, but its effects are more difficult to predict. In the contiguous United States, La Niña typically causes wetter-than-usual winters in the Pacific Northwest, warmer weather in the southeast, and drought conditions in the southwest. Atlantic hurricanes are stronger and more numerous during a La Niña event.

CONCEPT CHECK STOP

How are the Coriolis effect, prevailing winds, and surface ocean currents related?

What is the El Niño-Southern Oscillation (ENSO)? What are some of its global effects?

Human Impacts on the Ocean

LEARNING OBJECTIVES

Contrast fishing and aquaculture and relate the environmental challenges of each activity.

Identify human activities that contribute to marine pollution and describe their effects.

The ocean is so vast it's hard to imagine that human activities could harm it. Such is the case, however. Fisheries and aquaculture, marine shipping, marine pollution and eutrophication, coastal development, offshore mining, and global climate change and acidification all contribute to the degradation of marine environments. Scientists estimate that in 2008, less than 4 percent of the ocean remained unaffected by human activities, and 41 percent had experienced serious harm (**FIGURE 9.32A** on page 326).

MARINE POLLUTION AND DETERIORATING HABITAT

One of the great paradoxes of human civilization is that the same ocean that provides food to a hungry world is also used as a dumping ground. Coastal and marine ecosystems receive pollution from land, from rivers emptying into the ocean, and from atmospheric contaminants that enter the ocean via precipitation. Offshore mining and oil drilling pollute the neritic province with oil and other contaminants. Pollution increasingly threatens the world's fisheries. Events such as accidental oil spills and the deliberate dumping of litter pollute the water. The World Resources Institute estimates that about 80 percent of global ocean pollution comes from human activities on land. In 2003, the Pew Oceans Commission, composed of scientists, economists, fishermen, and other experts, verified the seriousness of ocean problems in a series of studies. Some of their findings are shown in **FIGURE 9.32B**.

WORLD FISHERIES

The ocean contains a valuable food resource. About 90 percent of the world's total marine catch is fishes, with clams, oysters, squid, octopus, and other mollusks representing an additional 6 percent of the total catch. Crustaceans, including lobsters, shrimp, and crabs, make up about 3 percent, and marine algae constitute the remaining 1 percent.

Fleets of deep-sea fishing vessels obtain most of the world's marine harvest. Numerous fishes are also captured in shallow coastal waters and inland waters. According to the Food and Agricultural Organization (FAO), the world annual fish harvest increased substantially from 19 million tonnes in 1950 to nearly 158 million tonnes in 2005, the latest year for which data are available.

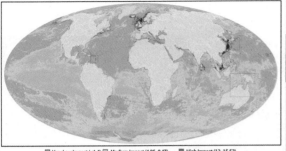

A Mapping human impacts. In 2008, an international team of marine scientists mapped effects of 17 human activities on the ocean. Almost no location remains unaffected, and 41 percent have been seriously altered by multiple activities.

VIEW THIS IN ACTION

B Major threats to the ocean.

Climate change
Coral reefs and polar seas are particularly vulnerable to increasing temperatures.

Aquaculture
Fish farms produce wastes that can pollute ocean water and harm marine organisms.

Invasive species
Organisms are transported and released in ballast water from ships containing foreign crabs, mussels, worms, and fishes.

Overfishing
Populations of many commercial fish species are severely depleted.

Nonpoint source pollution (runoff from land)
Agricultural runoff (fertilizers, pesticides, and livestock wastes) pollutes water.

Point source pollution
Passenger cruise ships dump sewage, shower and sink water, and oily bilge water.

Bycatch
Fishermen unintentionally kill dolphins, sea turtles, and sea birds.

Habitat destruction
Trawl nets (fishing equipment pulled along the ocean floor) destroy habitat.

Coastal development
Developers destroy important coastal habitat, such as salt marshes and mangrove swamps.

NATIONAL GEOGRAPHIC

Problems and challenges for the fishing industry
No nation lays legal claim to the open ocean. Consequently, resources in the ocean are more susceptible to overuse and degradation than land resources, which individual nations own and for which they feel responsible.

The most serious problem for marine fisheries is that many species, particularly large predatory fish, have been harvested to the point that their numbers are severely depleted. This generally causes a fishery to become unuseable for commercial or sport fishermen, let alone the other marine species that rely on it as part of the food web. Scientists have found that dramatically depleted fish populations recover only slowly. Some show no real increase in population size up to 15 years after the fishery has collapsed.

According to the FAO, at least 76 percent of the world's fish stocks are considered fully exploited, over-exploited, or depleted. Fisheries have experienced such pressure for two reasons. First, the growing human population requires protein in its diet, leading to a greater demand for fish. Second, technological advances allow us to fish so efficiently that every single fish is often removed from an area (see What a Scientist Sees).

Fishermen tend to concentrate on a few fish species with high commercial value, such as salmon, tuna, and

Modern Commercial Fishing Methods

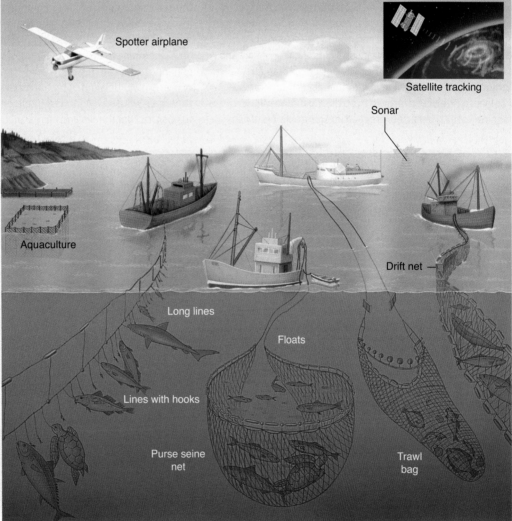

A A full fishnet is pulled on board a fishing vessel off the coast of Alaska.

B Scientific evidence indicates that modern methods of harvesting fish are so effective that many fish species have become rare. Sea turtles, dolphins, seals, whales, and other aquatic organisms are accidentally caught and killed in addition to the target fish. The depth of long lines is adjusted to catch open-water fishes such as sharks and tuna or bottom fishes such as cod and halibut. Purse seines catch anchovies, herring, mackerel, tuna, and other fishes that swim near the water's surface. Trawls catch cod, flounder, red snapper, scallops, shrimp, and other fishes and shellfish that live on or near the ocean floor. Drift nets catch salmon, tuna, and other fishes that swim in ocean waters.

In the summer of 1992, when northern cod biomass in the Canadian Atlantic fell to unprecedented levels, the Canadian government was forced to take drastic measures and close the fishery that had been the backbone of Canadian east coast communities for over 500 years. This closure resulted in the loss of over 40,000 jobs, an impact that local communities are still trying to recover from. The relationship between fishing communities and the northern cod resource was pushed to its breaking point. Fear over the economic impact that fishing quotas would have for these coastal communities resulted in a failure to address overexploited cod stocks and destructive fishing practices, which resulted in a collapse of the northern cod fishery and devastating ecological consequences.

Northern cod were traditionally found in abundance in these waters. However, when small crafts and traditional techniques gave way to new factory trawlers that came from thousands of miles away searching for herring, haddock, flatfish, capelin, redfish, and northwest Atlantic cod, the cod population began to dwindle. These ships hauled colossal nets and worked around the clock, even in terrible weather. The design of "draggers," enormous bag-like nets as long as football fields that scrape the bottom of the ocean, proved to be the most devastating technology of all. These nets dragged up entire schools of fish

and everything else in their way, causing enormous damage to target fish, nontarget fish, critical habitat, and bottom-dwelling communities, destabilizing the ecosystem of the northern cod.

The extent to which the fishery was mismanaged to the point of collapse served as a warning to governments around the world with similar scientific-based fisheries management programs. By the end of 1994 the northern cod population had declined to just 1700 tonnes, down from the 1990 estimated biomass of 400,000 tonnes. With so many people out of work, Newfoundland was an economic disaster area. The federal government paid close to $1 billion in social welfare payments to the province, but that amount did not even come close to covering the bill, which was estimated to be at least twice that.

The Canadian Atlantic fisheries collapse illustrates the devastating costs that resource overexploitation has on both the ecosystem and on human communities. These were tough lessons to learn about taking the health of the Earth's natural ecosystems seriously. But the conflict between biologists and fishermen still continues. Biologists around the world continue to indicate regions that are too heavily fished, warning of overexploitation and potential ecosystem damage, but fishermen whose livelihoods depend on each catch continue to push for liberal limits and increased quotas.

Traditional fishing practices were sustainable until technological improvements like this commercial trawler came along, which had devastating effects on the northern cod population and led to the eventual closure of east coast fisheries.

flounder, while other species, collectively called **bycatch**, are unintentionally caught and then discarded. The FAO reports that about 25 percent of all marine organisms caught—some 27 million tonnes—are dumped back into the ocean. Most of these unwanted animals are dead or soon die because they are crushed by the fishing gear or are out of the water too long. Canada and other countries are trying to significantly reduce the amount of bycatch and develop uses for the bycatch that remains.

In response to harvesting, many nations extended their limits of jurisdiction to 320 kilometres offshore. This action removed most fisheries from international use because more than 90 percent of the world's fisheries are harvested in relatively shallow waters close to land. This policy was supposed to prevent overharvesting by allowing nations to regulate the amount of fishes and other seafood harvested from their waters. However, many countries also have a policy of **open management**, in which all fishing boats of that country are given unrestricted access to fishes in national waters.

> **aquaculture**
> The growing of aquatic organisms (fishes, shellfish, and seaweeds) for human consumption.

Aquaculture: Fish farming

Aquaculture is more closely related to agriculture on land than it is to the fishing industry. Aquaculture is carried out both in fresh and marine water; the cultivation of marine organisms is sometimes called **mariculture**. According to the FAO, world aquaculture production has increased substantially, from 600,000 tonnes in 1950 to 59.4 million tonnes in 2004.

Aquaculture differs from fishing in several respects. First, although highly developed nations harvest more fishes from the ocean, developing nations produce much more seafood by aquaculture. Developing nations have an abundant supply of cheap labour, which is a requirement of aquaculture because it is labour intensive, like land-based agriculture. Another difference between fishing and aquaculture is that the limit on the size of a catch in fishing is the size of the natural population, whereas the limit on aquacultural production is primarily the size of the area in which organisms can be grown.

In aquacultural "fish farms," fish populations are concentrated in a relatively small area and produce higher-than-normal concentrations of waste that pollute the adjacent water and harm other organisms. Aquaculture also causes a net loss of wild fish because many of the fishes farmed are carnivorous. Sea bass and salmon, for example, eat up to 5 kilograms of wild fish to gain 1 kilogram of weight.

Some aquaculture operations are carried out in deep water (**FIGURE 9.33**). Opponents are concerned about the potential for pollution, the spread of disease, and the accidental release of caged species into the deep-water environment. Caged populations are more genetically homogenous than wild ones; if the two groups

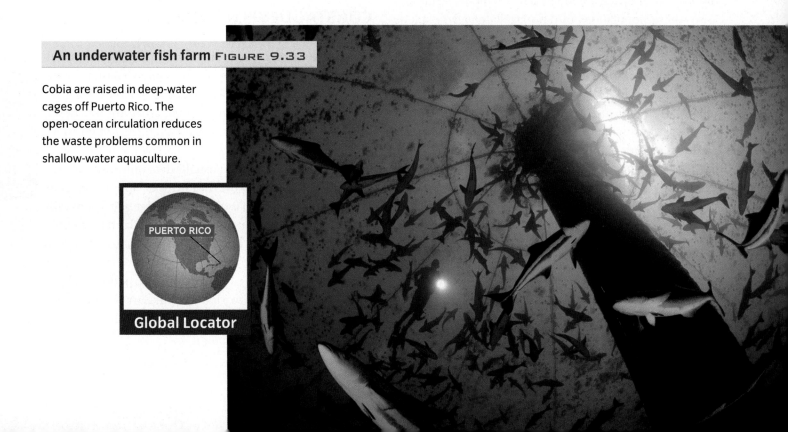

An underwater fish farm FIGURE 9.33

Cobia are raised in deep-water cages off Puerto Rico. The open-ocean circulation reduces the waste problems common in shallow-water aquaculture.

PUERTO RICO

Global Locator

interbreed, genetic diversity of wild populations could be diminished. The introduced organisms may also outcompete wild species.

SHIPPING, OCEAN DUMPING, AND PLASTIC DEBRIS

Millions of ships dump oily ballast and other wastes overboard in the neritic and oceanic provinces. The International Maritime Organization's International Convention for the Prevention of Pollution from Ships (MARPOL) bans marine pollution arising from the shipping industry. MARPOL regulations specifically address six types of marine pollution caused by shipping: oil, noxious liquids, harmful packaged substances, sewage, garbage, and air pollution. The 2004–2006 revisions to MARPOL regulations included stricter controls on oil tankers and added certain marine sites to the list of special protected areas. Unfortunately, MARPOL is not well enforced in the open ocean.

Several cities in Canada continue to dump their sewage sludge into the ocean. Disease-causing viruses and bacteria from human sewage may contaminate shellfish and other seafood and pose an increasing threat to public health. The Gulf of Mexico is now considered to have dead zones due to the nutrient enrichment and organic loading that contributes to escalated BOD and anoxic conditions. Harmful algal blooms such as red tide events may arise where populations of dinoflagellate algae explode and excrete elevated concentrations of neurotoxins that concentrate in shellfish, increase BOD in decomposition, and pose a significant human health risk if ingested.

Huge quantities of trash containing plastics are released into the ocean from coastal communities or, sometimes accidentally, from cargo ships. Plastics don't biodegrade; they photodegrade, which means they break down into smaller and smaller pieces yet still exist for an indefinite period. This trash collects in certain areas of the open ocean defined by atmospheric pressure systems. For example, in the north Pacific gyre, halfway between Hawaii and the U.S. mainland, researchers found a continuous array of floating plastics estimated at 1 million pieces per square mile that covered an area the size of Texas. (See the EnviroDiscovery box on the Great Pacific Garbage Dump on page 446 of Chapter 12.)

Plastic pollution in the ocean FIGURE 9.34

This bottlenose dolphin is unable to fully open its mouth because of the plastic entangled on its snout.

Not only are marine mammals and birds susceptible to being entangled and/or strangled by larger pieces of plastic (**FIGURE 9.34**), but the many filter-feeding organisms near the bottom of the ocean food chain constantly ingest the smaller degraded pieces. These plastic pieces may absorb and transport hazardous chemicals such as PCBs. Scientists have yet to determine whether these substances are incorporated into marine food webs when organisms ingest the plastic.

COASTAL DEVELOPMENT

Development of resorts, cities, industries, and agriculture along coasts alters or destroys many coastal ecosystems, including mangrove forests, salt marshes, sea grass beds, and coral reefs. Many coastal areas are overdeveloped, highly polluted, and overfished. Although more than 50 countries have coastal management strategies, their goals are narrow and usually deal only with the economic development of the thin strips of land that directly border the oceans. Coastal management plans generally don't integrate the management of both land and water, nor do they take into account the main cause of coastal degradation—sheer human numbers.

Perhaps as many as 3.8 billion people—about 60 percent of the world's population—live within 150 kilometres of a coastline. Demographers project

that three-fourths of all humans—perhaps as many as 6.4 billion—will live in that area by 2025. If the world's natural coastal areas aren't to become urban sprawl or continuous strips of tourist resorts during the 21st century, coastal management strategies must be developed that take into account projections of human population growth and distribution.

HUMAN IMPACTS ON CORAL REEFS

Coral formations, which are important ecosystems, are being degraded and destroyed. Approximately one-fourth of the world's coral reefs are at high risk. In some areas, silt washing downstream from clear-cut inland forests has smothered reefs. High salinity resulting from the diversion of fresh water to supply the growing human population may be killing Florida reefs. Overfishing (particularly the removal of top predators), damage by scuba divers and snorkellers, pollution from ocean dumping and coastal runoff, oil spills, boat groundings, fishing with dynamite or cyanide, hurricane damage, disease, coral bleaching, land reclamation, tourism, and the mining of corals for building material all take a heavy toll.

Since the late 1980s, corals in the tropical Atlantic and Pacific have suffered extensive **bleaching** (see What a Scientist Sees), in which stressed corals expel their zooxanthellae. Scientists suspect that several environmental stressors, particularly warmer seawater temperatures, contribute to coral bleaching. Many scientists attribute recent record sea temperatures and large die-offs of

Ocean Warming and Coral Bleaching

A Bleached coral off the coast of Panama. Scientists have linked coral bleaching to ocean warming. Warmer-than-usual temperatures stress the coral animals, causing them to lose their zooxanthellae. Without their algae, the corals can't get enough food and die.

B An evaluation of more than 5 million temperature profiles between 1950 and 1995 indicates that the ocean has warmed. Most warming has occurred in shallow waters where corals live.

What a Scientist Sees

corals (more than 70 percent losses in some areas) to El Niño effects, global climate change, or a combination of the two. Impaired coral growth in Australia was recently linked to increased ocean acidification associated with warmer water temperatures. Other potential stressors are pollution and coral diseases.

OFFSHORE EXTRACTION OF MINERAL AND ENERGY RESOURCES

Large deposits of minerals, including **manganese nodules**, lie on or below the ocean floor (**FIGURE 9.35**). Dredging manganese nodules from the ocean floor would adversely affect sea life, and the current market value for these minerals wouldn't cover the expense of obtaining them using existing technology. Furthermore, it isn't clear which countries have legal rights to minerals in international waters. Despite these concerns, many experts think that deep-sea mining will be technologically feasible in a few decades, and several industrialized nations have staked claims in a region of the Pacific known for its large number of nodules.

Manganese nodules on the ocean floor FIGURE 9.35

These potato-sized nodules have enticed miners, but it isn't yet commercially feasible to obtain them. Photographed in the Pacific Ocean.

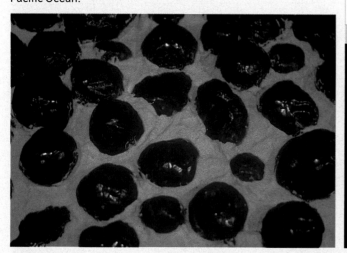

Offshore reserves of oil have long been tapped as a major source of energy. Additional reserves of methyl hydrate may become economical to produce in the future. However, obtaining oil and gas resources from the sea floor generally poses a threat to fishing. The environmental concerns associated with extracting offshore energy resources are discussed in Chapter 11.

CLIMATE CHANGE, SEA-LEVEL RISE, WARMER TEMPERATURES, AND ACIDIFICATION

Our understanding of global climate is so incomplete that unanticipated effects from a globally warmed world will undoubtedly occur. As the Arctic undergoes an overall warming as a result, scientists anticipate reductions in sea ice and glacial ice that could influence everything from patterns of oceanic circulation to sea level rise (see Chapter 10 for more detail).

The oceans are expected to absorb more carbon dioxide in equilibrium with atmospheric increases and become more acidified in the process. The change in pH is expected to alter the availability of carbonates and bicarbonates to shell-forming biota (clams and scallops; lobster and crayfish). Exoskeletons and shells will weaken and could threaten the viability of these species in the future.

CONCEPT CHECK STOP

What are some of the harmful environmental effects associated with the fishing industry? With aquaculture?

HOW does the widespread use of plastics contribute to ocean pollution?

HOW might the effect of global climate change on the ocean alter the global climate?

LIMNOLOGIST

Limnologists study the physical, chemical, and biological properties of lakes, rivers, and streams and play a vital role in protecting freshwater resources.

Imagine you are at the helm of a small boat, bobbing on the waves of a large lake about 50 metres from shore. You are a limnologist and you have spent the last few hours gathering water samples from different spots in the lake as part of an environmental assessment. You are here because a large petrochemical company wants to build a refinery on the lakeshore, but before construction can be approved, an environmental assessment must be carried out to determine the potential effects of the refinery on the surrounding environment. As an expert in aquatic ecosystems, you have joined the assessment team to investigate the impact the refinery could have on the lake's health.

As a limnologist, you have been part of many environmental assessments that have looked at the impact of industry on lakes and streams. Aquatic biota are sensitive to changes in their environment, making your assessment a critical component of the environmental review.

You begin by gathering baseline data to give you a better picture of the current status of the lake's biotic and abiotic characteristics. You use a temperature probe to measure water temperature at different depths. Temperature changes with depth in lakes, which produces stratified layers that affect the amount of oxygen and nutrients available, influencing where biota can live. You then link physical characteristics to water chemistry, which will be determined from laboratory analysis of the water samples you've collected. You also gather data on the aquatic communities, including fish, algae, plants, and zooplankton. You can use this information on species richness and abundance to construct a food web to predict the potential impact of an industrial development, spill, or climate change scenario.

All the baseline data you gather will be used to provide a picture of the water quality of the lake before the establishment of the refinery and will be included in the environmental assessment. From the environmental assessment it will be decided if the refinery project should go ahead and what the petrochemical company needs to do to ensure that the refinery's impacts are minimized.

Limnologists are scientists who study the physical, chemical, and biological properties of lakes, rivers, and streams. They study abiotic characteristics, such as stratification and water chemistry, as well as biotic elements, such as aquatic vegetation, algae, microbes, and invertebrates. Limnologists and their work play a vital role in protecting freshwater resources, and Canadian researchers are global leaders in the field.

For more information or to look up other environmental careers go to www.eco.ca.

SAVING LAKE WINNIPEG

Lake Winnipeg is the tenth largest body of fresh water in the world, covering an area of 24,514 square kilometres. Its watershed is the second largest in Canada, roughly 953,250 square kilometres, and spans four provinces and four U.S. states. Over 5.5 million people live within the watershed and nearly 20 million livestock are raised there.

There are four major user groups of Lake Winnipeg: commercial fisheries; Manitoba Hydro, which built a hydroelectric power facility on one of the lake's outflow rivers in the 1970s; recreation and tourism; and downstream communities on the Nelson River that rely on the lake for drinking water and indigenous fisheries. Each of these human-created demands places strain on this ecosystem.

In addition to these direct human uses of the lake, human activities from within the watershed are resulting in dramatic changes to the healthy balance of the ecosystem. One of the most notable effects is the introduction of large amounts of nutrients such as nitrogen and phosphorus that come from numerous landscape processes and activities including weathering and erosion, agricultural production, industrial manufacturing, and discharges from urban centres within the watershed. The excessive loading of nutrients into Lake Winnipeg has caused major changes to the quality and health of the lake.

Since 1924 scientists have been studying Lake Winnipeg, and the results of over 70 years of investigation reveal that Lake Winnipeg is approaching a deteriorating status that may affect the sustainability of the entire ecosystem. The continuous stresses on the lake have resulted in significant changes to water transparency, biological species population abundance, composition and distribution, and the lake's productivity and sediment chemistry. The lake is on a course of progressive eutrophication, resulting in re-occurring harmful "blooms" of algae growing and covering the lake (see photograph). As the plants decompose, they deplete the oxygen supply within the system, leaving no oxygen for remaining plant and animal life.

Several actions have been instituted to try to save Lake Winnipeg. In 1998, the Lake Winnipeg Research Consortium (LWRC) was founded, which brought together a group of interested agencies in a continued commitment to coordinate scientific and interdisciplinary research on Lake Winnipeg. In the years between 2002–2004, the LWRC facilitated three research surveys to study the biological, chemical, and physical attributes of Lake Winnipeg during the open-water season of the year.

In 2003, the Manitoba government announced the Lake Winnipeg Action Plan, a commitment to reduce nitrogen and phosphorus loads in Lake Winnipeg by 10 percent and 13 percent, respectively, to achieve pre-1970s conditions. This action plan acknowledges that human activities contribute to the nutrient loads found in the lake and that nutrient reductions will be needed across many sectors. The action plan includes improvements to onsite water system management regulations; nutrient limits for large municipal and industrial wastewater dischargers; regulation on application of nutrients to land (restrictions on phosphorus in lawn fertilizer, ensuring buffer zones, and restrictions in sensitive areas); and riparian tax credits to enhance riparian buffer zones and wetland restoration initiatives.

The Lake Winnipeg Action Plan identifies the collective responsibility we all share (see Table 9.3 on pages 320 and 321 for a list of things you can do to reduce water pollution). Without this collective effort, restoring the health of an ecosystem such as Lake Winnipeg is next to impossible to achieve.

The transparency of Lake Winnipeg water has been steadily declining for years. Shown here are harmful algal blooms and the effect these blooms have on water quality.

SUMMARY

1 The Importance of Water

1. In the hydrologic cycle, water continuously circulates through the abiotic environment. **Surface water** is precipitation that remains on the surface. **Runoff** is the movement of fresh water from precipitation and snowmelt to rivers, lakes, wetlands, and the ocean. **Groundwater** is the supply of fresh water that is stored in **aquifers**, underground reservoirs.

2. Water molecules are polar. The negatively charged (oxygen) end of one molecule is attracted to the positively charged (hydrogen) end of another molecule, forming a hydrogen bond. This is the basis for many of water's properties, including its high melting point, high boiling point, high heat capacity, and dissolving ability.

2 Water Issues

1. Overdrawing surface water causes wetlands to dry up and estuaries to become saltier. **Aquifer depletion** is the removal of groundwater faster than it can be recharged. **Salt water intrusion** is the movement of sea water into a freshwater aquifer near the coast. **Salinization** is the gradual accumulation of salt in soil, often due to improper irrigation.

2. **Water pollution** is any physical or chemical change in water that adversely affects the health of humans and other organisms. **Sewage** is the release of waste water from drains or sewers. It carries disease-causing agents and causes enrichment, the fertilization of a body of water due to high levels of nutrients. **Eutrophication** is the enrichment of a lake, estuary, or slow-flowing stream. **Cultural eutrophication** is enrichment of an aquatic ecosystem by stimulating primary productivity due to human activities. The **biochemical oxygen demand (BOD)**, the amount of oxygen that microorganisms need to decompose biological wastes, is raised by processes of eutrophication and raw sewage discharge. A high BOD decreases water quality by reducing levels of dissolved oxygen.

3. **Point source pollution** is water pollution that can be traced to a specific spot, such as waste water released from a factory or sewage treatment plant. **Nonpoint source pollution** includes pollutants that enter bodies of water over large areas, such as agricultural runoff, mining wastes, municipal wastes, construction sediments, and soil erosion.

3 Water Management

1. **Sustainable water use** is the wise use of water resources, without harming the hydrologic cycle or the ecosystems on which humans depend.

2. **Microirrigation** is an innovative type of irrigation that conserves

water by piping it to crops through sealed systems. Industries and cities can employ measures to recapture, purify, and reuse water in homes and buildings.

3. Most North American municipal water supplies are treated so the water is safe to drink. A chemical coagulant first clumps suspended particles, which then settle as sediment to the basin floor. Rapid or slow filtration removes remaining suspended materials and microorganisms, and disinfection by chlorination or UV exposure kills disease-causing agents.

4 The Global Ocean

1. Prevailing winds over the ocean generate **gyres**, large, circular ocean current systems that often encompass an entire ocean basin. The Coriolis effect is a force resulting from Earth's rotation that influences the paths of surface ocean currents, which move in a circular pattern, clockwise in the northern hemisphere and counterclockwise in the southern hemisphere.

2. The ocean and the atmosphere are strongly linked. The **El Niño-Southern Oscillation (ENSO)** event, which

is responsible for much of Earth's interannual climate variability, is a periodic, large-scale warming of surface waters of the tropical eastern Pacific Ocean that temporarily alters both ocean and atmospheric circulation patterns. A La Niña event occurs when surface water in the eastern Pacific Ocean becomes unusually warm. Its effects on weather patterns are less predictable than an ENSO event.

5 Human Impacts on the Ocean

1. The most serious problem for marine fisheries is the overharvesting of many species to the point that their numbers are severely depleted. Fishermen usually concentrate on a few fish species with high commercial value. In doing so, they also catch bycatch, fishes, marine mammals, sea turtles, seabirds, and other animals caught unintentionally in a commercial fishing catch and then discarded. **Aquaculture** is the growing of aquatic organisms (fishes, shellfish, and seaweeds) for human consumption. Aquaculture is common in developing nations with abundant cheap labour, and it is limited by the size of the space dedicated to cultivation. Aquaculture produces wastes that pollute the adjacent water and also causes a net loss of wild fish because many of the fishes farmed are carnivorous.

2. Marine pollution and garbage is generated by many human activities, including the release of trash and contaminants through commercial shipping, ocean dumping of sludge and industrial wastes, and discarding of plastics that are potentially harmful to marine organisms. Similar to fresh water, oxygen concentrations can be consumed in eutrophication that leads to dead zones devoid of biota. Marine environments are also deteriorated by coastal development and the extraction of offshore minerals.

Ocean acidification and sea level rise are consequences to the oceans expected as carbon dioxide concentrations continue to increase in the atmosphere. Oceans absorb carbon dioxide in equilibrium with the atmosphere and carbonic acid is formed. The acidification of the oceans influences the availability of bicarbonates and carbonates essential to shell-forming biota.

KEY TERMS

- **surface water** p. 289
- **runoff** p. 289
- **groundwater** p. 289
- **aquifers** p. 289
- **water withdrawal** p. 295
- **water consumption** p. 295
- **aquifer depletion** p. 296
- **salt water intrusion** p. 296
- **salinization** p. 298

- **water pollution** p. 298
- **sediment pollution** p. 300
- **sewage** p. 301
- **biochemical oxygen demand (BOD)** p. 301
- **eutrophication** p. 303
- **cultural eutrophication** p. 303
- **point source pollution** p. 305
- **nonpoint source pollution** p. 305

- **sustainable water use** p. 310
- **microirrigation** p. 311
- **primary treatment** p. 316
- **secondary treatment** p. 316
- **tertiary treatment** p. 317
- **stormwater rentention pond** p. 318
- **gyres** p. 322
- **El Nino-Southern Oscillation (ENSO)** p. 323
- **aquaculture** p. 329

CRITICAL AND CREATIVE THINKING QUESTIONS

1. Are our water supply problems largely the result of too many people? Give reasons for your answer.

2. Briefly describe the complexity of international water use, using the Rhine River or the Aral Sea as examples.

3. Outline a brief water conservation plan for your own daily use. How could you use water more sustainably?

4. Explain why untreated sewage may kill fish when it is added directly to a body of water.

5. Tell whether each of the following represents point or nonpoint source pollution: fertilizer runoff, thermal pollution from a power plant, urban runoff, sewage from a ship, and erosion

sediments from deforestation. Why is nonpoint pollution more difficult to control than point source pollution?

6. Describe the stages of wastewater treatment and indicate how these assist to reduce impacts to the receiving body of water.

7. Why has chlorination been replaced as a disinfection stage in water treatment?

8–9. The graph below reflects the monitoring of dissolved oxygen concentrations at six stations along a river. The stations are located 20 metres apart, with A the farthest upstream and F the farthest downstream.

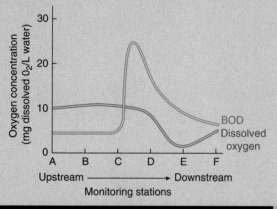

8. Where along the river did a sewage spill occur?

9. At which station would you most likely discover dead fish?

10. How do ocean currents affect climate on land?

11. Compare the different global effects of El Niño with those of La Niña. How are the two events similar? How are they different?

12. Explain how human activities impact coral reefs.

13. Imagine that you live in a small Atlantic coast community where a company wants to set up an aquaculture facility in a salt marsh. What are its benefits and its environmental drawbacks? Would you support or oppose this proposal? Explain your answer.

14. A 2003 study reports that in recent years tropical ocean waters have become saltier, whereas polar ocean waters have become less salty. What is a possible explanation for these changes?

What is happening in this picture ?

- Rain soaks the streets of Winnipeg. Where does the rainwater go?

- Many materials dissolve in water. What types of pollution might be dissolved in or carried by the rainwater?

- Is street runoff an example of point source pollution or nonpoint source pollution?

The Atmospheric Environment

10

LONG-DISTANCE TRANSPORT OF AIR POLLUTION

Winds distribute hazardous air pollutants globally Often these pollutants are volatile chemicals that easily evaporate in warmer southern latitudes and are then transported northward by air currents. It is through this *long-range transportation* that several persistent toxic compounds have begun to appear in the Canadian Arctic and accumulate to dangerously high concentrations. These chemicals are now being detected in the body fat of predators at the top of the food chain, like seals, whales, polar bears, as well as humans. Scientists have reported such high concentrations of PCBs in whale skin that they suggest one bite can expose an Inuit person to more PCBs than they think should be consumed in one week.

Air pollution does not remain in the country where it is produced, and therefore it becomes an *intercontinental* or *transboundary issue*. Environment Canada, for example, has tracked the atmospheric mercury detected in Canada's Arctic and found that it originated in Asia and Russia. Mercury is a heavy metal that undergoes biological magnification in food webs and is now found in top predators throughout the Arctic. Because of atmospheric transport, there is no location in Canada that is considered pristine and without accumulations of air pollutants.

In addition to the transport of air pollutants, Canada's Arctic faces rapid changes that are being driven by climate change. Warmer weather is accelerating sea ice melt and is challenging the ability of local and global biota to adapt rapidly or face extinction (see inset photograph). These changes to Arctic weather patterns are also forcing the Inuit peoples of northern Canada to adapt their cultural and traditional lifestyle activities (see photograph). Climate change "adaptation" refers to changes that people in various parts of the world are being forced to make at many levels to reduce impacts from the hazards that a changing climate may cause.

In this chapter we describe the features of the atmosphere that are vital to our lives. We consider the challenges and adaptation requirements from global atmospheric problems: air pollution, acid deposition, global climate change, and ozone depletion.

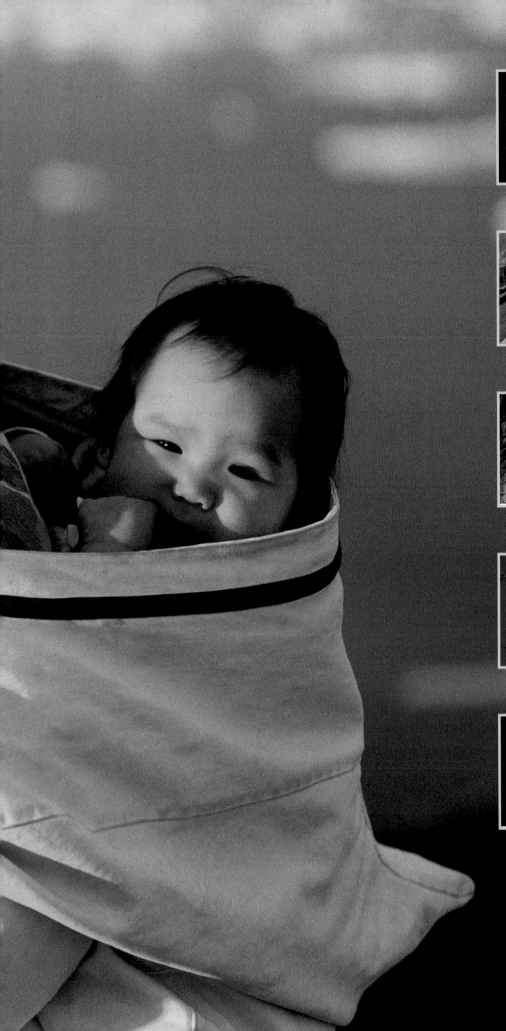

339

Features of the Atmosphere

LEARNING OBJECTIVES

Describe the layers of the atmosphere and their properties.

Discuss the factors that modify solar radiation input on Earth and that contribute to the Earth's energy budget.

Identify the factors responsible for global atmospheric and oceanic circulation.

Describe how the sun powers weather and climate.

he Earth's **atmosphere** is described as the ocean of air above the Earth composed of gases that extend from sea level to outer space some 350 to 500 kilometres in altitude. It is a large and dynamic system that interacts with not only the sun, by also the oceans, land masses, and ecosystems. Organisms depend on the atmosphere for existence, but they also interact with it, modifying its composition. The atmosphere screens out harmful incoming solar radiation and cosmic rays. It is through these interactions between solar energy, oceans, and the atmosphere that our weather and climate is generated.

atmosphere
The gaseous envelope surrounding Earth composed primarily of nitrogen and oxygen. Other gases are found in trace amounts.

Nitrogen (78 percent) and oxygen (21 percent) account for about 99 percent of dry air, with argon, carbon dioxide, and other trace gases (including air pollutants) making up the remaining 1 percent (**FIGURE 10.1**). While several of these gases are generally found in fixed quantities, others such as water vapour and pollutants are present in highly variable amounts depending on location and time of day or year.

Ulf Merbold, a German space shuttle astronaut, felt differently about the atmosphere after viewing it in space (**FIGURE 10.2**). "For the first time in my life, I saw the horizon as a curved line. It was accentuated by a thin seam of dark blue light—our atmosphere. Obviously, this wasn't the 'ocean' of air I had been told it was so many times in my life. I was terrified by its fragile appearance."

Composition of the atmosphere FIGURE 10.1

Nitrogen (78 percent) and oxygen (21 percent) are the largest contributors of gases to the atmosphere. Trace quantities of argon and other gases, including carbon dioxide, neon, helium, hydrogen, and xenon, have rather fixed concentrations that do not vary significantly. Conversely, water vapour and various pollutants (methane, ozone, nitrous oxide, particulates, and chlorofluorocarbons) may vary in concentration over time and between locations.

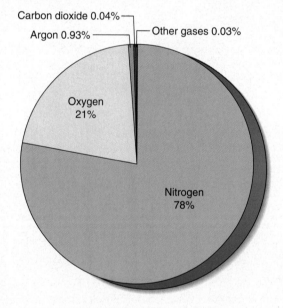

The atmosphere is composed of four major concentric layers: the troposphere, stratosphere, mesosphere, and thermosphere (**FIGURE 10.3** on page 346). These layers vary in altitude and temperature, depending on the latitude and season. A characteristic zigzag profile of

The atmosphere FIGURE 10.2

The "ocean of air" is a thin blue layer that separates the planet from the blackness of space.

temperature occurs with increasing altitude that helps to identify and distinguish between these layers. Between each layer there is a boundary (tropopause, stratopause, and mesopause) where temperature remains constant over a small range of altitudes. These boundaries hinder vertical mixing of gases and help isolate the atmospheric layers from one another. In each of these layers various interactions occur that alter incoming solar radiation. The troposphere is an important layer that distributes gases and energy around the planet and the formation of local and global climate regimes.

Even though the atmosphere seems rather thick at 500 kilometres above Earth's surface, about 80 percent of the atmospheric mass is actually located within the troposphere and about 50 percent within the first 6 kilometres altitude above sea level. This is because gravitational forces push the molecules and atoms of the atmosphere down toward Earth, compressing the molecules closer to one another, which causes more frequent collisions that exert more force and increases the **atmospheric pressure**. Atmospheric pressure declines exponentially with altitude.

Water vapour is one of the trace variable gases found in the atmosphere, ranging from about 0 to 4 percent of its composition, depending on the specific location or time of the year being considered. The amount of water vapour in the atmosphere changes on a regular basis. We use measurements of **relative humidity** to describe the ratio of measured water vapour detected in a volume of air to the amount that it could contain given the current temperature. It is typically described as a percentage. A reference to relative humidity of 50 percent means that the atmosphere contains only one-half of the water vapour that it could given the present temperature. Deserts typically have extremely low relative humidity while tropical regions have extremely high relative humidity.

> **atmospheric pressure**
> The force exerted by the column of air molecules located above a surface.

> **relative humidity**
> The amount of moisture present in the atmosphere in relation to the amount the atmosphere can hold given the present temperature.

SOLAR RADIATION AND ATMOSPHERIC CIRCULATION

The sun provides a constant input of energy to the planet. This energy has qualities of both waves and particles (photons). The quantity of energy in these waves and photons can be determined by a measure of their vibration frequency that exists along the **electromagnetic spectrum**.

High-frequency short wavelengths of energy are more energetic than lower frequency long wavelengths. High-energy forms include gamma, x-ray, and ultraviolet (UV) light. Low-energy forms include infrared (also termed thermal energy since it results in a rise in temperature), microwave, and radio waves. In the middle

> **electromagnetic spectrum**
> The entire range of electromagnetic radiation including gamma, x-ray, ultraviolet, visible, infrared, microwave, and radio.

Layers of Atmosphere

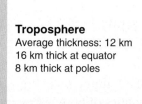

Mesopause

Stratopause

Ozone layer

Tropopause

Thermosphere
Extends from 80 to 500 km
above sea level

Mesosphere
Extends from 50 to 80 km
above sea level

Stratosphere
Extends from 11 to 50 km
above sea level

Troposphere
Average thickness: 12 km
16 km thick at equator
8 km thick at poles

Extremely low atmospheric density. Variable temperatures are observed. Gases in extremely thin air absorb x-rays and short-wave radiation, raising the temperature to 1000°C or more. The thermosphere is important in long-distance communication because it reflects outgoing radio waves back to Earth without the use of satellites. Auroras occur here.

Directly above the stratosphere. Temperatures drop to the lowest in the atmosphere—as low as –138°C. Meteors burn up from friction with air molecules in mesosphere.

Commercial jets fly here. Contains a layer of ozone that absorbs much of the sun's damaging ultraviolet (UV) radiation, which extends between 15 and 30 km above Earth's surface. Temperature increases with increasing altitude (up to –3°C) because absorption of UV radiation by ozone layer heats the air.

Layer of atmosphere closest to Earth's surface. Provides the air we breathe and contains most of the atmospheric mass. Temperature decreases with increasing altitude (to –52°C). Weather, including turbulent wind, storms, and most clouds, occurs here.

During a lightning flash, a negative charge moves from the bottom of the cloud to the ground, followed by an upward-moving charge along the same channel. The expansion of air around the lightning strike produces sound waves, or thunder.

An aurora in the northern hemisphere. Electrically charged particles from the sun collide with the gas molecules in the thermosphere releasing energy visible as light of different colours.

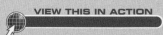
VIEW THIS IN ACTION

NATIONAL GEOGRAPHIC

is visible light that can be detected by the human eye when refracted by a prism, ranging from violet to blue to green to yellow to orange to red colours. About 43 percent of the incoming solar radiation peaks in the visible light wavelengths between 400 and 700 nm, and about 50 percent is radiated as infrared radiation. Gamma rays, x-rays, and most of the UV are absorbed by the stratosphere. Earth surfaces absorb portions of infrared radiation and ultraviolet light as well as visible light. Plants use portions of visible light in reactions of photosynthesis to produce glucose.

When considering the fate of incoming and outgoing energy from the atmosphere, we recognize that Earth has an energy budget that includes all gains through the absorption of incoming solar energy and all losses to outer space of outgoing radiation. About two-thirds of the incoming solar radiation is absorbed by the atmosphere, land, and oceans. The other one-third is reflected back to space by clouds, atmospheric particles (aerosols), snow, ice, and the oceans (**FIGURE 10.4**). The importance of this energy budget will be discussed further when we describe global climate change.

Incoming solar radiation is distributed on the planet by a large circulation process involving the atmosphere and oceans. This is influenced by a number of factors. One factor is that Earth is a sphere and its curvature affects the intensity of solar energy received on the

Earth's energy budget FIGURE 10.4

The Earth's energy budget describes the gains and losses of incoming solar radiation. Thirty-one percent is reflected back into outer space by the atmosphere as well as land surfaces, ice, and snow, and another 20 percent is absorbed in the atmosphere. The remaining 49 percent reaches the surface of the Earth as visible light and infrared radiation where it warms landscapes and oceans, powers the formation of weather and climate, and drives photosynthesis and ecosystem production. Warm objects, surfaces, and biota continuously emit infrared radiation into the atmosphere. This may dissipate into outer space or be absorbed by gases and warm the troposphere. When more infrared radiation is absorbed by tropospheric gases, positive radiative forcing results.

VIEW THIS IN ACTION

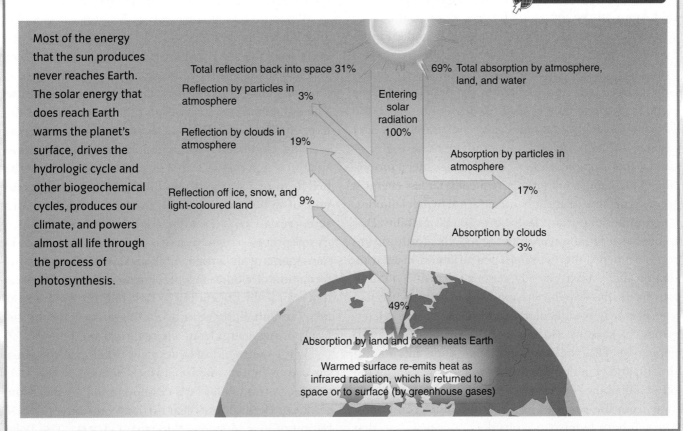

Most of the energy that the sun produces never reaches Earth. The solar energy that does reach Earth warms the planet's surface, drives the hydrologic cycle and other biogeochemical cycles, produces our climate, and powers almost all life through the process of photosynthesis.

Total reflection back into space 31%

Reflection by particles in atmosphere 3%

Reflection by clouds in atmosphere 19%

Reflection off ice, snow, and light-coloured land 9%

Entering solar radiation 100%

69% Total absorption by atmosphere, land, and water

Absorption by particles in atmosphere 17%

Absorption by clouds 3%

49%

Absorption by land and ocean heats Earth

Warmed surface re-emits heat as infrared radiation, which is returned to space or to surface (by greenhouse gases)

Process Diagram

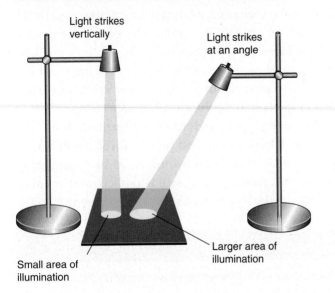

Light strikes
vertically

Light strikes
at an angle

Small area of
illumination

Larger area of
illumination

A The angle of incoming energy affects the area of illumination.

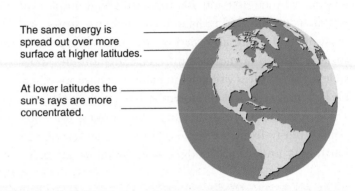

The same energy is
spread out over more
surface at higher latitudes.

At lower latitudes the
sun's rays are more
concentrated.

B Due to the curvature of the Earth, the angle of incoming solar radiation changes from being perpendicular at the equator to oblique at the Poles. The amount of energy received per unit of area at these higher latitudes is reduced as a result.

surface. At the equator, the angle of incoming solar radiation is perpendicular to the surface of the Earth and the sun appears directly overhead (represented by the desk lamp on the left at a 90 degree angle to the table; **FIGURE 10.5**). In contrast, at higher latitudes Earth's curvature means that incoming solar radiation is received at an oblique angle spreading the same amount of energy over a larger area (represented by the desk lamp on the right at an angle greater than 90 degrees to the desk). Radiation at higher latitudes must also travel a greater distance through the atmosphere where there is greater chance that air molecules will absorb and scatter a proportion of incoming sunlight. Consequently, far less energy is received per unit of area on the surface at higher latitudes.

Another important factor is seasonal variability. Due to the Earth's rotation around the sun, the northern and southern hemispheres experience variability in solar radiation that establishes our seasons (**FIGURE 10.6**). Earth's inclination on its axis (23.5 degrees from a line drawn perpendicular to the orbital plane) determines the seasons. During half the year (March 20 to September 22), the northern hemisphere tilts toward the sun and experiences warmer conditions. During the other half of the year (September 22 to March 20), it tilts away from the sun and experiences colder conditions. The southern hemisphere tilts the opposite way, so that

summer in the northern hemisphere corresponds to winter in the southern hemisphere. The equator is not affected tremendously by this seasonal variability.

When incoming solar energy is absorbed by ocean and land, temperatures at the surface begin to rise (**FIGURE 10.7A**). Energy is radiated to the air above and water evaporates. As the air above absorbs this heat, it becomes energized, expands, and becomes less dense. The lighter warmer air rises to higher altitudes. As this happens, the air begins to cool and moisture condenses, forming clouds and precipitation. Near to the troposphere, a high pressure area forms because the air at this height stops rising and upwelling continues by the column of air below. Air is forced to deflect toward low pressures created in places where cold air sinks to the surface. By this process a *convective cell* is created, characterized by movements of air between high and low pressure regions, formation of clouds and precipitation. Molecules of air move vertically (ascending or descending) and horizontally (at high altitudes or at Earth's surface) in the cell transporting energy from one region to another.

Large convection Hadley cells develop at the equator driven by the intense surface heating at this location. Ferrel and Polar cells develop in the northern and southern Hemispheres that further contribute to distribution of heat globally (**FIGURE 10.7B**). The Polar

Progression of seasons FIGURE 10.6

Earth's inclination on its axis remains the same as it travels around the sun. Thus, the sun's rays hit the northern hemisphere obliquely during its winter months and more directly during its summer. In the southern hemisphere, the sun's rays are oblique during its winter, which corresponds to the northern hemisphere's summer. At the equator, the sun's rays are approximately vertical on March 20 and September 22.

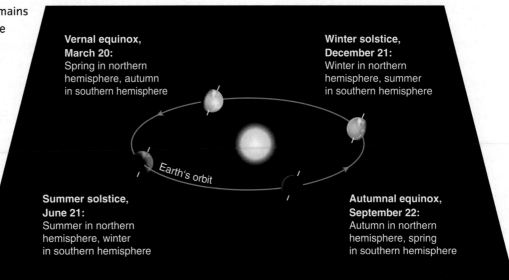

Vernal equinox, March 20:
Spring in northern hemisphere, autumn in southern hemisphere

Winter solstice, December 21:
Winter in northern hemisphere, summer in southern hemisphere

Earth's orbit

Summer solstice, June 21:
Summer in northern hemisphere, winter in southern hemisphere

Autumnal equinox, September 22:
Autumn in northern hemisphere, spring in southern hemisphere

Convective currents transfer heat FIGURE 10.7

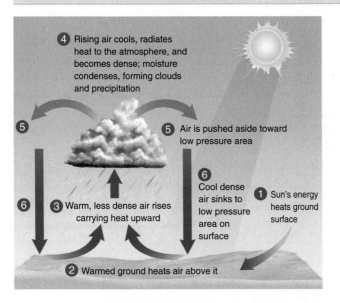

④ Rising air cools, radiates heat to the atmosphere, and becomes dense; moisture condenses, forming clouds and precipitation

⑤ Air is pushed aside toward low pressure area

⑥ Cool dense air sinks to low pressure area on surface

① Sun's energy heats ground surface

③ Warm, less dense air rises carrying heat upward

② Warmed ground heats air above it

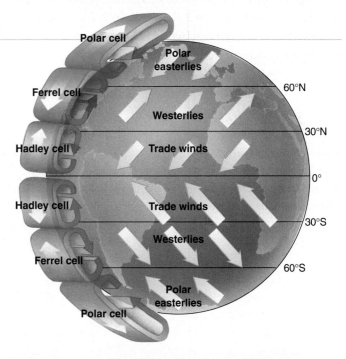

Polar cell

Polar easterlies

Ferrel cell

60°N

Westerlies

Hadley cell

30°N

Trade winds

0°

Hadley cell

Trade winds

30°S

Ferrel cell

Westerlies

60°S

Polar easterlies

Polar cell

A In convection, heating of the ground surface warms air above it resulting in an updraft of less dense, warm air (red arrows). As the air mass moves vertically and cools (blue arrows), moisture condenses as precipitation. Air moves toward low pressure created by the descending column of cold dense air. Warming and compression create a high pressure area at the surface of the Earth. Air then moves along the surface toward the low pressure region picking up moisture and heat. The cycle occurs repeatedly driven by the input of solar energy (numbers represent the sequence of events in the convection process).

B Six giant convection cells distribute heat and moisture over Earth's surface. The Hadley cells, between 0 and 30 degrees latitude, are the most intense due to incoming solar radiation. The convection cells distribute moisture to the equator and 60 degrees latitude. Descending air at 30 degrees and 90 degrees is drier and leads to desert and arid biomes. There are consistently three prevailing winds arising from convective processes in each hemisphere, diverted east or west by Earth's rotation and the Coriolis effect.

cells create high pressure regions at the poles because of descending cold air. The mid-latitude Ferrel cells involve additional factors contributing to the observed patterns of air currents. The cells contribute to the uneven distribution of moisture over the planet's surface leading to major precipitation at the equator and 60 degrees latitude, and to dry and arid conditions at 30 and 90 degrees latitude.

The atmosphere has three **prevailing winds**— major surface winds that blow more or less continually (Figure 10.7B). The direction of these winds is deflected east or west over different parts of the Earth by the *Coriolis effect*. To visualize the Coriolis effect, imagine that a rocket is launched from the North Pole toward New York (**FIGURE 10.8**). Prevailing winds from the northeast near the North Pole, or from the southeast near the South Pole, are called *polar easterlies*. Winds that blow in the middle latitudes from the southwest in the northern hemisphere or from the northwest in the southern hemisphere are called *westerlies*. Tropical winds blow from the northeast in the northern hemisphere or from the southeast in the southern hemisphere.

In addition to these giant global circulation patterns, the atmosphere features smaller-scale horizontal movements, or **winds**, that result partly from fluctuations in atmospheric pressure and partly from the planet's rotation. The pressure that is exerted by the atmospheric gases will depend on factors such as altitude, temperature, and humidity. This in turn will influence the direction and velocity of winds. Winds blow from areas of high atmospheric pressure to areas of low pressure, and we experience stronger winds when this pressure difference is increased.

These large-scale global circulation processes do not distribute only energy toward the poles. Pollutants, dust, and particulates within the air currents rise high into the atmosphere at lower latitudes or are deposited into the oceans. Through a process of long-range transport they are carried toward the poles where they may accumulate and enter the Arctic and Antarctic food web (see chapter opening vignette).

In the oceans, winds generate waves and currents. Gravity also contributes to oceanic circulation through the process of tides. In addition to these forces, there is

The Coriolis effect FIGURE 10.8

Viewed from the North Pole, the Coriolis effect appears to deflect ocean currents and winds to the right. From the South Pole, the deflection appears to be to the left.

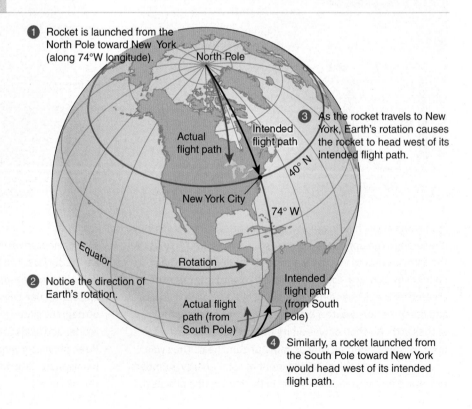

① Rocket is launched from the North Pole toward New York (along 74°W longitude).

North Pole

③ As the rocket travels to New York, Earth's rotation causes the rocket to head west of its intended flight path.

Intended flight path

Actual flight path

40° N

New York City

74° W

Equator

Rotation

② Notice the direction of Earth's rotation.

Actual flight path (from South Pole)

Intended flight path (from South Pole)

④ Similarly, a rocket launched from the South Pole toward New York would head west of its intended flight path.

Oceanic conveyor belt distributes heat energy FIGURE 10.9

The ocean conveyer belt transfers warm water from the Pacific Ocean to the Atlantic as a surface current. Cold, dense water sinks and slowly flows south from the Atlantic Ocean to the Pacific Ocean as a deep current. The Gulf Stream in the Atlantic Ocean represents one of the legs of the conveyor flowing northward that releases heat to the atmosphere influencing temperatures in western Europe.

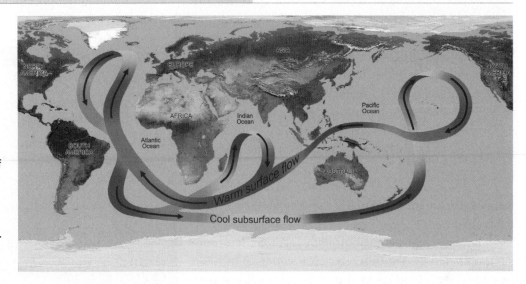

> **ocean conveyor belt (thermohaline gradient)** Deep-water circulation of the oceans due to a temperature salinity gradient. The flow of sea water transports vast amounts of absorbed energy from the equator to higher latitudes.

a large **ocean conveyor belt** (FIGURE 10.9) circulating water from the surface to the sea floor. The conveyor is important in the transport of heat from the equator to higher latitudes and therefore influences global atmospheric circulation. Norway, for example, is on average about 10°C warmer than Manitoba because of the influence of the conveyor, even though the two places are otherwise similar in latitude.

The ocean conveyor circulation is driven by density gradients in salt and temperature, and for this reason, it is often called the thermo- (temperature) haline (salt) gradient. Solar energy input at the equator causes surface waters to evaporate leaving behind dissolved salts that increase the density of sea water. When this sea water flows to higher latitudes, it radiates heat to the cooler atmosphere and further increases the density of the water. This forms the North American Deep Water (NADW) water mass around Labrador and Iceland. At particular locations, such as in the North Atlantic near Greenland, the heavy, dense, frigid salty water sinks to the sea floor at sites referred to as *deep-water formations*. Sea water travels at great depths along the sea floor toward the equator where it rises to the surface to complete the cycle. At these upwelling locations, nutrients that were entrained in the deep ocean surface with the ocean current. The full cycle can take a thousand years to complete.

As described in Chapter 9, El Niño and La Niña events arising in the Pacific Ocean alter global oceanic and atmospheric circulation patterns far from the tropical Pacific where the phenomenon originates (see Figure 9.30 on page 323). These events affect dynamics of energy and moisture transport delivering vast amounts of precipitation to some regions.

WEATHER AND CLIMATE

Unevenly distributed solar energy ultimately translates to the short-term weather and long-term climatic conditions we experience. Factors including air temperature, humidity, clouds, precipitation, air pressure, and winds influence what we can expect on a particular day but will further impact long-term conditions measured over a period of 30 years or more. Climatologists describe **weather** as the conditions in the atmosphere at a given place at a particular point in time and **climate** as the long-term averages (usually at least 30 years) of typical weather for a particular region over long periods of time. You can think of weather as what you see when you look outside today and climate as what you expect based on historical records.

There are many contributing factors that determine a region's climate. Latitude, elevation, topography, wind, humidity, fog, cloud cover, quantity and quality of vegetation, distance from the ocean, and geographic location all influence temperature, precipitation, and other

aspects of climate. Unlike weather, which changes rapidly, climate generally changes slowly, over hundreds or thousands of years.

CONCEPT CHECK **STOP**

Describe each of the layers of the atmosphere and provide descriptions of the major properties of each.

Describe the fate of incoming solar radiation in the atmosphere and the Earth's energy budget. How is solar intensity affected by Earth's curvature and axis?

How do the atmosphere and oceans distribute heat around the Earth?

What are the features that describe weather? How is climate distinguished from weather?

Air Quality and Air Pollution

The quality of the air we breathe deteriorates when **air pollution** develops. Pollutants include airborne substances such as chemicals, particulates, and/or biological agents that accumulate in the atmosphere creating conditions threatening the health for humans and biota alike. Air pollution contributes the hazy skyline over cities that obscures visibility, and

attacks and corrodes metals, plastics, rubber, and fabrics. Air currents can relocate these pollutants hundreds or even thousands of kilometres from their emission sites, making the problem of international significance. Examples of air pollution include tropospheric ozone and particulate matter, lead and heavy metals, acid precipitation, greenhouse gas emissions, and chemicals that deplete stratospheric ozone. Air quality is also altered when pollutants accumulate in sufficient concentrations indoors to affect human health.

Airborne chemicals of concern are categorized into either primary or secondary pollutants (**FIGURE 10.10**). Carbon oxides, nitrogen oxides, sulphur dioxide, most forms of particulate matter, and many volatile organic compounds represent the major

Primary air pollutants are emitted, unchanged, from a source directly into the atmosphere, whereas secondary air pollutants are produced from chemical reactions involving primary air pollutants.

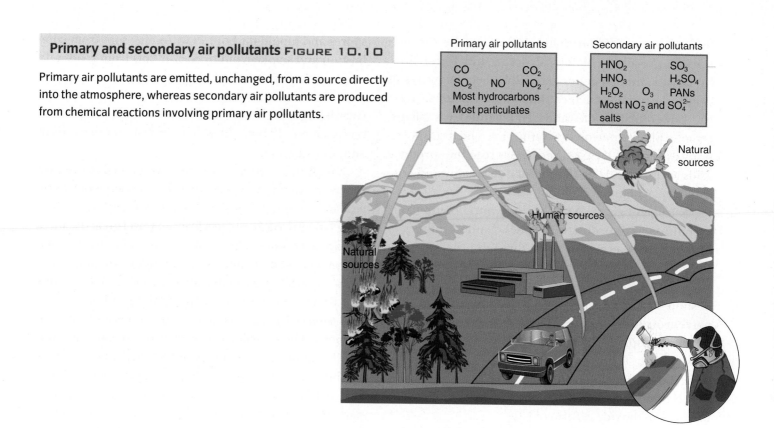

Primary air pollutants		Secondary air pollutants	
CO	CO_2	HNO_2	SO_3
SO_2 NO	NO_2	HNO_3	H_2SO_4
Most hydrocarbons		H_2O_2 O_3	PANs
Most particulates		Most NO_3^- and SO_4^{2-} salts	

> **primary air pollutants** Harmful chemicals that enter directly into the atmosphere from either a human or natural source.

> **secondary air pollutants** Harmful chemicals that form when primary air pollutants react in the atmosphere.

primary air pollutants. These are chemicals that are released directly into the atmosphere by either natural or human sources. When primary pollutants react in the atmosphere they form **secondary air pollutants** that lead to the development of fine particulate matter, tropospheric (ground-level) ozone, and acid precipitation. Sulphur dioxide, for example, reacts in the atmosphere to form sulphuric acid when it reacts with water.

MAJOR CLASSES OF AIR POLLUTANTS

The release of major pollutants in Canada are tracked and reported through the National Pollutant Release Inventory (NPRI). We focus on the most important classes from a regulatory perspective: particulate matter

(PM_{10} and $PM_{2.5}$), nitrogen oxides (NO_x), sulphur oxides (SO_x), carbon monoxide, volatile organic compounds (VOCs), tropospheric (ground-level) ozone, and air toxics. The health effects associated with exposure to these pollutants are summarized later in Table 10.1 on page 358.

Carbon oxides are primary pollutants including the gases carbon monoxide (CO) and carbon dioxide (CO_2). Carbon monoxide is a colourless, odourless, and tasteless gas produced in larger quantities than any other atmospheric pollutant except carbon dioxide. CO is a product of incomplete burning of hydrocarbon-based fossil fuels. Carbon monoxide is poisonous and reduces the blood's ability to transport oxygen. Carbon dioxide, also colourless, odourless, and tasteless, is produced through combustion and cellular respiration reactions. It is considered a greenhouse gas associated with climate change.

Sulphur oxides are compounds formed when sulphur and oxygen combine to form sulphur dioxide (SO_2) and sulphur trioxide (SO_3). Sulphur oxides form both primary and secondary pollutants. Volcanic eruptions can emit large volumes of sulphur dioxide into the atmosphere. Other natural sources arise through oceanic processes, biological decay, and forest fires. In Canada, human-generated sulphur emissions result from industrial activities including smelters, electric power plants, oil sands and natural gas plants, cement plants, petroleum refineries, and transportation (**FIGURE 10.11**). Secondary reactions produce acids and particulates that reduce air quality and contribute to acid precipitation downwind.

Nitrogen oxides (NO_x) include nitric oxide (NO) and nitrogen dioxide (NO_2) gases. Natural sources of nitrogen oxide emissions arise through processes such as soil bacteria that decays vegetation, lightning, and volcanic emissions. Transportation involving the combustion of fossil fuels accounts for approximately 60 percent of nitrogen oxides emission (**FIGURE 10.12**). In the atmosphere, NO_x undergo numerous secondary reactions to form various products—photochemical smog, fine particulates, and acid precipitation. The reddish-brown layer of air smog observed over cities is due to nitrogen dioxide, a form of NO_x.

Particulate matter (PM) refers to the matrix of solids and liquids that form in the air. It may also be referred to as aerosols. Common form include primary PM from volcanic eruptions; dust and dirt (and possibly the fungal spores and mould harboured within it); soot from combustion and smoke from fires; and salts from roads and ocean spray. Secondary PM is formed in the atmosphere from reactions with sulphur dioxide, nitrogen oxides, volatile organic compounds, and ammonia. Depending on its source, PM may also contain lead, mercury, iron, and/or cadmium.

Often these particulates are large enough to see, yet others are so tiny they require electron microscopy to reveal their presence. The size of particles impacts our health since larger particles are less likely to travel great distances and are likely filtered by our nose and throat. We describe particles based on diameter and monitor two groups: the *course (inhalable) particles* that are between 2.5 micrometres and 10 micrometres in diameter (PM_{10}); and *fine (respiratory) particles* that are less than 2.5 micrometres ($PM_{2.5}$) in diameter. These small particles provide an absorptive surface for numerous chemicals in the atmosphere such as sulphur oxides, nitrogen oxides, ammonia, hydrocarbons, and metals, and this contributes to the formation of fine particulate matter, or $PM_{2.5}$, that is a major component of smog. Nanoparticles (less than 100 nanometres in diameter) are the smallest and may penetrate the barrier of our lungs to enter the bloodstream and affect organs of the body.

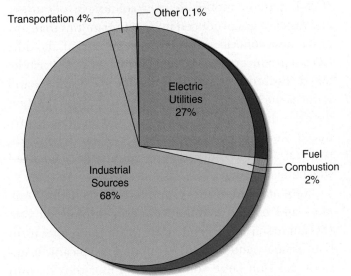

Sources of sulphur emissions in Canada, 2000 FIGURE 10.11

Transportation 4%
Other 0.1%
Electric Utilities 27%
Fuel Combustion 2%
Industrial Sources 68%

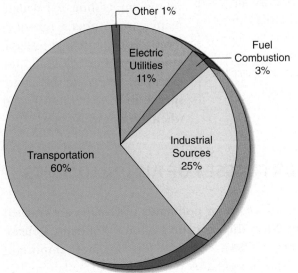

Sources of nitrogen oxide emissions, Canada 2000 FIGURE 10.12

Other 1%
Electric Utilities 11%
Fuel Combustion 3%
Transportation 60%
Industrial Sources 25%

Volatile organic compounds (VOCs) represent unburned hydrocarbon and other vapours that have a tendency to evaporate easily into the atmosphere. Natural sources originate from trees and plants and may account for two-thirds of the VOCs found in some locations (see What a Scientist Sees on page 352). Human sources include a wide array of products from various industries: gasoline fumes and solvents; benzene; cleansers and disinfectants; paints, strippers, preservatives, and sprays; pesticides; copiers and printers; glues and adhesives; and the list goes on. Because many household products contain these compounds, VOC concentrations are often 10 times higher indoors, contributing to air quality problems. VOCs are major contributors to primary pollutants but also undergo reactions with NO_x to form photochemical smog. Numerous VOCs are known to have or suspected of having direct toxic effects on humans, ranging from carcinogenesis (cancer causing) to neurotoxicity (affecting the nervous system). Benzene and dichloromethane, for example, are identified as toxic substances by Health Canada.

Unlike ozone that forms a protective barrier to UV penetration in the stratosphere, **tropospheric (ground-level) ozone (O_3)** results from secondary reactions of air pollutants and contributes to smog formation in our cities. While ground-level ozone pollution does form naturally (see What a Scientist Sees), a large amount is also produced by fossil fuel and automobile combustion. Tropospheric ozone is produced when NO_x and VOCs react in the presence of sunlight. Tropospheric ozone reduces visibility, causes serious health problems, and is a powerful greenhouse gas associated with climate change.

Tropospheric ozone is also known to adversely affect crop and forest vegetation (**FIGURE 10.13**). Ozone diffuses from the air into the leaves of plants and trees, and at higher concentrations it can be poisonous to plants. The impacts of air pollution on plant vegetation can include leaf injury, reduced plant growth, decreased yield, changes in quality, alterations in susceptibility to abiotic and biotic stresses, and overall decreased reproduction. Plants and trees weakened by ozone are therefore more susceptible to pests, drought, and disease.

Researchers in Wisconsin have confirmed these short-term effects of ozone on plants through a massive research project that has been initiated to examine the

Ozone damage FIGURE 10.13

A scientist measures the effects of ozone on the growth and productivity of plum trees. Plants exposed to ozone generally exhibit damaged leaves, reduced root growth, and reduced productivity.

effects of elevated levels of carbon dioxide and ozone (greenhouse gases) on the forest. This project, called Aspen Face, is feeding the trees the potential atmosphere of 50 years in the future. Researchers are monitoring the forest responses to these elevated levels of carbon dioxide and ozone. Results so far show that, at least in the short term, elevated levels of carbon dioxide result in a thicker, fuller tree canopy. Ozone, however, is showing the opposite results. Increased levels of ozone result in increased tree mortality, more disease, increased pest invasion, and higher percentages of leaf injury. The results of the Aspen Face research project provide only a snapshot of the effects of greenhouses gases on forests, but so far, the results of increased ozone are showing devastating effects on trees.

Most of the hundreds of other air pollutants—such as chlorine, lead, formaldehyde, radioactive substances, and fluorides—are present in very low concentrations, although it is possible to have high local concentrations of specific pollutants. Some of these air pollutants, known as **hazardous air pollutants**, or **air toxics**, are potentially harmful and may pose long-term health risks to people who live and work around chemical factories, incinerators, or other facilities that produce or use them.

Air Pollution from Natural Sources

A On a hot summer day in the Blue Ridge Mountains (part of the Appalachians) of Virginia, a bluish-purple haze hangs over the forested hills. This haze is caused by a group of volatile organic compounds, including isoprene and sesquiterpenes, emitted from the leaves of the trees in the forest. These chemicals evaporate into the air where they interact with other substances to produce tropospheric ozone contributing to the hazy skyline that scatters sunlight, leading to the blue appearance of the mountains.

B Many plants produce volatile organic compounds to attract pollinators, perhaps reduce predation, or even to deal with environmental stress such as too much heat. The sharp scent of white spruce comes from monoterpenes emitted from the needles.

C Scientists from the Institute of Arctic and Alpine Research at the University of Boulder, Colorado, have been conducting experiments on forests to study the emissions of VOCs from trees. The team is determining the rates of emissions of sesquiterpene from a variety of tree species. Scientists use Teflon bags to collect the VOC emitted and use the data to generate models predicting rates of emission from forests and specific tree species. This information can assist urban planners and foresters to identify best-suited tree species to offset air pollution.

SOURCES OF AIR POLLUTANTS

Air pollution originates through both natural and human sources. Erupting volcanoes, forest fires ignited by lightning, ocean spray and salt particulates, volatile organic compounds from coniferous forests and other vegetation represent large contributions from natural processes and activities (FIGURE 10.14A). Human activities add large quantities of pollutants that have the potential to impact both local and global air quality.

A significant contribution originates through the combustion of fossil fuels including coal, oil, and natural gas generated through transportation (mobile) and industrial (stationary) sources. Cars, trucks, tractors, and heavy construction equipment are examples of *mobile sources* (FIGURE 10.14B). *Stationary sources* include electric power plants and other industrial facilities (FIGURE 10.14C). Toxic air pollutants are released through activities and processes involved in chemical, metals, and paper industries.

Sources of air pollutants FIGURE 10.14

A Ocean spray is one example of a natural source of air pollution.

B Mobile sources of air pollution, like these trucks, produce particulate matter and other kinds of air pollution.

C This coal-fired electric power plant releases ash from the smokestacks and is an example of a stationary source of air pollution.

HEALTH EFFECTS OF MAJOR AIR POLLUTANTS

Generally speaking, exposure to low levels of pollutants irritates the nose, eyes, and throat and may lead to inflammation of the respiratory tract and premature aging of the lungs (TABLE 10.1). Many air pollutants also suppress the immune system, increasing susceptibility to infection. In addition, exposure to air pollution during respiratory illnesses may result in the development later in life of chronic respiratory diseases, such as emphysema and chronic bronchitis. In *emphysema,* the air sacs (alveoli) in the lungs become irreversibly distended, causing breathlessness and wheezy breathing. *Chronic bronchitis* is a disease in which the air passages (bronchi) of the lungs become permanently inflamed, causing breathlessness and chronic coughing.

Researchers now report that even short-term exposure to elevated concentrations of $PM_{2.5}$ can contribute to heart disease. With the smaller particles, the severity of illness may be greater. Nanoparticles can pass through cell membranes and migrate into other organs, including

Health effects of several major air pollutants TABLE 10.1

Pollutant	Source	Effects
Particulate matter (PM_{10}/$PM_{2.5}$) and soot	Industries, motor vehicles; photochemical smog formation	Causes irritation to eyes, nose, and throat; difficulties for individuals with asthma and other respiratory difficulties. Bronchitis may develop. May carry surface-absorbed carcinogenic compounds. Asbestos linked to lung cancer.
Sulphur dioxide (SO_2)	Coal-burning electric power plants, industrial emissions	Irritate respiratory tract; decrease lung function in asthmatics. Tightness in the chest and coughing. More harmful when particulates and other pollution concentrations high.
Nitrogen dioxide (NO_2)	Combustion engines of motor vehicles, industries, and heavily fertilized farmland	Irritate respiratory tack; aggravate respiratory conditions such as asthma and chronic bronchitis. Acute respiratory incidence in children when concentrations elevated.
Volatile organic compounds (VOCs)	Over 1000 products contribute volatile hydrocarbons: gasoline vapours, benzene, cleaning products, aerosols, paints, thinners, etc.	May lead to cancer, central nervous system disorders, and liver and kidney damage through prolonged exposure. Higher levels may cause mental impairment and birth defects.
Carbon monoxide (CO)	Incomplete fossil fuel combustion	Reduces blood's ability to transport oxygen; headache and fatigue at lower levels.
Tropospheric ozone (O_3)	Formed in atmosphere (secondary air pollutant)	Ozone irritates the lungs, increasing the symptoms of asthma and chronic bronchitis.
Lead and heavy metals		Harmful in small amounts, especially to unborn fetuses and young children. Exposure causes impairment in mental function, visual–motor performance, and causes neurological damage in children.

the brain, producing brain injuries similar to those found in Alzheimer patients. Particulates produced by diesel engines are commonly within the size of nanoparticles and also carry carcinogenic components. The shape of particulate matter is also of concern. Asbestos, which was once used widely in construction, may produce $PM_{2.5}$ dust, and these fibres may be inhaled and lodged deep in the lungs leading to the development of malignant lung cancer in many people. For this reason, asbestos must be removed with extreme caution when it is detected to ensure minimal contamination of dust.

According to the World Health Organization, the five worst cities in the world in terms of exposing children to air pollution are Mexico City, Mexico; Beijing, China; Shanghai, China; Tehran, Iran; and Kolkata (Calcutta), India. Respiratory disease is now the leading cause of death for children worldwide. More than 80 percent of these deaths occur in children under age five who live in cities in developing countries. The organization estimates that approximately 800,000 deaths per year worldwide (or 1.4 percent of all deaths) are attributed to outdoor air pollution. Lead pollution from heavily leaded gasoline is an especially serious problem in developing nations. The gasoline refineries in these countries are generally not equipped to remove lead from gasoline. In Cairo, Egypt, for example, children's blood lead levels are more than two times higher than the level considered at-risk in North America. Lead can retard children's growth and cause brain damage.

Health Canada has reported that approximately 5900 premature deaths occur in Canada annually as a result of exposure to air pollution. In response, the government of Canada, along with provincial and municipal governments and health and environmental nongovernmental organizations, developed the Air Quality Health Index (AQHI). The AQHI provides the general public with information about air pollution in Canada, including levels of ground-level ozone, particulate matter ($PM_{2.5}/PM_{10}$), and nitrogen dioxide.

Through the AQHI, air quality is reported on a scale from 1 to 10+ (most severe) and ranked low to high in health risk. The AQHI is also reported on an hourly basis and forecasts are given for the following day. In addition, Health Canada provides information to identify those "most at risk" and provides suggestions on how activity levels can be adjusted to limit short-term exposure.

Environment Canada is also working to inform the public about the ways Canadians can reduce their personal contributions to air pollution. To find out more about the AQHI, contact the Ministry of Environment air monitoring branch, your local lung association, or visit Health Canada at www.hc-sc.gc.ca.

MANAGING AIR POLLUTION

Air pollution is an international problem because pollution does not stay in the place where it is created (recall the discussion of long-range transport of pollutants in Chapter 4 and in the opening vignette to this chapter). Pollution regularly crosses borders, boundaries, or any other artificial divisions humans have created on Earth. To combat this problem, nations have to work together to control and reduce pollution for the sake of the planet as a whole.

Emissions trading, also referred to as "cap and trade," is an administrative approach used to provide economic incentives for achieving reductions in the emissions of pollutants. The overall goal of emissions trading is to control pollution. An administrative body or government sets a cap on the amount of pollution that can be emitted within a specified area or nation. Companies and other groups are issued an emission permit with a specific number of credits (allowances) that represent how much pollution they are allowed to emit. A company's credits cannot exceed this cap. Companies that need to increase their emissions allowance buy (or trade) credits from those facilities that pollute less. The cap is usually lowered over time, aiming toward a national emissions reduction target.

Although many countries have implemented emissions trading within their own borders, geographically close nations can also implement these programs as a group. The European Union Emissions Trading System (EU ETS) is one of the largest multinational programs for greenhouse gases. This program is a major aspect of the European Union's climate policy, and when the Kyoto Protocol came into effect in 2005 the EU ETS was already operational. The EU ETS covers more than 10,000 installations in the energy and industrial sectors, which collectively represent close to half of the EU's emissions of carbon dioxide and 40 percent of their total greenhouse gas emissions. The EU predicts an overall reduction of greenhouse gases for the sector of 21 percent of their 2005 emissions levels by 2020.

The close proximity of Canada and the United States has opened up discussions of the possibility of an emissions trading program being implemented between these two countries. In 1991, Canada and the United States established the Canada–U.S. Air Quality Agreement to address the issue of transboundary air pollution. In 2003, Canada and the United States undertook a joint feasibility and research study to explore the cross-border trading of capped emissions of nitrogen oxides and sulphur dioxides, known pollutants that contribute to smog and acid rain and that are associated with human health risks. The focus of this study has been on electricity generators that burn fossil fuels and emit nitrogen oxides and sulphur dioxides. The U.S. Acid Rain Program, established in 1995, has successfully reduced sulphur dioxide emissions by 32 percent from 1990 levels as facilities have had to pursue innovative technological methods of reducing sulphur dioxide emissions. A joint national study made a number of findings:

- Canada and the United States share three principal transboundary air quality problems (ground-level ozone, particulate matter, and nitrogen oxide and sulphur dioxide).

- Canada and the United States would need legislative and regulatory changes to ensure that credits (allowances) issued by each country are equivalent and are recognized for trade to be used in either country.

- Electricity generators that burn fossil fuels are the best candidates to participate in future cross-border emissions trading programs; however, additional exploration of how other large pollution contributors within the sector could participate should be investigated.

- An emissions monitoring and reporting system would need to be designed, implemented, and consistent in both countries.

- Trading rules would need to be defined consistently in both countries.

The Kyoto Protocol also uses emissions trading for the six major greenhouse gases. Emissions quotas were agreed upon by each participating nation to reduce their overall emissions by 5.2 percent of their 1990 levels by the end of 2012. Under this treaty, nations that emit less than their quota will be able to sell emissions credits to nations that exceed their quota.

Carbon credits are also traded. One carbon credit is equivalent to one tonne of carbon. Companies sell their carbon credits to commercial and individual customers. This program uses market mechanisms to drive industrial and commercial processes in the direction of low emissions and less carbon intensive approaches to production. There are two types of carbon credits:

1. Carbon offset credits (COCs), which consist of clean forms of energy production (wind, solar, hydro, and biofuels)

2. Carbon reduction credits (CRCs), which consist of collected and stored carbon from the atmosphere through reforestation, forestation, and soil collection and storage efforts.

Canada's response to air pollution

Because air pollution causes many health and environmental problems, most highly developed nations and many developing nations have established air quality standards for numerous air pollutants. In Canada, the **Clean Air Act** refers to a number of pieces of legislation established to reduce emissions of both air pollutants and greenhouse gases. The Clean Air Act permits the government of Canada to ensure industries' strict compliance with clean, consistent, and comprehensive national standards. The legislation takes a collaborative, all-inclusive approach to both air pollutants and greenhouse gases. In the early 1970s, the first of two Clean Air Acts was passed, which regulated the release of four air pollutants: asbestos, lead, mercury, and vinyl chloride.

There has been some noted opposition to Clean Air Act regulations. Critics argue it has seriously affected corporate profits, resulting in business being taken to outside sources, while other opponents suggest the legislation is not tough enough to actually control air pollution. While the federal government has the authority to sign international air protection treaties, Parliament is unable to put these regulations into action without consent from the provinces since the federal and provincial governments share the legal jurisdiction over air pollution. Incentive-based regulation is currently being used to stimulate the economy while encouraging companies to apply innovative and environmentally sustainable production technologies.

In 1997, Canada signed the Kyoto Protocol, but Kyoto was not supported by the Conservative government that was elected in 2006, and Canada's Kyoto obligations were subsequently dropped. As a result, Canada introduced the new set of regulations in a second Clean Air Act. These regulations set out a plan to reduce the greenhouse gas emission levels of 2003 by 45 to 65 percent by 2050. This new plan places regulations on vehicle fuel consumption for 2011 and a target for ozone and smog levels for 2025.

CONCEPT CHECK STOP

What is the difference between primary and secondary air pollutants?

What are the main classes of air pollutants, and what are some of their effects?

What adverse health effects have been related to air pollution?

What are some of the ways the international community has responded to air pollution?

Effects of Urban Air Pollution

LEARNING OBJECTIVES

Distinguish between industrial smog and photochemical smog.

Explain how weather and topography influence the dispersion of smog.

Describe the processes in the formation and deposition of acid precipitation.

Relate some of the effects of acid precipitation and ways to combat these effects.

Describe sick building syndrome.

n the atmosphere, secondary reactions take place between primary pollutants that contribute to the formation of smog and acid rain. Once in the atmosphere, wind currents transport pollutants hundreds and even thousands of kilometres, which often leads to transboundary dispersal. For this reason, air pollution needs to be treated as an international issue to attempt to resolve it.

INDUSTRIAL AND PHOTOCHEMICAL SMOG

The term **smog** was coined at the beginning of the 20th century for the smoky fog that prevailed over London during the Industrial Revolution when sulphur-rich coal was burned for residential homes and in industry. Through this process, **industrial smog** is produced containing large quantities of carbon dioxide and carbon monoxide along with soot and other forms of particulate matter. Sulphur found in the coal is also released to form sulphur dioxide and sulphur trioxide. Secondary reactions with water produce sulphuric acid and reactions with particulates form fine particulate matter. Because of air quality laws and pollution-control devices, industrial smog is generally not a significant problem in highly developed nations.

Industrial smog still affects developing nations, however. For example, many cities and towns in China

industrial smog
A whitish-grey haze resulting from sulphur-rich coal combustion. Sulphuric acid, particulate matter, carbon dioxide, and carbon monoxide are major end products of the reaction.

Air pollution in China FIGURE 10.15

Coal smoke pollutes the air in a small town in Shanxi Province, China. All forms of pollution are increasing as China becomes industrialized.

CHINA

Global Locator

burn tremendous quantities of coal to heat homes, producing thick layers of industrial smog that permit residents to see the sun only a few weeks of the year. The rest of the time residents are choked in a haze of orange-coloured coal dust (**FIGURE 10.15**). China's low air quality even affected the 2008 Summer Olympics held in Beijing.

photochemical smog A brownish-orange haze formed by reactions of nitrogen oxides, volatile organic compounds, and particulate matter, typically on a sunny day. Tropospheric ozone and fine particulate matter are major end products of the reaction.

A second form of noxious smog is **photochemical smog**, which results in the formation of tropospheric ozone (O_3) and fine particulate matter ($PM_{2.5}$). Photochemical smog is typically associated with large cities where a high number of vehicles operate.

Photochemical smog production is typically worst during the summer months when ample sunlight is available to drive the atmospheric reactions (**FIGURE 10.16**). Driving your vehicle generates the precursors to photochemical smog: nitrogen oxide is released from the tailpipe along with volatile organic compounds like unburned gasoline. Sunlight adds the energy and heat necessary to activate photochemical reactions to form tropospheric ozone. As well, other chemicals, including nitrogen dioxide, nitric acid, ammonium nitrates, $PM_{2.5}$, and peroxyacyl nitrates (PANs) are formed.

The growing number of automobiles is the most significant contributing factor to the formation of urban photochemical smog. In developing nations, many vehicles are 10 or more years old and have no pollution control devices. Motor vehicles produce about 60 to 70 percent of the air pollutants in urban areas of Central America, and they produce 50 to 60 percent in urban areas of India. The most rapid proliferation of motor vehicles worldwide is currently occurring in Latin America, Asia, and eastern Europe.

Many cities in Canada also experience the problems of photochemical smog. Windsor, Toronto, Montreal, and Vancouver have experienced smog conditions that exceed air quality targets an average of 10 or more days

Tropospheric photochemical smog chemistry FIGURE 10.16

Photochemical smog involves reactions of nitrogen oxides such as nitric oxide with volatile organic compounds in the presence of sunlight. Through the light-dependent reactions, tropospheric ozone is produced. Further reactions generate additional chemicals including peroxyacyl nitrates (PANs), nitric acid, and organic compounds like formaldehyde. The formation of nitrogen dioxide contributes to the brownish haze that forms over urban centres.

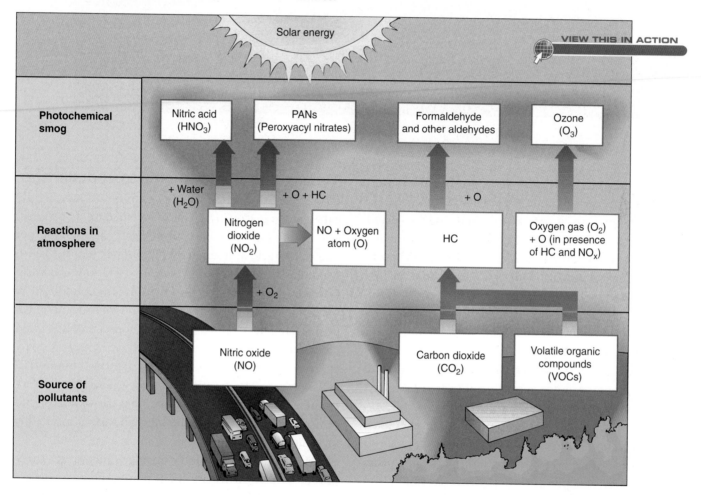

each year. Between 2000 and 2007, Environment Canada reported a 13 percent increase in tropospheric ozone concentrations from photochemical smog.

How weather and topography affect smog problems

Ozone and PM levels vary seasonally, daily, and even hourly depending on the weather conditions and where the wind and air is coming from. Smog weather is usually associated with light or stagnant winds and warm temperatures conducive for a temperature inversion. Recall that, normally, air temperature declines with increasing altitude above Earth.

This drives a process of vertical mixing (*convection cells*) so that warmer surface air rises to higher altitudes, creating a low pressure area that cooler air moves into. This vertical mixing is an important factor in the dilution and dispersion of pollutants upward and downwind. However, when a **temperature inversion** occurs, a warm air mass presses down on the column of cool air below. Air circulation is impeded and the particulate matter and polluting gases are

temperature inversion A layer of cold air temporarily trapped near the ground by a warmer upper layer.

Temperature inversion FIGURE 10.17

Cooler air

Cool air

Warm air

Normal pattern

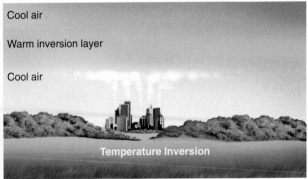

Cool air

Warm inversion layer

Cool air

Temperature Inversion

trapped and accumulate close to the ground, often over a rather short period of time (**FIGURE 10.17**).

A temperature inversion resulted in the deadly smog over London in 1952 that killed an estimated 12,000 people. A similar temperature inversion was responsible for the smog in Donor, Pennsylvania, in 1948 that took the lives of 20 residents and lead to the development of clean air legislation in the United States.

Temperature inversions usually last for only a few hours before solar energy warms the air near the ground. Sometimes, however, a stalled high-pressure air mass allows a temperature inversion to persist for several days, causing atmospheric stagnation.

Certain types of topography increase the likelihood of temperature inversions. Cities located in valleys or near a coast, hills, or mountains are prime candidates for temperature inversions. In valleys sunlight is blocked by nearby slopes and the air remains cool close to the ground increasing the frequency of temperature inversions. The Los Angeles Basin, for example, is notorious for encountering extreme smog conditions (see the EnviroDiscovery box). The city lies between the Pacific Ocean on the west and mountains to the north and east. During the summer the sunny climate produces a layer of warm dry air at upper elevations. However, a region of upwelling occurs just off the Pacific coast, bringing cold ocean water to the surface and cooling the ocean air. As this cool air blows inland over the basin, the mountains block its movement further. Thus, a layer of warm, dry air overlies cool air at the surface, producing a temperature inversion.

Making the commitment to reduce photochemical smog

Smog continues to be a significant challenge from a regulatory perspective. Crop losses due to smog are estimated to be nearly $70 million per year in Ontario and about $9 million per year in British Columbia's Lower Fraser Valley. In 1990, the Canadian Council of Ministers of the Environment committed to reductions in smog precursor emissions of NO_x and VOCs through the development of regulations, guidelines, and educational initiatives. These targets have largely been focused on automobile emission standards and commercial emission restrictions. Newer designs of catalytic converters in automobiles, for example, help treat exhaust and remove NO_x and other pollutants like carbon monoxide and the VOCs that contribute to smog.

Many public education campaigns that target transportation alternatives are held in an attempt to reduce smog across Canada (**FIGURE 10.18**). Ottawa and Carleton, Ontario, have introduced the Transportation Environment Action Plan, a public education program

Public transportation FIGURE 10.18

Many cities in Canada are promoting public transit as a way to "go green" and reduce the incidence of smog in urban centres.

Los Angeles, California, has some of the worst smog in the world. Its location, in combination with its sunny climate, is conducive to the formation of stable temperature inversions that trap photochemical smog near the ground, sometimes for long periods. Passenger vehicles, heavy-duty trucks, and buses are the source of more than half of the smog-producing emissions in Los Angeles.

In 1969 California became the first state in the United States to enforce emission standards on motor vehicles, largely because of the air pollution problems in Los Angeles. Today Los Angeles has stringent smog controls that regulate everything from low-emission alternative fuels (such as compressed natural gas) for buses to lawn mower emissions to paint vapours. Using the cleanest emission-reduction equipment available significantly reduces emissions from large industrial and manufacturing sources, including oil refineries and power plants. After several decades devoted to improving its air quality, Los Angeles now has the cleanest skies it has had since the 1950s. Despite the impressive progress, however, Los Angeles still exceeds federal air quality standards on more days than almost any other metropolitan region in the United States.

Photochemical smog

Photographed in Los Angeles, California, on a day when air pollution exceeded federal air quality standards.

to teach people about alternative transportation techniques. This program has developed innovative ways to show how commuting to work can be easy through public transit, walking, cycling, and ridesharing. In Toronto, the Healthy City Office is developing a public education program to promote the use of public transit, make cycling to work safer, easier, and more convenient, and provide partners for employees who want to rideshare. Calgary is initiating the world's first voluntary automobile emissions testing program. The program, called SMOG FREE, is an acronym for Save Money On Gas From Reduced Exhaust Emissions. Under the program, Calgary residents can have their vehicle emissions measured at local service outlets, and sponsors offer coupons for a $10 reduction on service costs to improve or repair exhaust systems.

ACID PRECIPITATION

Precipitation is a natural way to cleanse the atmosphere of chemicals and particulates. Carbon dioxide, for example, is removed when carbonic acid is formed in rainwater. The problem of acid precipitation, however, lies with strong acids formed when precursor chemicals such as sulphur dioxide and nitrogen oxides are emitted into the atmosphere. **Acid precipitation** results when

■ **acid precipitation**
A group of secondary air pollutants that include sulphuric, nitrous, and nitric acids formed by reactions of sulphur dioxide and nitrogen oxides with water in the atmosphere (wet deposition) or following contact with water (dry deposition).

these chemicals combine with water to form secondary pollutants including sulphuric acid (H_2SO_4), nitrous acid (HNO_2), and nitric acid (HNO_3). Robert Angus Smith, a British chemist, coined the term *acid rain* in 1872 after he noticed that buildings in areas with heavy industrial activity were almost dissolving away, worn down by years of acidic rainfall. Acid precipitation affects aquatic and terrestrial ecosystems and erodes and corrodes buildings and structures. The major culprit is sulphur dioxide, contributing to about 60 to 70 percent of the acid precipitation observed.

Acid precipitation is deposited on landscapes, often hundreds of kilometres downwind from the emission source in one of two ways (**FIGURE 10.19**). **Wet deposition** occurs when pollutants react with water in the atmosphere in its various forms to create an acidic solution that precipitates onto the landscape. In areas where the weather is dry, acid chemicals may attach to particles and fall to the ground through **dry deposition**. The addition of rainfall, fog, sleet, hail, or snow dissolves the oxides to form acidic compounds of sulphuric and nitric acids. About half of the acidity in global ecosystems is derived from dry deposition.

Acidity and pH

How acidic is acid precipitation? The acidity of a solution is determined from a measure of the concentration of hydrogen protons (H+), or pH. pH values range along a logarithmic scale from 0 (acidic) to 14 (basic or alkaline; **FIGURE 10.20**). A solution is neutral if the pH is 7, which means it is neither acidic nor alkaline.

Because the scale is logarithmic, small changes in the pH scale translate into substantial changes in the actual acidity. A change from pH 7 to pH 6 is really a 10-fold change. Unpolluted rainwater is slightly acidic with a pH of around 5.7 due to the formation of carbonic acid. Acid precipitation is below a pH of 5.0 with most in the range between pH 4.3 to 5.0. In regions of Ontario, pH is 4.5, or 40 times more acidic than normal.

Impacts of acid precipitation

Acid precipitation corrodes metals, building materials, and statues (**FIGURE 10.21**). It eats away at important monuments and ancient ruins. When acid rain reacts with the calcium carbonates found in the limestone of the building, calcium sulphate is formed, which is soluble and

Acid deposition FIGURE 10.19

The pollutants sulphur dioxide (SO_2) and nitrogen oxide (NO_x) are emitted into the atmosphere. Wet deposition may occur when these pollutants combine with water vapour to form acids that precipitate to the landscape with rainfall. Dry deposition may also occur when particulates laced with pollutants return to the surface and later combine with water to form acids.

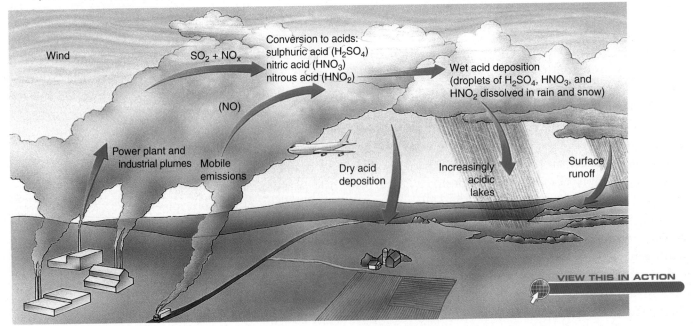

The pH Scale FIGURE 10.20

Acidity is measured in terms of pH, on a scale that runs from zero, the most acidic, to 14, the most alkaline. A change of one unit on the pH scale represents a 10-fold change in acidity. Typically, organisms live under conditions hovering around pH 7 and function less successfully toward either end of the scale. By a pH of 4.2, most aquatic invertebrates and vertebrates are killed.

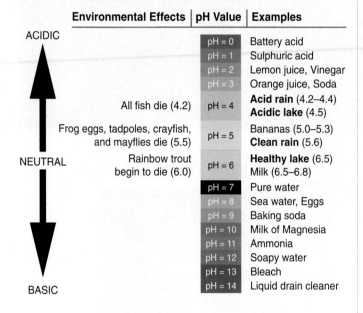

Environmental Effects	pH Value	Examples
	pH = 0	Battery acid
	pH = 1	Sulphuric acid
	pH = 2	Lemon juice, Vinegar
	pH = 3	Orange juice, Soda
All fish die (4.2)	pH = 4	**Acid rain** (4.2–4.4) **Acidic lake** (4.5)
Frog eggs, tadpoles, crayfish, and mayflies die (5.5)	pH = 5	Bananas (5.0–5.3) **Clean rain** (5.6)
Rainbow trout begin to die (6.0)	pH = 6	**Healthy lake** (6.5) Milk (6.5–6.8)
	pH = 7	Pure water
	pH = 8	Sea water, Eggs
	pH = 9	Baking soda
	pH = 10	Milk of Magnesia
	pH = 11	Ammonia
	pH = 12	Soapy water
	pH = 13	Bleach
	pH = 14	Liquid drain cleaner

ACIDIC

NEUTRAL

BASIC

washed away with rainfalls. In Greece the Parthenon marbles have been moved indoors to prevent further damage by acid precipitation. In Canada, headstones that are over 90 years old at the National Military Beechwood Cemetery in Ottawa are deteriorating in the face of acid precipitation. Affected buildings and statues can be costly to repair requiring millions of dollars of investment.

In the 1970s, researchers began to question how acid precipitation affects aquatic ecosystems. The Canadian

Acid rain damage FIGURE 10.21

Acid rain has destroyed much of the detail of this stone angel's face through chemical reactions of calcium carbonate and sulphuric acid. Photographed in London, U.K.

Department of Fisheries and Oceans began a series of whole-lake acidification experiments on Lake 223 at the Experimental Lakes Area in northwestern Ontario. Starting in 1976, the lake was increasingly acidified under controlled manipulations to gradually reduce the pH from 6.7 to nearly 5.0 over a period of five years. Numerous plant and animal species disappeared from the lake, and it became apparent that overall biological diversity declined. As various populations of plants and animals were lost, the entire food chain was affected by the acidic conditions (FIGURE 10.22 on page 364). A recovery was initiated in 1984, and pH slowly returned to its previous condition by increasing the alkalinity through controlled manipulations. Some species, such as the white sucker and lake trout, recovered quickly while others are now extirpated from the lake or were slow to return.

Other studies have confirmed that acid precipitation causes a decrease in biodiversity. Researchers have documented reductions in prey including populations of zooplankton, phytoplankton, clams, and crayfish. Fish eggs degrade and young are not able to develop. In those water bodies with pH 4.0, there may be no fish found at all. Without healthy populations of fish to feed upon, loons and other water birds dwindle in numbers. Decomposition rates also slow in acidic lakes and the undecayed matter accumulates at the bottom of the lake.

Some species benefit from the acidity. Benthic plants and mosses, for example, can perform well under these conditions. Blackfly larvae also seem to do well. The Canadian government has estimated that about 800,000 lakes of various sizes in eastern Canada now fit into the acidic category.

Acid precipitation also causes serious damage to forests. Through cumulative effects from several stressors including acid precipitation, insect infestation, tropospheric ozone, UV radiation, and drought, there may be an overall weakening of trees and the inability to withstand adverse conditions (FIGURE 10.23A on page 364). Acid precipitation may be more severe and cause direct damage to the protective waxy surface of leaves and needles resulting in defoliation or even death (FIGURE 10.23B).

Acid precipitation alters soil chemistry and reduces the availability of essential nutrients. Through a process called *cation-exchange*, the solubility of calcium, magnesium, and potassium is enhanced in acidic soil water and nutrients are leached. This means that nutrients are

Lake trout in Experimental Lake 223 FIGURE 10.22

Shown here is a healthy lake trout. In the acidified Lake 223, the acidic conditions caused populations at lower trophic levels, including white sucker, to crash, reducing available food sources for lake trout and causing them to starve. During recovery, lake trout biomass increased as the trout's food sources started to come back. Not all species at lower trophic levels recovered to the same extent, though.

Acid precipitation's effects on forests FIGURE 10.23

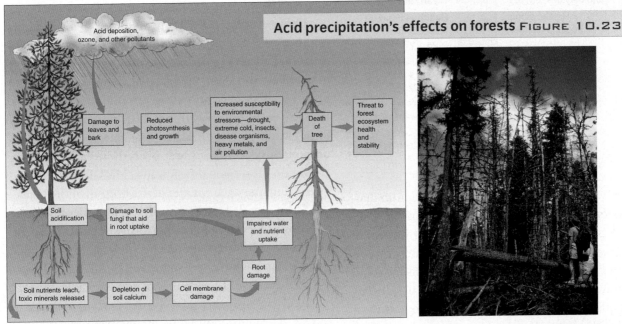

A The impacts of acid precipitation to forests include increased susceptibility to environmental stress, injury to leaves and needles, and even death. Soil chemistry is altered accelerating nutrient leaching and aluminum toxicity.

B Trees exposed to acid precipitation are more susceptible to pests and disease resulting in the death of many of the trees in a region.

washed deeper into the soil or washed out of the topsoil altogether. Soils are then devoid of essential nutrients and trees become nutrient deficient, slowing in growth. Soil mineral deficiencies are now found in forests in Ontario, Quebec, and the Atlantic provinces in Canada as well as in the northeastern United States. Projections suggest that upward of 50 percent of Canada's eastern boreal forests may be affected by acid rain in the future without further cuts in emissions.

Combatting the effects of acid precipitation The basic concept of acid precipitation control seems rather straightforward: if we reduce sulphur emissions we can curb acid precipitation. Canada has committed to reducing sulphur dioxide emissions through the development of a series of agreements over the past 15 years. The Eastern Canadian Acid Rain Program was adopted in 1985 and resulted in a cut of 43 percent to pre-1980 sulphur dioxide emissions across several provinces.

The goal was to reduce depositions by 50 percent to a target of 20 kilograms per hectare per year.

Even with these cuts, it was impossible to sufficiently reduce sulphur emissions and meet targets without the cooperation of the United States, since much of the air pollution originated south of the Canadian border. In 1991, the Canada–United States Air Quality Agreement was signed, which led to radically cutting sulphur dioxide emissions by 55 percent. Canada continues to strive for further cuts of 30 to 50 percent with the development of the Canada-Wide Acid Rain Strategy for Post-2000.

The significant reductions made by Canada and the United States have been achieved in part by installing scrubbers in the smokestacks of coal-fired power plants and using clean-coal technologies to burn coal without excessive sulphur emissions (see Chapter 11). Rainfall in parts of central Ontario and the U.S. Midwest, northeast, and mid-Atlantic regions is less acidic today than it was a few years ago as a result of cleaner-burning power plants and the use of reformulated gasoline. Many power plants in Quebec and Ontario switched from high-sulphur to low-sulphur coal. However, solving one environmental problem often creates others. While the move to low-sulphur coal reduced sulphur emissions, it contributed to the problem of global climate change. Because low-sulphur coal has a lower heat value than high-sulphur coal, more of it must be burned—and more CO_2 emitted—to generate a given amount of electricity. Low-sulphur coal also contains higher levels of mercury and other trace metals, so burning it adds more of these hazardous pollutants to the air.

Even if the target acid loading of between 15 to 20 kilograms per hectare per year is achieved, scientists have noted that sensitive ecosystems continue to be vulnerable to the influence of acidity. This is because different regions of Canada vary in their buffering or **acid neutralizing capacity (ANC)**. In those areas of the country where there is plenty of acid neutralizing limestone rock present, ecosystems are better able to buffer against acid precipitation and may not show immediate signs of acid stress. The high alkalinity of the soils and rock provides bases including calcium and magnesium salts that neutralize against the acid and keep pH stable near or at the pre-exposure condition. Conversely, in areas of Canada dominated by

granite, such as the Canadian and Precambrian Shield, there is little or no calcium or magnesium carbonates present and therefore little buffer against acid precipitation. As a result pH is reduced and the impacts of acid rain are more immediately observed. This situation describes provinces in eastern Canada including Ontario, Quebec, and Nova Scotia. As well, lakes in northeastern Alberta, northern Saskatchewan and Manitoba, parts of western British Columbia, Nunavut, and the Northwest Territories found on the Canadian Shield are susceptible should acid precipitation loading increase in the future.

It is imperative that we identify whether current emissions exceed the ability of an ecosystem to naturally neutralize acid precipitation. Scientists have determined this by measuring critical loads for regions across the country. **Critical loads** represent estimates of the amount of acid precipitation (e.g., wet sulphate deposition) that a region can receive and not be damaged. Critical loads are high for regions that have high alkalinity and high acid neutralizing capacity and low for regions when the opposite occurs. For eastern Canadian lakes, critical loads for wet sulphate deposition to maintain a pH of at least 6.0 range from more than 20 kilograms of sulphate per hectare per year in areas with high acid neutralizing capacity to less than 8 kilograms per hectare per year in the most sensitive regions.

In many regions in eastern Canada, current wet sulphate depositions exceed critical loads and are projected to continue in the future. This equates to 25 percent of lakes and 38 percent of forests in eastern Canada. Environment Canada estimates that an additional 75 percent cut beyond the current emission targets are necessary to bring wet sulphate deposition levels below the critical loads for virtually all ecosystems in eastern Canada. Even then impacted lakes and forests will likely be slow to recover since the past 30 or more years of acid precipitation have profoundly altered soil chemistry, depleting ecosystems of essential calcium and magnesium. Because soils take hundreds or even thousands of years to develop, it may be decades or centuries before they recover from the effects of acid rain.

Many scientists are convinced that ecosystems will not recover until substantial cuts in nitrogen oxide emissions also occur. However, these emissions are hard to control because

acid neutralizing capacity (ANC) The ability of water or soil to neutralize acids that have been added.

critical load The maximum amount of pollution that an ecosystem can tolerate without being damaged.

motor vehicles produce large amounts of nitrogen oxides. Engine improvements such as the catalytic converter help reduce nitrogen oxide emissions, but as the human population continues to grow, the increasing number of motor vehicles will probably offset these gains. Dramatic cuts in nitrogen oxide emissions will require a reduction in high-temperature energy generation, especially in gasoline and diesel engines.

INDOOR AIR POLLUTION

If you are reading this chapter indoors, you are probably inhaling air pollution. The air in enclosed places, such as automobiles, homes, schools, and offices, may have significantly higher levels of air pollutants than the air outdoors. Indoor air pollution is of particular concern to urban residents because they may spend as much as 95 percent of their time in enclosed places.

Because illnesses from indoor air pollution usually resemble common ailments such as colds, influenza, or upset stomachs, which are collectively known as **sick building syndrome**, they are often not recognized. The most com-

> **sick building syndrome**
> Eye irritations, nausea, headaches, respiratory infections, depression, and fatigue caused by indoor air pollution.

mon contaminants of indoor air are radon, cigarette smoke, carbon monoxide, nitrogen dioxide (from gas stoves), formaldehyde (from carpet, fabrics, and furniture), household pesticides, cleaning solvents, ozone (from photocopiers), and asbestos. In addition, viruses, bacteria, fungi (yeasts, moulds, and mildews), dust mites, pollen, and other organisms are often found in heating, air conditioning, and ventilation ducts.

Indoor air pollution is a particularly serious health hazard in developing countries, where many people burn fuels such as firewood or animal dung indoors to cook and heat water (**FIGURE 10.24**). Smoke from indoor cooking contains carbon monoxide, particulates, hydrocarbons, and hazardous air pollutants such as formaldehyde and benzene. Women and children are harmed the most by indoor cooking, which can contribute to acute lower respiratory infections, pneumonia, and lung cancer. The World Health Organization estimates that smoke from indoor cooking kills 1.6 million people each year.

Indoor air pollution FIGURE 10.24
Cooking indoors with open fires or traditional cooking stoves results in dangerous levels of indoor air pollution.

Radon The most serious indoor air pollutant is probably radon, a colourless, tasteless, odourless radioactive gas produced naturally during the radioactive decay of uranium in Earth's crust. Radon seeps through the ground and enters buildings, where it sometimes accumulates to dangerous levels (**FIGURE 10.25**). Although radon is also emitted into the atmosphere, it gets diluted and dispersed and is of little consequence outdoors.

Only ingested or inhaled radon harms the body. The National Research Council of the National Academy of

How radon infiltrates a house FIGURE 10.25

Cracks in basement walls or floors, openings around pipes, and pores in concrete blocks provide some of the entries for radon.

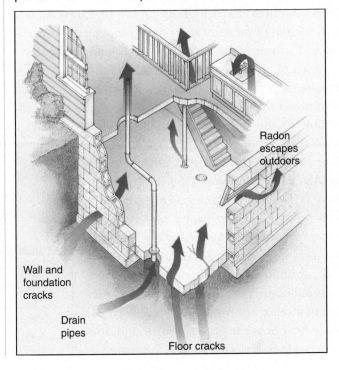

Radon escapes outdoors

Wall and foundation cracks

Drain pipes

Floor cracks

Sciences estimates that residential exposure to radon causes 12 percent of all lung cancers—between 15,000 and 22,000 cases of lung cancer annually in the United States. Cigarette smoking exacerbates the risk from radon exposure; about 90 percent of radon-related cancers occur among current or former smokers.

Ironically, efforts to make our homes more energy efficient have increased the hazard of indoor air pollutants, including radon. Drafty homes waste energy but allow radon to escape outdoors so it does not build up inside. Every home should be tested for radon because levels vary widely from home to home, even in the same neighbourhood. Testing and corrective actions are reasonably priced.

FIGURE 10.26 summarizes many possible sources of air pollution in homes.

CONCEPT CHECK STOP

What is the difference between industrial and photochemical smog?

How does topography affect smog?

What are some of the effects of acid precipitation?

How have energy conservation efforts contributed to indoor air pollution and the sick building syndrome?

Indoor air pollution FIGURE 10.26

Homes may contain higher levels of air pollutants than outside air, even near polluted industrial sites.

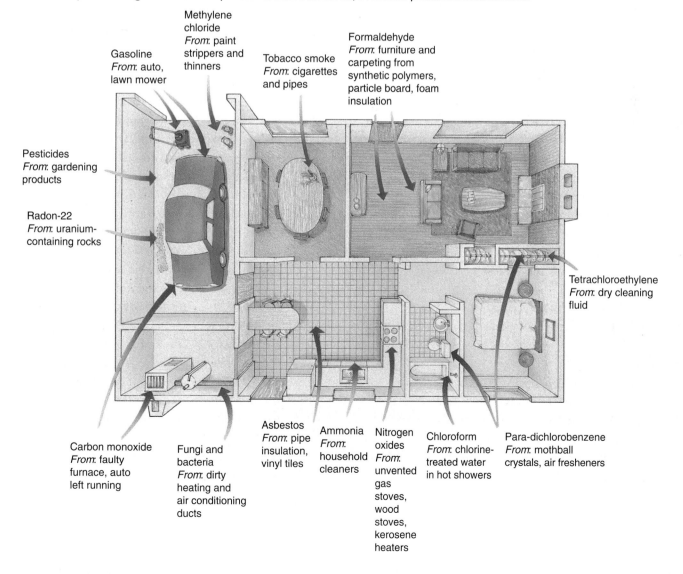

Gasoline
From: auto, lawn mower

Methylene chloride
From: paint strippers and thinners

Tobacco smoke
From: cigarettes and pipes

Formaldehyde
From: furniture and carpeting from synthetic polymers, particle board, foam insulation

Pesticides
From: gardening products

Radon-22
From: uranium-containing rocks

Tetrachloroethylene
From: dry cleaning fluid

Carbon monoxide
From: faulty furnace, auto left running

Fungi and bacteria
From: dirty heating and air conditioning ducts

Asbestos
From: pipe insulation, vinyl tiles

Ammonia
From: household cleaners

Nitrogen oxides
From: unvented gas stoves, wood stoves, kerosene heaters

Chloroform
From: chlorine-treated water in hot showers

Para-dichlorobenzene
From: mothball crystals, air fresheners

Global Climate Change

eather and climate are driven by the energy of the sun, which heats the Earth unevenly. Then, through a system of interactions with the atmosphere, lithosphere, hydrosphere, and biosphere, this heat is redistributed around the globe. Weather includes not only temperature, but winds, atmospheric pressure, humidity, and precipitation that change over time to influence the conditions we experience today, next week, next year, or in the future. Change that occurs in weather over long periods of time, such as from one century to another or between hundreds of thousands of years, is called **climate change**. Changes in climate that have taken place during Earth's history have included natural events—continental drift, solar radiation, Earth's orbit, and volcanic emissions—but more recently humans have affected the climate system by enhancing the concentration of greenhouse gases in the atmosphere.

NATURAL AND ENHANCED GREENHOUSE EFFECT

The sun powers the Earth's **energy budget**, which is the amount of incoming radiation that is absorbed and the outgoing infrared radiation emitted back to space. Recall that about one-third of the energy budget is reflected back to outer space by clouds, particulates, and the planet, and the other two-thirds is absorbed by the atmosphere and clouds, as well as the planet's land and oceans (refer to Figure 10.4 on page 343). Absorbed energy runs the hydrologic cycle, powers winds and atmospheric and ocean circulation, enters food webs through photosynthesis, and is used to generate biomass and warm the planet. Ultimately, this energy is radiated back to space as outgoing long-wave infrared (or thermal) radiation.

Earth's energy budget may change through a concept referred to as **radiative forcing**. *Positive radiative forcing* results from additional incoming energy to the budget and generally results in a warming influence. In *negative radiative forcing*, a cooling influence typically results when reflection is increased or the amount of energy radiated to outer space is enhanced. Because climate results from a series of interactions between atmosphere, oceans, and land processes, radiative forcing triggers a number of feedback processes, many of which we are only beginning to uncover and appreciate.

One way that radiative forcing can arise is when the amount of incoming solar radiation is changed either because of Earth's orbit or the activity of the sun itself. Solar flares, for example, result during periods of heightened solar activity and in turn provide additional quantities of energy input to Earth. Changes in Earth's rotation around the sun may alter the distribution of radiation over the planet's surface that may lead to the occurrence of glaciation periods.

climate change
A change in the long-term weather patterns observed over a period from decades to millions of years.

energy budget
The gains and losses of solar radiation to Earth. When in equilibrium, there are no changes in overall energy amount for the planet.

radiative forcing
The imbalance in Earth's energy budget that results when either the gains through energy absorbed or losses through the amount of energy radiated to outer space is changed through either natural or human influences. Positive forcing results in a warming influence and negative forcing leads to a cooling influence.

Natural and enhanced greenhouse warming FIGURE 10.27

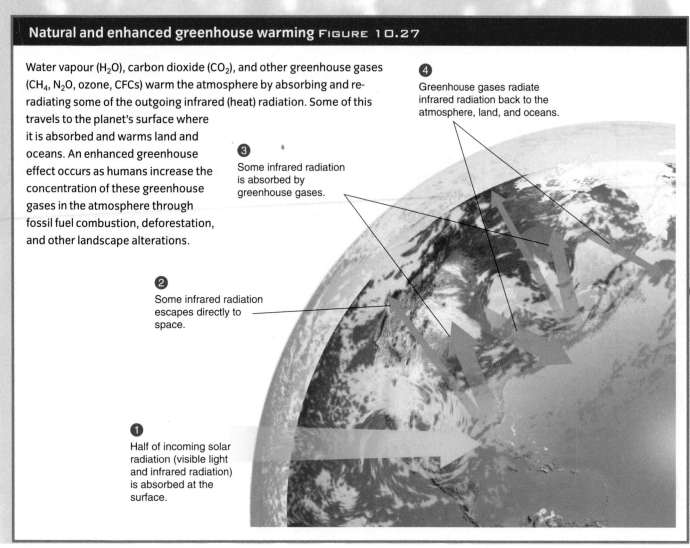

Water vapour (H$_2$O), carbon dioxide (CO$_2$), and other greenhouse gases (CH$_4$, N$_2$O, ozone, CFCs) warm the atmosphere by absorbing and re-radiating some of the outgoing infrared (heat) radiation. Some of this travels to the planet's surface where it is absorbed and warms land and oceans. An enhanced greenhouse effect occurs as humans increase the concentration of these greenhouse gases in the atmosphere through fossil fuel combustion, deforestation, and other landscape alterations.

4 Greenhouse gases radiate infrared radiation back to the atmosphere, land, and oceans.

3 Some infrared radiation is absorbed by greenhouse gases.

2 Some infrared radiation escapes directly to space.

1 Half of incoming solar radiation (visible light and infrared radiation) is absorbed at the surface.

Another way to induce positive radiative forcing is by altering the quantity of long-wave radiation emitted into outer space. Through a phenomenon commonly called the *greenhouse effect*, gases in the troposphere intercept and absorb infrared radiation as it travels to outer space from the surface of the planet. Some of this infrared radiation travels down to the planet's surface where it is absorbed by land and oceans and heats the surface in the process (FIGURE 10.27).

The gases that create the greenhouse effect are commonly called **greenhouse gases**. The most significant of these radiative gases are water vapour (H$_2$O) and

greenhouse gases Gases that absorb infrared radiation, including water vapour, carbon dioxide, methane, nitrous oxide, chlorofluorocarbons, and tropospheric ozone.

carbon dioxide (CO$_2$), which are responsible for most of the *natural* heating influence to Earth's troposphere and surface. Other trace gases that contribute to the greenhouse effect include methane (CH$_4$), nitrous oxide (N$_2$O), ozone (O$_3$), as well as entirely human-made forms including carbon tetrachloride (CCl$_4$) and chloro-flourocarbons (CFCs). Earth's natural temperature and energy budget is actually set around 15°C on average due to these greenhouse gases, about 33°C higher than otherwise would be the case. Without this natural warming influence, the average temperature on the planet would be −18°C and

Greenhouse gas	Natural Sources	Anthropogenic Sources	Pre-Industrial (pre-1750) Concentration	Present Concentration	Average Atmospheric Residence Time (yrs)	Warming Potential
Carbon dioxide (CO_2)	Natural processes of plant and animal life decay	Burning of fossil fuels, deforestation, plant burning	278 ppm	393 ppm	100–120	1
Methane (CH_4)	Natural decay of matter in an oxygen-free environment (wetlands, rice paddies, animal digestive processes)	Gas production and pipelines, natural gas leaks, landfills, agricultural practices, coal production	700 ppb	1721 ppb	12–18	23
Nitrous oxide (N_2O)	Soils and oceans	Fossil fuel combustion, fertilizers, soil cultivation, livestock wastes, nylon production	275 ppb	311 ppb	114–120	310
Chlorofluorocarbons	N/A	Air conditioners, refrigerators, plastic foams	0 ppt	503 ppt	11–20	6200–7100
Hydrochlorofluorocarbons	N/A	Air conditioners, refrigerators, plastic foams	0	105 ppt	9–39 (avg. 15 years)	470–2000

water would freeze, making it impossible for life to exist as we know it.

However, measurements of greenhouse gases indicate that concentrations are generally increasing, mostly due to human activities (TABLE 10.2). Evidence now points to an imbalance in the Earth's energy budget through what has been called the **enhanced greenhouse effect**. Essentially, human activities are turning up the Earth's "thermostat," leading to positive radiative forcing and a number of climate system responses, discussed later in the chapter.

enhanced greenhouse effect The additional warming that results from increased levels of CO_2 and other gases that absorb infrared radiation. Carbon dioxide is the gas of most concern in causing the enhanced warming influence.

Causes of the enhanced greenhouse effect
Not all greenhouse gases behave the same when they are in the atmosphere. Two factors that are important in determining the radiative impact of a greenhouse gas include the amount of gas emitted and the specific properties of the gas. The length of time the gas resides in the atmosphere, or *average residence time*, is important as this will determine whether the gas has the opportunity to become well mixed in the troposphere. Short residence times mean that the gas may not be homogenously mixed in

the air and may not have long to interact and influence radiative forcing. As well, the efficiency or effectiveness in absorbing infrared radiation (or **global warming potential, GWP**) is another important factor since some greenhouse gases are better able to absorb energy than others. Carbon dioxide molecules, for example, remain in the troposphere anywhere from 100 to 120 years, but these molecules are actually very poor absorbers of infrared radiation and are therefore assigned a GWP value equal to 1. If we contrast this with methane molecules, which remain in the atmosphere for only 12 to 18 years but are very effective absorbers of heat energy, the GWP is much higher. Essentially, with some gases a little goes a long way in contributing to a warming influence in the atmosphere. **TABLE 10.2** summarizes features of the greenhouse gases, their sources, and GWP. Water vapour is not affected significantly by human activities and is therefore excluded from the table.

Even though carbon dioxide has a relatively low GWP, it is the greenhouse gas of most concern since it is found in greatest concentration of all greenhouse gases found in the atmosphere. Atmospheric levels of carbon dioxide have been monitored through a number of methods. Researchers can examine air from the ancient atmosphere by studying ice cores that have trapped air bubbles from the time in them (**FIGURE 10.28A** and **B** on page 372).

Since 1958, researchers have been collecting information at the famous Mauna Loa Observatory in Hawaii and more recently by taking measurements at several thousand land-based meteorological stations around the world, collecting data from weather balloons, orbiting satellites, transoceanic ships, and hundreds of sea-surface buoys with temperature sensors. Records confirm that carbon dioxide concentrations have steadily increased from pre-Industrial Revolution concentrations of 280 parts per million measured in 1750 (**FIGURE 10.28C**). In 2010, concentrations were reported to be 393 parts per million, close to 40 percent above the concentration observed prior to the start of the Industrial Revolution. Comparing current concentrations with bubbles of the ancient atmosphere

acquired from ice cores taken from Vostok, Antarctica, reveals that the present CO_2 concentration is much higher than any peaks recorded in the past 400,000 years.

Variability in CO_2 concentrations in the Vostok cores also highlights the strong correlation between CO_2 and temperature. As our modern atmospheric CO_2 concentrations rise rapidly, it is interesting to note that the National Oceanic and Atmospheric Administration (NOAA) has indicated that 11 of the 12 years from 1995 to 2006 were among the warmest years since the mid-1800s.

Recall from Chapter 5 (refer to Figure 5.9 on page 140) that the carbon cycle has numerous sources and sinks moving carbon into and out of the atmosphere. CO_2 dissolves in water and therefore the oceans represent a major sink. Another major sink includes forests and vegetation that utilize CO_2 in photosynthesis to form glucose and organic matter. During summer months in the northern hemisphere, atmospheric CO_2 concentrations reach their annual low because high rates of photosynthesis reduce the atmospheric pool of CO_2. In the winter, concentrations reach their annual peak. This leads to the appearance of the zigzag jaggedness of the line in Figure 10.28C. Organic matter may be stored in soils or sequestered for longer periods in the form of fossil fuels. Respiration returns CO_2 to the atmosphere. We humans release CO_2 back to the atmosphere when we burn these fossil fuels, remove forests, or burn the biomass of forests. We can devise a budget that accounts for all of the sources (or emissions) of CO_2 to the atmosphere and similarly all of the sinks that remove CO_2 from the atmosphere. If in equilibrium, there will be no net buildup of the atmospheric pool.

The Global Carbon Project has done just that. From 2000 to 2008 researchers assessed major CO_2 emissions against natural sinks. Their findings indicated that the natural sinks were only able to reduce the atmospheric concentration by about 57 percent, leaving the excess in the atmosphere to contribute to positive radiative forcing. CO_2 emissions resulted from coal combustion in China, the United States, and India and represented a whopping increase of 29 percent between 2000 and 2008. Land conversion, such as when tracts of tropical forests are logged or burned, represented 12 percent of the total human emissions in 2008, and primarily involved Brazil and Indonesia. The project report reinforced the findings of other researchers who suggested that while oceans are able to hold 50 times more

A Scientists use ice cores taken from glaciers to identify concentrations of CO_2 in the ancient atmosphere.

B A thin cut of an ice core shows the bubbles of ancient atmosphere trapped in dark bubbles between ice grains.

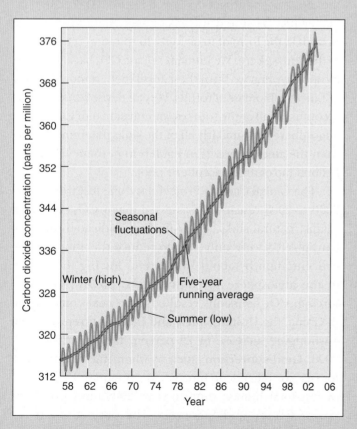

C Note the steady increase in the concentration of atmospheric CO_2 since 1958, when measurements began at the Mauna Loa Observatory in Hawaii. This location was selected because it is far from urban areas where factories, power plants, and motor vehicles emit CO_2. Seasonal fluctuations correspond to rates of photosynthesis in the northern hemisphere. Winter corresponds to elevated levels of CO_2 when vegetation in the northern hemisphere lies dormant, and summer corresponds to high rates of photosynthesis when CO_2 is taken out of the atmosphere. Concentrations have risen steadily over the past 40 years and have reached 393 ppm in 2010.

NATIONAL GEOGRAPHIC

Your carbon footprint is the amount of carbon that is emitted by your activities and lifestyle. It relates to the amount of greenhouse gas that is produced in our daily lives through the burning of fossil fuels for electricity, heating, and transportation. Your carbon footprint is made up of the sum of two parts:

1. Your **primary carbon footprint** is the measure of your direct emissions of carbon dioxide from the burning of fossil fuels including energy consumption and transportation (cars, plane rides, etc.). We have direct control over our primary footprint.

2. Your **secondary carbon footprint** is the measure of the indirect carbon dioxide emissions from the entire lifecycle of the products we use. Simply stated, the more products we buy, the more emissions are produced.

The pie chart shows the main elements that make up a typical person's carbon footprint in a developed country. Take time to think about the ways you contribute to the emissions of greenhouse gases. Here are some ideas of ways you can reduce your carbon footprint:

- Turn off lights, TVs, computers, DVD players, and other electronics when they are not being used. Limit the amount of time in a day that you use electronics.
- Turn down your central heating system and adjust the timing to lower the heating when you are away from the house (at work each day).
- Turn down your water heater.
- Buy energy-efficient appliances.
- Buy energy-efficient light bulbs.
- When possible, install thick insulation in your loft and walls.

- Buy locally produced food.
- Reduce your consumption of meat.
- Recycle grey water.
- Travel less by car. When possible walk, ride a bike, take a bus, or carpool.
- Sign up with a green energy supplier to have your energy supplied from renewable sources.
- Hang clothes out to dry.
- RECYCLE!

To calculate your personal carbon footprint, visit the carbon footprint website at www.carbonfootprint.com

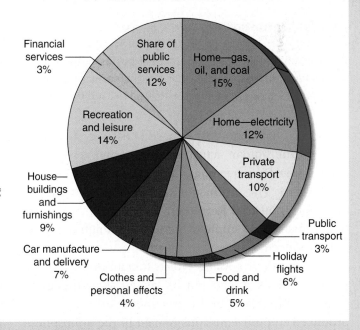

CO_2 than the atmosphere, the current rates of emissions exceed slowing rates of absorption, resulting in a build up of CO_2 in the atmospheric pool.

Cooling of the atmosphere
Another influencing factor on the Earth's energy budget relates to the dynamics of solar energy reflection. When more light is reflected back into space, it creates a general cooling influence. **Aerosols** are fine particles that are suspended in the atmosphere. These aerosols influence climate in one of two ways. First, they may contribute to a scattering effect on sunlight, enhancing reflection directly back into space and preventing this energy from reaching the surface of the Earth. Second, in the lower atmosphere aerosols can modify the size of cloud particles, changing how the clouds reflect and absorb sunlight.

Aerosols are created by both natural and human activities. Volcanic eruptions and desert dust contribute

PHILIPPINES

Global Locator

Volcanic eruption FIGURE 10.29

The eruption of Mount Pinatubo in the Philippines in 1991 injected massive amounts of sulphur into the atmosphere. Because sulphur haze reduces the amount of sunlight reaching the surface, this eruption caused Earth to cool temporarily. Compared with temperatures during the rest of the 1990s, 1992 and 1993 global temperatures were relatively cool.

aerosols to the lower atmosphere (**FIGURE 10.29**). Some of the human activities contributing aerosols include smoke from burning tropical forests and fossil fuel combustion. The production of aerosols has increased since the Industrial Revolution and we now recognize that human-made sulphate aerosols outweigh the naturally produced sources.

Despite their cooling effect, human-produced sulphur emissions should not be viewed as a strategy to counteract increases in the greenhouse effect. Human-produced sulphur emissions remain in the atmosphere for only days, weeks, or months. They do not disperse globally and may cause only regional cooling. Also, as described earlier, many primary and secondary pollutants are created through sulphur oxides, and therefore most nations are trying to reduce their sulphur emissions, not maintain or increase them.

> **albedo** The extent to which an object diffusely reflects light from the sun.

Albedo is a measure of a surface's ability to reflect incoming solar radiation. For example, lighter-coloured areas of the Earth's surface, such as those covered by ice and snow, will generally reflect upward of 80 to 90 percent of this incoming energy. In contrast, darker-coloured areas of the Earth's surface, such as those covered by forests, grass, cities, and oceans, tend to absorb more energy than they reflect, and thus have a much lower albedo (**FIGURE 10.30**).

Human activities such as deforestation, logging, and farming alter the Earth's albedo by altering the amount of absorption by the landscape. For example, the development or expansion of a city contributes to more heating in the urban environment and small heat islands are created. Retreating glaciers and shrinking sea ice reduces the albedo as well. Researchers study the ice-albedo effect to determine the overall influence of retreating sea ice to the radiative forcing on the planet. (See the EnviroDiscovery on page 377.)

EFFECTS OF GLOBAL CLIMATE CHANGE

Global climate change is not simply an increase in temperature as a result of Earth's energy imbalance. The climate system involves numerous interrelated components, meaning that positive radiative forcing results in changes observed in the atmosphere, hydrosphere (and crysosphere), lithosphere, and biosphere through positive and negative feedback loops.

In response to the growing consensus that the Earth is undergoing a rapid and accelerated change in climate,

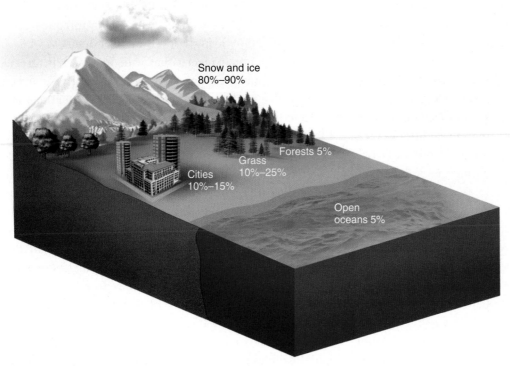

Snow and ice
80%–90%

Forests 5%

Grass
10%–25%

Cities
10%–15%

Open
oceans 5%

The albedo of snow and ice is highest, meaning that these surfaces reflect a large amount of solar radiation back into space. As the ice shelves in the Arctic and Antarctica melt, the amount of solar radiation that is absorbed by the surface of the Earth is increasing.

governments around the world organized the Intergovernmental Panel on Climate Change (IPCC). With input from hundreds of climate experts, the IPCC provides the definitive scientific statement about global climate change.

In 2007, the IPCC projected a 1.8°C to 4°C increase in global temperature by the year 2100. Canada may experience upward of a 10°C increase as a result of enhanced greenhouse gases and radiative forcing. The IPCC predicts higher temperatures and an increase in the number of hot days over nearly all land areas, fewer frost days, and fewer cold days.

But temperature is not the only climatic factor expected to be altered by the enhanced greenhouse effect. The additional input of energy is expected to result in more frequent and severe weather patterns that will impact precipitation, with numerous consequences to human well-being and other organisms as well. Computer models of future weather conditions associated with global climate change predict that precipitation patterns will change, causing some areas to have more frequent and severe droughts and other areas to have heavier snow and rainstorms, which may cause more frequent flooding. Coastal regions may experience stronger hurricanes, accelerated coastal erosion, and possible breaching of levies and dykes.

Models predict that the availability of fresh water will be impacted by changes in precipitation, particularly in areas that are currently arid or semi-arid, such as the Sahel region just south of the Sahara Desert. With less annual snowfall and a rapid melt in the spring, the availability of proven water resources may be of increasing concern even in Canada where our current supplies are abundant.

Changes in Antarctica and the Arctic

At the Poles, researchers are studying the dynamics of sea ice to predict how these polar regions may respond to climate change. One aspect scientists are examining in particular is the ice-albedo effect. Ice has a high albedo and reflects most of the solar energy emitted from the sun. The conversion of sea ice to barren rock or open ocean, however, changes the surface from one that is largely reflective to an expanse of heat absorbing surfaces. Computer models suggest that the overall temperature increase may be 4°C to over 7°C in the Arctic by 2100, depending on greenhouse gas emissions. This would be a more rapid and pronounced warming than found in other regions of the planet and has been called **polar amplification**. Polar amplification arises through positive feedbacks where the climate forcing that initiated the process of sea ice retreat

is reinforced by increased energy absorption due to the loss of the reflective ice surfaces.

Over the course of the past 30 years, scientists have been monitoring the expanse of sea ice in the Arctic to further understand the dynamics of sea ice thinning and loss (**FIGURE 10.31**). The worst year on record for measured annual sea ice minimum was 2007. While scientists had hoped that recovery was being seen in sea ice development in 2008 and 2009, they are finding that the perennial ice that typically remains year after year regardless of the summer melt has been wasted and exists only as a much thinner sheet, and that it is easily broken apart by ocean turbulence.

Another issue associated with Antarctica and the Arctic is the melting of the polar ice caps. One of the anticipated effects of increased temperatures in the Arctic and Antarctica is a rise in sea level. Records indicate that during the 20th century the sea level rose between 10 to 20 centimetres, and researchers estimate that it could rise an additional 48 centimetres by 2100. Such a rise could have devastating effects on sensitive coastal regions in Atlantic Canada and parts of the Beaufort Sea coast. Storm surges could result in flooding along Prince Edward Island, Newfoundland and Labrador, and Nova Scotia, and dykes constructed to protect communities could be breached by high tides. Coastal erosion rates are expected to increase since storms along the coast will occur in greater frequency and duration. Small island nations such as the Maldives in the Indian Ocean are considered highly vulnerable to a rise in sea level, as storm surges could easily sweep over entire islands. Other vulnerable countries—such as Bangladesh, Egypt, Vietnam, and Mozambique—have dense populations living in low-lying river deltas.

There is also concern that with increases in greenhouse gas concentrations and loss of sea ice in the Arctic there will be both a warming influence on oceanic water and a reduction in saltiness of the northern North Atlantic Ocean. Computer models depict a weakening or complete stop to the process of the ocean conveyor belt formation that is important in the transport of energy from the equator to the higher latitudes. The measurements of the level of freshness observed in the North Atlantic Ocean over the last 40 years indicate that the region has indeed gradually been showing a reduction in saltiness. If this continues and the oceanic conveyor belt stalls, the average temperatures in northern Europe may drop by about 5°C. To illustrate the significance of this decline in temperature, we can compare this same difference in temperature to that which occurred between the global average today and the temperature during the last Ice Age when Canada was covered in glaciers over 3000 metres thick.

Effects on agriculture Global climate change will also increase problems for agriculture. The rise in sea level will inundate some river deltas, which are fertile agricultural lands. Certain agricultural pests and

Dwindling Arctic sea ice FIGURE 10.31

A Minimum sea ice concentration for the year 1979

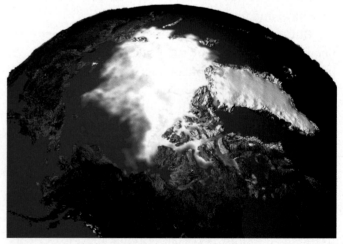

B Minimum sea ice concentration for the year 2003

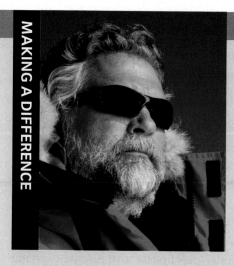

DAVID BARBER

David Barber, a sea ice specialist, has made enormous contributions to the research, protection, and knowledge of Canada's Arctic environment. Barber is a lead investigator on the Canadian Arctic Shelf Exchange Study (CASES), which brings together over 400 scientists from 11 different countries to study the arctic ecosystem and the thinning and disappearing ice of the Arctic Ocean resulting from climate changes.

Barber's research involves the effects of circumpolar flaw leads (CFL) that are created when arctic ice packs move away from coastal ice, leaving areas of open water. During the winter in northern polar regions of the world, the sea ice is covered by snow. In the summer, the sun is higher in the sky and the temperature warms the snow, causing it to melt. The melting snow exposes bare sea ice, leads (cracks), and the ocean water, which all tend to enhance energy absorption, resulting in a decrease in albedo and more solar energy being absorbed. Through positive feedback, the process results in further snow and ice melt. This positive feedback and polar amplification is one of the primary reasons that the Arctic is more sensitive to changes in temperature than other geographic areas.

Barber's efforts have contributed to the establishment of remote observatories as a part of an initiative known as ArcticNet, a collaborative collection of scientists, Inuit people, northern communities, and governments studying the impacts of climate change on coastal Canadian Arctic (see the EnviroDiscovery box below on ArcticNet). Dr. Barber is the Canada Research Chair in Arctic Systems Science and is a member of the Research Management Committee for ArcticNet.

EnviroDiscovery   ARCTICNET

In 2002, an iceberg roughly twice the size of Prince Edward Island broke off from the Antarctic Peninsula. The Antarctic Peninsula has experienced a 3°C temperature warming over the last century, an unprecedented increase in such a short period of time. This temperature increase has led to the retreat and collapse of several ice shelves with at least six ice shelves lost completely. Climate change is suggested to be causing these massive ice shelves to disintegrate and break apart at a far more rapid rate than predicted (see graph). Studies of the Arctic have revealed similar findings. Since the 1970s the ice-covered ocean in the Arctic has retreated, and the remaining ice has thinned rapidly, losing 40 percent of its volume in less than three decades.

Research is ongoing to study the impact of climate change in the Arctic and Antarctic. In Canada, this research has taken the shape of the multidisciplinary ArcticNet team. ArcticNet brings together the knowledge and expertise of dozens of scientists, managers, northern communities, and government leaders with linkages extending to the international community. Through such research, we are learning more about the impacts of climate change on the biological, ecological, geological, economical, political, and sociological world in which we live. It is hoped that such findings will contribute to the establishment of policies and strategies in Canada and internationally to mitigate and adapt to climate change.

Visit www.arcticnet.ulaval.ca to learn more about this important Arctic research program.

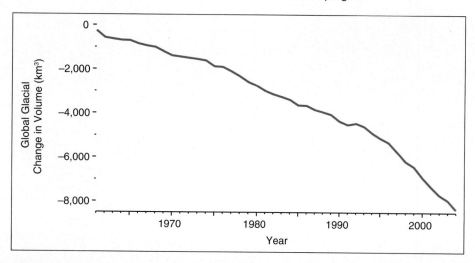

disease-causing organisms will probably proliferate and reduce crop yields. The likely increase in the frequency and duration of droughts will be a particularly serious problem for countries with limited water resources. Lack of water for drinking and agriculture may force millions of people to relocate.

Increasingly warmer temperatures in Canada may lengthen the growing season and may also introduce production possibilities in northern areas previously unsuitable for agriculture. Coupled with these changes are a few predicted adverse effects, such as reduced moisture and increased rates of evaporation. These factors could result in severe droughts, and hot temperatures will create favourable conditions for a variety of weeds and insects. Overall, these predicted changes indicate lower crop yields. In addition, the changes will also negatively impact livestock production. Insects and disease previously not found in prairie regions would move into these areas with climate change. Livestock may experience less cold stress during winter months, but they will experience greater heat stress during the summer.

Warmer and drier temperatures will force the development of drought-resistant crop varieties. Agriculture, which is currently the largest consumer of water in the Canadian Prairies, will be pressured to use more water to respond to changing rainfall, evaporation, runoff, and soil moisture storage conditions that are predicted to accompany climate change in the future. The water crisis already apparent in the world will be further stressed by these predicted changes. Agriculture and Agri-Food Canada estimates that agriculture accounts for close to 10 percent of Canada's total greenhouse gas emissions with livestock and manure accounting for 58 percent of these gases and crops accounting for 37 percent of these gases.

Ecological effects

Virtually all species on Earth have been or will be affected by global climate change. Species such as weeds, insect pests, and disease-carrying organisms already common in a wide range of environments will come out as winners, with greatly expanded numbers and ranges. Canadian forests have already suffered from one insect infestation, the mountain pine beetle. The mountain pine beetle was previously found only in the southern portion of British Columbia, since other regions further north were climatically unsuitable for the beetles. However, they have now infested forests in Alberta, parts of Saskatchewan, and the western United States (**FIGURE 10.32**). The British

Mountain pine beetle damage in British Columbia FIGURE 10.32

The dark brown trees are those that have been killed by mountain pine beetle infestation, and you can see that close to the entire forest area has been devastated.

Columbia Ministry of Forests and Range aerial survey detected 9.2 million hectares of pine beetle infestation in 2006. The implications of mountain pine beetle infestation across Canadian forests would be devastating to the economy, aesthetics, and environment, as well as to all of the interrelated species within the ecosystem that rely on these trees.

Even human health appears to be affected by global climate change. Currently, most evidence linking climate warming to disease outbreaks is circumstantial, so scientists are reluctant to ascribe cause-and-effect relationships. Nonetheless, data linking climate warming and human health problems are accumulating. More frequent and more severe heat waves during summer may increase the number of heat-related illnesses and deaths. Climate warming may also affect human health indirectly. Mosquitoes and other disease carriers could expand their range into the newly warm areas and spread malaria, dengue fever, schistosomiasis, and yellow fever (FIGURE 10.33). According to the World Health Organization, during 1998, the second warmest year on record, the incidence of malaria, Rift Valley fever, and cholera surged in developing countries.

An increasing number of studies report measurable changes in the biology of plant and animal species as a result of climate warming. Climate change also affects populations, communities, and ecosystems. In "Visualizing the Effects of Global Climate Change" (on page 380), we report on the results of several of the hundreds of studies conducted thus far.

Rising temperatures in the waters around Antarctica have led to a decline in the population of shrimp-like krill, and Antarctic silverfish, major food sources for Adélie penguins. This decline has contributed to a reduction in Adélie penguin populations (FIGURE 10.34A on page 380). Warmer temperatures also cause higher rates of reproductive failure in these penguins by producing wet, snowy conditions that kill the developing chick embryos at egg-laying sites.

Worldwide, many frog populations have plummeted; these include Puerto Rico's national symbol, the tiny tree frog known as coqui (FIGURE 10.34B). Warmer temperatures and more frequent dry periods have stressed the coqui, making them more vulnerable to infection by a lethal fungus.

Ecosystems considered at greatest risk of climate-change loss are polar seas, coral reefs, mountain ecosystems, coastal wetlands, and tundra. Water temperature increases of 1°C to 2°C cause coral bleaching, which contributes to the destruction of coral reefs (FIGURE 10.34C). In 1998, when tropical waters were some of the warmest ever recorded, about 10 percent of the world's corals died.

As atmospheric CO_2 levels continue to rise, the level of CO_2 that dissolves in the ocean increases. Dissolved CO_2 produces carbonic acid, which acidifies the water (FIGURE 10.35 on page 380). This acidification could be disastrous for shelled sea animals, particularly zooplankton at the base of the marine food web; the acid would alter the availability of shell building minerals such as calcium carbonate, making shells thinner and weaker.

Sick with malaria FIGURE 10.33

In a climate-warmed world, the mosquito that spreads malaria could expand into temperate areas. Photographed in Ariquemes, Brazil, where most of the town's inhabitants suffer from malaria.

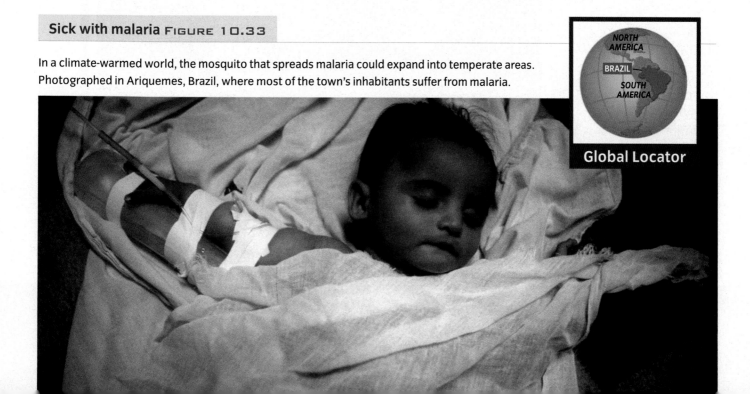

Global Locator

Global climate change affects organisms FIGURE 10.34

A Warmer temperatures in Antarctica threaten the Adélie penguin's food supply and reduce its reproductive success.

B A coqui tree frog in Puerto Rico. These once ubiquitous little frogs have become rarer, an indirect casualty of climate change.

C Ocean warming stresses corals, causing them to become bleached. Photographed near the Maldives in the Indian Ocean.

Acidification of the ocean FIGURE 10.35

Scientific models project that ocean water will become increasingly acidic if human-produced CO_2 levels continue to rise. Shown are computer models for 1995 and 2100 from NOAA Pacific Marine Environmental Laboratory.

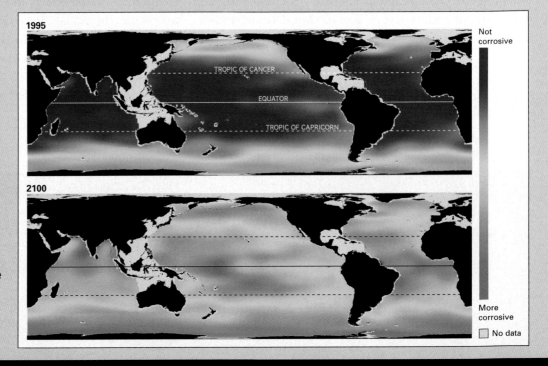

1995

TROPIC OF CANCER

EQUATOR

TROPIC OF CAPRICORN

2100

Not corrosive

More corrosive

☐ No data

NATIONAL GEOGRAPHIC

SPECIFIC WAYS TO DEAL WITH GLOBAL CLIMATE CHANGE

All greenhouse gases will have to be dealt with as we develop strategies to address global climate change, but we focus on CO_2 because it is produced in the greatest quantity and has the largest total effect. Carbon dioxide has an atmospheric lifetime of more than a century, so emissions produced today will still be around in the 22nd century. The extent and severity of global climate change will depend on the amount of additional greenhouse gas emissions we add to the atmosphere. Many studies make the assumption that atmospheric CO_2 will stabilize at 550 ppm, which is roughly twice the concentration of atmospheric CO_2 in the preindustrial world and almost 50 percent higher than the CO_2 currently in the atmosphere.

There are basically two ways to manage global climate change: mitigation and adaptation. *Mitigation* is the moderation or postponement of global climate change through measures that buy us time to further our understanding of global climate change and to pursue more permanent solutions. *Adaptation* is a response to changes caused by global climate change. Adopting adaptation strategies for climate warming implies that global climate change is unavoidable.

Mitigation of global climate change

Climate warming is essentially an energy issue. Developing alternatives to fossil fuels offers a solution to warming caused by CO_2 emissions and addresses the dilemma of dwindling supplies of fossil fuels. Alternatives to fossil fuels include solar energy and nuclear energy.

Increasing the energy efficiency of automobiles and appliances—thus reducing the output of CO_2—would help mitigate global climate change. Energy-pricing strategies, such as carbon taxes and the elimination of energy subsidies, are other policies that could mitigate global climate change. Most experts think using current technologies and developing such policies would significantly reduce greenhouse gas emissions with little cost to society. One promising avenue of mitigation is the development of renewable energy technologies, discussed further in Chapter 11.

Planting and maintaining forests also mitigates global climate change. Like other green plants, trees remove carbon dioxide from the air and incorporate the carbon into organic matter through photosynthesis. Reasonable estimates suggest that trees could remove 10 percent to 15 percent of the excess CO_2 in the atmosphere, but only through enormous plantings, so such efforts should not be considered a substitute for cutting emissions of greenhouse gases.

Many countries are investigating **carbon management**. Several power plants currently capture the CO_2 in their flue gases, but the technology is new. Technological innovations that more efficiently trap CO_2 from smokestacks would help mitigate global climate change and yet allow us to continue using fossil fuels (while they last) for energy. The carbon could be sequestered in geologic formations or in depleted oil or natural gas wells on land. **FIGURE 10.36** on page 382 summarizes several ways to mitigate global climate change.

> **carbon management**
> Ways to separate and capture the CO_2 produced during the combustion of fossil fuels and then sequester (store) it.

Trees are another important tool in the fight against climate change. Carbon is fixed in a variety of tree species and they are extremely effective in sequestering greenhouse gases. A mature poplar tree, as an example, traps 266 kilograms of carbon. Following shelterbelt spacing for trees means that 106 tonnes per kilometre of carbon can be sequestered by a poplar tree shelterbelt. On top of this, carbon is also sequestered in the trees roots, which may equal close to 50 to 75 percent of these amounts. In addition, shelterbelts reduce energy usage by protecting homes from the wind and reducing heating bills by as much as 25 percent!

Researchers are also examining ways to reduce greenhouse gas emissions on the farm. While reducing soil tillage activity reduces carbon emissions from machinery, studies have also shown that soils will benefit from increased organic matter, which acts as a sink for carbon dioxide and ultimately contributes to an increase in carbon storage. Other interesting benefits come from the use of perennial forage crops, enhanced efficiency of fertilizer usage, and livestock feeding that may reduce methane production.

In December 2009 the United Nations Climate Change Conference was held in Copenhagen, Denmark.

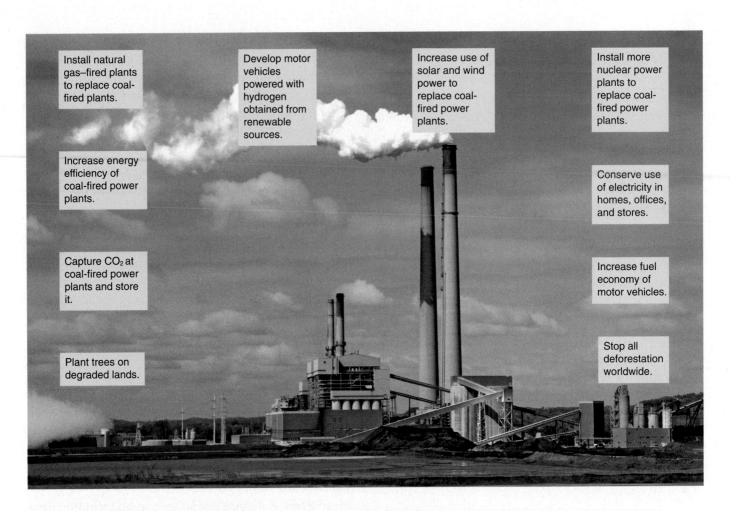

Install natural gas–fired plants to replace coal-fired plants.

Develop motor vehicles powered with hydrogen obtained from renewable sources.

Increase use of solar and wind power to replace coal-fired power plants.

Install more nuclear power plants to replace coal-fired power plants.

Increase energy efficiency of coal-fired power plants.

Conserve use of electricity in homes, offices, and stores.

Capture CO_2 at coal-fired power plants and store it.

Increase fuel economy of motor vehicles.

Plant trees on degraded lands.

Stop all deforestation worldwide.

Some of the many strategies suggested for mitigation of global climate change FIGURE 10.36

To be effective these strategies will have to be enacted on a global scale. For example, replacing coal-fired power plants with those powered by natural gas (upper-left corner) would have to involve 1400 or more coal-fired plants around the world.

The conference, commonly referred to as the Copenhagen Summit, gathered representatives from over 170 countries with the hope that a framework for climate change mitigation beyond 2012 would be agreed upon. Canada's commitment under the Copenhagen Accord was a 17 percent reduction of greenhouse gases from 2005 levels. Unfortunately, the Copenhagen Summit was widely criticized for creating a document that does not contain any legally binding commitments for reducing CO_2 emissions. Many Canadians voiced their frustration about the agreement and their fears that the commitments to emissions reductions are not only inadequate but will likely not even be met.

Adaptation to global climate change Because the overwhelming majority of climate experts think human-induced global climate change is inevitable, government planners and social scientists are developing strategies to help various regions and sectors of society adapt to climate warming. One of the most pressing issues is rising sea level. People living in coastal areas could be moved inland, away from the dangers of storm surges, although the societal and economic costs would be great. Another extremely expensive alternative is the construction of massive sea wells to protect coastal land. Rivers and canals that spill into the ocean could be channelled to prevent salt water intrusion into fresh water and agricultural land.

We must also adapt to shifting agricultural zones. Countries with temperate climates are evaluating semi-tropical crops to determine the best substitutes for traditional crops as the climate warms. Large lumber companies are developing heat- and drought-resistant strains of trees that will be harvested when global climate change may be well advanced. Evaluating such problems and finding and implementing solutions now will ease future stresses of climate warming.

CONCEPT CHECK STOP

What is the enhanced greenhouse effect and how does it affect global climate?

What is the sea-ice albedo effect and why does it contribute to an acceleration of climate change?

What are two examples of each of the approaches to global climate change: mitigation and adaptation?

Ozone and Ozone Loss in the Stratosphere

LEARNING OBJECTIVES

Describe the importance of the stratospheric ozone layer.

Explain how stratospheric ozone is created and describe ozone-depleting reactions.

Relate the harmful effects of ozone depletion to humans and ecosystems.

Tropospheric ozone (O_3) has been described thus far as a human-made pollutant. About 90 percent of total ozone, however, actually exists as a naturally produced component in the stratosphere, found in greatest concentrations from about 20 to 40 kilometres above Earth's surface. The popularly named "ozone layer" is essential to the existence of life as it shields Earth's surface from the high-energy harmful incoming **ultraviolet (UV) radiation** that causes mutations and cataracts and impairs immunological defences in exposed biota.

Ozone is continually produced through reactions of oxygen and ultraviolet light in the stratosphere. Oxygen molecules (O_2) are spliced by ultraviolet radiation in the initial step of the reaction producing individual oxygen atoms (O). This reactive O atom binds with oxygen molecules (O_2) to produce **ozone** molecules ($O + O_2 \rightarrow O_3$) in the second stage of the reaction. In terms of total production, the highest rates

ozone Three oxygen atoms bonded together through reactions driven by the presence of ultraviolet radiation in the stratosphere. Ozone occurs naturally in the stratosphere as the "ozone layer" shielding the Earth's surface from UV radiation.

occur at the equator in the location of highest incoming solar intensity. Atmospheric winds, however, redistribute stratospheric ozone and this accumulates more at the mid to high latitudes of the South and North Poles.

A slight ozone thinning occurs naturally over Antarctica for a few months each year. In 1976, however, the thinning was observed to be greater than it should have been if natural causes were the only factor contributing to the loss. In 1983, the Antarctic thinning lead scientists to believe their instruments must have been broken as ozone levels dipped to very low levels. By 2000, the thinning reached a record size of 29.2 million square kilometres (larger than the North American continent). The phenomenon has since been coined the "ozone hole" (**FIGURE 10.37A** on page 384). A smaller thinning has also been detected in the stratospheric ozone layer over the Arctic, and ozone levels over Europe and North America have dropped by about 10 percent since the 1970s (**FIGURE 10.37B**).

CAUSES OF OZONE DEPLETION

The chemicals responsible for **ozone depletion** in the stratosphere are a group of industrial and commercial human-made halogen compounds containing reactive chlorine (Cl) and chlorine monoxide (ClO) as well as bromine (Br) and bromine monoxide (BrO). The most common of these ozone-depleting substances are **chlorofluorocarbons (CFCs)**, which were first produced in 1928 as refrigerants. Freon has been used as a propellant for aerosol cans and coolants in air conditioners and refrigerators. Other CFCs have been used as solvents and as foam-blowing agents for insulation and packaging (Styrofoam, for example). Bromine is produced in

> **ozone depletion**
> The destruction of stratospheric ozone by human-produced chlorine and bromine containing chemicals.

the oceans but a significant amount was used to make fumigants. Methyl bromide is also released through biomass burning either naturally or through human activities.

These chemicals were easily and cheaply produced, chemically stable, and versatile. First used in the 1960s, their popularity grew and about 320,000 tonnes of CFCs were used worldwide in 1998. When scientists revealed their ozone-destroying qualities, the products were banned from further use. Additional ozone-depleting substances include *halons,* used as fire retardants, and *methyl chloroform* and *carbon tetrachloride,* used as industrial solvents.

Chemists began to research the reactions that were destroying the ozone to discover the major processes. Once in

Visualizing

Ozone depletion FIGURE 10.37

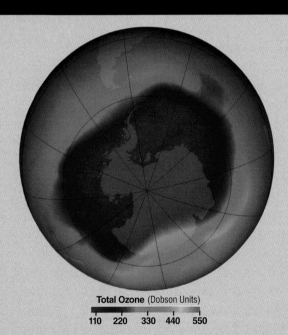

Total Ozone (Dobson Units)

110   220   330   440   550

A A computer-generated image of part of the southern hemisphere, taken in December 2007, reveals ozone thinning (the purple area over Antarctica) following destructive reactions that occur during the Antarctic polar spring in October. The ozone-thin area is not stationary but moves about as a result of air currents.

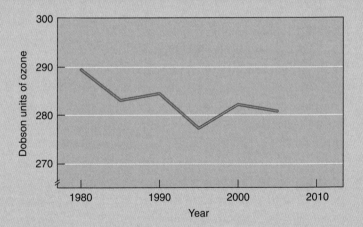

B World levels of stratospheric ozone (measured as Dobson units), 1980 to 2005.

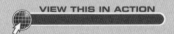
VIEW THIS IN ACTION

NATIONAL GEOGRAPHIC

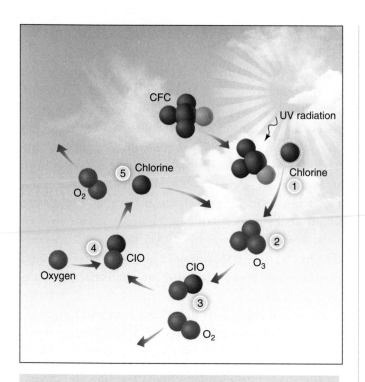

Chemistry of ozone destruction Figure 10.38

UV radiation breaks off a chlorine atom (1), which then breaks up an ozone molecule (2) releasing an ordinary oxygen molecule and a chlorine monoxide molecule (3). The chlorine monoxide molecule bonds with a free oxygen atom (4) releasing the chlorine atom (5), which is now free to break up another ozone atom.

the atmosphere, CFCs appear to drift slowly into the stratosphere, and the chlorine is released by ultraviolet radiation. Chlorine (or bromine) then reacts with ozone to produce chlorine monoxide (ClO) (or bromine monoxide; BrO) and an oxygen molecule ($Cl + O_3 \rightarrow ClO + O_2$). This is shortly followed by the reaction of ClO (or BrO) with an oxygen atom (O) producing an additional oxygen molecule ($ClO + O \rightarrow Cl + O_2$). The end result of this **catalytic reaction** is the destruction of an ozone molecule (O_3), the production of oxygen (O_2), and the reformation of the catalyst chlorine (or bromine) atoms (**Figure 10.38**). This reaction is termed *catalytic* since the reactive chlorine (or bromine) is reformed at the end of the reaction and available to undergo continued ozone-destroying reactions repeatedly in the stratosphere.

catalytic reaction Following the destructive reactions of ozone with reactive chlorine (or bromine), the halogen is reformed and continues to be available to undergo reactions. Ultimately one chlorine molecule can destroy hundreds of ozone molecules in the process.

Scientists believe that one chlorine (or bromine) atom has the potential to undergo hundreds of reactions before reacting with other gases that remove it from the stratosphere.

Researchers were not sure why the depletion was consistently observed and destruction so severe in Antarctica in spring following the polar night. We now recognize that there are unique weather properties over the South Pole that facilitate ozone hole formation. During the polar winter, extremely frigid temperatures ($-78°C$ or lower) that prevail for upward of six months produce ice bearing *polar stratospheric clouds* (PSCs). These PSCs represent highly reactive surfaces for the formation of chlorine monoxide during the polar winter. Following the polar night and the reappearance of solar energy in spring, the chlorine monoxide undergoes repeated ozone-destroying reactions. A polar vortex assists in the formation of the ozone hole by isolating the air mass and preventing the mixing of ozone transported from tropical latitudes for an extended period of time.

EFFECTS OF OZONE DEPLETION

With depletion of the ozone layer, higher levels of UV radiation reach the surface of Earth. Increased levels of UV radiation may disrupt ecosystems. For example, the productivity of Antarctic phytoplankton, the microscopic drifting algae that are the base of the Antarctic food web, has declined from increased exposure to UV radiation. (The UV radiation inhibits photosynthesis.)

Biologists have documented direct UV damage to natural populations of Antarctic fish. A possible link also exists between the widespread decline of amphibian populations and increased UV radiation. Because organisms are interdependent, the negative effect on one species has ramifications throughout the entire ecosystem.

Excessive exposure to UV radiation is linked to several health problems in humans, including eye cataracts, skin cancer (**Figure 10.39** on page 386), and weakened immunity. Malignant melanoma, the most dangerous type of skin cancer, is increasing faster than any other type of cancer.

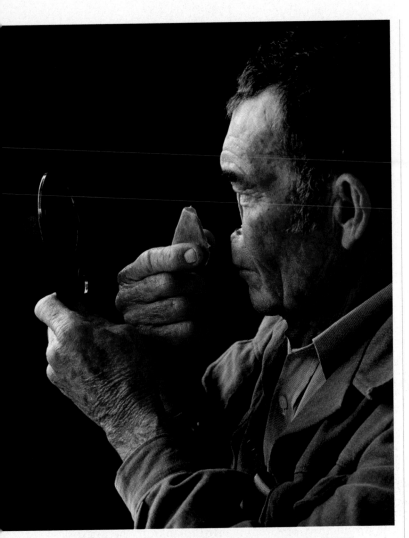

Exposure to UV radiation and skin cancer

FIGURE 10.39

This Australian worked outdoors throughout his career, and it was perhaps inevitable that he contracted skin cancer. Here he attaches an artificial nose where doctors removed a large melanoma.

In 1992, Environment Canada developed a method to predict the strength of the sun's UV radiation known as the UV index. The UV index is now an international standard of measurement identifying the sun's UV radiation for specific locations and on specified dates. This value ranges from 0 (low risk) to 11+ (extreme risk) and is reported regularly. Public health organizations generally promote that people should protect themselves any time the UV index records a rating of 3 or higher.

RECOVERY OF THE OZONE LAYER

In 1987, representatives from numerous countries met in Montreal to sign the **Montreal Protocol**, an agreement that originally stipulated a 50 percent reduction of CFC production by 1998. Despite this effort, stratospheric ozone continued to thin over the heavily populated mid-latitudes of the northern hemisphere, and the Montreal Protocol was modified to include even stricter limits on CFC production. By 1996 there were 161 countries participating in the Protocol, which grew to 191 countries by the 20th anniversary of the agreement in 2007. The Montreal Protocol has been hailed as perhaps the most successful international treaty negotiated to date.

Industrial companies that manufacture CFCs quickly developed substitutes, such as hydrofluorocarbons (HFCs) and hydrochlorofluorocarbons (HCFCs). HFCs do not attack ozone, although they are potent greenhouse gases. HCFCs attack ozone but are less destructive than the chemicals they are replacing.

CFC, carbon tetrachloride, and methyl chloroform production was almost completely phased out in Canada and other developed countries in 1996 and in 2005 in developing nations. HCFCs will be phased out in 2030.

Unfortunately, CFCs are extremely stable and will probably continue to deplete stratospheric ozone for several decades since they have an average residence time of 100 years or longer in the atmosphere. Human-exacerbated ozone thinning will reappear over Antarctica each year, although the area and degree of thinning will gradually decline over time until full recovery, which is anticipated to take place sometime after 2050.

CONCEPT CHECK STOP

Why is the stratospheric ozone layer important to us?

How is the ozone layer destroyed?

Why is ozone depletion a problem for humans and ecosystems?

Most environmental research focuses on one single issue, such as acid deposition, global climate change, or ozone depletion. In the past few years, however, some researchers have been exploring the interactions of all three problems simultaneously. One recent study of such interactions found that North American lakes may be more susceptible to damage from UV radiation than the thinning of the ozone hole would indicate. The reason? The organic matter in the lakes, which absorbs some UV radiation and protects the lakes' plant and fish life, is affected by acid deposition and global climate change. Acid deposition reacts with organic matter in lakes, causing it to settle to the lake floor, where it does not absorb as much of the UV radiation as it once did. And a warmer climate increases evaporation, which reduces the amount of organic matter washed into lakes by streams.

Several studies report a link between human-caused climate warming and polar ozone depletion. Greenhouse gases that warm the troposphere also contribute to stratospheric cooling, presumably because heat trapped in the troposphere acts like a blanket preventing heat to radiate up in altitude. The more greenhouse gases, the thicker the insulating blanket preventing warm air from entering into the stratosphere. The stratospheric temperature has been cooling for the past several years, and these lower temperatures provide better conditions for polar stratospheric clouds to form that contributes to ozone-destroying reactions. Record ozone holes over Antarctica are attributed to cooler stratospheric temperatures. Some scientists speculate that if the cooling trend in the stratosphere continues, recovery of the ozone layer may be delayed. This means climate warming could prolong ozone depletion in the stratosphere despite the success of the Montreal Protocol.

Scientists now know environmental problems can't be studied as separate issues because they often interact in surprisingly subtle ways. As global climate change, ozone depletion, and acid deposition are studied further, it is likely that other interactions will be discovered.

ECO CANADA | **CAREER FOCUS**

CLIMATOLOGIST

Imagine you are standing at the top of a ridge in one of the windiest places in Canada. You are a climatologist and you are here as part of a team of consultants working on a project for an energy development company. This company is interested in building a wind farm in the area that will generate electricity for residential and industrial consumers in the southern part of the province. Before it starts building million-dollar turbines and generators, the company needs to know if this location is windy enough for a farm. It's hired you to analyze long-term climate data from the area and determine if, where, and how the farm should be built.

As a climatologist, you have studied all kinds of weather phenomena, but your specialty is wind. In this case, you're looking for the kind of wind that can be used to produce electricity. You have heard from area residents just how windy the area can be, but you will also use long-term data and computer models to evaluate the wind's quantitative properties. For example, weather satellites and a nearby meteorological station will provide data that will help you determine how many days a year on average the wind blows at a speed high enough to turn the blades of a large turbine. Since the meteorological station has records for the last 65 years, you can use that information to detect trends and variability in the wind. You will also use computer models based on this data to determine exactly where the turbines should be placed on the ridge so that they receive the strongest wind. Finally, since these huge turbines don't spin like your average garden windmill, you must also determine which direction

Climatologists use long-range weather data to study trends, understand causes, and make predictions about climate change.

they should face in order to maximize effectiveness. After several months of research and data analysis, you will produce a report for the energy company on the feasibility of its wind farm based on the occurrence, strength, and prevailing direction of local winds.

Climatology is the study of weather patterns and the processes that cause them. Climatologists use long-term meteorological data such as temperature, wind speed, and precipitation to study trends, understand causes, and make predictions. Often confused with meteorologists, who study short-term weather patterns, climatologists forecast weather changes over the span of years rather than days. Considering that long-term climate change affects every aspect of our lives and our environment, the work of climatologists is crucial to sustaining and preserving the world's ecosystems.

For more information or to look up other environmental careers go to www.eco.ca.

INTERNATIONAL HUMAN DIMENSIONS PROGRAMME ON GLOBAL ENVIRONMENTAL CHANGE (IDHP)

Management of the atmospheric environment is challenging since it requires management on a global scale. Programs such as the International Human Dimensions Programme on Global Environmental Change (IDHP), an international, interdisciplinary science program, promotes the scientific and social research on global environmental change. The IDHP investigates issues such as climate change, the global carbon cycle, the role of institutions in global environmental change, human security, sustainable productivity, consumption issues, and urbanization.

Initiated in the 1990s and extending into the beginning of the 21st century, the Institutional Dimensions of Global Environmental Change (IDGEC), a long-term international research project developed under the IDHP, investigated the role of institutions in the management of the environment. The IDGEC has promoted and inspired research investigating the ways in which institutions cause and alleviate global environmental problems. Institutions shape the behaviours and outcomes of human activities,

and extensive analysis has proven that institutions can make a significant difference to climate change and other environmental factors, proving that governance without government can be effective and favourable in certain circumstances.

Decision making and policy creation at multiple levels is difficult; however, there is an increasing awareness of global climate change and overall loss of global biodiversity, and institutional governance is necessary to determine the cause and confront the changes necessary to address global environmental issues. Adaptive co-management is the combination of efforts that promotes multi-level institutional linkages and shared responsibility between a diversity of players and knowledge sources, and is promoted as a global management strategy in governance of the environment. Public, private, government, and corporate partnerships are necessary between states, provinces, and countries in multilateral agreements to increase the activity and importance of policy to manage the environment at all political levels.

SUMMARY

1 Features of the Atmosphere

1. The troposphere, the layer of atmosphere closest to Earth's surface, extends to a height of approximately 12 kilometres. Temperature decreases with increasing altitude, and weather occurs in the troposphere. In the stratosphere, ozone absorbs much of the sun's UV radiation. The mesosphere, directly above the stratosphere, has the lowest temperatures in the atmosphere. The thermosphere has steadily rising temperatures and gases that absorb x-rays and short-wave UV radiation. The thermosphere reflects outgoing radio waves back toward Earth without the aid of satellites.

2. Sunlight is the primary (almost sole) source of energy available in the biosphere. The sun's energy runs the hydrologic cycle, drives winds and ocean currents, powers photosynthesis, and warms the planet. Of all the solar energy that reaches Earth, 31 percent is immediately reflected away, and the remaining 69 percent is absorbed. Ultimately, all absorbed solar energy is radiated into space as infrared radiation, electromagnetic radiation with wavelengths longer than those of visible light but shorter than microwaves; it is perceived as invisible waves of heat energy. When the absorbed energy is the same as re-radiated energy, there are no changes in the Earth's energy budget.

3. Convective cells develop through a process of surface heating that radiates energy to the air directly above. Warmer air is less dense and rises in altitude. At the tropopause, air is deflected to areas of low pressure caused when cold dense air descends toward the planet's surface. At the surface, air moves toward the low pressure created by the ascending column of warm air. Through this process, heat is radiated to the atmosphere and moisture is transported to other locations.

4. **Weather** and **climate** represent short- and long-term trends in temperature, air pressure, precipitation, solar input, cloudiness, and humidity. The climate system consists of the sun as the engine that powers the system; the oceans that heat and cool and transport energy; and the hydrosphere, lithosphere, biosphere, and atmosphere.

2 Air Quality and Air Pollution

1. **Air pollution** consists of various chemicals (gases, liquids, or solids) present in the atmosphere in levels high enough to harm humans, other organisms, or materials. **Primary air pollutants** are harmful chemicals that enter the atmosphere directly due to either human activities or natural processes; examples include carbon oxides, nitrogen oxides, sulphur dioxide, particulate matter, and hydrocarbons. **Secondary air pollutants** are harmful chemicals that form in the atmosphere when primary air pollutants react chemically with each other or with natural components of the atmosphere; ozone and sulphur trioxide are examples.

2. Major classes of air pollutants include carbon monoxide, sulphur oxides, nitrogen oxides, particulates, tropospheric (ground-level) ozone, volatile organic compounds, and toxins like lead and mercury.

3. Exposure to low levels of air pollutants irritates the eyes and causes inflammation of the respiratory tract. Many air pollutants suppress the immune system, increasing susceptibility to infection. Exposure to air pollution during respiratory illnesses may result in the development of chronic respiratory diseases, such as emphysema and chronic bronchitis.

4. Air pollution is an international issue. One way that nations have tried to address this issue is through emissions trading, which is an administrative approach used to provide economic incentives for achieving reductions in the emissions of pollutants. Companies and other groups are issued an emission permit with a specific number of credits (allowances) that represent how much pollution they are allowed to emit. Companies that need to increase their emissions allowance buy (or trade) credits from those facilities that pollute less. Carbon credits are also traded.

3 Industrial and Photochemical Smog

1. **Photochemical smog** is produced when nitrogen oxides react with volatile organic compounds under sunny conditions. Temperature inversions and topography affect the incidence and severity of smog. Los Angeles is notorious for developing extreme smog conditions in North America. **Temperature inversions** result from a mass of warm air that develops at higher altitudes and prevents the convective mixing of air below. Mountains and coastal dynamics can also impact the development of temperature inversions.

2. **Acid precipitation** is a type of air pollution that includes sulphuric and nitric acids in precipitation as well as dry acid particles that settle out of the air. Wet deposition occurs when pollutants dissolve in water, forming an acid that precipitates onto the landscape with rainfall. Dry deposition occurs when dust laced with these pollutants settles onto the landscape. Once the dust dissolves in water, the acid properties of the pollutants begin to appear.

3. The **sick building syndrome** includes eye irritations, nausea, headaches, respiratory infections, depression, and fatigue caused by indoor air pollution.

4 Global Climate Change

1. **Greenhouse gases** are gases that absorb infrared radiation; they include carbon dioxide, methane, nitrous oxide, chlorofluorocarbons, and tropospheric ozone. The **enhanced greenhouse effect** is the additional warming that may be produced by human-increased levels of gases that absorb infrared radiation. These gases have increased considerably since the Industrial Revolution.

2. Global **climate change** will probably cause a rise in sea level, changes in precipitation patterns, extinction of many species, and problems for agriculture. It could result in the displacement of millions of people, thereby increasing international tensions. The Arctic and Antarctic are expected to be among the first locations to show dramatic changes in climate from greenhouse gas emissions.

3. Mitigation (slowing down the rate of global climate change) and adaptation (making adjustments to live with climate change) are two ways to address climate change. Mitigation includes developing alternatives to fossil fuels; increasing energy efficiency of automobiles and appliances; planting and maintaining forests; and instigating **carbon management**, by finding ways to separate and capture the CO_2 produced during the combustion of fossil fuels and then sequester it. Adaptation includes developing strategies to help various regions and sectors of society adapt to climate warming.

5 Ozone and Ozone Loss in the Stratosphere

1. **Ozone** (O_3) is a human-made pollutant in the troposphere but a naturally produced, essential component in the stratosphere. The stratosphere contains a layer of ozone that shields the surface from much of the sun's ultraviolet (UV) radiation, that part of the electromagnetic spectrum with wavelengths just shorter than those of visible light; UV radiation is a high-energy form of radiation that can be lethal to organisms at high levels of exposure.

2. Ozone formation involves oxygen molecules that absorb ultraviolet radiation forming oxygen atoms. One oxygen atom combines with an oxygen molecule to form ozone. Natural destructive reactions means there is no net buildup of ozone over time. **Ozone depletion** is the natural and human-caused removal of ozone from the stratosphere. The primary chemicals responsible for ozone thinning in the stratosphere are chlorofluorocarbons (CFCs), human-made organic compounds that contain chlorine and bromine. CFCs are now banned because they attack the stratospheric ozone layer.

3. Ozone depletion causes excessive exposure to UV radiation, which may increase cataracts, weaken immunity, and cause skin cancer in humans. Increased levels of UV radiation may also disrupt ecosystems.

KEY TERMS

CRITICAL AND CREATIVE THINKING QUESTIONS

1. The atmosphere of Earth has been compared to the peel covering an apple. Explain the comparison.

2. What basic forces determine the circulation of the atmosphere? Describe the general directions of atmospheric circulation.

3. Distinguish between primary and secondary air pollutants. Give examples of each that you are likely to encounter.

4. These graphs represent air pollutant measurements taken at two different locations. Which location is indoors, and which is outdoors? Explain your answer.

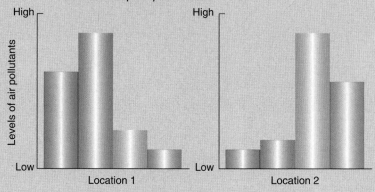

Formaldehyde   Radon   Sulphuric acid   Methane

5. One of the most effective ways to reduce the threat of radon-induced lung cancer is to quit smoking. Explain.

6. During a formal debate on the hazards of air pollution, one team argues that ozone is helpful to the atmosphere, and the other team argues that it is destructive. Explain why they are both correct.

7. The graph to the right shows air pollutant levels in a city in the northern hemisphere, measured throughout a year. Is this city likely to be found in a developing country or in a developed country? Why?

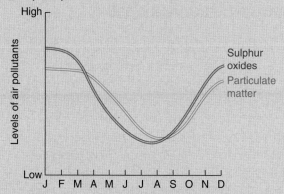

8. On the basis of what you know about the nature of science, do you think we can say with absolute certainty that the increased production of greenhouse gases is causing global climate change? Why or why not?

9. Biologists who study plants growing high in the Alps found that plants adapted to cold-mountain conditions migrated up the peaks as fast as 3.7 metres per decade during the 20th century, apparently in response to climate warming. Assuming that warming continues during the 21st century, what will happen to the plants if they reach the top of the mountains.

10. Some environmentalists contend that the wisest "use" of fossil fuels is to leave them in the ground. How would this affect air pollution? Global climate change? Energy supplies?

11. What is the Montreal Protocol?

12. Discuss some of the possible impacts of forest decline. How might these factors interact to speed the rate of decline?

What is happening in this picture ?

These scientists have drilled into a glacier in Greenland to remove an ice core. Do you think the ice deep within the glacier is old or relatively young? Explain your answer.

Some of the deeper samples were laid down thousands of years ago, when the climate was much cooler. The ice contains bubbles of air. Based on what you have learned in this chapter, do you think the level of carbon dioxide in the air bubbles in the oldest ice is higher or lower than the level in today's atmosphere? Explain your answer.

If the scientists compared CFCs in the air bubbles in the oldest ice with today's levels of CFCs, what do you think they would find?

Energy Resources 11

CANADA'S ENERGY ADDICTION

Canada is the fifth-largest energy producer in the world, generating close to 6 percent of the total global energy supplies. Canada is also the world's leading producer of hydroelectricity, contributing 13 percent to the world's total production, and the tenth-largest producer of coal. As well, Canada is a considerable producer of petroleum and natural gas. Today, roughly 98 percent of Canada's total energy is exported to the United States, although Canada also imports a large amount of energy (a majority of Canada's major coal and oil fields are located in Western Canada, so the eastern provinces such as Ontario and Quebec still import coal and oil).

Not only does Canada produce a lot of energy resources, we also consume a large amount of energy. In fact, David Suzuki has stated that Canada consumes as much energy as the entire continent of Africa. Our lifestyle is dependent on many energy sources, particularly oil (see photo). Oil, the world's largest single source of energy, is cheap (relative to most other energy sources), easy to transport and use, and extremely versatile. But it is not infinite, and it is not environmentally friendly. Oil drilling is hazardous and energy intensive, as are transportation and refining, and burning oil (usually as diesel or gasoline) releases a variety of pollutants. However, until viable alternatives are discovered and implemented, Canada's (and most of the world's) reliance on nonrenewable resources like oil will continue.

Canada's current energy policy has been developed and influenced by the many international energy agreements that deal with trade issues and emissions associated with the production and use of energy. Canada's energy policy is guided by several main principles, such as market orientation, jurisdictional authority, the role of provinces, pipeline regulations for health and safety, and environmental sustainability. Because Canada's Constitution places natural resources under provincial jurisdiction, the regulation of oil and natural gas activities is provincially managed by utility boards. Royalties and taxes on oil and natural gas production, drilling incentives, grant permits, and licensing are also all regulated provincially. For more information on Canada's energy policy, visit Natural Resources Canada's website at www.nrcan-rncan.gc.ca.

CHAPTER OUTLINE

Energy Consumption

Human society depends on energy. We use it to warm our homes in winter and cool them in summer; to grow, store, and cook our food; to light our homes; to extract and process natural resources for manufacturing items we use daily; and to power various forms of transportation. Many of the conveniences of modern living depend on a ready supply of energy. But use of energy brings along with it responsibilities and obligations of the users as well as the producers.

NONRENEWABLE AND RENEWABLE ENERGY

Energy can be classified into two basic forms—nonrenewable or renewable.

A **nonrenewable resource** is a natural resource, often existing in limited amounts, that once consumed cannot be produced or regenerated on a human time scale that can sustain its consumption rate. These resources are always retrieved by processes that require energy. Some examples of nonrenewable resources are aluminum, tin, copper, and fossil fuels such as coal, petroleum, and natural gas.

A **renewable resource** is a natural resource that, if consumed, can be replaced by natural processes at a rate that is comparable or faster than its rate of consumption by humans. Resources such as solar radiation, tides, winds, and hydroelectricity are

nonrenewable resources Natural resources that are present in limited supplies and are depleted as they are used.

renewable resources Resources that are replaced by natural processes and that can be used forever, provided they are not overexploited in the short term.

continual resources available for repeated use. Resources such as wood, paper, fish, and leather are resources that can be harvested sustainably to ensure a continuous supply. Similarly, geothermal power, fresh water, and biomass must be managed carefully to avoid overexploitation in the short term, exceeding the world's capacity to replenish these resources. The rapid growth in the human population may threaten the sustainability of many renewable resources. Specifically in developing countries, renewable resources are often overexploited for export to developed countries. Renewable resources must be managed properly to actually be renewable.

ENERGY CONSUMPTION AROUND THE WORLD

A conspicuous difference in per capita energy consumption exists between highly developed and developing nations (**FIGURE 11.1**). As you might expect, highly developed nations consume much more energy per person than developing nations—approximately eight times as much. In Canada, oil comprises the largest source of energy to support our energy demands, but natural gas and hydropower are not far behind. In contrast to other nations, Canada's demand for coal is only a small contributor to its total energy consumption. Currently, only about 1 percent of Canada's energy demands are met through the use of alternative renewable energy sources.

However, world energy consumption has increased every year since 1982, with most of the increase occurring in developing countries. From 2003 to 2004, for example, energy consumption increased worldwide by about 2.5 percent, most of it in China and India. One of the goals of developing countries is to improve their standards of living. One way to achieve this is through economic development, a process usually accompanied by a rise in per capita energy consumption. Furthermore, the world's

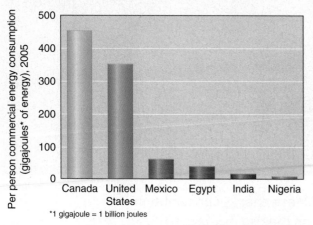

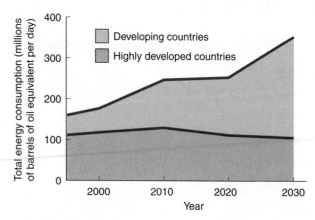

A Annual per capita commercial energy consumption in selected countries.

B Projected total energy consumption, to 2030.

energy requirements will continue to increase during the 21st century as the human population becomes larger, particularly in developing countries.

In contrast, the population in highly developed nations is more stable, and many energy experts think that their per capita energy consumption may be at or near saturation. Additional energy demands in highly developed nations may be met by increasing the energy efficiency of things like appliances, automobiles, and home insulation.

CONCEPT CHECK STOP

What is the difference between a nonrenewable energy source and a renewable energy source?

How does per capita energy consumption compare in highly developed and developing countries?

Coal, Oil, and Natural Gas

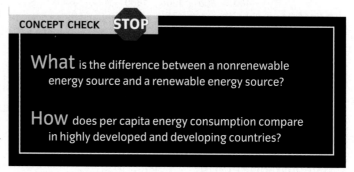

LEARNING OBJECTIVES

Describe two technologies that make coal a cleaner fuel.

Describe existing reserves of oil and natural gas.

Summarize the environmental problems caused by using coal, oil, and natural gas.

COAL

Coal, the most abundant fossil fuel in the world, is found primarily in the northern hemisphere. The largest coal deposits are found in the United States, Russia, China, Australia, India, Germany, and South Africa. China ranks at the top of these nations in terms of both

production and consumption (**FIGURE 11.2** on page 396). According to the World Resources Institute, known world coal reserves could last for more than 200 years at the present rate of consumption. Coal resources currently too expensive to develop have the potential to provide enough coal to last for 1000 or more years at current consumption rates.

Top ten coal producing countries worldwide, 2008 FIGURE 11.2

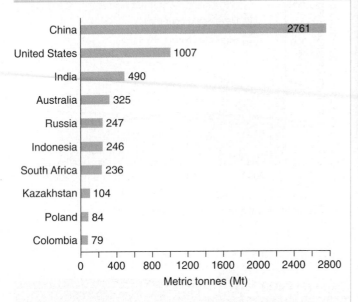

Country	Metric tonnes (Mt)
China	2761
United States	1007
India	490
Australia	325
Russia	247
Indonesia	246
South Africa	236
Kazakhstan	104
Poland	84
Colombia	79

Chinese energy consumption is on the rise FIGURE 11.3

Workers in Shanxi Province load coal onto trucks. Coal provides 65 percent of China's energy. Consumption of coal in China, the highest in the world, may double within 20 years as its economy grows.

Utility companies use coal to produce electricity, and heavy industries use coal for steel production. Different grades of coal (lignite to anthracite) have different industrial uses. Coal consumption has surged in recent years, particularly in the rapidly growing economies of China and India (**FIGURE 11.3**).

Coal mining

The two basic types of coal mines are surface and subsurface (underground) mines. If the coal bed is within 30 metres (100 feet) or so of the surface, **surface mining** is usually done. In **strip mining**, one type of surface mining, a trench is dug to extract the coal, which is scraped out of the ground and loaded into railroad cars or trucks. Surface mining is used to obtain approximately 60 percent of the coal mined.

When the coal is deeper in the ground or runs deep into the ground from an outcrop on a hillside, it is mined underground. **Subsurface mining** accounts for approximately 40 percent of the coal mined.

> **surface mining**
> The extraction of mineral and energy resources near Earth's surface by first removing the soil, subsoil, and overlying rock strata.

> **subsurface mining** The extraction of mineral and energy resources from deep underground deposits.

Surface mining has several advantages over subsurface mining: it is usually less expensive, safer for miners, and generally allows a more complete removal of coal from the ground. However, surface mining disrupts the land much more extensively than subsurface mining and has the potential to cause serious environmental problems.

Environmental impacts of coal

Coal mining, especially surface mining, has significant impacts on the environment (**FIGURE 11.4**). Prior to the mid-1970s, before the application of environmental regulations, abandoned surface coal mines were usually left as large open pits or trenches. Acid and toxic mineral drainage from such mines, along with the removal of topsoil, which was buried or washed away by erosion, prevented most plants from naturally re-colonizing the land. Streams were polluted with sediment and **acid mine drainage**, which is produced when rainwater seeps through iron sulphide minerals exposed in mine wastes. Landslides occurred on hills unstable from the lack of vegetation.

> **acid mine drainage** Pollution caused when sulphuric acid and dangerous dissolved materials such as lead, arsenic, and cadmium wash from coal and metal mines into nearby lakes and streams.

Surface coal mine FIGURE 11.4

The overlying vegetation, soil, and rock are stripped away, and then the coal is extracted out of the ground.

Dead trees enveloped in acid fog FIGURE 11.5

Acid deposition contributes to forest decline. Forest decline was first documented in Germany and eastern Europe. Later it was observed in eastern North America, particularly at higher elevations.

Generally, development companies are required to rehabilitate coal mine sites according to differing legislative standards in specific local jurisdictions. In many cases, rehabilitation attempts to mimic the historical condition of the land base. Ecosystem management approaches (as described in Chapters 5 and 6) are increasingly being applied to ensure efficient and effective results.

Coal burning generally contributes more of the common air pollutants than burning either oil or natural gas. In the United States, coal-burning electric power plants currently produce one-third of all airborne mercury emissions. Some coal contains sulphur and nitrogen that, when burned, are released into the atmosphere as sulphur oxides (SO_2 and SO_3) and nitrogen oxides (NO, NO_2, and N_2O), many of which form acids when they react with water. These reactions result in **acid deposition**, which was historically particularly prevalent downwind from coal-burning electric power plants (FIGURE 11.5), although recent technological innovations have reduced this hazard. In addition, these pollutants can contribute to the production of smog and reduce air quality. These are discussed in greater detail in Chapter 10.

Burning any fossil fuel releases carbon dioxide (CO_2) and contributes to greater concentrations of greenhouse gases in the atmosphere, which ultimately impacts global climate. It is interesting to note that when compared with other fossil fuels, such as oil and natural gas, coal combustion contributes more CO_2 gas to the atmosphere because it releases more CO_2 per unit of heat energy produced. This is of great concern as there is pressure by developing nations such as China and India to extract large quantities of coal in pursuit of energy to drive their increasing industrialization.

Making coal cleaner

Sulphur emissions associated with the combustion of coal can be reduced by **scrubbers**, or desulphurization systems, that clean the power plants' exhaust. As the polluted air passes through a scrubber, chemicals in the scrubber react with the pollution and cause it to precipitate, or settle out.

Clean coal technologies are methods of burning coal that reduce air pollution. **Fluidized-bed combustion** mixes crushed coal with limestone particles in a strong air current during combustion (**FIGURE 11.6**). This clean coal technology produces fewer nitrogen oxides and removes sulphur from the coal. It produces more heat from a given amount of coal, thereby reducing CO_2 emissions per unit of electricity produced.

OIL AND NATURAL GAS

Oil and natural gas supply approximately 55 percent of the energy used in Canada. In 2007, Canada consumed 69.6 percent, or 2.34 million barrels per day (bbl/d), of its 3.36 million bbl/d total oil production. In 2006, Canada produced 6.5 trillion cubic feet (Tcf) of natural gas, of which we consumed 3.3 Tcf, approximately 50 percent, that same year (**FIGURE 11.7**). Globally, oil and natural gas provide 61.5 percent of the world's energy.

Petroleum, or **crude oil**, is a liquid composed of hundreds of hydrocarbon compounds. During petroleum refining, the compounds are separated into different products—like gases, jet fuel, heating oil, diesel, and asphalt—based on their different boiling points (**FIGURE 11.8**). Oil is also used to produce **petrochemicals**, compounds used to make products like fertilizers, plastics, paints, pesticides, medicines, and synthetic fibres.

Process Diagram

Fluidized-bed combustion of coal FIGURE 11.6

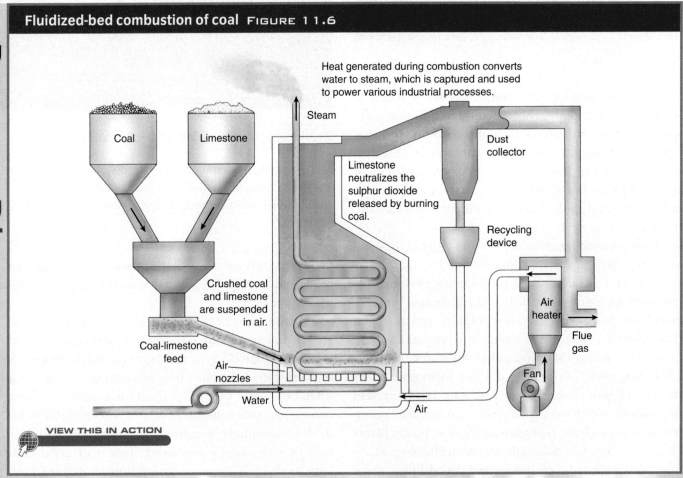

Heat generated during combustion converts water to steam, which is captured and used to power various industrial processes.

Steam

Coal

Limestone

Limestone neutralizes the sulphur dioxide released by burning coal.

Dust collector

Recycling device

Crushed coal and limestone are suspended in air.

Coal-limestone feed

Air nozzles

Water

Air

Air heater

Flue gas

Fan

VIEW THIS IN ACTION

Canada's commercial energy sources FIGURE 11.7

Note the overwhelming importance of oil, natural gas, hydroelectric power, and coal as commercial energy sources. "Other Renewables" include geothermal, solar, wind, and wood electric power.

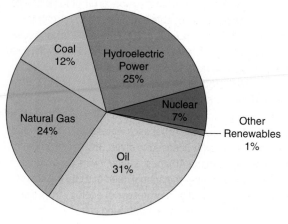

Compared to petroleum, **natural gas** contains only a few hydrocarbons: methane and smaller amounts of ethane, propane, and butane. Propane and butane are separated from the natural gas, stored in pressurized tanks as a liquid called **liquefied petroleum gas**, and used primarily in rural areas as fuel for heating and cooking. Methane is used to heat residential and commercial buildings, to generate electricity in power plants, and for a variety of purposes in the organic chemistry industry.

Natural gas use is increasing in four main areas: home heating, electricity generation, transportation, and commercial cooling. An increasingly popular use of natural gas is **cogeneration**, a clean and efficient process in which natural gas is used to produce both electricity and steam; the heat of the exhaust gases provides energy to make steam for water and space heating. Cogeneration systems that use natural gas provide electricity cleanly and efficiently.

As a fuel for trucks, buses, and cars, natural gas offers significant environmental advantages over gasoline or diesel: Natural gas vehicles emit 80 percent to 93 percent fewer hydrocarbons, 90 percent less carbon monoxide, 90 percent fewer toxic emissions, and almost no soot. Natural gas efficiently fuels residential and commercial

Petroleum refining FIGURE 11.8

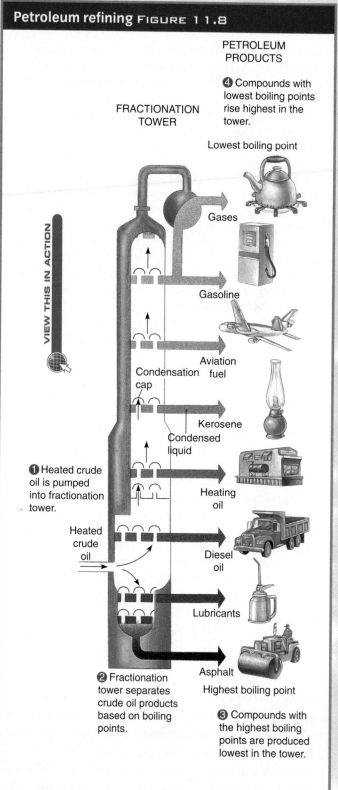

Process Diagram

PETROLEUM PRODUCTS

FRACTIONATION TOWER

❹ Compounds with lowest boiling points rise highest in the tower.

Lowest boiling point

VIEW THIS IN ACTION

Gases

Gasoline

Aviation fuel

Condensation cap

Kerosene

Condensed liquid

❶ Heated crude oil is pumped into fractionation tower.

Heating oil

Heated crude oil

Diesel oil

Lubricants

❷ Fractionation tower separates crude oil products based on boiling points.

Asphalt

Highest boiling point

❸ Compounds with the highest boiling points are produced lowest in the tower.

In northeastern Alberta, beneath 140,200 square kilometres of land, an area larger than the state of Florida, lie three major areas (Athabasca, Peace River, and Cold Lake) of bitumen deposits known as the Alberta oil sands, or the "tar sands," named from its bitumen-coated sand source. Bitumen is a heavy molasses-like crude oil that will flow only if it is diluted with lighter hydrocarbons or if it is heated. These bitumen deposits of the Alberta oil sands are the second-largest deposits in the world; only traditional oil reserves in Saudi Arabia are bigger. The oil sands deposits contain close to 1.7 trillion barrels of bitumen.

Although European fur traders first heard about the oil sands back in 1719, it wasn't until 1967 that the first commercially viable operation began. Today there are three main mines in operation, and in 2008 they produced a combined total of 1.2 million barrels per day (bbl/d) of crude oil. Future predictions made by the Canadian Association of Petroleum Producers suggest the Canadian oil sands production will grow to close to 3.3 million bbl/d by 2020, and potentially 5 million bbl/d by 2030. This would position Canada as one of the top five oil-producing countries in the world and would situate Alberta as a global energy leader.

The advances made on the Alberta oil sands have been a victory of energy research and technological innovation; however, many Canadians remain opposed to the oil sands production and suggest that government and industry have failed to adequately address the environmental and health risks that are posed by this large-scale mining operation. Although the Alberta government requires companies to restore the land destroyed by mining to "equivalent land capacity," many argue that these lands are not restored to support previous uses. In some cases, mining land has been converted to agricultural or pasture land rather than to the original forest land. And others note that the amount of land that has been reclaimed is only a fraction of what has been damaged.

As well, the processing of bitumen into synthetic crude oil requires energy that is currently generated by burning natural gas: For every barrel of synthetic crude oil that is produced in Alberta, more than 80 kilograms of greenhouse gases are released into the atmosphere and 2000 to 4000 barrels of waste water are dumped into tailing ponds that now occupy 130 square kilometres of previous boreal forest. The magnitude of greenhouse gas emissions from the Alberta oil sands is a large reason why Canada was not able to fulfill the commitments made under the Kyoto Protocol; in 2002 Canada's greenhouse gas emissions had increased by 24 percent of 1990 levels, even though our Kyoto target was 6 percent *below* 1990 levels. The Alberta oil sands production contributed 3.4 percent of Canada's greenhouse gas emissions in 2003.

Another significant concern is the influence that the Alberta oil sands are having on the Athabasca watershed. The oil sands operations are currently licensed to divert 359 million cubic metres of water from the Athabasca River, twice the volume of water required to meet the annual need of a city the size of Calgary. Dr. David Schindler, a professor of ecology from the University of Alberta, warns that continued water extraction from the Athabasca River will have devastating consequences. (Read more about Dr. Schindler in the Chapter 10 Making a Difference feature.) Schindler suggests that the water removed is not replaced because it is too contaminated, leaving the river with lower water volume and large amounts of toxic waste to dispose of, all factors that have serious impacts for the entire ecosystem.

Several concerned groups have tried to curtail the oil sands operations over the years. In 2009, the Pembina Institute, an environmental nongovernmental organization, filed an affidavit with the Alberta Energy Resources Conservation Board (ERCB) and the government of Canada requesting a public hearing on a projected mine expansion. For more information on Alberta's oil sands visit the government of Alberta website at www.alberta.ca/home/, or visit the Pembina Institute website at www.pembina.org/

---

air-cooling systems. One example is the use of natural gas in a desiccant-based (air-drying) cooling system, which is ideal for restaurants and supermarkets, where humidity control is as important as temperature control.

The main disadvantage of natural gas is that deposits are often located far from where the energy is used. Because it is a gas and less dense than a liquid, natural gas costs four times more to transport through pipelines than crude oil. To transport natural gas over long distances, it must first be compressed to form *liquefied natural gas (LNG)*, after which it must be returned to the gaseous state at regasification plants before being piped to where it will be used.

A series of pipelines have been constructed to establish a North American grid to transport natural gas across Canada and the United States. Perhaps the largest is the TransCanada pipeline (**FIGURE 11.9**), originally constructed in 1958. It carries natural gas through Alberta, Saskatchewan, Manitoba, Ontario, and Quebec.

The TransCanada pipeline FIGURE 11.9

A network of pipelines has been built to transport natural gas throughout Canada. The largest one in the network is the TransCanada pipeline (shown here in green).

CentraGas Ltd.
Pacific Northern Gas Ltd.
BC Gas Ltd.
WestCoast Energy Inc.
Alliance Pipeline Inc.
TransCanada Alberta (NGTL)
Foothills Pipeline Ltd.
TransGas Ltd.
TransCanada pipeline
Union Gas Ltd.
Gaz Metropolitan Inc.
TransQuébec & Maritimes Pipeline
Maritimes & NorthEast Pipeline

There is considerable attention given to the Arctic for further development of natural gas reserves. The MacKenzie Valley Pipeline is a proposed pipeline project to transport natural gas from the Beaufort Sea through the Northwest Territories to tie into gas lines already located in northern Alberta. This project was first proposed in 1970 and again in 2004. Alaska is also a source of natural gas that is currently being brought to market. The Alaska Pipeline Project stretches 2737 kilometres from the North Slope in Alaska through the Yukon to the British Columbia–Alberta border near Boundary Lake. The construction of the pipeline will be difficult due to harsh climate, challenging terrain, three mountain ranges, active fault lines, hundreds of rivers and streams, and natural migratory routes of caribou and moose—not to mention the numerous conservation and environmental groups that have voiced opposition to the pipeline's construction because of how it will impact numerous other wildlife species, including deer and elk. It is suspected, though, that in light of political goals for cleaner fuel sources, North American natural gas demand, and a desire to be less dependent on oil from overseas, both the MacKenzie Valley Pipeline and the Alaska Pipeline will be required to meet North American energy demands.

Reserves of oil and natural gas

Oil and natural gas deposits exist on every continent, but their distribution is uneven. More than half of the world's

Distribution of oil deposits FIGURE 11.10

The Persian Gulf region contains huge oil deposits in a relatively small area, whereas other regions have few.

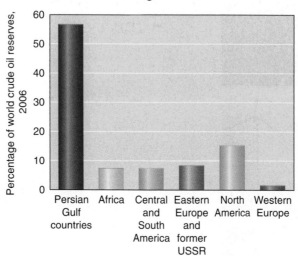

total estimated reserves are situated in the Persian Gulf region, which includes Iran, Iraq, Kuwait, Oman, Qatar, Saudi Arabia, Syria, United Arab Emirates, and Yemen (**FIGURE 11.10**). Major oil fields also exist in Canada, Venezuela, Mexico, Russia, Kazakhstan, Libya, and the United States (in Alaska and the Gulf of Mexico). **FIGURE 11.11** on pages 402 and 403 outlines global supply and production of energy sources.

Global energy supply and consumption FIGURE 11.11

▶ ENERGY CONSUMPTION

The use and availability of primary energy resources are unequally distributed around the globe. More than 86 percent of energy consumed globally is from nonrenewable fossil fuels—coal, oil, and natural gas. Consumption of these fuels is greatest in industrialized nations, with the United States using up nearly one-quarter. Developing countries, especially those in sub-Saharan Africa, rely on more traditional sources of energy, such as firewood and dung.

HYDROPOWER

NUCLEAR

SOLAR

WIND

GEOTHERMAL

ALTERNATIVE ENERGIES

Hydropower provides nearly 18 percent of the world's electricity, but it is limited to countries with adequate water resources, and it poses threats to local watersheds. **Nuclear energy** makes up 17 percent of Earth's electricity, but few countries have adopted it because of potential environmental risks and waste disposal issues. **Solar** and **wind energy** are inexhaustible and are the focus of new energy technologies and research. **Geothermal energy** is efficient but limited to countries with ready sources of hot groundwater, such as Iceland.

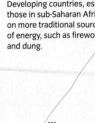

Annual energy consumption, in trillions of British thermal units (BTUs)
- More than 25,000
- 10,001–25,000
- 1001–10,000
- 101–1000
- 10–100
- Less than 10
- No data available

Major energy deposit
- Coal
- Natural gas
- Oil
- Oil transit chokepoint

▼ RENEWABLE ENERGY

Renewable sources of energy—geothermal, solar, and wind— make up a small percentage of the world's energy supply. They have a significant impact, however, on local and regional energy supplies, especially for electricity, in places such as the United States, Japan, and Germany. These sources of energy can be regenerated or renewed in a relatively short time, whereas fossil fuels form over geologic time.

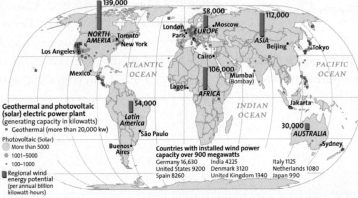

Geothermal and photovoltaic (solar) electric power plant (generating capacity in kilowatts)
- ■ Geothermal (more than 20,000 kw)

Photovoltaic (Solar)
- ● More than 5000
- ● 1001–5000
- · 100–1000
- ■ Regional wind energy potential (per annual billion kilowatt-hours)

Countries with installed wind power capacity over 900 megawatts
Germany 16,630	India 4225
United States 9200	Denmark 3120
Spain 8260	United Kingdom 1340
Italy 1125	
Netherlands 1080	
Japan 990	

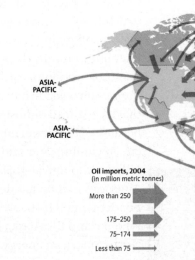

Oil imports, 2004 (in million metric tonnes)
- More than 250
- 175–250
- 75–174
- Less than 75

▶ FLOW OF OIL WORLDWIDE

Major oil reserves are clustered in a handful of countries, more than half of which are in the Middle East, whereas the greatest demand for oil is in the United States, Europe, Japan, and China. Other major oil exporters include the Russian Federation, Norway, Venezuela, and Mexico.

ENERGY

Energy enables us to cook our food, heat our homes, move about our planet, and run industry. Every day the world uses some 320 billion kilowatt-hours of energy—equivalent to each person burning 22 lightbulbs nonstop, and over the next century, demand may increase threefold. Consumption is not uniform around the globe. People in industrialized counties consume far greater amounts of energy than those in developing countries. The world's energy supply is still fossil-fuel based, despite advances in alternative energy sources. To meet demand, many countries must import fuels, making the trade of energy a critical, often volatile global political issue. Instability where most oil is found—the Persian Gulf, Nigeria, and Venezuela—make this global economic power line fragile. Insatiable demand where most energy is consumed—the United States, Japan, China, India, and Germany—makes national economies increasingly dependent. Furthermore, extraction and use of fossil fuels have serious environmental effects, such as air pollution and climate warming. The challenge for the future? Reducing reliance on fossil fuels, developing alternative energies to meet demand, and mediating the trade-offs between the environment and energy.

GeoBytes

LACK OF ACCESS
More than 2 billion people, mostly in the developing world, do not have access to electricity. Increasingly, small-scale wind and solar projects bring power to poor rural areas.

WINDS OF CHANGE
Worldwide, wind supplies less than 1 percent of electric power, but it is the fastest-growing source, especially in Europe. Denmark gets 20 percent of its electricity from wind.

POWER OF THE SUN
Near Leipzig, Germany, some 33,000 photovoltaic panels produce up to 5 megawatts of power. It is one of the world's largest solar arrays.

GOING NUCLEAR
France gets 78 percent of its electricity from nuclear power. Developing nations, such as China and India, are building new reactors to reduce pollution and meet soaring energy demands.

GROWING PAINS
China is fuelling its economic growth with huge quantities of coal, and it suffers from energy-related environmental problems. China is second only to the United States in greenhouse gas emissions that contribute to global warming.

◄ **WORLD OIL SUPPLY**
The world's hunger for oil is insatiable, but the supply is finite and unequally distributed, making it one of the world's most valuable commodities. It is the leading source of energy worldwide, and in industrialized countries it accounts for more than one-third of all energy consumed. Pressure on the world's oil supply continues to mount as both industrialized and developing countries grow more dependent on it to meet increasing energy needs.

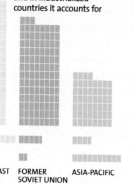

Large oil deposits also exist under the **continental shelves**, the relatively flat underwater areas that surround continents, and in deep-water areas adjacent to the continental shelves. Despite problems like storms at sea and the potential for oil spills, many countries engage in offshore drilling. Continental shelves off the coasts of western Africa and Brazil are also promising potential sources of oil. Environmentalists and coastal industries like fishing generally oppose opening the continental shelves for oil and natural gas exploration because of the threat of a major oil spill.

Almost half of the world's proven recoverable reserves of natural gas are located in two countries, Russia and Iran. But Canada's natural gas reserves are substantial too. In 2006, Canada's natural gas production was 6.5 trillion cubic feet while Canadian consumption that year totalled 3.3 trillion cubic feet. In January 2008, Canada's proven natural gas reserves totalled 57.9 trillion cubic feet. Canada is the second-largest producer of natural gas in the western hemisphere. Natural gas is increasingly being produced in Canada and the U.S. from "tight" shale deposits using new and advanced technologies.

Another major reserve of natural gas is found in methane hydrates. Trapped beneath the Arctic permafrost and in the deep ocean floor at depths greater than 1600 feet, the natural gas can be released with an increase in temperature or a decrease in pressure. The most potentially recoverable deposits are found in the permafrost conditions in the Arctic; the ocean deposits are found scattered throughout the world and are thinly deposited, making them harder to recover. There are three identified methods that are being explored to release natural gas from its methane hydrate form:

1. Heat the hydrate reservoir with hot water or steam resulting in vaporization. This method uses a significant amount of energy and is considered to be cost-ineffective.

EnviroDiscovery   CANADIAN OFFSHORE OIL PRODUCTION

Many oil reserves are found under the ocean. To access these, permanent or mobile oil platforms are built to extract this oil. Hibernia is a petroleum oil field located 315 kilometres off the coast of Newfoundland. Hibernia was discovered in 1979 and is the fifth-largest oil field ever discovered in Canada. It consists of two principal reservoirs, the Hibernia and Ben Nevis-Avalon, which are located at average depths of 3700 and 2400 metres, respectively. The Terra Nova and White Rose fields off Newfoundland and the Sable Island gas field off Nova Scotia are also likely to be significant in the future.

Hibernia is the world's largest oil platform and was built by a consortium of private companies and the Canadian government. Created in Bull Arm, Newfoundland, this massive structure, designed to withstand the force of sea ice and icebergs, sits on the ocean floor and is 111 metres high with a storage capacity for 1.3 million barrels of crude oil. Hibernia has proven to be one of the most productive oil fields in Canada. The Hibernia oil platform began operations in November 1997. It is designed to hold 185 crew members at a time, and has a capacity of producing 230,000 barrels of crude oil per day.

Despite the sheer magnitude and strength of these offshore drilling structures, shocking accidents have occurred in Canadian ocean waters. The *Ocean Ranger*, a semi-submersible mobile offshore drilling unit located 267 kilometres east of St. John's, sank with 84 crew members aboard. During a vicious storm in February 1982, the $120 million *Ocean Ranger* toppled over and went missing in the ocean waters. The accident left both the Canadian and United States governments wondering how the "unsinkable" *Ocean Ranger* could capsize, leaving not a single survivor. Investigation later pointed to poor training and design flaws as the reasons for the devastating accident.

The Hibernia oil platform began operations in November 1997. It is designed to hold 185 crew members at a time, and has a capacity of producing 230,000 barrels of crude oil per day.

2. Inject an inhibitor that reduces the stability of the hydrates. This method is also considered cost-ineffective since it requires a significant amount of inhibitor over a large area.

3. Reduce pressure allowing the hydrate to vaporize in the drill well. This method has the most promise.

To harvest natural gas from methane hydrates, however, is challenging and will require infrastructure to be built, reliable technologies to be implemented, and pipelines and storage facilities to be designed and developed. There is potential to harvest natural gas from methane hydrates in places such as Alaska's North Slope where pipelines and roads are already implemented. If methane hydrates can be harvested in the future it may potentially change the global energy market, since an abundance of natural gas has been located in traditionally importing nations (United States and the United Kingdom) that would now become large exporters of their new supply of natural gas.

How long will oil and natural gas supplies last? We cannot predict how many reserves will be discovered, whether technological breakthroughs will allow us to extract more fuel from each deposit, or whether world consumption of oil and natural gas will increase, remain the same, or decrease. Even with technological advances, most industry analysts predict that global oil production will peak between 2050 and 2100. Natural gas is more plentiful than oil. Experts estimate that, at current rates of consumption, readily recoverable reserves of natural gas will keep production rising for at least 10 years after conventional supplies of petroleum have begun to decline.

As the prices for oil and natural gas rise in the global marketplace, production of previously uneconomic reserves becomes profitable. Thus, the life expectancy of these energy reserves is tied to human demand for the product. The issue is not really that oil or natural gas will run out, but rather that $75 per barrel oil will someday run out, and then $100 per barrel oil, then $150 per barrel oil, and so on. At some point the remaining reserves will become so expensive relative to alternatives that demand will likely shift to those alternatives for purely economic reasons.

Environmental impacts of oil and natural gas
The environmental problems associated with oil and natural gas result from burning and transporting

them. As with coal, the burning of oil and natural gas produces CO_2 that can contribute to global warming. Every litre of gasoline your car or truck burns releases about 45 kilograms of CO_2 into the atmosphere. Burning oil also leads to acid deposition and the formation of photochemical smog. Oil combustion produces almost no sulphur oxides, but it does produce nitrogen oxides. In fact, gasoline combustion contributes about half the nitrogen oxides released into the atmosphere by human activities. (Coal combustion contributes the rest.)

Natural gas, on the other hand, is a relatively clean, efficient source of energy that contains almost no sulphur and releases far less CO_2, fewer hydrocarbons, and almost no particulate matter, compared with oil and coal.

One risk of oil and natural gas production relates to their transport, often over long distances by pipelines or ocean tankers. A serious spill along the route creates an environmental crisis, particularly in aquatic ecosystems, where the oil slick can travel great distances. One of the most well-known disasters in North America was the *Exxon Valdez* supertanker hitting Bligh Reef in 1989 and spilling 260,000 barrels of crude oil into Prince William Sound along the coast of Alaska (**FIGURE 11.12**). According to the U.S. Fish and Wildlife Service and the Alaska Department of Environmental Conservation, more than 30,000 birds (sea ducks, loons, cormorants, bald eagles, and other species) and between 3500 and 5500 sea otters died as a result of the spill. The area's killer whale and harbour seal populations declined, salmon migration was disrupted, and the fishing season in the area was halted that year. Although Exxon declared the cleanup "complete" in late 1989, it left behind many problems such as contaminated shorelines, continued damage to some species of birds, fishes, and mammals, and a reduced commercial salmon catch. The BP oil spill that occurred in April 2010 is discussed in the EnviroDiscovery box on page 406.

The *Exxon Valdez* oil spill FIGURE 11.12
An aeriel view of the massive oil slick at the southwest end of Prince William Sound, Alaska, 1989.

NATIONAL GEOGRAPHIC

The Deepwater Horizon oil spill, also known as the BP Oil Spill, occurred in the Gulf of Mexico in April 2010 and is now considered to be the largest offshore oil spill in U.S. history. Deepwater Horizon, under lease to BP until September 2013, is a semi-submersible mobile offshore drilling unit with the capability of drilling down over 9000 metres. Beginning on April 20, 2010, 210 kilometres to the southeast of New Orleans, a wellhead blowout located 1500 metres below the ocean surface exploded, killing 11 platform workers and injuring 17 others and initiating a massive oil gusher. The gusher was estimated to have been releasing anywhere between 25,000 to 200,000 barrels (4 million to 32 million litres) of oil per day into the Gulf of Mexico. The resulting oil slick covered a surface area estimated to be at least 4000 square kilometres, depending on the day and weather conditions of measurement, although scientists report that there were massive underwater oil plumes that were not visible from the surface.

Experts fear that the oil spill will result in the worst environmental disaster in U.S. history. There are more than 400 species that live in the islands and marshlands that are at risk, some of which are endangered. As of June 7, 2010, there were already hundreds of dead birds, sea turtles, dolphins, and other mammals, presumably killed as a result of the spill.

Ecosystem recovery from such a spill will likely take years if not decades. Coupled with these concerns is the possibility that the Gulf Stream sea currents may spread the oil out into the Atlantic Ocean, potentially affecting marine life between the water surface and the ocean floor.

Fisheries and tourism have also been seriously affected by the spill. On May 24, 2010, the U.S. federal government declared a fisheries disaster for the states of Alabama, Mississippi, and Louisiana, with cost estimates to the fishing industry in the order of $2.5 billion. Tourists reacted to the spill by cancelling their vacations, fearing the arrival of the oil on the coastal beaches. The impacts to tourism along Florida's Paradise Coast are estimated at $3 billion.

Containment and cleanup of the oil spill began almost immediately with mixed success. Many strategies were used and many of these failed. By early June 2010, a cap was finally attached, capturing some of the oil. The cap only captured a portion of the oil being released, however, and long-term containment strategies were still necessary at the time of writing. Long-term efforts include drilling relief wells into the original well to allow them to block the original well flow. As of June 2010 the relief wells were slated to be finished by August 2010, at a cost of approximately US $100 million per well.

The oil slick as seen from space by NASA's Terra satellite on May 24, 2010.

A brown pelican covered in oil from the BP oil spill found along the Louisiana coast line on June 3, 2010.

One positive outcome of the *Exxon Valdez* disaster was passage of the U.S. Oil Pollution Act of 1990. This legislation establishes liability for damages to natural resources resulting from a catastrophic oil spill, including a trust fund that pays to clean up spills when the responsible party cannot, and requires by 2015 double hulls on all oil tankers that enter U.S. waters. In Canada, considerable emphasis was placed on oil and natural gas contaminant legislation, particularly at the federal level and in northern environments. In addition, Canada focused efforts on determining the most appropriate mitigative and containment measures in the event of a spill occurring.

Another massive oil spill occurred in 1991 during the Persian Gulf War, when about 6 million barrels (960 million litres) of crude oil—more than 20 times the amount of the *Exxon Valdez* spill—were deliberately dumped into the Persian Gulf. Many oil wells were set on fire, and lakes of oil spilled into the desert around the burning oil wells. In 2001 Kuwait began a massive remediation project to clean up its oil-contaminated desert. Progress is slow, and it may take a century or more for the area to completely recover.

CONCEPT CHECK STOP

Which type of coal mining—surface or subsurface mining—is more land-intensive?

How long are reserves of oil and natural gas projected to last?

What are three environmental problems associated with mining and burning coal, and using oil and natural gas as energy resources?

Nuclear Energy

LEARNING OBJECTIVES

Define *nuclear energy* and describe a typical nuclear power reactor.

Discuss the pros and cons of electric power produced by nuclear energy versus coal.

Describe safety issues associated with nuclear power plants and risks associated with the storage of radioactive wastes.

---

 ll atoms are composed of positively charged protons, negatively charged electrons, and electrically neutral neutrons. Protons and neutrons, which have approximately the same mass, are clustered in the centre of the atom, making up its nucleus. Electrons, which possess little mass in comparison with protons and neutrons, orbit the nucleus in distinct regions.

As a way to obtain energy, nuclear processes are fundamentally different from the combustion that produces energy from fossil fuels. Combustion is a chemical reaction. In chemical reactions, atoms of one element do not change into atoms of another element, nor does any of their mass (matter) change into energy. The energy released in chemical reactions comes from changes in the chemical bonds that hold the atoms together. Chemical bonds are associations between electrons, and chemical reactions involve the rearrangement of electrons.

In contrast, **nuclear energy** involves changes in the nuclei of atoms; small amounts of matter from the nucleus are converted into large amounts of energy. There are two different nuclear reactions that release energy: fission and fusion. In nuclear **fission**, the process nuclear power plants use, energy is released when a single neutron crashes into

nuclear energy
The energy released by nuclear fission or fusion.

fission The splitting of an atomic nucleus into two smaller fragments, accompanied by the release of a large amount of energy.

Nuclear fission FIGURE 11.13

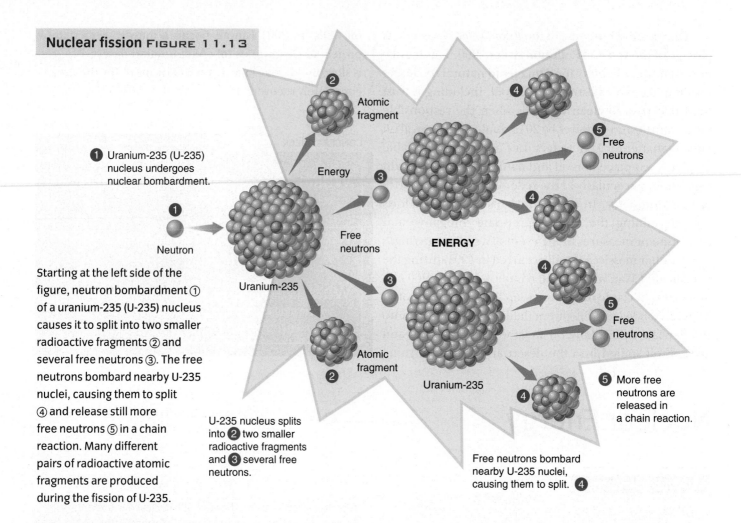

1 Uranium-235 (U-235) nucleus undergoes nuclear bombardment.

Neutron

Uranium-235

2 Atomic fragment

Energy

3 Free neutrons

4

5 Free neutrons

ENERGY

3

2 Atomic fragment

Uranium-235

4

4

5 Free neutrons

5 More free neutrons are released in a chain reaction.

Free neutrons bombard nearby U-235 nuclei, causing them to split. **4**

U-235 nucleus splits into **2** two smaller radioactive fragments and **3** several free neutrons.

Starting at the left side of the figure, neutron bombardment ① of a uranium-235 (U-235) nucleus causes it to split into two smaller radioactive fragments ② and several free neutrons ③. The free neutrons bombard nearby U-235 nuclei, causing them to split ④ and release still more free neutrons ⑤ in a chain reaction. Many different pairs of radioactive atomic fragments are produced during the fission of U-235.

fusion The joining of two lightweight atomic nuclei into a single, heavier nucleus, accompanied by the release of a large amount of energy.

a large atom of one element, like uranium, and splits it into two smaller atoms of different elements (**FIGURE 11.13**). In **fusion**, the process that powers the sun and other stars, two small atoms are combined, forming one larger atom of a different element.

CONVENTIONAL NUCLEAR FISSION

Uranium ore, the mineral fuel used in conventional nuclear power plants, is a nonrenewable resource present in limited amounts in sedimentary rock in Earth's crust. Uranium ore contains three isotopes, U-238 (which makes up 99.28 percent of uranium), U-235 (0.71 percent), and U-234 (less than 0.01 percent). Because U-235, the isotope used in

conventional fission reactions, is such a minor part of uranium ore, uranium ore must be refined after mining to increase the concentration of U-235 to about 3 percent. This refining process, called **enrichment**, requires a great deal of energy.

After enrichment, the uranium is processed into small pellets of uranium dioxide; each pellet contains the energy equivalent to a tonne of coal (**FIGURE 11.14A**). The pellets are then placed in closed pipes, often as long as 3.7 metres, called **fuel rods**. The fuel rods are then grouped into square **fuel assemblies**, generally made up of 200 rods each (**FIGURE 11.14B**). A typical **nuclear reactor** contains about 250 fuel assemblies.

enrichment
The process by which uranium ore is refined after mining to increase the concentration of fissionable U-235.

nuclear reactor
A device that initiates and maintains a controlled nuclear fission chain reaction to produce energy for electricity.

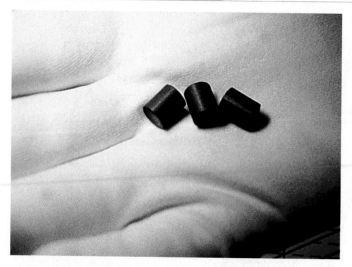

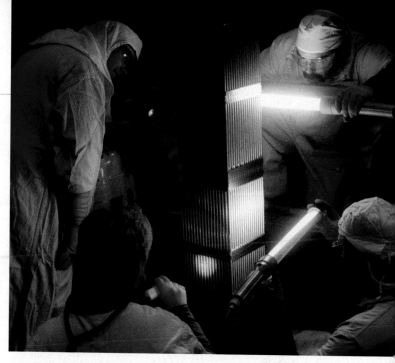

A Uranium dioxide pellets, held in a gloved hand, contain about 3 percent uranium-235, the fission fuel in a nuclear reactor. Each pellet contains the energy equivalent of one tonne of coal.

B The uranium pellets are loaded into long fuel rods, which are grouped into square fuel assemblies like the one being inspected here by technicians.

The fission of U-235 releases an enormous amount of heat, which is used to transform water into steam. The steam, in turn, is used to generate electricity. The production of electricity is possible because the fission reaction is controlled. Operators of a nuclear power plant can start or stop and increase or decrease the fission reactions in the reactor to produce the desired amount of heat energy.

A typical nuclear power plant has four main parts: the reactor core, the steam generator, the turbine, and the condenser (**FIGURE 11.15** on page 410). Fission occurs in the **reactor core**, and the heat produced by nuclear fission is used to produce steam from liquid water in the **steam generator**. The **turbine** uses the steam to generate electricity, and the **condenser** cools the steam, converting it back to a liquid.

CANADA'S NUCLEAR ENERGY PRODUCTION

As of June 2008, there were 439 nuclear power reactors operating in 35 countries around the world producing 15 percent of the world's electricity. There are currently 22 nuclear reactors in Canada that are owned by three provincial electric utilities. Ontario Power Generation

Inc. owns 20 of the power reactors, while Hydro-Québec and New Brunswick Power each own one reactor. All nuclear reactors are of the CANDU type (short for CANada Deuterium Uranium; see **FIGURE 11.16** on page 410). The CANDU reactor was developed in Canada in the late 1950s and 1960s, and in 1962 it produced the first nuclear-generated electricity in Canada. What makes the CANDU reactor different from other water-moderated reactors is its use of heavy water for neutron moderation. CANDU is the most efficient of all reactors, using approximately 15 percent less uranium than a pressurized water reactor for each megawatt of electricity produced. The use of natural uranium makes fuel fabrication easier and there is no need for uranium enrichment facilities, which provides significant fuel cost savings. Offsetting these potential savings, however, is the initial one-time cost of the heavy water, but the heavy water used in CANDU reactors is readily available and can be produced locally, which provides additional advantages.

Canada is the world's largest exporter of uranium and has the second-largest proven uranium reserves. Most of Canada's vast supply of uranium is in highly concentrated deposits in northern Saskatchewan, giving the province the nickname "the Saudi Arabia of the uranium industry"; in 2007, the uranium mines in Saskatchewan

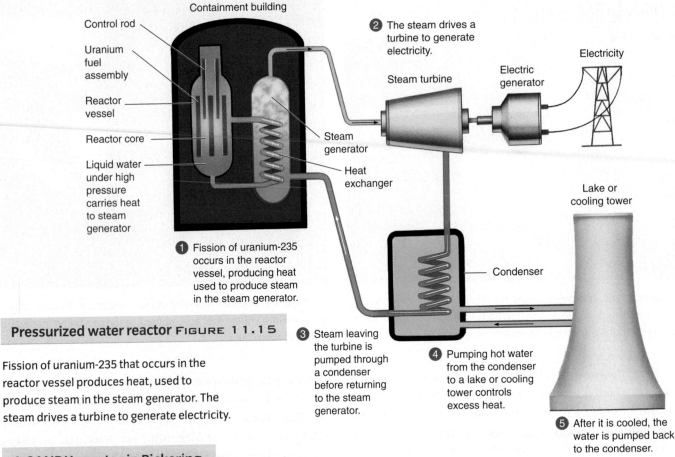

Containment building

Control rod

Uranium fuel assembly

Reactor vessel

Reactor core

Liquid water under high pressure carries heat to steam generator

Steam generator

Heat exchanger

2 The steam drives a turbine to generate electricity.

Steam turbine

Electric generator

Electricity

Lake or cooling tower

Condenser

1 Fission of uranium-235 occurs in the reactor vessel, producing heat used to produce steam in the steam generator.

3 Steam leaving the turbine is pumped through a condenser before returning to the steam generator.

4 Pumping hot water from the condenser to a lake or cooling tower controls excess heat.

5 After it is cooled, the water is pumped back to the condenser.

Pressurized water reactor FIGURE 11.15

Fission of uranium-235 that occurs in the reactor vessel produces heat, used to produce steam in the steam generator. The steam drives a turbine to generate electricity.

A CANDU reactor in Pickering, Ontario FIGURE 11.16

The CANDU reactor was initially intended as a proof-of-concept design. Today, there are 29 CANDU reactors in use around the world and a total of 48 heavy water–moderated reactors based on the CANDU design in operation, under construction, or under refurbishment worldwide.

accounted for roughly 22 percent of the world's total uranium production. Canada also supplies 75 percent of the world's cobalt-60, used to sterilize the world's single-use medical supplies. Currently, Canada produces 15 percent of its own electricity at nuclear power plants. The province of Ontario is Canada's primary nuclear energy producer; it has 18 operating reactors, plus two reactors undergoing refurbishment, which provide close to 50 percent of the province's electricity. Overall in Canada, the industry employs 21,000 people directly and 10,000 people indirectly. In 2007, Canada's uranium mining industry produced sales of $1.1 billion.

Many nations are looking to nuclear power to reduce the harmful emissions that are produced when burning fossil fuels. Electricity that is produced from uranium does not release combustible gases into the atmosphere. From 1971 to 2007, the nuclear generating stations in Ontario prevented the release of over 1.6 billion tonnes of carbon dioxide, 30 million tonnes of acid gas, and 73 million tonnes of ash into the atmosphere that would have resulted if the electricity had been produced by the burning of coal.

Because of this the support for nuclear energy is growing. International commitments to reduce greenhouse gas emissions are persuading countries to consider the environmental advantages of nuclear energy. In Canada, a public opinion poll conducted in 2008 by Ipsos Reid on behalf of the Canadian Nuclear Association showed that 73 percent of Canadians believe that nuclear energy is reliable with 68 percent indicating they believe that nuclear power will be an important part of our future energy production. In 2001, Canada became a member of Generation 1V, an international forum created to develop new generation nuclear power reactors by 2030. In 2001, a charter for collaborative research toward the goals of sustainability, reliability, economics, and safety was signed by Canada and 10 other countries.

IS NUCLEAR ENERGY CLEANER THAN COAL?

One of the reasons supporters of nuclear energy argue for its widespread adoption is that nuclear energy impacts the environment less than fossil fuels like coal (TABLE 11.1).

The combustion of coal to generate electricity is responsible for more than one-third of the air pollution in North America and contributes to acid precipitation and global warming. In comparison, nuclear energy emits few pollutants into the atmosphere. Nuclear energy also provides power without producing climate-altering CO_2.

However, nuclear energy generates radioactive waste in the form of **spent fuel**. Nuclear power plants also produce radioactive coolant fluids and gases in the reactor. These radioactive wastes are extremely dangerous, and the hazards of their health and environmental impacts mean that special measures must be taken to ensure their safe storage and disposal.

> **spent fuel** Used fuel elements that were irradiated in a nuclear reactor.

Opponents of nuclear energy contend that the replacement of coal-burning power plants with nuclear power plants does not significantly lessen the threat of global warming because only 15 percent of greenhouse gases come from power plants. Auto emissions and industrial processes produce most greenhouse gases, and neither is affected by nuclear power. Also, uranium mining, reactor construction, and disposal of nuclear wastes require the combustion of fossil fuels. This means that

Comparison of environmental impacts of 1000-MWe coal and conventional nuclear power plants*
TABLE 11.1

Impact	Coal	Nuclear (conventional fission)
Land use	17,000 acres	1900 acres
Daily fuel requirement	8000 tonnes (of coal)/day	3 kg (of enriched uranium)/day
Availability of fuel, based on present economics	A few hundred years	100 years, maybe longer
Air pollution	Moderate to severe, depending on pollution controls	Low
Climate change risk (from CO_2 emissions)	Severe	No risk
Radioactive emissions, routine	1 curie	28,000 curies
Water pollution	Often severe	Potentially severe at nuclear waste disposal sites
Risk from catastrophic accidents	Short-term local risk	Long-term risk over large areas
Link to nuclear weapons	No	Yes
Annual occupational deaths	0.5 to 5	0.1 to 1

* Impacts include extraction, processing, transportation, and conversion. Assumes coal is strip-mined. (A 1000-MWe utility, at a 60 percent load factor, produces enough electricity for a city of 1 million people.)

nuclear power indirectly contributes a small amount to the greenhouse effect—about 2 to 6 grams of carbon per kilowatt-hour. This is two orders of magnitude lower than fossil fuels.

SAFETY AND ACCIDENTS IN NUCLEAR POWER PLANTS

Although conventional nuclear power plants cannot explode like atomic bombs, accidents do happen in which dangerous levels of radiation are released into the environment and result in human casualties. At high temperatures, the metal encasing uranium fuel can melt, releasing radiation; this is called a **meltdown**. The meltdown is caused by a loss of control of the fission reaction in the nuclear core. Temperatures become not just high, but excessively high, since the cooling system can no longer cope, and meltdown occurs. Also, the water used in a nuclear reactor to transfer heat can boil away during an accident, contaminating the atmosphere with radioactivity.

Although the nuclear industry considers the probability of a major accident low, the consequences of such accidents are drastic and life threatening, both immediately and long after the accident has occurred.

The world's worst nuclear power plant accident took place in 1986 at the Chernobyl plant, located in the former Soviet republic of the Ukraine. One or possibly two explosions ripped apart a nuclear reactor and expelled large quantities of radioactive material into the atmosphere (**FIGURE 11.17**). The effects of this accident were not confined to the area immediately surrounding the power plant: Significant amounts of radioisotopes quickly spread across large portions of Europe. The Chernobyl accident affected and will continue to affect many nations.

Although cleanup in the immediate vicinity of Chernobyl is finished, the people in Ukraine face many long-term problems. Ultimately, more than 170,000 people had to permanently abandon their homes. Much of the farmland and forests are so contaminated that they cannot be used for more than a century. Inhabitants of many areas of Ukraine cannot drink the water or consume locally produced milk, meat, fish, fruits, or vegetables. Mothers do not nurse their babies because their

Chernobyl, Ukraine FIGURE 11.17

The arrow indicates the site of the explosion. The upper part of the reactor was completely destroyed.

milk is contaminated by radioactivity. The frequency of birth defects and mental retardation in newborns has increased in affected areas, and children exposed to the Chernobyl fallout experience higher incidences of leukemia, thyroid cancer, abnormalities of the immune system, and other birth defects (**FIGURE 11.18**).

Health consequences of Chernobyl FIGURE 11.18

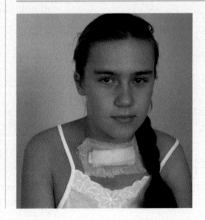

Shown is a 14-year-old who is recovering from thyroid cancer. A significant (25-fold) increase in thyroid cancer in children and adolescents occurred within a few years after the accident.

RADIOACTIVE WASTES

Radioactive wastes are classified as either "low-level" or "high-level." **Low-level radioactive wastes** include glassware, tools, paper, clothing, and other items contaminated by radioactivity. They are produced by nuclear power plants, university research labs, nuclear medicine departments in hospitals, and industries. In Canada, low-level radioactive waste is shipped to Atomic Energy of Canada Limited (AECL) for disposal. Uranium mine and mill tailings are a type of low-level waste that is generated during the mining and milling of uranium ore and during the production of uranium concentrate. Uranium ore concentrate is used to make fuel for Canadian and foreign power reactors. These mine tailings are often placed in specified containment areas such as lakes, valleys, or pits close to the mine or mill site. Some tailings may be used as backfill in the underground or open pit mine.

High-level radioactive wastes produced during nuclear fission include the reactor metals (fuel rods and assemblies), coolant fluids, and air or other gases found in the reactor. High-level radioactive wastes are also generated during the reprocessing of spent fuel. Produced by nuclear power plants and nuclear weapons facilities, high-level radioactive wastes are among the most dangerous human-made hazardous wastes.

Radioactive waste has been produced in Canada since the 1930s when the first uranium mine began operating at Port Radium in the Northwest Territories. Radioactive waste is currently managed in accordance with the requirements set out by the Canadian Nuclear Safety Commission (CNSC), an independent nuclear regulator. Waste inventory projections indicate that by 2033 Canada will have 15,000 m³ of nuclear fuel waste, 2.6 million m³ of low-level radioactive waste, and 222 million tonnes of uranium mill tailings.

As the radioisotopes in spent fuel decay, they produce considerable heat, are extremely toxic to organisms, and remain radioactive for thousands of years. Their dangerous level of radioactivity requires special handling. Secure storage of these materials must be guaranteed for thousands of years, until the materials decay sufficiently to be safe. The safe disposal of radioactive wastes is one of the main difficulties that must be overcome if nuclear energy is to realize its potential in the 21st century.

What are the best sites for the long-term storage of high-level radioactive wastes? Many scientists recommend storing the wastes in stable rock formations deep in the ground. People's reluctance to have radioactive wastes stored near their homes complicates the selection of these sites. Meanwhile, radioactive wastes continue to accumulate. Many commercially operated nuclear power plants store their spent fuel in huge indoor pools of water or in storage casks on-site. None of these plants were designed for long-term storage of spent fuel (**Figure 11.19** on page 414).

Yucca Mountain in Nevada has been identified as the only candidate in North America for a permanent underground storage site for high-level radioactive wastes from commercially operated power plants (see What a Scientist Sees on page 414). In 2002 it was chosen by the federal government as the U.S. nuclear-waste repository, despite controversy and the state of Nevada's opposition.

Whether or not nuclear waste is eventually stored in Yucca Mountain, the scientific community generally agrees that storage of high-level radioactive waste in deep underground repositories is the best long-term option. An underground waste facility is far safer than storing high-level nuclear waste as we do now, at many different commercial nuclear reactors, which pose a greater risk of terrorist attacks, theft, and, possibly, human health problems.

low-level radioactive wastes Solids, liquids, or gases that give off small amounts of ionizing radiation.

high-level radioactive wastes Radioactive solids, liquids, or gases that initially give off large amounts of ionizing radiation.

DECOMMISSIONING NUCLEAR POWER PLANTS

As nuclear power plants age, certain critical sections, like the reactor vessel, become brittle or corroded. At the end of their operational usefulness, nuclear power plants are not simply abandoned or demolished, because many parts have become contaminated with radioactivity.

A On-site storage casks at the Prairie Island nuclear power plant in Minnesota. Each cask holds 40 spent fuel assemblies (16 tonnes).

Storage casks for spent fuel FIGURE 11.19

Cask length: 5.2 m (17 ft)

- Protective cover
- Lid with metallic seals
- Storage cask (neutron shield)
- Canister of steel
- Spent fuel rods

B Details of a storage cask. Each cask, designed to last at least 40 years, is monitored and will be replaced if leakage occurs.

Yucca Mountain

A This tunnel provides access to nuclear waste storage at Yucca Mountain, an arid, sparsely populated area of Nevada.

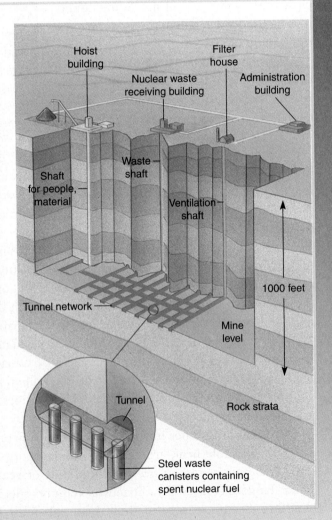

Hoist building

Nuclear waste receiving building

Filter house

Administration building

Shaft for people, material

Waste shaft

Ventilation shaft

Tunnel network

Mine level

1000 feet

Tunnel

Rock strata

Steel waste canisters containing spent nuclear fuel

B There is a huge complex of interconnected tunnels located in dense volcanic rock 305 metres beneath the Yucca Mountain crest. Canisters containing high-level radioactive waste can be stored in the tunnels.

Over the past three decades, Soviet (and now Russian) practices for radioactive waste disposal have often violated international standards:

- Billions of litres of liquid radioactive wastes were pumped directly underground, without being stored in protective containers. Russian officials claim that layers of clay and shale at the sites prevent leakage but admit that more leaks than expected have occurred.

- Highly radioactive wastes were dumped into the ocean in amounts double that of dumped wastes from 12 other nuclear nations combined.

- Both underground injection and underwater dumping of radioactive wastes continue because Russia lacks alternatives for nuclear waste processing. Potential health and environmental hazards associated with these wastes are unknown because so little data exist for these types of long-term storage.

Murmansk nuclear waste site

Three options exist when a nuclear power plant is closed: storage, entombment, or decommissioning. If an old plant is put into storage, the utility company guards it for 50 to 100 years while some of the radioactive materials decay, making it safer to dismantle the plant later. Accidental leaks during the storage period are an ongoing concern.

Most experts do not consider entombment, permanently encasing the entire power plant in concrete, a viable option because the tomb would have to remain intact for at least 1000 years. Accidental leaks would probably occur during that time, and we cannot guarantee that future generations would inspect and maintain the site.

The third option for the retirement of a nuclear power plant is to **decommission** the plant immediately after it closes. Advances in robotics may make it feasible to tear down sections of old plants that are too "hot" (radioactive)

decommission
To dismantle a nuclear power plant after it closes.

for workers to safely dismantle. As the plant is torn down, small sections of it are transported to a permanent storage site.

According to the International Atomic Energy Agency, 107 nuclear power plants worldwide were permanently retired as of 2004, and many nuclear power plants are nearing retirement age. During the 21st century, we may find we are paying more in our utility bills to decommission old plants than we are to construct new ones.

CONCEPT CHECK STOP

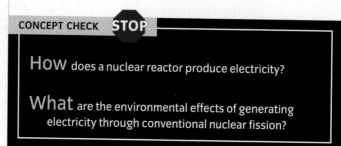

How does a nuclear reactor produce electricity?

What are the environmental effects of generating electricity through conventional nuclear fission?

Solar, Biomass, Wind, Hydroelectric, Geothermal, and Tidal Energy

LEARNING OBJECTIVES

Distinguish between active and passive solar heating and describe how each is used.

Contrast the advantages and disadvantages of solar thermal electric generation and photovoltaic solar cells in converting solar energy into electricity.

Define *biomass* and outline how it is used as a source of energy.

Compare the potential of wind energy and hydropower.

Describe geothermal and tidal energy.

DIRECT SOLAR ENERGY

The sun produces a tremendous amount of energy, and most of it dissipates into space. Only a small portion is radiated to Earth. Solar energy is different from fossil and nuclear fuels because it is always available; we will run out of solar energy only when the sun's nuclear fire burns out. To make solar energy useful, however, we must collect it.

active solar heating A system of putting the sun's energy to use in which a series of collectors absorb the solar energy, and pumps or fans distribute the collected heat.

In **active solar heating**, a series of collection devices mounted on a roof or in a field is used to gather solar energy. The most common solar collection device is a panel or plate of black metal, which absorbs the sun's energy (**FIGURE 11.20**). Active solar heating is used primarily for heating water, either for household use or for swimming pools. The heat absorbed by the solar collector is transferred to

Process Diagram

Active solar water heating FIGURE 11.20

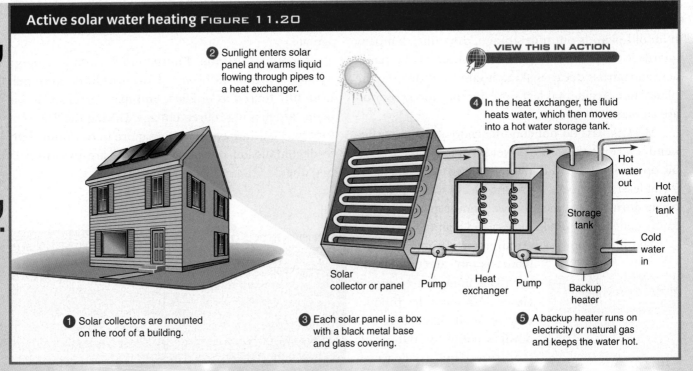

❷ Sunlight enters solar panel and warms liquid flowing through pipes to a heat exchanger.

VIEW THIS IN ACTION

❹ In the heat exchanger, the fluid heats water, which then moves into a hot water storage tank.

Hot water out

Hot water tank

Storage tank

Cold water in

Solar collector or panel

Pump

Heat exchanger

Pump

Backup heater

❶ Solar collectors are mounted on the roof of a building.

❸ Each solar panel is a box with a black metal base and glass covering.

❺ A backup heater runs on electricity or natural gas and keeps the water hot.

a fluid inside the panel, which is then pumped to the heat exchanger, where the heat is transferred to water that will be stored in the hot water tank. Solar domestic water heating can provide a family's hot water needs year-round.

Active solar energy is not used for space heating as commonly as it is used for heating water, but it may become more important when diminishing supplies of fossil fuels force gas and oil prices higher.

In **passive solar heating**, solar energy heats buildings without the need for pumps or fans to distribute the heat. Certain design features are incorporated into a passive solar heating system to warm buildings in winter and help them remain cool in summer (**FIGURE 11.21**). In the northern hemisphere, large south-facing windows receive more total sunlight during the day than windows facing other directions. Sunlight entering through the windows provides heat, which is then stored in floors and walls made of concrete or stone, or in containers of water. This stored heat is transmitted throughout the building naturally by **convection**, the circulation that occurs because warm air rises and cooler air sinks. Buildings with passive solar heating systems must be well insulated so that accumulated heat doesn't escape. Depending on the building's design and location, passive heating can save as much as 50 percent of heating costs.

> **passive solar heating** A system of putting the sun's energy to use without requiring mechanical devices to distribute the collected heat.

Passive solar heating FIGURE 11.21

A Behind the greenhouse glass of this passive solar home, rooms remain at steady temperatures during winter.

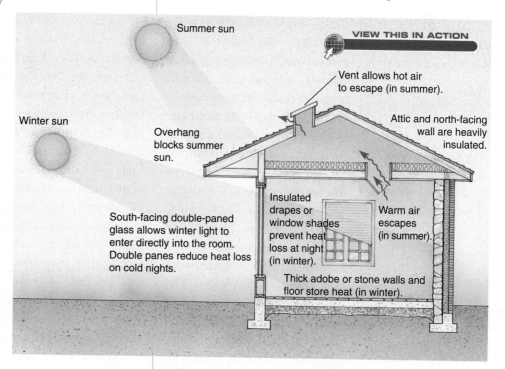

Summer sun

VIEW THIS IN ACTION

Winter sun

Overhang blocks summer sun.

Vent allows hot air to escape (in summer).

Attic and north-facing wall are heavily insulated.

South-facing double-paned glass allows winter light to enter directly into the room. Double panes reduce heat loss on cold nights.

Insulated drapes or window shades prevent heat loss at night (in winter).

Warm air escapes (in summer).

Thick adobe or stone walls and floor store heat (in winter).

B Several passive designs are incorporated into this home.

Solar, Biomass, Wind, Hydroelectric, Geothermal, and Tidal Energy 417

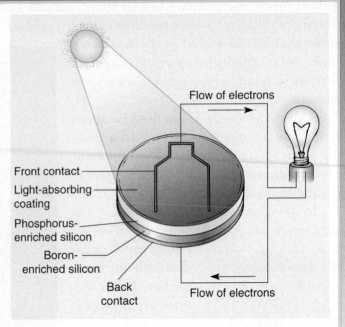

Photovoltaic Cells

A A student seeing the roof of the Intercultural Center of Georgetown University probably knows that it has arrays of photovoltaic (PV) cells to collect solar energy. The PV system supplies about 10 percent of the school's electricity.

Flow of electrons

Front contact
Light-absorbing coating
Phosphorus-enriched silicon
Boron-enriched silicon
Back contact

Flow of electrons

B A scientist looking at those arrays knows that photovoltaic cells contain silicon and other materials. Sunlight excites electrons, which are ejected from silicon atoms. Useful electricity is generated when the ejected electrons flow out of the PV cells through a wire.

Photovoltaic solar cells

Photovoltaic (PV) solar cells can convert sunlight directly into electricity (see What a Scientist Sees). They are usually arranged on large panels that absorb sunlight even on cloudy or rainy days.

PV cells generate electricity with no pollution and minimal maintenance. They can be used on any scale, from small portable modules attached to camping lanterns to large, multi-megawatt power plants, and can power satellites, uncrewed airplanes, highway signals, wristwatches, and calculators. The cells' widespread use to generate electricity is currently limited by their low efficiency at converting solar energy to electricity, and by the amount of land needed to hold the number of solar panels required for large-scale use.

> **photovoltaic (PV) solar cell** A wafer or thin film of solid state materials, such as silicon or gallium arsenide, that are treated with certain metals in such a way that the film generates electricity—that is, a flow of electrons—when solar energy is absorbed.

In remote areas not served by electric power plants, like the rural areas of developing countries, it is more economical to use PV cells for electricity. Photovoltaics generate energy that can pump water, refrigerate vaccines, grind grain, charge batteries, and supply rural homes with lighting. According to the Institute for Sustainable Power, more than 1 million households in the developing countries of Asia, Latin America, and Africa have installed PV solar cells on the roofs of their homes. A PV panel the size of two pizza boxes supplies a rural household with enough electricity for five lights, a radio, and a television.

Utility companies can purchase PV devices in modular units, which can become operational in a short period. Rather than committing a billion dollars or more and a decade or more to build a new power plant, they can increase generating capacity in small increments. The PV units can provide the additional energy, for example, to power irrigation pumps on hot, sunny days.

Future technological progress may make PVs economically competitive with electricity produced by conventional energy sources. The production of "thin-film"

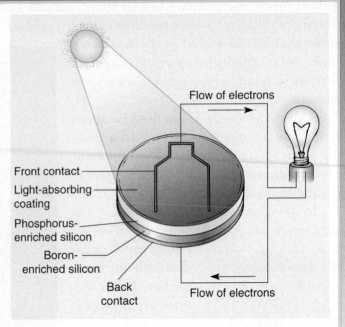

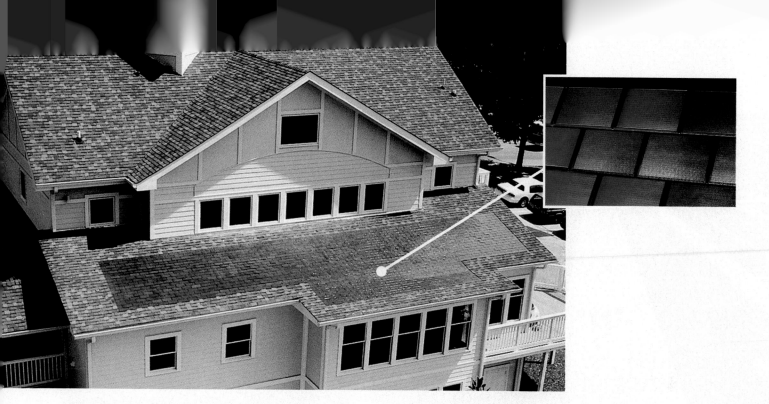

Solar shingles FIGURE 11.22

These thin-film solar cells look much like conventional roofing materials.

solar cells (FIGURE 11.22), which are much cheaper to manufacture than standard PVs, has decreased costs for PVs. More than 120,000 Japanese homes have installed PV solar-energy roofing in the past few years. The Million Solar Roofs initiative, sponsored by the U.S. government, plans to have solar roofing on 1 million buildings by 2010. Another technological advance that shows promise is dye-sensitized solar cells, which can be produced at about one-fifth the cost of conventional silicon panels.

Solar thermal electric generation

In **solar thermal electric generation**, electricity is produced by several different systems that collect sunlight and concentrate it using a combination of mirrors or lenses to heat a working fluid to high temperatures.

In one such system, computer-guided trough-shaped mirrors track the sun for optimum efficiency, centre sunlight on nearby oil-filled pipes, and heat the oil to 390°C (FIGURE 11.23 on page 420). The hot oil is circulated to a water storage system and

solar thermal electric generation
A means of producing electricity in which the sun's energy is concentrated by mirrors or lenses onto a fluid-filled pipe; the heated fluid is used to generate electricity.

used to boil water into super-heated steam, which turns a turbine to generate electricity.

Solar thermal systems often have a backup—usually natural gas—to generate electricity at night and during cloudy days when solar power isn't operating. The world's largest solar thermal system of this type currently operates in the Mojave Desert in southern California.

Solar thermal energy systems are inherently more efficient than other solar technologies because they concentrate the sun's energy. With improved engineering, manufacturing, and construction methods, solar thermal energy may become cost-competitive with fossil fuels (TABLE 11.2 on page 420). In addition, the environmental benefits of solar thermal plants are significant: They don't produce air pollution or contribute to acid rain or global warming.

Solar-generated hydrogen Increasingly, people think of hydrogen as the fuel of the future, as it is abundant as well as easily produced. Electricity generated by photovoltaics or wind energy can split water into the gases oxygen and hydrogen, though this process isn't yet economical. Hydrogen can also be produced using conventional energy sources such as fossil fuels and nuclear power. However, using fossil fuels or nuclear energy to create hydrogen results in the same

Solar thermal electric generation FIGURE 11.23

A

A A solar thermal plant in California uses troughs to focus sunlight on a fluid-filled tube, as shown in **B**. The heated oil is pumped to a water tank where it generates steam used to produce electricity. For simplicity, arrows show sunlight converging on several points; sunlight actually converges on the pipe throughout its length.

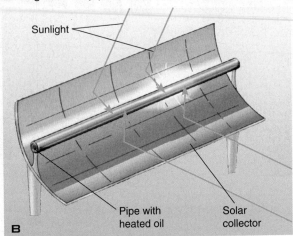

Sunlight

Pipe with heated oil

Solar collector

B

Generating costs of electric power plants TABLE 11.2

Energy source	Generating costs (cents per kilowatt-hour)*
Hydropower	4–10
Biomass	6–8
Geothermal	3–8
Wind	4–5
Solar thermal	10–15
Photovoltaics	15–25
Natural gas	4–5
Coal	4–5
Nuclear power	10–15

*Electricity production and consumption are measured in kilowatt-hours (kWh). As an example, one 50-watt light bulb that is on for 20 hours uses one kilowatt-hour of electricity ($50 \times 20 = 1000$ watt-hours = 1 kWh).

serious environmental problems we discussed previously. We therefore limit our discussion to hydrogen fuel production using solar electricity, which is sustainable but not yet cost-efficient.

Hydrogen is a clean fuel; it produces water and heat as it burns and produces no sulphur oxides, carbon monoxide, hydrocarbon particulates, or CO_2 emissions that are contributing factors to most of our environmental challenges today. It does produce some nitrogen oxides, though in amounts that are fairly easy to control. Hydrogen has the potential to provide energy for transportation (in the form of hydrogen-powered automobiles) as well as for heating buildings and producing electricity.

It may seem wasteful to use electricity generated from solar energy to make hydrogen that will then be used to generate electricity. However, the electricity generated by existing photovoltaic cells must be used immediately, whereas hydrogen offers a convenient way to store solar energy as chemical energy. It can be transported by pipeline, possibly less expensively than electricity is transported by wire.

Production of hydrogen from PV electricity currently has a relatively low efficiency (perhaps 10 percent), which means that very little of the solar energy absorbed by the PV cells is actually converted into the chemical energy of hydrogen fuel. Low efficiency translates into high costs. Scientists are working to improve this efficiency because decreased costs could make solar-generated hydrogen fuel commercially viable.

Other challenges besides high costs face us if we are to replace gasoline with hydrogen as a transportation fuel. First, we would need to develop a complex infrastructure (like hydrogen pipelines) to provide hydrogen to service stations. Another challenge is developing **fuel cells** for motor vehicles that are inexpensive, safe, and can drive a long distance without the need to be re-fuelled. A fuel cell is an electrochemical cell similar to a battery (**FIGURE 11.24**).

Fuel cells produce power as long as they are supplied with fuel, whereas batteries store a fixed amount of energy. The major car

fuel cell A device that directly converts chemical energy into electricity without producing steam that runs a turbine and generator; the fuel cell requires hydrogen and oxygen from the air.

biomass Plant and animal material used as fuel.

manufacturers are now developing automobiles and buses powered by hydrogen fuel cells.

BIOMASS ENERGY

Some renewable energy sources indirectly use the sun's energy. Combustion of **biomass** (organic matter) is an example of indirect solar energy, because plants use solar energy for photosynthesis and store the energy in biomass.

Biomass is one of the oldest fuels known to humans, and it consists of materials like wood, fast-growing plant and algal crops, crop wastes, sawdust and wood chips, and animal wastes. Biomass contains chemical energy that comes from the sun's radiant energy, which photosynthetic organisms use to form organic molecules. Biomass is a renewable form of energy if managed properly.

Biomass fuel, which may be a solid, liquid, or gas, is burned to release its energy. Solid biomass fuels like wood, charcoal (wood turned into coal by partial burning), animal dung, and peat (partly decayed plant matter found in bogs and swamps) supply a substantial portion of the world's energy. At least half of the human population relies on biomass as their main source of energy. In developing countries, wood is the primary fuel for cooking and heat (**FIGURE 11.25**).

Fuel cells in a laboratory FIGURE 11.24

A These fuel cells combine hydrogen and oxygen to create electricity.

Flow of negatively charged electrons provides electricity

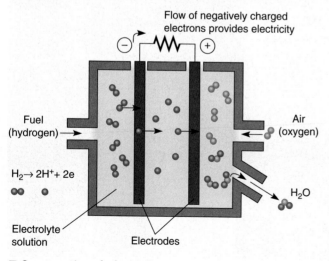

Fuel (hydrogen) →

Air (oxygen)

$H_2 \rightarrow 2H^+ + 2e$

H_2O

Electrolyte solution

Electrodes

B Cross-section of a fuel cell.

Biomass FIGURE 11.25
Firewood is the major energy source for most of the developing world. Photographed in Nepal.

Nepal

Global Locator

It is possible to convert biomass, particularly animal wastes, into **biogas**. Biogas is produced through anaerobic digestion whereby bacteria decompose organic matter under conditions where there is no oxygen available. The end product is a gas mainly composed of methane (60 percent) and carbon dioxide and is comparable in many ways to natural gas. Unlike fossil fuel combustion, biogas production has the opportunity to be carbon neutral, and when recovered properly it does not emit additional greenhouse gases into the atmosphere. In addition, this digestion process provides the advantage of treating organic waste and reducing the environmental impact of these wastes. Biogas collected can be used as an energy source for generators, boilers, burners, dryers, or any equipment using propane, gas, or diesel. These types of equipment require minor adjustments to run on biogas.

The technology to create anaerobic digesters is well established in European countries such as Denmark and Germany where over 1000 MW of electricity are produced from waste and farm manure. In India and China, several million family-sized biogas digesters use microbial decomposition of household and agricultural wastes to produce biogas for cooking and lighting (**Figure 11.26**). In 2007, agreement was reached in Montreal to develop biogas anaerobic digester power plants in Canada. As well, biofuels generated at landfills are used to heat buildings on-site and in adjacent areas, as demonstrated in Montreal.

Biomass can also be converted into liquid fuels, especially **methanol** (methyl alcohol) and **ethanol** (ethyl alcohol), which can be used in internal combustion engines. Mixing gasoline with 10 percent ethanol (usually produced from corn) produces a cleaner-burning

Biogas digester in India FIGURE 11.26

This small-scale biogas digester is being evaluated at a research centre. Animal manure placed in the digester decomposes, releasing methane gas that can be used as cooking fuel.

Global Locator

mixture called *gasohol*. *Biodiesel*, made from plant or animal oils, is becoming more popular as an alternative fuel for diesel engines in trucks, farm equipment, and boats. The oil is often refined from waste oil produced at restaurants (like the oil used to make french fries); biodiesel burns much cleaner than diesel fuel.

Although some North American energy companies convert sugarcane, corn, or wood crops to alcohol, others are interested in the commercial conversion of agricultural and municipal wastes into ethanol. Currently, the profitability of ethanol is only possible because of government subsidies that reduce ethanol's cost. However, as more car companies introduce ethanol-friendly vehicles, these subsidies may cease to be necessary.

Biomass is attractive as a source of energy because it reduces dependence on fossil fuels and often makes use of waste products, thereby reducing our waste disposal problem. Biomass combustion is not completely free of the pollution problems of fossil fuels, but biomass combustion produces levels of sulphur and ash that are lower than those that coal produces.

Some problems associated with obtaining energy from biomass include the use of land and water that might otherwise be dedicated to agriculture. Shifting the agricultural balance toward energy production might decrease food production; contribute to greater challenges to feed an ever-expanding human population; exploit marginal lands and reduce the important ecological services these provide in their natural state; increase carbon emissions rather than be a carbon neutral process; and contribute to higher food prices at the supermarket. Also, as mentioned earlier, at least half of the world's population relies on biomass as its main source of energy. Unfortunately, in many areas people burn wood faster than they replant trees. Intensive use of wood for energy has resulted in severe damage to the environment, including soil erosion, deforestation and desertification, air pollution, and degradation of water supplies.

Excessive use of crop biomass can also harm soil quality. Crop residues such as cornstalks are increasingly being used for energy. Crop residues left in and on the ground prevent erosion by holding the soil in place; their removal would eventually deplete the soil of minerals and reduce its productivity.

Food energy, another biologically based energy source, is described in more detail in Chapter 5.

WIND ENERGY

During the 1990s and early 2000s, wind became the world's fastest growing source of energy. Wind results from the sun warming the atmosphere.

> **wind energy**
> Electric energy obtained from surface air currents caused by the solar warming of air.

Wind energy is also an indirect form of solar energy: The radiant energy of the sun is transformed into mechanical energy through the movement of air molecules. Wind is sporadic over much of Earth's surface, varying in direction and magnitude. Like direct solar energy, wind energy is a highly dispersed form of energy. Harnessing wind energy to generate electricity has great potential.

New wind turbines are huge—100 metres tall—and have long blades designed to harness wind energy efficiently (**FIGURE 11.27A** and **11.27B** on page 424). As turbines have become larger and more efficient, costs for wind power have declined rapidly—from $.40 per kilowatt-hour in 1980 to a current cost of $.04 to $.05 per kilowatt-hour. Wind power is cost-competitive with many forms of conventional energy. Advances such as turbines that use variable-speed operation may make wind energy an important global source of electricity during the first half of the 21st century.

Denmark, one of the world leaders in wind power, currently generates 21 percent of its electricity using wind energy. Much of this power is generated offshore because ocean winds are strong. Other leading wind energy countries include Germany, which produces about one-third of the world's total wind power, Spain, the United States, and India.

Harnessing wind energy is most profitable in rural areas that receive fairly continual winds, like islands, coastal areas, mountain passes, and grasslands. The world's largest concentration of wind turbines is currently located in the Tehachapi Pass at the southern end of the Sierra Nevada mountain range in California. In Canada, some of the best locations for large-scale electricity generation from wind energy are on the Great Plains. All three Prairie provinces have been developing wind energy operations. In November 2008, Manitoba's government and the province's hydro utility signed a deal to develop what would be the largest wind farm in Canada near St. Joseph in southern Manitoba.

A Wind energy represents a promising alternative that could diminish our dependence on fossil fuels.

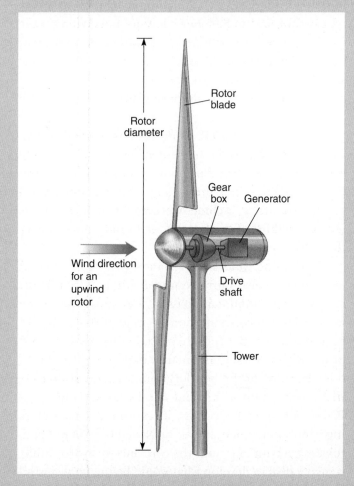

Rotor diameter

Rotor blade

Gear box

Generator

Wind direction for an upwind rotor

Drive shaft

Tower

B This basic wind turbine design has a horizontal axis (horizontal refers to the orientation of the drive shaft). Airflow causes the turbine's blades to turn 15 to 60 revolutions per minute (rpm). As the blades turn, gears within the turbine spin the drive shaft. This spinning powers the generator, which sends electricity through underground cables to a nearby utility. (The tower isn't drawn to scale and is much taller than depicted.)

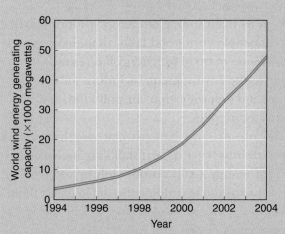

C Global wind energy has shown record growth in recent years.

NATIONAL GEOGRAPHIC

The project is estimated to cost $800 million and would generate 300 megawatts of electricity from 130 turbines. With construction underway in 2011, the wind farm will generate $70 million in lease payments to local landowners, increase opportunities for local jobs, and bring in almost $500,000 to municipal coffers, which constitutes a third of the total budget. The wind farm will displace 800,000 tonnes of greenhouse gas emissions annually, the equivalent of taking 145,000 cars off the road.

Wind produces no waste and is a clean source of energy. It produces no emissions of sulphur dioxide, carbon dioxide, or nitrogen oxides. Every kilowatt-hour of electricity generated by wind power rather than fossil fuels prevents 2 to 4.5 kilograms of the greenhouse gas CO_2 from entering the atmosphere.

The use of wind power doesn't cause major environmental problems, although one concern is reported bird and bat kills, and disturbances to flight patterns and nesting behaviours of various avian species. The California Energy Commission estimated that several hundred birds turned up dead in the vicinity of the 7000 turbines at Altamont Pass in California during a two-year study; most had collided with the turbines. Studies later determined that Altamont Pass is a major bird migration pathway. In response to this concern, wind power sites have implemented technical "fixes" like painted blades and anti-perching devices, or have shut down operations during peak bird migration periods. Developers of future wind farm sites currently conduct voluntary wildlife studies and try to locate sites away from bird and bat routes. Locating wind power sites strategically can minimize the effects of this energy source into the future.

Not all people welcome wind power projects. The Maple Ridge Wind Farm in upstate New York has almost 200 windmills. Many local residents appreciate the extra money that wind-farm leases provide to the local economy. However, others think the wind turbines ruin their view of the Adirondack Mountains. A wind farm in Innisfil, Ontario, along Highway 400 near Toronto, is being planned with the intent to build five 120-metre-high turbines on 80 hectares of agricultural land along the highway's east side. However, local residents and stakeholders voiced opposition to the placement of the wind farm. Many feel that the location is too close to residential homes and would occupy valuable commercial and retail space along the highway. Once built, local residents fear that nothing will stop many additional wind farms from being developed in the area. Knowing that once they are constructed the turbines will be there forever, local residents say they should not have to put up with plummeting property values when they are likely never to see any benefits of the electricity from the wind turbines.

HYDROELECTRIC ENERGY

The sun's energy drives the hydrologic cycle, which includes precipitation, evaporation, transpiration, and drainage and runoff. As water flows from higher elevations back to sea level through rivers and streams, dams can harness and make use of its energy. The potential energy of water held back by a dam is converted to kinetic energy as the water turns turbines to generate electricity (**FIGURE 11.28** on page 426). **Hydropower** is more efficient than any other energy source for producing electricity—about 90 percent of available hydropower energy is converted into consumable electricity.

> **hydropower**
> A form of renewable energy that relies on flowing or falling water to generate electricity.

Hydropower generates approximately 19 percent of the world's electricity, making it the most widely used form of renewable energy. The four countries with the greatest hydroelectric production are Canada, the United States, Brazil, and China. In Canada, there are a multitude of reservoirs and dams that produce 60 percent of the country's electricity. In 1999, Canada generated more than 340 billion kilowatt-hours of power, far outdistancing other major producers. Many of the developed countries have already built dams at most of their potential sites. However, in countries in Africa and South America, hydropower represents a great potential source of electricity.

Traditional hydropower technology was only suited for large dams with rapidly flowing water, such as those found on the Columbia River and its tributaries in British Columbia, the Nelson and Churchill Rivers in Manitoba, and the La Grande River in northwestern Quebec. New designs allow modern turbines to harness electricity from large, slow-moving rivers or from streams with small flow capacities. These new technologies have the potential to increase the amount of hydroelectric power generated by existing dams.

Hydroelectric power FIGURE 11.28

A The Mactaquac hydro dam in New Brunswick, like other hydro dams in Canada, generates electricity as the water moves through the dam.

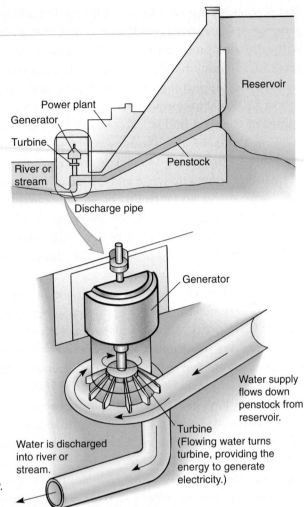

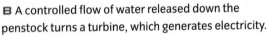

B A controlled flow of water released down the penstock turns a turbine, which generates electricity.

Problems with hydroelectric power

While hydroelectric power has major advantages, such as eliminating the need to import fuel, lower operating costs, and the lack of carbon dioxide production, there are a number of downfalls as well. Building a dam changes the natural flow of a river: Water backs up, flooding large areas of land and forming a reservoir, which destroys plant and animal habitats. Native fishes are particularly susceptible to dams because the original river ecosystem is so altered, which also affects the migration of spawning fish. Below the dam, the once-powerful river is reduced to a relative trickle. The natural beauty of the countryside is affected, and certain forms of wilderness recreation are made impossible or less enjoyable, although the dams permit water sports in the reservoir. In Canada, the Center of Expertise on Hydropower Impacts on Fish and Fish Habitat (CHIF) was created to research the impacts of hydroelectricity on fish and fish habitat and to promote cooperation and collaboration between scientists, managers, and developers within the industry in Canada. CHIF research focuses on reservoir construction and management, fish passage at and near hydro facilities, and river flow modification.

At least 200 large dams around the world are associated with earthquakes that may occur during and after the filling of a reservoir behind a dam. The larger the reservoir and the faster it is filled, the greater the intensity of seismic activity. An area need not be seismically active to have earthquakes induced by reservoirs.

If a dam breaks, people and property downstream may be affected. In addition, waterborne diseases may spread through the population. *Schistosomiasis* is a tropical disease caused by a parasitic worm. As much as half the population of Egypt suffers from this disease, largely as a result of the Aswan Dam, built on the Nile River in 1902 to control flooding but used since 1960 to provide electric power. The large

The James Bay Hydroelectric Project is a series of constructed hydroelectric power stations in the La Grande watershed in northwestern Quebec. In May 1972, the James Bay Hydroelectric Project began building a series of dams, dikes, reservoirs and power stations, altering the flow of water in an area covering over 350,000 square kilometres, approximately the size of the state of New York. The development occurred in three distinct phases and resulted in the largest hydroelectric power development in Canada, and one of the largest hydroelectric systems in the world. Once completed the project had the generating capacity of more than three times the capacity of all of the power stations at Niagara Falls, eight times the power of the Hoover Dam, and over twice the power of all eight CANDU units at the Bruce Nuclear Generating Station, the largest in North America.

During the first phase of the James Bay Hydroelectric Project three major rivers, the Caniapiscau, Eastmain, and Opinaca, were diverted into reservoirs on La Grande Rivière. The diversion of these rivers nearly doubled the power potential, and power stations were built to channel these massive amounts of hydroelectric power. In 1972, the Cree and Inuit of James Bay opposed the project, although they eventually gave up their claim to certain land in northern Quebec in exchange for a $225 million settlement. By 1985, the first phase of the

project was complete with a total cost of development close to $14 billion.

The second phase of construction on the James Bay Hydroelectric Project began in 1989, redirecting the Eastmain, Laforge, and Caniapiscau Rivers into La Grande Rivière. In 1994, with the second phase of the development almost complete, the project was halted over environmental concerns. It was not until 2002 that an agreement with the Cree of James Bay cleared the way for the completion of phase two of the development.

The third and final phase of the James Bay project, known as James Bay 2 or the Great Whale Hydroelectric Project, was set to begin in March 1989. However, public concern over the ecological impacts of the development hindered construction. This phase of the project was set to affect a watershed area the size of France. In November 1994, a joint effort between federal and provincial governments and Aboriginal representatives ordered Hydro-Québec to revise the environmental impact study it had originally prepared for the project. Hydro-Québec announced the following day its decision to halt phase three construction indefinitely. Increasing energy demands and world energy prices may reawaken the drive to continue development of the James Bay Hydroelectric Project in the future.

reservoir behind the dam provides habitat for the worm, which spends part of its life cycle in the water. The worms infect humans during bathing, swimming, or walking barefoot along water banks, or by drinking infected water.

In arid regions, the creation of a reservoir results in greater evaporation because it has a larger surface area in contact with the air than the stream or river did. As a result, serious water loss and increased salinity of the remaining water may occur.

When an area behind a dam is flooded, the trees and other vegetation die and are decomposed, releasing the large quantities of carbon that were tied up in organic molecules in the plant bodies as carbon dioxide and methane. These gases absorb infrared radiation and are therefore associated with global warming.

The construction of large dams also involuntarily displaces people from lands flooded by reservoirs. The Three Gorges Project, the largest dam ever built, spans the Yangtze River in China (**FIGURE 11.29**). This dam created a reservoir 663 kilometres long. The tops of as many as 100 mountains became small islands, fragmenting habitat and threatening 57 endangered species, and almost 2 million people were displaced, the largest number for any dam project.

The environmental and social impacts of a dam may not be acceptable to the people living in a particular area. Laws prevent or restrict the building of dams in certain locations. In the United States, the Wild and Scenic Rivers Act prevents the hydroelectric development of certain rivers, although this law protects less than 1 percent of the nation's total river systems. Other countries, such as Norway and Sweden, have similar laws.

Dams cost a great deal to build but are relatively inexpensive to operate. A dam has a limited life span, usually 50 to 100 years, because over time the reservoir fills

The Three Gorges Dam in China FIGURE 11.29

in with silt until it cannot hold enough water to generate electricity. This trapped silt, which is rich in nutrients, is prevented from enriching agricultural lands downstream. Egypt, for example, historically had rich fertile soils constantly recharged by sediments deposited through the annual flooding of the Nile River. Now Egypt must rely on heavy applications of chemical fertilizer downstream from the Aswan Dam to maintain the fertility of the Nile River Valley.

GEOTHERMAL ENERGY

Geothermal energy, the natural heat within Earth that steadily increases as a gradient away from the surface, arises from Earth's core, from friction along continental plate boundaries, from gravitational compression, and from the decay of radioactive elements. The amount of geothermal energy is enormous. Scientists estimate that just 1 percent of the heat contained in the uppermost 10 kilometres of Earth's crust is equivalent to 500 times the energy contained in all of Earth's oil and natural gas resources.

> **geothermal energy** The use of energy from Earth's interior for space heating or generation of electricity.

Geothermal energy is also associated with volcanism. Large underground reservoirs of heat exist in areas of geologically recent volcanism. As groundwater in these areas travels downward, it is heated, becomes buoyant, and then rises until it is trapped by an impermeable layer in Earth's crust, forming a **hydrothermal reservoir**. Hydrothermal reservoirs contain hot water and possibly steam, depending on the temperature and pressure of the fluid. Some of the hot water or steam may escape to the surface, creating hot springs or geysers. Hot springs have been used for thousands of years for bathing, cooking, and heating buildings. Drilling a well brings the hot fluid to the surface, where a power station may use it to supply heat directly to consumers or to generate electricity (**FIGURE 11.30A**). The electricity these power stations generate is inexpensive and reliable.

The United States is the world's largest producer of geothermal electricity. Other important producers of geothermal energy include the Philippines, Italy, Japan, Mexico, Indonesia, New Zealand, and Iceland, with Canada beginning to experiment with its use (**Figure 11.30b**).

Iceland, a country with minimal oil and natural gas resources, is located on the mid-Atlantic ridge, a boundary between two continental plates. Iceland is therefore an island of intense volcanic activity with considerable geothermal resources (as evidenced by the massive volcano eruption in April 2010). Iceland uses geothermal energy to generate electricity, to heat two-thirds of its homes, and to grow most of the fruits and vegetables required by the people of Iceland in geothermally heated greenhouses.

Is geothermal energy renewable? As a source of heat for geothermal energy, the planet is inexhaustible on a human time scale. However, the water used to transfer the heat to the surface isn't inexhaustible. Some geothermal applications recirculate all the water back into the underground reservoir, ensuring many decades of heat extraction from a given reservoir.

Geothermal energy is considered environmentally benign because it emits only a fraction of the air pollutants released by conventional fossil fuel–based energy technologies. The most common environmental hazard is the emission of hydrogen sulphide (H_2S) gas, which comes from the very low levels of dissolved minerals and salts found in the steam or hot water. Hydrogen sulphide smells like rotten eggs and is toxic to humans at high concentrations. A lesser concern is that the surrounding land may subside, or sink, as the water from hot springs and their connecting underground reservoirs is removed.

Scientists are studying how to economically extract some of the vast amount of geothermal energy stored in hot, dry rock. Such a technology could greatly expand the extent and use of geothermal resources.

Heating and cooling buildings with geothermal energy

Geothermal heat pumps (GHPs) take advantage of the difference in temperature between Earth's surface and subsurface (at depths from 1 metre to about 100 metres). In an underground arrangement of pipes containing circulating fluids, GHPs extract natural heat in winter, when Earth acts as a heat source, and transfer excess heat underground in summer, when Earth acts as a heat sink. Though available for many years, GHPs aren't

Geothermal power plant Figure 11.30

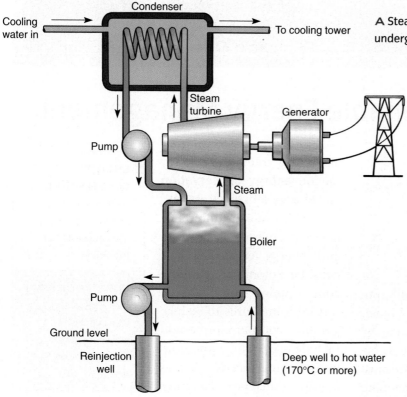

A Steam separated from hot water pumped from underground turns a turbine and generates electricity.

B A geothermal power plant in Iceland.

widely used because their installation is expensive. However, with the growth of *green architecture* and rising fuel costs, commercial and residential use of GHPs is on the rise. The system's benefits include low operating costs and high efficiency.

TIDAL ENERGY

Tides, or the rise and fall of the surface waters of the ocean and seas that occur twice each day, are the result of the gravitational pull of the moon and the sun. A dam built across a bay can harness the energy of large tides to generate electricity. As the tide falls, water flowing back to the ocean over the dam's spillway turns a turbine and generates electricity through **tidal energy**.

■ **tidal energy**
A form of renewable energy that relies on the ebb and flow of the tides to generate electricity.

Power plants using tidal power are in operation in France, Russia, China, and Canada. Tidal energy can't become a significant resource worldwide because few areas have large enough differences in water level between high and low tides to make power generation feasible. The most promising locations for tidal power in North America include the Bay of Fundy in Nova Scotia, Passamaquoddy Bay in Maine, Puget Sound in Washington State, and Cook Inlet in Alaska.

Other problems associated with tidal energy include the high cost of building a tidal power station and potential environmental problems associated with tidal energy in estuaries, coastal areas where river currents meet ocean tides. Fishes and countless invertebrates migrate to estuaries to spawn. Building a dam across the mouth of an estuary would prevent these animals from reaching their breeding habitats.

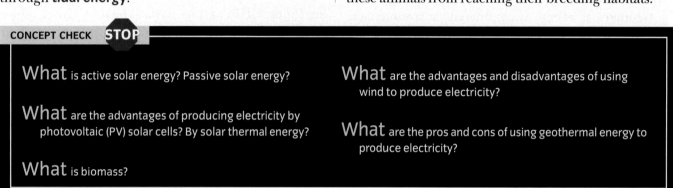

CONCEPT CHECK STOP

What is active solar energy? Passive solar energy?

What are the advantages of producing electricity by photovoltaic (PV) solar cells? By solar thermal energy?

What is biomass?

What are the advantages and disadvantages of using wind to produce electricity?

What are the pros and cons of using geothermal energy to produce electricity?

Approaches to Sustainable Energy Management

LEARNING OBJECTIVES

Distinguish between energy conservation and energy efficiency and give examples of each.

Define *cogeneration*.

Human requirements for energy will continue to increase, if only because the human population is growing. In addition, energy consumption continues to increase as developing countries raise their standards of living. We must therefore place a high priority not only on developing alternative sources of energy but on **energy conservation** and **energy efficiency** as well.

To illustrate the difference between energy conservation and energy efficiency, let's consider automobile gasoline consumption. Energy conservation measures to reduce gasoline consumption would include carpooling and lowering driving speeds, whereas energy efficiency measures

■ **energy conservation**
Using less energy by reducing energy use and waste, for example.

■ **energy efficiency**
Using less energy to accomplish a given task by using new technology, for example.

would include designing and manufacturing more fuel-efficient automobiles. Both conservation and efficiency accomplish the same goal—saving energy.

Many energy experts consider energy conservation and energy efficiency the most promising energy "sources" available, because they save energy for future use and buy us time to explore new energy alternatives. Developing technologies for energy conservation and efficiency costs less than developing new sources or supplies of energy; the technologies also improve the economy's productivity.

Energy-efficient technologies and greater efforts at conservation would also provide important environmental benefits by reducing air pollution. CO_2 emissions that contribute to global warming, acid precipitation, and other environmental problems are related to large quantities of energy production and consumption.

ENERGY-EFFICIENT TECHNOLOGIES

The development of more efficient appliances, automobiles, buildings, and industrial processes has helped reduce energy consumption in highly developed countries. Compact fluorescent light bulbs produce light of comparable quality to that of incandescent light bulbs but require only 25 percent of the energy and last up to 15 times longer, although they can be the source of certain toxic wastes. Although relatively expensive, the energy-efficient bulbs more than pay for themselves in energy savings. Standard long-tube fluorescent bulbs have become more efficient. New condensing furnaces require approximately 30 percent less fuel than conventional gas furnaces. Heat exchangers are also becoming more prevalent as sustainable energy management options for residential dwellings and commercial buildings. "Superinsulated" buildings use 70 to 90 percent less heat energy than buildings insulated by standard methods (FIGURE 11.31).

Most industrialized countries have set appliance efficiency standards for refrigerators, freezers, washing machines, clothes dryers, dishwashers, room air conditioners, and ranges and ovens (including microwaves). For example, refrigerators built today consume 75 percent less energy than comparable models built in the mid-1970s.

Superinsulated buildings FIGURE 11.31

A Some of the characteristics of a superinsulated home, which is so well insulated and airtight it doesn't require a furnace in winter. Heat from the inhabitants, light bulbs, and appliances provides almost all the necessary heat.

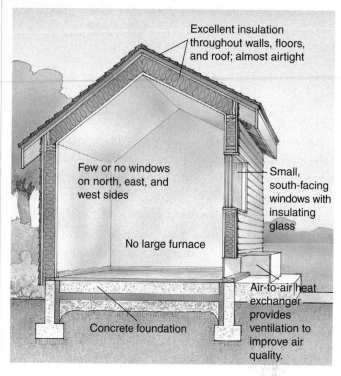

Excellent insulation throughout walls, floors, and roof; almost airtight

Few or no windows on north, east, and west sides

Small, south-facing windows with insulating glass

No large furnace

Air-to-air heat exchanger provides ventilation to improve air quality.

Concrete foundation

B A superinsulated office building in Toronto has south-facing windows with insulating glass. The building is so well insulated it uses no furnace.

Cogeneration FIGURE 11.32

In this example of a cogeneration system, fuel combustion generates electricity in a generator. The electricity produced is used in-house or sold to a local utility. The waste heat (leftover hot gases or steam) is recovered for useful purposes, such as industrial processes, heating buildings, hot water heating, and generating additional electricity.

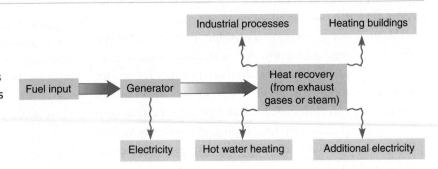

Energy costs often account for 30 percent of a company's operating budget. It makes good economic sense for businesses in older buildings to invest in energy improvements, which often pay for themselves in a few years. Energy improvements may be as simple as fine-tuning existing heating, ventilation, and air conditioning systems or as major as replacing all the windows and lights. Either way, both the environment and the company's bottom line can benefit.

Automobile efficiency has improved dramatically since the mid-1970s as a result of the use of lighter materials and designs that reduce air drag. The average fuel efficiency of new passenger cars doubled between the mid-1970s and the mid-1980s. It declined after that, as larger vehicles became more popular, but high gasoline prices are reversing the trend: Sales of compact and subcompact cars have increased over the last few years, and sales of trucks and SUVs have dropped. The number of hybrid models available now is indicative of the further gains that can be made using current technology.

Cogeneration
One energy technology with a bright future is **cogeneration**, or **combined heat and power (CHP)**. Cogeneration involves the production of two useful forms of energy from the same fuel. A CHP system first generates electricity, and then the steam produced during this process is used rather than wasted. The system's overall conversion efficiency (that is, the ratio of useful energy produced to fuel energy used)

cogeneration (CHP) An energy technology that involves recycling "waste" heat.

is high because some of what is usually waste heat is incorporated into the process.

Cogeneration can be cost-effective on both a small and a large scale. Modular CHP systems enable hospitals, hotels, restaurants, factories, and other businesses to harness steam, which would otherwise be wasted, to heat buildings, cook food, or operate machinery before it cools and gets pumped back into the boiler as water (**FIGURE 11.32**).

ELECTRIC POWER COMPANIES AND ENERGY EFFICIENCY

Changes in the regulations that govern electric utilities allow these companies to make more money by generating less electricity. Such programs provide incentives for energy conservation and thereby reduce power plant emissions that contribute to environmental problems.

Electric utilities can often avoid the massive expenses of building new power plants or purchasing additional power by helping electricity consumers save energy. Some utilities support energy conservation and efficiency by offering cash awards to consumers who install energy-efficient technologies. Other utilities give customers energy-efficient compact fluorescent light bulbs, air conditioners, or other appliances. They then charge slightly higher rates or a small leasing fee, but the greater efficiency results in savings for both the utility company and the consumer. The utility company makes more money from selling less electricity because it does not have to invest in additional power generation to meet increased demand. The consumer saves because the efficient light bulbs or appliances use less energy, which more than offsets the higher rates.

Electric power plants are themselves an important target for improved energy efficiency. Much heat is lost during the generation of electricity. If all this wasted energy were harnessed, for example, by cogeneration, it could be used productively, thereby conserving energy.

Another way to increase energy efficiency would be to improve our electric grids because about 10 percent of electricity is lost during transmission. To accomplish this, some energy experts envision that future electricity will be generated far from population centres, converted to supercooled hydrogen, and transported through underground superconducting pipelines. The technology to build such conduits has not yet been developed.

ENERGY CONSERVATION AT HOME

The average household spends $1500 each year on utility bills. This cost could be reduced considerably by investing in energy-efficient technologies (**FIGURE 11.33**). Although a more energy-efficient house might cost more up front depending on the technologies employed, the improvements usually pay for themselves in two or three years. Energy efficiency has become an essential element of design codes nationwide.

Some energy-saving improvements, such as thicker wall insulation, are easier to install while a home is being built. Other improvements can be made to older homes to reduce heating and cooling costs and enhance energy efficiency. Examples include installing thicker attic insulation, installing storm windows and doors, caulking

Some energy-saving measures for the home FIGURE 11.33

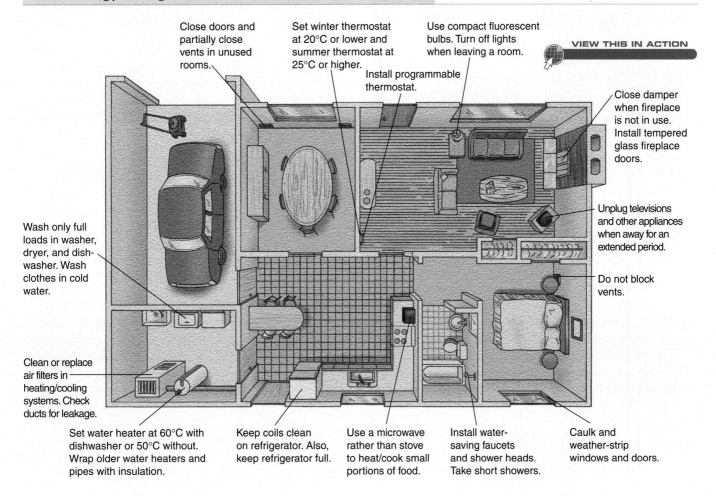

Close doors and partially close vents in unused rooms.

Set winter thermostat at 20°C or lower and summer thermostat at 25°C or higher.

Install programmable thermostat.

Use compact fluorescent bulbs. Turn off lights when leaving a room.

VIEW THIS IN ACTION

Close damper when fireplace is not in use. Install tempered glass fireplace doors.

Unplug televisions and other appliances when away for an extended period.

Do not block vents.

Wash only full loads in washer, dryer, and dishwasher. Wash clothes in cold water.

Clean or replace air filters in heating/cooling systems. Check ducts for leakage.

Set water heater at 60°C with dishwasher or 50°C without. Wrap older water heaters and pipes with insulation.

Keep coils clean on refrigerator. Also, keep refrigerator full.

Use a microwave rather than stove to heat/cook small portions of food.

Install water-saving faucets and shower heads. Take short showers.

Caulk and weather-strip windows and doors.

cracks around windows and doors, replacing inefficient furnaces and refrigerators, and adding heat pumps.

Many news articles provide these and more suggestions every year. Most local utility companies will also perform an energy audit on a home for little or no charge.

CONCEPT CHECK **STOP**

What is the difference between energy conservation and energy efficiency?

What is cogeneration?

ECO CANADA   **CAREER FOCUS**

ENERGY AUDITOR

Energy auditors measure, record, and evaluate the flow of energy in homes, commercial buildings, and industrial plants, looking for ways energy can be used more efficiently.

Imagine you are wearing your bright white hard hat, steel-toed boots, and safety glasses. With clipboard in hand, you begin your tour of a massive bottling factory. You are an energy auditor and you have been hired by the factory to perform an energy audit of the facility.

The factory's owners are concerned with rising energy costs and have asked you to evaluate their operation and find ways to reduce energy consumption. The owners also want to demonstrate their commitment to a progressive environmental policy to the facility's staff and the public. Making the factory more energy efficient is one way to do so. You and your auditing team will spend several days at the facility reviewing processes and operating data and observing workers, looking for ways the facility can reduce the amount of energy it uses and decrease energy costs.

As an energy auditor, you are an expert on how facilities like this bottling plant can cost-effectively manage their energy

consumption. One of the first things you look at is the monthly energy costs on the factory's utility bill. You will also look at the energy rating of the equipment in the plant, for example, the motors, boilers, and heating furnaces, because the energy rating tells you the maximum joules of energy each machine uses per hour. In this case, the factory's machinery is only a few years old, so its equipment already has a very efficient energy rating.

In addition to the energy rating, you will review the plant production logs and maintenance records for potential opportunities. For example, the bottling plant has hundreds of metres of conveyer belts: If the bearings on the conveyor belts are not greased regularly, the belts do not move as smoothly as they should and require more energy input to keep them moving. You'll also check if the furnace or any other machinery has night settings that will stop them from using as much energy when no one is on-site. Next you will spend considerable time interviewing workers about their practices, looking for ways individuals can help reduce energy consumption. The result of your energy audit will be a report for the bottling facility's owner detailing current energy usage and recommending economical opportunities for making the facility more energy efficient.

Energy auditors use a systematized approach to measure, record, and evaluate the flow of energy, determining if it is being used efficiently and pinpointing where it is being wasted. Energy audits usually fall into one of three categories: homes, commercial buildings, and industrial plants. They also range in complexity, from a quick walk-through inspection to a comprehensive analysis of the implications of alternative energy efficiency measures. Once an energy audit has been conducted, the energy auditor works with a team of professionals to analyze the results and produce a technical report for the client that reveals areas where energy efficiency can be improved.

For more information or to look up other environmental careers go to www.eco.ca.

THE HYDROGEN ECONOMY

The hydrogen economy is an energy-based system that uses fuel cells for motive power (such as cars and boats) and produces energy for buildings and electronics that are powered by hydrogen rather than by fossil fuels. The energy in a hydrogen economy is produced by separating hydrogen ions from their electrons and combining the ions back with oxygen molecules to form water as the waste product.

First called "the hydrogen economy" in 1970 by John Bockris, the system proposes to overcome some of the negative impacts of using hydrocarbon fuels where carbon is released into the atmosphere. Those in favour of a hydrogen economy suggest that the use of hydrogen can be an environmentally cleaner source of energy with significantly less carbon dioxide pollutant released into the atmosphere than other systems. In addition, the hydrogen economy offers a potential solution to the world's reliance on oil. Reducing world dependence on oil means less drilling and potentially fewer oil spills, pipeline explosions, and oil well fires, all of which have devastating impacts on the environment.

Despite the identified benefits associated with the hydrogen economy, implementation across the globe will not be easy because of many complex obstacles. First, hydrogen must be separated from its parent compounds to become useable, and the current methods of doing this are unsafe, expensive, and inefficient. Second, the storage and transportation of hydrogen poses a problem as hydrogen is difficult to store as a liquid or as a compressed gas. Extensive research is underway to address this challenge. In addition, technology has to develop and produce vehicles able to use this hydrogen energy. Combatting these issues requires a massive investment of funds in power plants, pipeline construction, distribution networks, and automobile development and production.

The hydrogen economy offers many potential benefits but still faces many challenges. It will be interesting to see what the future holds for the use of hydrogen energy to power many of the things we use in our everyday lives.

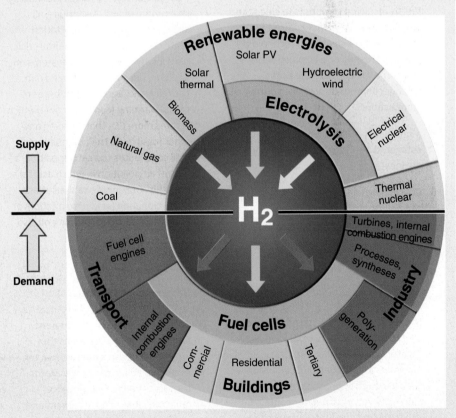

This diagram shows the many different types of inputs into the hydrogen economy as well as the numerous uses of this type of energy.

SUMMARY

1 Energy Consumption

1. A **nonrenewable resource** is a natural resource, often existing in limited amounts, that once consumed cannot be produced or regenerated on a human time scale that can sustain its consumption rate. These resources are always retrieved by processes that require energy. A **renewable resource** is a natural resource that, if consumed, can be replaced by natural processes at a rate that is comparable or faster than its rate of consumption by humans.

2. The per capita energy consumption in highly developed nations is eight times higher than that in developing nations.

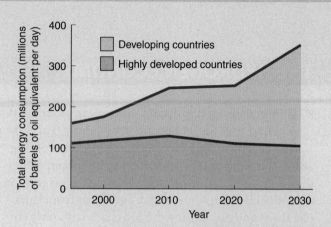

2 Coal, Oil, and Natural Gas

1. Power plants can make coal a cleaner fuel by installing scrubbers to clean the power plants' exhaust. Fluidized-bed combustion is a clean-coal technology in which crushed coal is mixed with limestone to neutralize the acidic sulphur compounds produced during combustion.

2. More than half of the world's total estimated oil and natural gas reserves are located in the Persian Gulf region.

3. Coal mining can lead to landslides and can pollute streams with sediment and **acid mine drainage,** pollution caused when sulphuric acid and dangerous dissolved materials such as lead, arsenic, and cadmium wash from coal and metal mines into nearby lakes and streams. A serious spill along an oil or gas transportation route creates an environmental crisis, particularly in aquatic ecosystems. In addition, burning coal, oil, and natural gas releases CO_2 and contributes to global warming (especially coal, which creates more CO_2 than burning any other fossil fuel). The combustion of coal contributes to acid deposition, a type of air pollution in which acid falls from the atmosphere to the surface as precipitation (acid precipitation) or as dry acid particles. Burning oil also

leads to acid deposition by producing nitrogen oxides. Natural gas contains almost no sulphur and produces much less CO_2 and other pollutants compared with oil and coal.

3 Nuclear Energy

1. **Nuclear energy** is the energy released by nuclear **fission** or **fusion**. A **nuclear reactor** is a device that initiates and maintains a controlled nuclear fission chain reaction to produce energy for electricity. A typical reactor contains a reactor core, where nuclear fission occurs; a steam generator; a turbine; and a condenser.

2. Generating electric power through nuclear energy emits few pollutants (such as CO_2) into the atmosphere, compared with the combustion of coal, but generates highly dangerous radioactive waste, such as **spent fuel,** the used fuel elements that were irradiated in a nuclear reactor. Accidents at nuclear power plants can release dangerous levels of radiation into the environment and result in human casualties. The safe storage of radioactive wastes is another concern associated with nuclear energy. **Low-level radioactive wastes** are radioactive solids, liquids, or gases that give off small amounts of ionizing radiation. **High-level radioactive wastes** are radioactive solids, liquids, or gases that initially give off large amounts of ionizing radiation. Radioactive wastes must be isolated securely for thousands of years. One option for the retirement of an aging nuclear power plant is to **decommission** it, which entails dismantling the old nuclear power plant after it closes.

4 Solar, Biomass, Wind, Hydroelectric, Geothermal, and Tidal Energy

1. **Pasive solar heating** is a system of putting the sun's energy to use without requiring mechanical devices to distribute the collected heat in order to heat buildings. **Active solar heating** is a system of putting the sun's energy to use; in this system, a series of collectors absorb the solar energy, and pumps or fans distribute the collected heat. Active solar heating is used primarily for heating water.

2. A **photovoltaic (PV) solar cell** is a wafer or thin film of solid state materials, such as silicon or gallium arsenide, that are treated with certain metals in such a way that they generate electricity—that is, a flow of electrons—when solar energy is absorbed. PVs generate electricity with no pollution and minimal maintenance but are limited by their low efficiency and by the amount of land needed for their large-scale use. **Solar thermal electric generation** is a means of producing electricity in which the sun's energy is concentrated by mirrors or lenses onto a fluid-filled pipe; the heated fluid is used to generate electricity. Solar thermal energy systems are efficient and provide significant environmental benefits, but they are not cost-competitive with fossil fuels.

3. **Biomass** is plant and animal material used as fuel. Biomass fuel—materials such as wood, fast-growing plant and algal crops, crop wastes, sawdust and wood chips, and animal wastes—is burned to release its energy.

4. **Wind energy** is electric energy obtained from surface air currents caused by the solar warming of air. Restricted primarily to rural areas that receive fairly continual winds, wind power is a clean and cost-effective source of energy, but wind turbines can kill birds and bats. **Hydropower** is a form of renewable energy that relies on flowing or falling water to generate electricity. Hydropower is highly efficient, but the dams built

in traditional hydropower projects can greatly disrupt the natural environment and displace local residents.

5. **Geothermal energy** is the use of energy from Earth's interior for either space heating or generation of electricity. **Tidal energy** is a form of renewable energy that relies on the ebb and flow of the tides to generate electricity.

5 Approaches to Sustainable Energy Management

1. **Energy conservation** is using less energy—for example, by reducing energy use and waste. **Energy efficiency** is using less energy to accomplish a given task—for example, with new technology. Examples of energy conservation measures that reduce gasoline consumption include carpooling and lowering driving speeds; energy efficiency measures include designing and manufacturing more fuel-efficient automobiles.

2. **Cogeneration** is an energy technology that involves recycling "waste" heat. It produces two useful forms of energy from the same fuel: it generates electricity and captures and uses the steam released.

KEY TERMS

- **nonrenewable resource** p. 394
- **renewable resource** p. 394
- **surface mining** p. 396
- **subsurface mining** p. 396
- **acid mine drainage** p. 396
- **nuclear energy** p. 407
- **fission** p. 407
- **fusion** p. 408
- **enrichment** p. 408

- **nuclear reactor** p. 408
- **spent fuel** p. 411
- **low-level radioactive wastes** p. 413
- **high-level radioactive wastes** p. 413
- **decommission** p. 415
- **active solar heating** p. 416
- **passive solar heating** p. 417
- **photovoltaic (PV) solar cell** p. 418
- **solar thermal electric generation** p. 419

- **fuel cell** p. 421
- **biomass** p. 421
- **wind energy** p. 423
- **hydropower** p. 425
- **geothermal energy** p. 428
- **tidal energy** p. 430
- **energy conservation** p. 430
- **energy efficiency** p. 430
- **cogeneration (CHP)** p. 432

CRITICAL AND CREATIVE THINKING QUESTIONS

1. Distinguish between coal, oil, natural gas, and nuclear energy, and compare the environmental impacts of each.

2. In your estimation, which fossil fuel has the greatest potential for the 21st century? Why?

3. Which major consumer of oil is most vulnerable to disruption in the event of another energy crisis: electric power generation, motor vehicles, heating and air conditioning, or industry? Why?

4. Can we prevent catastrophic accidents at nuclear power plants in the future? Why or why not?

5. Are you in favour of Canada developing additional nuclear power plants to provide us with electricity in the 21st century? Why or why not?

6. On the basis of what you have learned about coal, oil, and natural gas, which fossil fuel do you think Canada should exploit during the next 20 years? Explain your answer.

7. Examine the bar graph to the right, which shows the six countries with the greatest natural gas deposits. Is most of the world's natural gas located in North and South America, or in Europe and the Persian Gulf countries?

8. Some scholars think the Industrial Revolution may have been concentrated in Europe and North America because coal is located there. Explain the connection between coal and the Industrial Revolution.

9. Explain the following statement: Unlike fossil fuels, solar energy is not resource-limited but is technology-limited.

10. One advantage of the various forms of renewable energy, such as solar, thermal, and wind energy, is that they cause no net increase in atmospheric carbon dioxide. Is this true for biomass? Why or why not?

11. Give an example of how one or more of the alternative energy sources discussed in this chapter could have a negative effect on each of the following aspects of ecosystems:
    a. Soil preservation
    b. Natural water flow
    c. Foods used by wild plant and animal populations
    d. Preservation of the diversity of organisms found in an area

12. Explain how energy conservation and efficiency are major "sources" of energy.

13. Evaluate which forms of energy, other than fossil fuels and nuclear power, have the greatest potential where you live.

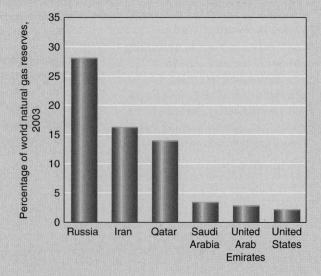

These hoses suck methane, which is used as fuel, from decomposing trash.

What type of renewable energy is being provided?

What would be some advantages and disadvantages of this type of energy production?

Wastes in the Environment

REUSING AND RECYCLING OLD AUTOMOBILES

Each year in North America, close to 12 million vehicles are discarded *(see photo)*. Although 75 percent of a car can be reused as secondhand parts or be recycled, the remaining 25 percent—glass, metals, plastics, fabrics, rubber, foam, and leather—usually ends up in landfills.

The automobile is the most recycled consumer product in the world today. It takes about 45 seconds to shred the average automobile into fist-sized pieces for recycling. Roughly 37 percent of the iron and steel scrap reprocessed in North America comes from old cars. On average, Canadians discard between 116,000 and 232,000 tonnes of household scrap metal every year. About 42 percent of all new steel in North America comes from recycled metal. According to the U.S. Environmental Protection Agency (EPA), recycling scrap iron and steel produces 86 percent less air pollution and 76 percent less water pollution than mining and refining an equivalent amount of iron ore.

Recycling plastic, which automakers use because it is lightweight and improves fuel efficiency, is one of the biggest challenges in auto recycling. No industry standards for plastic parts currently exist, so the kinds and amounts used in cars vary a great deal.

Auto manufacturers worldwide have begun to address the challenge of recycling old cars. Toyota developed a way to recycle urethane foam and other shredded materials. Chrysler is developing a concept vehicle with completely recyclable body sections. And many other automakers have started to design cars with completely recyclable or reconditionable parts (see inset of recyclable plastic and composite parts on a Toyota vehicle). The European Union has mandated that by 2015, 95 percent of each discarded car must be recoverable. When they design new models, European auto manufacturers now take into account the car's entire life cycle.

This kind of *product stewardship* encourages optimal reuse and recycling when products are returned to manufacturers. But despite these initiatives, there currently is no legislation governing the management of end-of-life vehicles in North America.

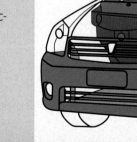

441

Solid Waste

Canada and the United States generate more solid waste per capita than any other country (U.S. is first; Canada is a close second). In North America, an average individual will throw away 600 times that person's adult body weight in garbage in his or her lifetime. In 2002, each individual in the United States generated 1.4 tonnes of solid waste, roughly 2 kilograms per person per day. That same year in Canada, 31 million tonnes of municipal, commercial, and industrial waste was generated, roughly 2.7 kilograms of waste for each Canadian per day! In Canada, there are over 10,000 landfill sites with a price tag of more than $1.5 billion per year to manage. These landfill sites account for close to 38 percent of the total methane emissions in Canada. This is in part caused by our consumption habits; by six months of age, the average Canadian has consumed the same amount of resources as the average person in a developing country does in a lifetime. The problem worsens each year as the North American population increases.

Waste generation is an unavoidable consequence of the prosperous, high-technology, industrial economies of highly developed nations. Many products that would be repaired, reused, or recycled in less affluent nations are simply thrown away. Nobody likes to think about solid waste, but it is certainly a concern of modern society—we keep producing it, and places to dispose of it safely are dwindling in number (**FIGURE 12.1**).

This problem has been acutely felt in large cities like Toronto and Vancouver in years when civic workers have gone on strike, causing garbage to pile up in the streets (**FIGURE 12.2**). In Vancouver in 2007, residential garbage pickup came to a halt for nearly three months as outside municipal workers went on strike for 88 days during the months of July to October. Residents were left to contend with smelly garbage pileups and were forced to find solutions for the garbage accumulating in private residences. The city of Vancouver organized garbage disposal sites in city parks. The smell and mess that resulted from the workers' strike remind us that trash removal is an essential aspect of modern city life.

Recycling FIGURE 12.1

Most municipalities in Canada have recycling programs in place, but some recyclable material is still thrown into landfills.

Garbage pileup FIGURE 12.2

When garbage doesn't get conveniently picked up from our curbsides, the result is often quite unpleasant. During the 2009 civic workers' strike in Toronto, household waste piled up on the streets, creating a smelly, ugly mess that lasted for months.

TYPES OF SOLID WASTE

Municipal solid waste consists of the combined residential and commercial waste produced in a municipal area. Municipal solid waste is a heterogeneous mixture composed primarily of paper and paperboard; yard waste; food waste; plastics; metals; rubber, leather, and textiles; glass; and wood (FIGURE 12.3). Canada produces about 791 kilograms per capita of municipal waste each year, which has been steadily increasing since 1980. The proportions of the major types of solid waste in this mixture change over time. Today's solid waste contains more paper and plastics, but less glass and steel than it did in the past.

municipal solid waste Solid materials discarded by homes, offices, stores, restaurants schools, hospitals, prisons, libraries, and other facilities.	
nonmunicipal solid waste Solid waste generated by industry, agriculture, and mining.	

Municipal solid waste is actually only a small proportion—less than 2 percent—of the total solid waste produced each day. **Nonmunicipal solid waste**, which includes mining, agricultural, and industrial wastes, is produced in substantially larger amounts. Most solid waste generated in North America is from nonmunicipal sources. Some of the impacts of mine tailings to water quality are discussed in greater detail in Chapter 9.

Municipal solid waste FIGURE 12.3

A Composition of total solid waste.

B Composition of municipal solid waste.

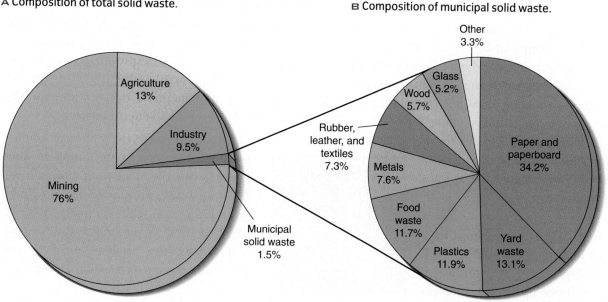

DISPOSAL OF SOLID WASTE

Solid waste has been traditionally regarded as material that is no longer useful and that should be disposed of. Statistics Canada reported some interesting trends in waste disposal for Canadian residents and industries in 2006:

- Canadians produced over 1000 killograms of waste per person, which was up by 8 percent from 2004. This amounted to an additional 2 million tonnes of waste going to landfills in 2006.

- Of this total, only 237 kilograms were diverted from landfills.

- A greater proportion of waste, some 22 million tonnes, came from nonresidential sources, with only 13 million tonnes from residential sources.

- Only Prince Edward Island did not show an overall increase in waste disposal to landfill. The largest increase was found in Alberta, which contributed almost 37 percent to that increase. Ontario and Quebec contributed 31 percent and 18 percent respectively.

- Materials processed for recycling increased 9 percent to just over 7.7 million tonnes in 2006.

How do we deal with this waste material? Essentially, we can get rid of it in four ways: dump it, bury it, burn it, or recycle and compost it. We consider the benefits and consequences of each of these methods in greater detail below.

Open dumps

The old method of solid waste disposal was dumping. Open dumps, which are now illegal, were unsanitary, malodorous places where disease-carrying vermin like rats and flies proliferated. Methane gas was released into the surrounding air as microorganisms decomposed the solid waste, and fires polluted the air with acrid smoke. Liquid oozed and seeped through the heaps of solid waste, often leaching hazardous materials that then contaminated soil, surface water, and groundwater.

Sanitary landfills

Open dumps have been replaced by **sanitary landfills**, which receive about 55 percent of the municipal solid waste generated in North America

sanitary landfill
The most common method of disposing of solid waste, by compacting it and burying it under a shallow layer of soil.

today. Sanitary landfills differ from open dumps in that the solid waste is placed in a hole, compacted, and covered with a thin layer of soil every day (FIGURE 12.4). This process reduces the number of rats and other vermin usually associated with solid waste, lessens the danger of fires, and decreases the amount of odour. If a sanitary landfill is operated in accordance with solid waste management–approved guidelines, it does not pollute local surface and groundwater. Safety is ensured by layering compacted clay and plastic sheets at the bottom of the landfill, which prevent liquid waste from seeping into groundwater.

Newer landfills possess a double liner system (plastic, clay, plastic, clay) and use sophisticated systems to collect **leachate** (liquid that seeps through the solid waste) and methane gas that forms during decomposition.

The location of an "ideal" sanitary landfill is based on a variety of factors, including the geology of the area, soil drainage properties, and the proximity of nearby bodies of water and wetlands. The landfill should be far enough away from centres of dense population so it is inoffensive but close enough so as not to require high transportation costs.

Although the operation of sanitary landfills has improved over the years with the passage of stricter guidelines, few landfills are ideal. Most sanitary landfills in operation today do not meet current legal standards for new landfills and encounter a variety of problems:

- Methane gas, produced by microorganisms that decompose organic material anaerobically (in the absence of oxygen), may seep through the solid waste and accumulate in underground pockets, creating the possibility of an explosion—even in basements of nearby homes. Landfill operators typically collect and burn off methane, but a growing number use the methane for gas-to-energy projects.

- Leachate that seeps from unlined landfills or through cracks in the lining of lined landfills can potentially contaminate surface water and groundwater. Because even household trash contains toxic chemicals like heavy metals, pesticides, and organic compounds, the leachate must be collected and treated to neutralize its negative effects.

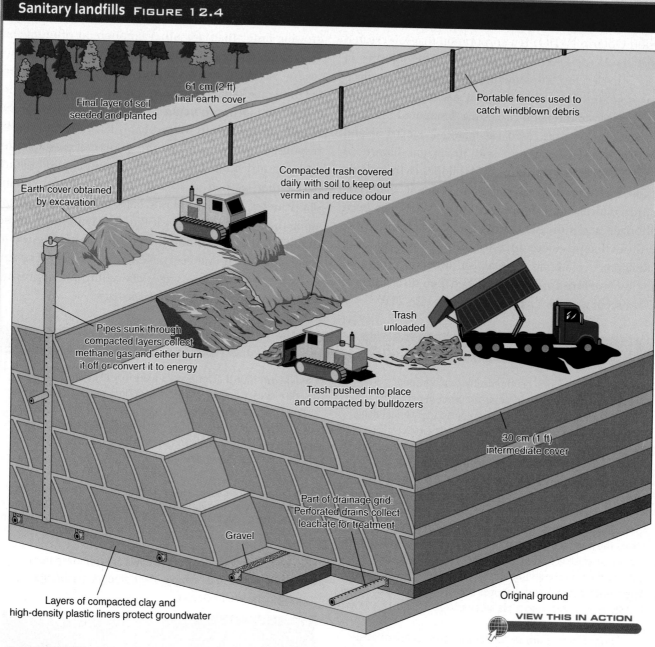

Sanitary landfills FIGURE 12.4

Final layer of soil seeded and planted

61 cm (2 ft) final earth cover

Portable fences used to catch windblown debris

Earth cover obtained by excavation

Compacted trash covered daily with soil to keep out vermin and reduce odour

Pipes sunk through compacted layers collect methane gas and either burn it off or convert it to energy

Trash unloaded

Trash pushed into place and compacted by bulldozers

30 cm (1 ft) intermediate cover

Part of drainage grid: Perforated drains collect leachate for treatment

Gravel

Layers of compacted clay and high-density plastic liners protect groundwater

Original ground

VIEW THIS IN ACTION

Sanitary landfills are more than just dumps. They are constructed today with protective liners of compacted clay and high-density plastic and sophisticated leachate collection systems that minimize environmental problems like groundwater contamination. Solid waste is spread in a thin layer, compacted into small sections called "cells," and covered with soil.

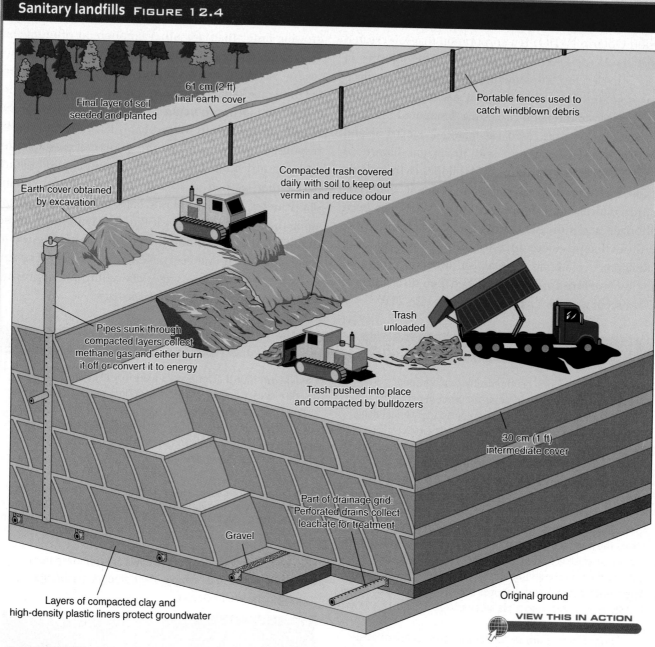

Process Diagram

- Landfills, by their nature, fill up. They are not a long-term remedy for waste disposal. Many have reached their capacity, and others do not meet provincial or federal environmental standards. Fewer new sanitary landfills are being opened; many desirable sites are already taken, and people are usually adamantly opposed to the construction of a landfill near their homes.

Waste diversion is a process of diverting waste from a landfill through programming to recycling, reusing, re-giving, or through biological treatments such as anaerobic digestion, composting, or incineration. An example of a waste diversion program is found in Ontario. Waste Diversion Ontario (WDO) is a corporation that was created under the Waste Diversion Act of 2002. The objective of WDO is to develop, implement, and operate waste diversion programs for a wide range of materials. Currently, WDO has waste diversion programs designated for bluebox waste, used tires, used oil material, electrical and electrical equipment waste, and municipal hazardous or special waste. In the future, WDO hopes to include diversion for household special waste, organic materials, pharmaceuticals, and fluorescent tubes. WDO has been authorized under the Act to work cooperatively with those industries that produce and distribute products that result in designated materials to establish diversion programs. The city council of Toronto established their diversion goal of 70 percent waste diversion to be achieved by 2010.

The special problem of plastic The amount of plastic in our solid waste, more than half of it from packaging, is growing faster than any other component of municipal solid waste. Most plastics are chemically stable and do not readily decompose. This characteristic, although essential in the packaging of products like food and medicine, causes long-term problems. Most plastic debris disposed of in sanitary landfills will probably last for centuries. In response to concerns about the volume of plastic waste, some countries, like Taiwan, India, Australia, and Ireland, have either banned or started taxing the use of certain plastic items, especially plastic bags, in hopes of discouraging their use. In 2007, Leaf Rapids, Manitoba, became the first Canadian community to ban plastic bag use in retail stores. San Francisco, New York City, and Los Angeles are also in various stages of drafting and approving legislation to ban the use of plastic bags.

EnviroDiscovery   GREAT PACIFIC GARBAGE DUMP

In 1997, Charles Moore, an American oceanographer, accidentally stumbled upon a massive amount of plastic waste in the Pacific Ocean, which he coined the "Great Pacific Garbage Dump." Currently, this plastic trash vortex is twice the size of the continental United States and growing rapidly. Moore suggests that there are over 90 million tonnes of flotsam circulating in this region, creating an island of plastic garbage, commonly called "plastic soup." The plastic soup is held together by underwater currents and stretches from approximately 500 nautical miles off of the California coast across the northern Pacific, past Hawaii, almost as far as Japan. The plastic trash is translucent and lies just below the ocean water surface, making it undetectable by satellite photographs. When the large trash vortex moves close to land (as it does near the Hawaiian archipelago), it leaves garbage behind, resulting in a beach covered in plastic.

Modern plastics are so durable that every piece of plastic manufactured in the past 50 years that has made its way to the ocean is still out there somewhere. There are currently more than 2 million plastic bottles disposed of every five minutes. Moore warns that unless consumers cut back on their use of disposable plastics, the plastic waste will double in size over the next decade, having a devastating impact on marine ecosystems.

Plastic debris causes the death of more than a million seabirds and over 100,000 marine animals each year. Countless birds have mistaken these tiny plastics for food and been found with syringes, cigarette lighters, toothbrushes, and other such plastics in their stomachs. The UN Environment Programme estimated in 2006 that for every square mile of ocean there are 46,000 pieces of floating plastic. The millions of plastic pellets act like chemical sponges attracting human-made chemicals such as hydrocarbons and DDT. These chemicals enter into the food chain and the animals and sea fish that end up on our dinner plates. The threat to human health is real!

To see an informative video produced by *Good Morning America* on the Garbage Dump, go to www.youtube.com/watch?v=uLrVCl4N67M.

The plastic that washes up on beaches or is removed from the ocean gives a small sample of the amount of garbage floating in the Pacific Ocean.

Special plastics that degrade or disintegrate have been developed. Some of these are **photodegradable**—that is, they break down after being exposed to sunlight—which means they will not break down when these are buried deep under layers of waste in a sanitary landfill. Other plastics are **biodegradable**—they are decomposed by microorganisms like bacteria. Whether biodegradable plastics actually break down under the conditions found in a sanitary landfill is not yet clear, although preliminary studies indicate that they probably do not. (Other waste management options for plastic are discussed later in this chapter.)

Legislation on plastic bags Close to 60,000 plastic bags are discarded every five seconds in Canada. In Ontario, almost 80 plastic bags are used every second, meaning 7 million plastic bags are used each day. The Ontario government, relevant industry associations, and environmental organizations have formed a partnership to find creative solutions for disposing of plastic bags. This partnership has agreed to reduce the use of single-use plastic bags across Ontario over the next five years by 50 percent. The partnership has also agreed to the following:

- Expansion of use of reuseable bags and increasing reuseable bag options for consumers

- Increasing their commitment to the amount of recycled content in plastic bags

- Incentives to encourage consumers to reduce single-use plastic bag demand

- Incentives to encourage consumers to expand use of reuseable bags

- Options to encourage consumers to recycle plastic bags

- Increasing education of Ontarians on issues related to plastic bags

- Annual reporting on plastic bag use and recycling efforts

- Increasing monitoring efforts on plastic bag use reduction, recycling, and reuse

Environment Canada is funding a number of research initiatives that will provide a more in-depth understanding of the environmental issues associated with plastic bags and plastic packaging. The results of this research will assist the government in creating a comprehensive plan to deal with plastics in the future.

Some stores have made significant efforts, such as Ikea and Wholesale Foods, to promote the use of reuseable bags. Loblaw Companies, a Canadian supermarket chain, was able to divert close to 150 million kilograms of plastic bags from landfills in 2008, and their Superstore in Milton, Ontario, became the first in Canada to not offer plastic bags at all. Loblaw Companies aims to divert more than 1 billion plastic bags and close to 70 percent of the store-generated waste from landfills.

Incineration

Canada uses incineration to deal with about 5 percent of the solid waste generated. When solid waste is incinerated, two positive things occur. First, the volume of solid waste is reduced by up to 90 percent: Ash is more compact than unburned solid waste. Second, incineration produces heat that can make steam to warm buildings or generate electricity. In 2004 the United States had 89 waste-to-energy incinerators, which produce substantially less carbon dioxide emissions than power plants that burn fossil fuels (**FIGURE 12.5**). (Recall from Chapter 10 that carbon

Carbon dioxide emissions per kilowatt-hour of electricity generated FIGURE 12.5

Waste-to-energy incinerators release less carbon dioxide into the atmosphere than do equivalent power plants that burn fossil fuels.

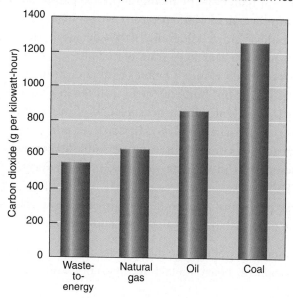

This mountain of tires will be burned to produce electricity. For every 1000 kilograms of chipped-up tires, 476 MJ of electricity is produced.

dioxide is a potent greenhouse gas.) While Canada is currently treating only a small part of disposed waste using energy-from-waste (EFW) facilities, Japan treats about 75 percent of its waste by EFW.

The best materials for incineration are paper, plastics, and rubber, all of which produce a lot of heat. Paper burns readily, and 1 kilogram of plastic waste yields almost as much heat as a kilogram of fuel oil.

Tires produce as much heat as coal and often generate less pollution. Some electric utilities in Canada and the United States burn tires instead of or in addition to coal (FIGURE 12.6). In 2001, 42 percent of all discarded tires were incinerated. In 2005, 20 percent of all scrap tires in Canada were used as tire-derived fuel (TDF).

Most problems associated with incineration arise from the potential for environmental contamination:

- Incinerators pollute the air with carbon monoxide, particulates, heavy metals like mercury, and other toxic materials, unless expensive air pollution control devices are used.

- Incinerators produce large quantities of ash, which must be disposed of properly. **Bottom ash**, or slag, is the ash left at the bottom of the incinerator when combustion is completed. **Fly ash** is the ash from the flue (chimney) that is trapped by air pollution control devices. Fly ash usually contains more toxic materials, including heavy metals and possibly dioxins, than bottom ash. Both types of incinerator ash are best disposed of in specially licensed hazardous waste landfills (discussed later in this chapter).

- Like sanitary landfills, site selection for incinerators is controversial. People may recognize the need for an incinerator, but they do not want it near their homes.

- Incinerators are expensive to run. Prices have escalated because costly pollution control devices are now required. Economic factors have also restricted construction of new plants.

The three types of incinerators are mass burn, modular, and refuse-derived fuel. Most **mass burn incinerators** are large and designed to recover the energy produced from combustion (FIGURE 12.7). **Modular incinerators** are smaller incinerators that burn all solid waste. Assembled at factories, they are less expensive to build. **Refuse-derived fuel incinerators** burn the combustible portion of solid waste. First, noncombustible waste such as glass and metals are removed by machine or by hand. The remaining solid waste, including plastic and paper, is shredded or shaped into pellets and burned.

mass burn incinerator A large furnace that burns all solid waste except for unburnable items such as refrigerators.

Composting

Yard waste, such as grass clippings, branches, and leaves, is a substantial component of municipal solid waste (see Figure 12.3). As space in sanitary landfills becomes more limited, other ways to dispose of yard waste are being developed and implemented.

One of the best recovery methods for organic waste is to convert it into soil conditioners like compost or

Mass burn, waste-to-energy incinerator FIGURE 12.7

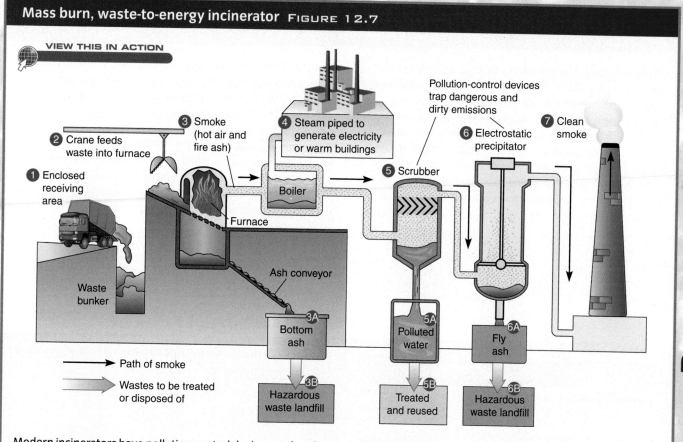

VIEW THIS IN ACTION

① Enclosed receiving area

② Crane feeds waste into furnace

③ Smoke (hot air and fire ash)

④ Steam piped to generate electricity or warm buildings

⑤ Scrubber

⑥ Electrostatic precipitator

⑦ Clean smoke

Pollution-control devices trap dangerous and dirty emissions

Boiler

Furnace

Waste bunker

Ash conveyor

3A Bottom ash

5A Polluted water

6A Fly ash

→ Path of smoke

⇒ Wastes to be treated or disposed of

3B Hazardous waste landfill

5B Treated and reused

6B Hazardous waste landfill

Modern incinerators have pollution control devices such as lime scrubbers and electrostatic precipitators to trap dangerous and dirty emissions like those produced by burning rubber tires.

mulch (FIGURE 12.8). Food scraps, sewage sludge, and agricultural manure are other forms of solid waste that can be used to make compost. Compost and mulch are used for landscaping in public parks and playgrounds or as part of the daily soil cover at sanitary landfills. Compost and mulch are also sold to gardeners.

Composting as a means of managing solid waste first became popular in Europe. Many municipalities in North America have composting facilities as part of their comprehensive solid waste management plans, and many have banned yard waste from sanitary landfills. This trend is likely to continue, making composting even more desirable.

Home composting of household waste FIGURE 12.8

These two different composter designs are both used to create natural garden mulch in an environmentally friendly way by recycling kitchen waste, garden prunings, and grass cuttings.

There are a number of correctional facilities in Canada that compost food waste by placing food discards into "food only" containers, which are then placed in compost containers. Kingston Penitentiary, a maximum security prison located in Kingston, Ontario, is an example of a Canadian prison taking recycling and composting seriously. In addition, CORCAN, also located in Kingston, Ontario, is a giant central composting facility run by a special operating agency of the Correctional Service of Canada that is having a huge impact on municipal and institutional composting. As a result of CORCAN, the community

is now at 50 percent waste diversion from the local landfill. This carefully managed operation is located at a minimum security institution near Kingston. The compost centre is situated in an enclosed building with six organic composting bays where organic waste is received and then disturbed mechanically through channels 66 metres long. CORCAN is an all-around winner: It provides prison inmates with the opportunity to learn work skills, it helps the Kingston municipality solve its garbage problem and save money, and it enhances the environment through waste diversion and organic soil enrichment.

CONCEPT CHECK STOP

How do municipal and nonmunicipal solid waste differ?

What are some features of sanitary landfills? What problems are associated with them?

What are the main features of a mass burn incinerator? What problems are associated with incineration?

How do composters work?

Reducing Solid Waste

LEARNING OBJECTIVES

Define *source reduction.*

Summarize how source reduction, reuse, and recycling help reduce the volume of solid waste.

Define *integrated waste management.*

G iven the problems associated with sanitary landfills and incinerators, it makes sense to do whatever we can to reduce the wastes we generate. The three goals of waste prevention, in order of priority, are to (1) reduce the amount of waste as much as possible, (2) reuse products as much as possible, and (3) recycle materials as much as possible.

Reducing the amount of waste includes purchasing products that have less packaging and that last longer or

that are repairable (**FIGURE 12.9**). Consumers can also decrease their consumption of products to reduce waste. Before deciding to purchase a product, a consumer should ask, "Do I really *need* this product, or do I merely *want* it?"

SOURCE REDUCTION

The most underutilized aspect of waste management is **source reduction**. Source reduction is accomplished in

> **source reduction**
> An aspect of waste management in which products are designed and manufactured in ways that decrease the amount of solid and hazardous waste in the solid waste stream.

a variety of ways, including the use of raw materials that introduce less waste during manufacturing, and reusing and recycling wastes at the plants where they are generated. Innovation and product modifications play a key role. Consider aluminum cans: They are 35 percent lighter now than in the 1970s because less material is introduced into their manufacture. Dry-cell batteries are another example: They contain much less mercury today than they did in the early 1980s.

Dematerialization, the progressive decrease in the size and weight of a product as a result of technological improvements, is an example of source reduction only if the new product is at least as durable as the one it replaces. If the smaller, lighter product has a shorter life span and must be replaced more often, source reduction is not achieved.

REUSING PRODUCTS

One example of reuse is refillable glass beverage bottles. Years ago, refillable beverage bottles were commonly used in Canada and the United States. However, for a glass bottle to be reused it must be considerably thicker (and heavier) than one-use bottles. Because of the increased weight, transportation costs are higher. The centralization of bottling facilities makes it economically difficult to go back to the days of refillable bottles. Several countries still reuse glass, including Canada, Japan, Ecuador, Denmark, Finland, Germany, the Netherlands, Norway, Sweden, and Switzerland.

RECYCLING MATERIALS

Many materials found in solid waste can be collected and reprocessed into new products. Recycling is preferred over landfill disposal because it conserves our natural resources and is more environmentally benign. It is estimated that every tonne of recycled paper saves 17 trees, 118,000 litres of water, 4100 kilowatt-hours of energy, and 3 cubic metres of landfill space. Recycling also has a positive effect on the economy by generating jobs and revenue from selling the recycled materials. Recycling does have environmental costs: Like all human activities, it uses energy and generates pollution. For example, the de-inking process in paper recycling requires energy and produces a toxic sludge that contains heavy metals.

The many different materials in municipal solid waste must be separated before recycling. The

Many research initiatives have looked at a number of different products asking the question of whether reuseable products are always better than disposables. It would appear that in some instances they may not be.

Disposable versus cloth diapers

Research conducted by the U.S. Natural Resources Defense Council suggests reuseable diapers may be as bad for the environment as disposables. The studies, which looked at the complete path of cloth diapers from cotton to diaper to washing machine, and disposable diapers from plastic factory to diaper to dump, suggest that reuseable diapers may not be as good for the environment as originally thought. Disposable diapers consume more raw materials and produce more solid waste; however, cloth diaper production uses more water and energy and produces more atmospheric emissions and waste water effluent.

A study conducted by an advisory board to the U.K. Environmental Agency drew a similar conclusion. This study looked at the entire lifecycle of both products, specifically analyzing the materials, chemicals, energy consumed during production, usage, and disposal, and measured these against environmental impacts such as climate change, ozone depletion, human toxicity, acidification, fresh water, aquatic toxicity, terrestrial toxicity, photochemical oxidant formation, and eutrophication. The reuseable cloth diapers save the landfills, yet load washing machines and sewage systems. However, disposable diapers will always likely be very damaging to the environment, but there are opportunities to minimize the impacts of renewable cloth diapers.

Reuseable versus disposable cups

Surprising results were also found concerning the lifecycle energy efficiency of reusable versus disposable cups. Dr. Martin Hocking at the University of Victoria studied three types of reuseable drinking cups (ceramic, glass, and reuseable plastic) and compared them with two types of disposable cups (paper and polystyrene foam). The energy used to manufacture reuseable cups is vastly greater than the energy needed to manufacture disposable cups. The efficiency of the dishwasher and of the energy system that powers the dishwasher made a substantial difference when studying the energy of the lifecycle of each cup. Assuming that a new commercial dishwasher running on Canadian electricity is used, the total amount of energy per use is a combination of the wash energy plus the appropriate fraction of manufacturing energy in relation to the cup's lifetime. The lifetime needed for the energy per use of a reuseable cup to become less than that for a disposable cup depends on the lifetime of the reuseable cup. In order for a reuseable cup to be an improvement over a disposable one on the basis of energy, the reuseable cup must be used multiple times to "cash in" on the energy investment made in the cup. If the reuseable cup is used 10 times, then each use is "charged" one-tenth of the manufacturing energy. In addition, research indicated that by choosing even a slightly less efficient dishwasher, the reuseable cup would never be as efficient as even the foam cup. In situations where reusable cups are likely to be lost or broken, and thus have a shorter lifetime, then disposable cups are the preferred option.

Rechargeable versus disposable batteries

In the debate over disposable batteries compared with rechargeable batteries, research overwhelming suggests that rechargeable batteries are the best option in terms of the environmental and economic advantages.

separation of materials in items with complex compositions is difficult. Some food containers are composed of thin layers of metal foil, plastic, and paper, and trying to separate these layers is a daunting prospect.

The number of communities with recycling programs increased remarkably during the 1990s but levelled off in the early 2000s. In 2002, over 80 percent of Canadian households recycled metal cans, plastics, and paper. By 2006, 93 percent of Canadian households had access to recycling programs for at least one recyclable material, and close to 90 percent of all Canadian households reported using some type of recycling program.

In all, close to 17 million Canadians have access to recycling.

Many communities, particularly in areas with sparser populations, have drop-off centres where people can take their separated recyclables. Recyclables are usually sent to a materials recovery facility where they are sorted and prepared for remanufacturing. Currently, 80 percent of the municipal solid waste in Canada is sent to landfills, and the remaining 20 percent is either recycled or disposed of through resource recovery or by incineration (FIGURE 12.10A on page 454).

Most people think recycling involves merely separating certain materials from the solid waste stream, but that is only the first step. For recycling to work, there must be a market for the recycled goods, and the recycled products must be used in preference to virgin products. Prices paid by processors for recyclable materials vary significantly from one year to the next, depending largely on the demand for recycled products. In some places, recycling—particularly curbside collection—is not economically feasible.

Recycling paper

Many highly developed countries have relatively high recycling rates for paper and paperboard. Denmark, for example, recycles 97 percent of its paper. Legislation sometimes increases the demand for recycled paper. In Toronto, city council passed a law requiring daily newspapers that are sold in vending boxes on city streets to contain at least 50 percent recycled fibre. Federally, the Office of Greening Government Operations (OGGO) was created to enhance recycling and purchase of greener products including recycled paper (see the Making a Difference feature on page 455). Increasingly, public institutions as well as private sector companies are selecting "greener" products in their operations, and recycled paper is a good example of that trend.

Currently, roughly one-third of the waste in Canada is paper and paperboard. About 450 grams of newspaper can be recycled to make six cereal boxes, six egg cartons, or 2000 sheets of writing paper. Recycling 1 tonne of newspaper saves 19 trees, 3 cubic metres of landfill space, 4000 kilowatt hours of energy, 29,000 litres of water, and 30 kilograms of air pollution.

Recycling glass

Each year Canadians throw away 6 million tonnes of glass. In 1992, 17 percent of glass was recycled in Canada. Recycled glass costs less than glass made from virgin materials. Glass food and beverage containers are crushed to form cullet, which glass manufacturers can use to make new products. For example, cullet from a mixture of different colours is sometimes added to asphalt to make an attractive roadway (FIGURE 12.10B).

Recycling metals

The recycling of aluminum is one of the best success stories in North American recycling, largely because of economic factors (FIGURE 12.10C). Making a new aluminum can from a recycled one requires a fraction of the energy it would take to make a new can from raw metal. According to the U.S. Environmental Protection Agency (EPA), in 2003 almost 44 percent of aluminum beverage cans were recycled.

Other recyclable metals include lead, gold, iron and steel, silver, and zinc. According to the Institute of Scrap Recycling Industries, new steel products contain an average of 60 percent recycled scrap steel. One of the obstacles to recycling metal products discarded in municipal solid waste is that their metallic compositions are often unknown. It is also difficult to extract metal from products like stoves that contain materials besides metal (plastic, rubber, or glass, for instance).

Recycling plastic

In Canada, plastic accounts for 7 percent of the total weight (30 percent of the total volume) in a typical landfill. Each week, Canadians take home more than 55 million plastic bags. About 5 percent of plastic is recycled. Depending on the economic situation, it is sometimes less expensive to make plastic from raw materials (petroleum and natural gas) than to recycle it. Some governments support or require the recycling of plastic. Polyethylene terephthalate (PET), the plastic used in soda bottles, is recycled more than any other plastic. According to the EPA, almost 9 percent of plastic bottles and packaging is recycled to make such diverse products as carpeting, automobile parts, tennis ball felt, and polyester cloth. One of the challenges associated with recycling plastic is that there are many different kinds. Forty-six different plastics are common in consumer products, and many products contain multiple kinds that must be separated or sorted.

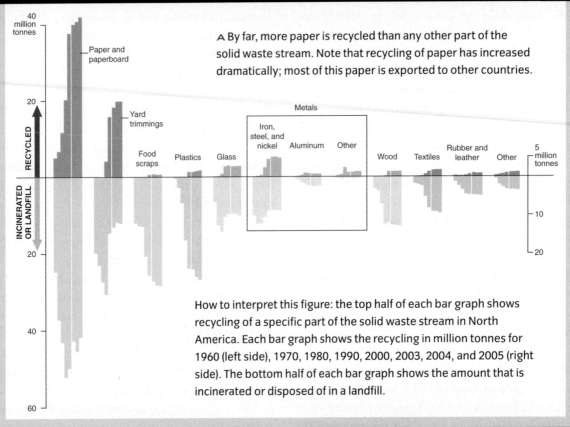

A By far, more paper is recycled than any other part of the solid waste stream. Note that recycling of paper has increased dramatically; most of this paper is exported to other countries.

How to interpret this figure: the top half of each bar graph shows recycling of a specific part of the solid waste stream in North America. Each bar graph shows the recycling in million tonnes for 1960 (left side), 1970, 1980, 1990, 2000, 2003, 2004, and 2005 (right side). The bottom half of each bar graph shows the amount that is incinerated or disposed of in a landfill.

B A bin of green glass that has been sorted and crushed at a recycling plant. Separating the different colours of glass in a cost-effective manner is challenging. Note in A that the amount of glass recycling has not increased since 1990.

C Bales of crushed aluminum beverage cans are ready for processing. Nearly 55 percent of every aluminum can is made from recycled aluminum. Canada recycles 63 percent of its aluminum cans. The recycling of plastics and aluminum uses only 5 percent to 10 percent as much energy as making new plastic or smelting aluminum. It is estimated that every three months the United States throws away enough aluminum to completely rebuild its commercial air fleet.

OFFICE OF GREENING GOVERNMENT OPERATIONS (OGGO)

The government of Canada, with annual spending of over $13 billion on products and services, has an enormous impact on the market economy, and it has been making efforts to "go green." The Office of Greening Government Operations (OGGO), created in April 2005 within Public Works and Government Services Canada (PWGSC), was established to provide advice, support, and guidance on greening government activities and operations and developing policy that relates to green government. The OGGO mandate is to accelerate the greening of government operations by working with federal departments such as the Treasury Board, Secretariat, and Environment Canada. The OGGO provides an opportunity to establish consistent government priorities, accountabilities, targets, timelines, and reporting requirements that will assist the government in achieving environmentally friendly operations. As of April 2005, all new government office buildings must be certified by Leadership in Energy and Environmental Design (LEED) gold standards. The activities of the OGGO include:

- Lowering energy resource consumption
- Lowering greenhouse gases and other air polluting emissions
- Acquiring green policy
- Green procurement
- Remediating contaminated sites
- Improving waste management
- Increasing efficiency of government vehicles and use
- Sustainable office building design
- Environmentally responsible e-waste programming

For more information on the Office of Greening Government Operations, please visit the Public Works and Government Services Canada website at www.pwgsc.gc.ca.

Provincial and municipal governments have also implemented green procurement. The city of Calgary, for example, has implemented the Sustainable Environmental and Ethical Procurement Policy (SEEPP) that provides the municipal government with guidelines to procurement of activities such as the purchase of products and services considering environmental labelling as well as factors such as energy efficiency, minimal packaging, and other sustainable aspects over the entire lifetime of a product or service. Calgary spends over $2 billion annually on procurement. The guidelines established by SEEPP are consistent with both national and international environmental standards and guidelines.

Recycling tires According to the EPA, almost 36 percent of the rubber in tires is currently recycled to make other products. Scrap tires in stock piles range between 300 million to 500 million tires. The United States is set to add roughly another 280 million, and Canada is set to add close to 28 million more each year. Tires in stockpiles are subject to fires and also represent a significant health concern due to standing water buildup at tire stockpiles. Yet, recycled tires are used for only a few products: retread tires; playground equipment; trashcans, garden hoses, and other consumer products; reef building on coastlines; and rubberized asphalt for pavement. Tire recycling programs are common throughout North America. As noted earlier in this chapter, tires are also burned in waste-to-energy incinerators to provide electricity.

Recycling electronic and electrical equipment In Canada, it is a provincial responsibility to regulate electronic waste. The first Canadian program was established in 1994 with British Columbia Post-Consumer Paint Stewardship Regulation. The Provincial Extended Producer Responsibility (EPR) Program provides residents with access to an electronic "take-back" program. Currently, this program exists in British Columbia, Alberta, Saskatchewan, Ontario, and Nova Scotia, and approval is pending in Manitoba. Only those items that cannot be reused are allowed. The EPR is an environmental policy approach in which it is a producer's responsibility, either physical and/or financial, to see a product through to the post-consumer stage of the product's life cycle. The benefits of such programs are reducing waste, increasing reuse and recycling, and providing a framework for industry responsibility and action.

Despite all of these efforts, there are still massive problems with e-waste in Canada. For example, if you drop off your old computer at a recycling centre in an attempt to be environmentally responsible, do you know if it is soundly disposed of? Or is it sent to southern China, where the majority of worldwide e-waste ends up?

Recycling centres export used electronics to places like Guiyu, China, where the poor community makes a living by breaking down any parts of old electronics that can be used again (**FIGURE 12.11**). The electronics are burned and dumped in acid to expose the precious metals that are contained inside the parts. The old, useless portions are then dumped into the local river. The health risks are severe. Children and adults living in these communities are exposed to numerous toxins and chemicals that cause skin lesions, neurological disorders, and cancer. The earth and water are contaminated beyond redemption, and residents are unable to drink the water. Clearly, further legislation, monitoring, and education on e-waste are needed to prevent Canadian electronics from ending up in Guiyu.

INTEGRATED WASTE MANAGEMENT

The most effective way to deal with solid waste is with a combination of techniques. In **integrated waste management**,

> **integrated waste management**
> A combination of the best waste management techniques into a consolidated program to deal effectively with solid waste.

Guiyu, China, has become the waste dump for the world's electronics FIGURE 12.11

a variety of waste minimization methods, including the 3Rs of waste prevention (reduce, reuse, and recycle), are incorporated into an overall waste management plan (FIGURE 12.12). Even on a large scale, recycling and source reduction can substantially reduce, but not entirely eliminate, the need for disposal facilities such as incinerators and landfills.

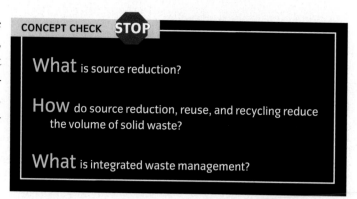

CONCEPT CHECK STOP

What is source reduction?

How do source reduction, reuse, and recycling reduce the volume of solid waste?

What is integrated waste management?

Integrated waste management FIGURE 12.12

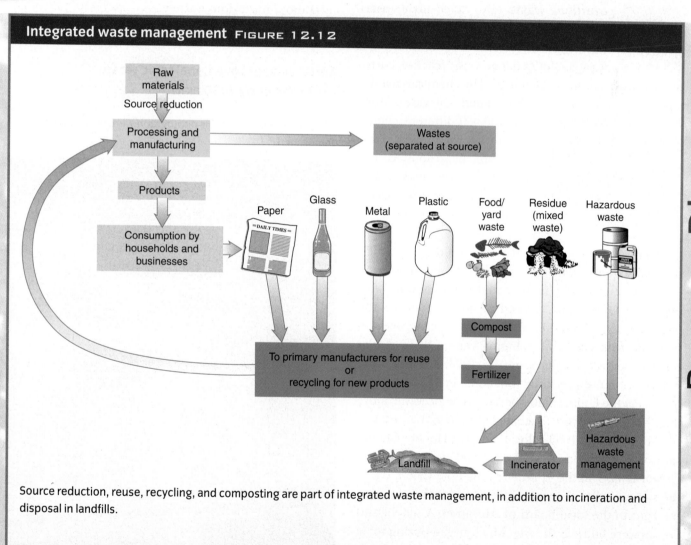

Source reduction, reuse, recycling, and composting are part of integrated waste management, in addition to incineration and disposal in landfills.

Process Diagram

Hazardous Waste

LEARNING OBJECTIVES

Define *hazardous waste.*

Briefly characterize two types of hazardous waste: dioxins and PCBs.

Hazardous waste (also called **toxic waste**) accounts for about 1 percent of the solid waste stream in North America. Hazardous waste includes dangerously reactive, corrosive, explosive, or toxic chemicals. The chemicals may be solids, liquids, or gases. More than 700,000 different chemicals are known to exist. How many are hazardous is unknown because most have never been tested for toxicity, but without a doubt, there are thousands.

> **hazardous waste**
> Any discarded chemical that threatens human health or the environment.

Hazardous waste has held attention since 1977 when it was discovered that toxic waste from an abandoned chemical dump had contaminated homes and possibly people in **Love Canal**, a small neighbourhood on the edge of Niagara Falls, New York. Love Canal became synonymous with chemical pollution caused by negligent hazardous waste management. In 1978, it became the first location ever declared a national emergency disaster area in the United States because of toxic waste; more than 700 families were evacuated (**FIGURE 12.13**).

From 1942 to 1953, a local industry, Hooker Chemical Company, disposed of about 20,000 tonnes of toxic chemical waste in the 914-metre-long Love Canal. When the site was filled, Hooker added topsoil and donated the land to the local board of education. A school and houses were built on the site, which began oozing toxic waste several years later. Over 300 chemicals, many of them carcinogenic, have been identified in Love Canal's toxic waste, and the synergistic effects of the various chemicals reacting and interacting with each other are largely unknown.

In 1990, after almost 10 years of cleanup, the EPA and the New York Department of Health declared the area safe for resettlement. Today, the canal is a 16-hectare mound covered by clay and surrounded by a chain-link fence and warning signs. The Love Canal episode resulted in the passage of the federal Superfund law, which held polluters accountable for the cost of cleanups and generated immediate and ongoing concern about hazardous wastes.

Aerial view of Love Canal toxic waste site in the early 1980s FIGURE 12.13

All the homes shown in this photograph were evacuated and demolished.

TYPES OF HAZARDOUS WASTE

Hazardous chemicals include a variety of acids, dioxins, abandoned explosives, heavy metals, infectious waste, nerve gas, organic solvents, polychlorinated biphenyls (PCBs), pesticides, and radioactive substances (**TABLE 12.1**). Many of these chemicals are discussed in other chapters; see Chapters 4, 8, 9, 10, and 11, which examine endocrine disrupters, pesticides, water pollution, air pollution, and radioactive waste. Here we discuss dioxins and PCBs.

Dioxins **Dioxins** are a group of 75 similar chemical compounds formed as by-products during the combustion of chlorine compounds. Incineration of medical and municipal wastes accounts for 70 to 95 percent of known human emissions of dioxins. Some other known sources of dioxins are iron ore mills, copper smelters, cement kilns, metal recycling, coal combustion, pulp and paper plants that use chlorine for bleaching (**FIGURE 12.14** on page 460), and chemical accidents. Motor vehicles, barbecues, and cigarette smoke emit minor amounts of dioxins. Forest fires and volcanic eruptions are natural sources of dioxins. Dioxins also form during the production of some pesticides.

Dioxins are emitted in smoke and then settle on plants, the soil, and bodies of water; from there they are incorporated into the food web. When humans and other animals ingest dioxins—primarily in contaminated meat, dairy products, and fish—they store and accumulate the dioxins in their fatty tissues (see the bioaccumulation and biomagnification discussion in

Examples of hazardous waste TABLE 12.1

Hazardous material	Some possible sources
Acids	Ash from power plants and incinerators, petroleum products
CFCs (chlorofluorocarbons)	Coolant in air conditioners and refrigerators
Cyanides	Metal refining; fumigants in ships, railway cars, and warehouses
Dioxins	Emissions from incinerators and pulp and paper plants
Explosives	Old military installations
Heavy metals	Paints, pigments, batteries, ash from incinerators, sewage sludge with industrial waste, improper disposal in landfills
Arsenic	Industrial processes, pesticides, additives to glass, paints
Cadmium	Rechargeable batteries, incineration, paints, plastics
Lead	Lead-acid storage batteries, stains and paints, TV picture tubes and electronics discarded in landfills
Mercury	Coal-burning power plants, paints, household cleaners (disinfectants), industrial processes, medicines, seed fungicides
Infectious waste	Hospitals, research labs
Nerve gas	Old military installations
Organic solvents	Industrial processes, household cleaners, leather, plastics, pet maintenance (soaps), adhesives, cosmetics
PCBs (polychlorinated biphenyls)	Older appliances (built before 1980), electrical transformers and capacitors
Pesticides	Household products
Radioactive waste	Nuclear power plants, nuclear medicine facilities, weapons factories

Air pollution from a paper mill FIGURE 12.14

Dioxins may be in these emissions.

Chapter 4). Because dioxins are so widely distributed in the environment, virtually everyone has dioxins in their body fat.

Dioxins cause several kinds of cancer in laboratory animals, but the data conflict on their cancer-causing ability in humans. A 2001 EPA report suggests that dioxins probably cause several kinds of cancer in humans and likely affect the human reproductive, immune, and nervous systems. Because human milk contains dioxins, nursing infants are considered particularly at risk.

Dioxins have been shown to accumulate easily in the human body. Parents and researchers are concerned that dioxins have been found in disposable diapers that are manufactured using a chlorinated bleaching process. This raises frightening questions about whether these dioxins, which are stored in fatty tissues in the human body, may cause cancer or immune system depression. In Europe, changes have already been made and chlorine-free diapers are common. Some countries, such as Sweden, Germany, Australia, and Japan, are making similar changes as well.

PCBs Polychlorinated biphenyls (PCBs) are a group of 209 industrial chemicals composed of carbon, hydrogen, and chlorine. PCBs were manufactured between 1929 and 1979 for a wide variety of uses: as cooling fluids in electrical transformers, electrical capacitors, vacuum pumps, and gas-transmission turbines; and in hydraulic fluids, fire retardants, adhesives, lubricants, pesticide extenders, inks, and other materials. Prior to their ban in the 1970s, PCBs were dumped in large quantities into landfills, sewers, and fields. Such improper disposal is one of the reasons PCBs are still a threat today.

The dangers of PCBs first became evident in Japan in 1968, where hundreds of people ate rice bran oil accidentally contaminated with PCBs and consequently experienced serious health problems, including liver and kidney damage. A similar mass poisoning tied to PCBs occurred in Taiwan in 1979. Since then, toxicity tests conducted on animals have indicated that PCBs harm the skin, eyes, reproductive organs, and gastrointestinal system. PCBs are endocrine disrupters: They interfere with hormones released by the thyroid gland. Several studies have demonstrated that in utero exposure to PCBs can lead to certain intellectual impairments in children. PCBs may be carcinogenic; they are known to cause liver cancer in rats, and studies in Sweden and the United States have shown a correlation between high PCB concentrations in the body and incidences of certain cancers.

In 1985, near Kenora, Ontario, a transport truck spilled PCBs on the Trans-Canada Highway. Traffic was detoured for hundreds of kilometres as the cleanup took place. The asphalt was stripped and the highway was repaved. In response to this spill, the federal and provincial government regulations on the transport of dangerous materials were strengthened. Federally, under the Transport of Dangerous Goods Act, transportation of PCBs now requires detailed and precise documentation for international and interprovincial movement of PCBs. In addition, rigid, leak-proof containers must be used. All the provincial governments also have implemented legislation controlling the movement of PCBs through their provinces.

Although high-temperature incineration is one of the most effective ways to destroy PCBs in most solid waste, it is too costly to be used for the removal of PCBs

Studying bacteria that break down PCBs in contaminated soil FIGURE 12.15

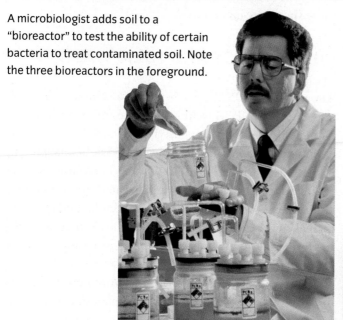

A microbiologist adds soil to a "bioreactor" to test the ability of certain bacteria to treat contaminated soil. Note the three bioreactors in the foreground.

that have leached into soil and water. One way to remove PCBs from soil and water is to extract them with solvents. But this method is undesirable because the solvents themselves are hazardous chemicals, and these extraction methods are also costly.

More recently, researchers have discovered several bacteria that degrade PCBs at a fraction of the cost of incineration. Additional research is needed to make the biological degradation of PCBs practical (FIGURE 12.15).

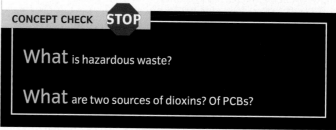

CONCEPT CHECK **STOP**

What is hazardous waste?

What are two sources of dioxins? Of PCBs?

Managing Hazardous Waste

LEARNING OBJECTIVES

Describe some of the public policy initiatives in Canada designed to reduce Canadians' exposure to hazardous waste.

Explain how environmental chemistry is related to source reduction.

We have the technology to manage toxic waste in an environmentally responsible way, but it is extremely expensive. Although great strides have been made in educating the public about the problems of hazardous waste, we have only begun to address the issues of hazardous waste disposal. No country currently has an effective hazardous waste management program, but several European countries have led the way by producing smaller amounts of hazardous waste and by using fewer hazardous substances. In Canada, the definition of toxic and hazardous wastes is a provincial matter, and definitions tend to differ from province to province.

CHEMICAL ACCIDENTS

When a chemical accident occurs in an industrialized country, whether at a factory or during the transport of hazardous chemicals, a response team is generally notified. In Canada, each province has a "hazmat" (hazardous material) team that deals with these types of situations on a jurisdictional basis (FIGURE 12.16 on page 462). Most chemical accidents reported involve oil, gasoline, or other petroleum spills. The remaining accidents involve some 1032 other hazardous chemicals, such as PCBs, ammonia, sulphuric acid, and chlorine.

Chemical safety programs have traditionally stressed accident mitigation and adding safety systems to existing procedures. More recently, industry and government agencies have stressed accident prevention through

Cleaning up hazardous waste FIGURE 12.16

In Canada and other industrialized countries, chemical spills are dealt with by a "hazmat" team. Removal and destruction of the wastes are often complicated by the fact that usually nobody knows what chemicals are present.

the **principle of inherent safety**, in which industrial processes are redesigned to involve less toxic materials so dangerous accidents are less likely to occur in the first place.

In December 1984, at the Union Carbide pesticide plant in Bhopal, India, a subsidiary of the U.S. multinational Dow Chemical Company, an accidental spill of 38 tonnes of toxic methyl isocyanate gas occurred, exposing more than 500,000 people to the toxic gas. The immediate death toll was 2259 people, but within 72 hours of the spill the number had risen to 10,000. Since then, an additional 25,000 have died from gas-related diseases.

Bhopal is the world's worst industrial disaster. A number of factors have been identified that led to this disaster, including the storage of chemicals in large tanks instead of small individual steel drums, the corrosion of materials in pipelines, poor maintenance, failure of several safety systems (due to poor maintenance and regulations), and safety systems that were dismantled to save money. In 1993, the International Medical Commission on Bhopal was established to respond to the long-term health effects of the disaster. These included eye problems, respiratory difficulties, immune system and neurological disorders, cardiac failure, lung injury, female reproductive difficulties, birth defects, and the risk of having stillborn babies. The spill left a sea of toxic and volatile compounds that severely affected the land, water, and air. The effects of the damage of the spill are astounding, and researchers fear the spill will result in decades of slow poisoning. In 1986, the factory was closed and the pipes, tanks, and drums were cleaned and sold. The water in the area was polluted and strategies were created to supply residents near the factory with clean drinking water. In 2008, the Madhya Pradesh High Court determined that the toxic waste should be incinerated.

PUBLIC POLICY AND TOXIC WASTE

The small-scale PCB spill in Kenora, Ontario, in 1985 and the 1988 PCB warehouse fire in Saint-Basile-le-Grand, Quebec, both focused public and government attention on the issue of hazardous waste management. Canadians produce 7.3 million tonnes of hazardous waste each year, but only 40 percent is treated; the remaining amount ends up in landfill sites or is discharged into municipal sewers. The lack of adequate treatment and disposal sites in Canada has led to illegal dumping practices (FIGURE 12.17).

Several public policies have been instituted at all levels of government to reduce Canadians' exposure to hazardous waste. For example, the Transportation of Dangerous Goods Act, implemented cooperatively by the provincial and federal governments, is designed to protect Canadians and the environment from toxic or hazardous waste spills that may occur during transport. A major aspect of this program is actually controlling the transport of dangerous goods. In addition, federal and provincial guidelines provide detailed instruction on the storage of hazardous wastes in Canada and require storage facilities to have in-plant surveillance, inspection, and leak detectors in place.

Toxic waste in deteriorating drums FIGURE 12.17

Old toxic dump sites like this one are more commonplace than we'd like to think. The metal drums in which this waste is stored have corroded and started to leak.

At the federal level, there are 24 departments that have responsibility over 30 different statutes addressing various aspects of controlling toxic substances. In 1989, a Waste Management Branch was created within Environment Canada. Legislation such as the Canadian Environmental Protection Act, the Clean Air Act, the Canada Water Act, the Ocean Dumping Control Act, the Department of the Environment Act, among others, assist the federal government in safeguarding the environment and human health from polluting substances.

Provincial governments have substantial authority to deal with hazardous waste disposal; however, provincial legislation, similar to federal legislation, has focused on air and water discharges. Each province relies on its own set of regulations and legislation to govern waste management.

Basel Convention

Each year millions of tonnes of waste are generated in Canada as by-products of industrial activities. Almost 5.5 million tonnes of this waste contains toxic chemicals that are considered to be hazardous to human health. Public awareness over the transboundary movement of hazardous waste was intensified in the late 1970s and early 1980s. The major concern was the exporting from industrialized, developed countries into developing countries for cheap disposal at inadequately prepared sites. The concerns resulted in a sense of urgency to address this transboundary movement of waste by developing and implementing international controls.

The **Basel Convention**, a global convention under the United Nations, is a framework for harmonizing hazardous waste disposal among nations. By October 2006, there were 168 countries, including those of the European Union, that were a part of the Convention. Canada participated in the development and was one of the original signatories of the Basel Convention, which was ratified on August 28, 1992. The key objectives of the Convention are to:

- Minimize the generation of hazardous waste
- Minimize the generation of hazardous recyclable materials
- Ensure the disposal of hazardous waste occurs in an environmentally sound manner

- Ensure hazardous waste disposal occurs as close to the source as possible
- Minimize the international movement of hazardous waste and hazardous recyclable materials

Under the Basel Convention, parties cannot carry out or authorize the transboundary movement of hazardous waste or hazardous recyclable materials to the following:

- Any party not a part of the Convention
- Antarctica
- Any prospective location that has prohibited such imports
- Any location where the disposal sites for such waste are available in the country of origin
- Any location where there is reason to believe that environmentally sound management/disposal options are not available within the location of destination

Canada, in response to its commitment to the goals of the Convention, created the Export and Import of Hazardous Waste and Hazardous Recyclable Materials Regulations. These regulations ensure that hazardous wastes and hazardous recyclable materials are managed safely and in a manner that protects the environment and human health.

Environment Canada's Waste Management Division continues to represent Canada and participate in Basel Convention activities. Environment Canada is also regulating the national framework for the international movement of nonhazardous waste. For more information on the Basel Convention, please visit the Basel Convention website at www.basel.int.

MANAGING TOXIC WASTE PRODUCTION

As with municipal solid waste, the most effective approach to managing toxic waste is source reduction—that is, using less hazardous or nonhazardous materials in industrial processes. Source reduction relies on the increasingly important field of **environmental chemistry**, or *green chemistry*.

environmental chemistry
A subdiscipline of chemistry in which commercially important chemical processes are redesigned to reduce environmental harm.

For example, chlorinated solvents are widely used in electronics, dry cleaning, foam insulation, and industrial cleaning. To accomplish source reduction, it is sometimes possible to substitute a less hazardous water-based solvent for the toxic chlorinated one. Substantial source reduction of chlorinated solvents can also be procured by reducing solvent emissions. Installing solvent-saving devices benefits the environment and also saves money because smaller amounts of chlorinated solvents must be purchased. No matter how efficient source reduction becomes, however, it will never entirely eliminate hazardous waste.

The second-best way to deal with hazardous waste is to reduce its toxicity by chemical, physical, or biological means, depending on the nature of the waste. High-temperature incineration, for example, reduces dangerous compounds like pesticides, PCBs, and organic solvents to safe products such as water and carbon dioxide. The resulting ash is hazardous and must be disposed of in a landfill designed for hazardous materials. Incineration using a *plasma torch* produces such high temperatures (up to 10,000°C, five times higher than conventional incinerators) that hazardous waste is almost completely converted to harmless gases.

Hazardous waste that is not completely detoxified must be placed in long-term storage. Hazardous waste landfills are subject to strict environmental criteria and design features. They are located as far as possible from aquifers, streams, wetlands, and residences. These landfills include several layers of compacted clay and high-density plastic liners at the bottom of the landfill to prevent leaching of hazardous substances into surface water and groundwater (**FIGURE 12.18**). Leachate

Cutaway view through a hazardous waste landfill FIGURE 12.18

The bottom of this hazardous waste landfill has two layers of compacted clay, each covered by a high-density plastic liner. (Some hazardous waste landfills have three layers of compacted clay.) A drain system located above the plastic and clay liners allows liquid leachate to collect in a basin where it can be treated, and a leak detection system is installed between the clay liners. Barrels of hazardous waste are placed above the liners and covered with soil.

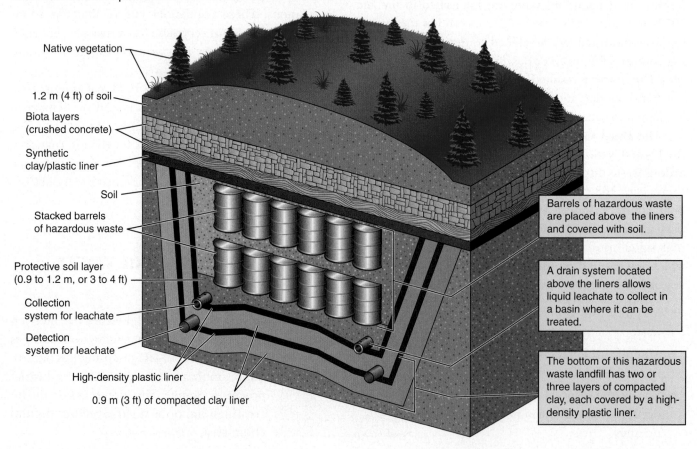

Native vegetation

1.2 m (4 ft) of soil

Biota layers (crushed concrete)

Synthetic clay/plastic liner

Soil

Stacked barrels of hazardous waste

Protective soil layer (0.9 to 1.2 m, or 3 to 4 ft)

Collection system for leachate

Detection system for leachate

High-density plastic liner

0.9 m (3 ft) of compacted clay liner

Barrels of hazardous waste are placed above the liners and covered with soil.

A drain system located above the liners allows liquid leachate to collect in a basin where it can be treated.

The bottom of this hazardous waste landfill has two or three layers of compacted clay, each covered by a high-density plastic liner.

(liquid that percolates through the landfill) is collected and treated to remove contaminants. The entire facility and nearby groundwater deposits are carefully monitored to make sure there is no leakage.

Few facilities are certified to handle toxic waste, although many larger companies are now licensed to treat their hazardous waste on-site. As a result, much hazardous waste is still dumped in sanitary landfills, burned in incinerators that lack the required pollution control devices, or discharged into sewers.

CONCEPT CHECK STOP

What is the focus of legislation in Canada dealing with hazardous waste?

What is the Basel Convention?

How is environmental chemistry applied to reducing sources of hazardous waste?

ECO CANADA   **CAREER FOCUS**

HAZARDOUS WASTE TECHNICIAN

Hazardous waste technicians are responsible for handling, processing, packaging, and tracking hazardous waste for shipment, treatment, and disposal.

Imagine you are dressed in a white protective suit complete with gloves and slippers for your shoes. You are a hazardous waste technician and you are unloading a minivan full of household hazardous waste containers. This is the city's annual household hazardous waste roundup, where for 10 days the city campaigns to have residents bring their old paint cans, solvents, motor oil, and batteries to your waste-handling facility for proper disposal. The roundup is part of a public campaign to educate residents on the proper disposal of household hazardous wastes so they don't end up in landfills or down the drain. Instead, they are brought to you and you ensure that these common household chemicals are properly collected and disposed of without harm to the environment.

As a hazardous waste technician, you have years of experience handling and disposing of hazardous substances and you know how

important it is to follow proper procedures. When residents or industrial traffic arrives at your facility, you carefully remove the hazardous waste containers from the vehicle and check the inventory list that should accompany the containers. When receiving hazardous waste, it is important that you can accurately identify what kind of waste has come in, where it has come from, and in what quantity.

Once you have confirmed everything is clearly marked and the necessary paperwork has been filled out, you sort the containers. First you identify which wastes can be recycled, for example, used motor oil or antifreeze. Those that cannot be recycled are inventoried according to their active ingredient and packed in plastic-lined drums to be shipped to specialized treatment and disposal facilities.

When the drums are ready for shipping, you must complete a shipping manifest indicating the volume of waste being transported and the name of the company hauling it. Copies of the shipping manifest are sent with the transport company with the drums. When the drums reach their destination, a delivery confirmation notice will be sent to you. Once you have the delivery confirmation, you can be confident the hazardous waste is being treated and disposed of properly and not ending up in landfills or draining into watersheds.

Hazardous waste technicians are responsible for handling, processing, packaging, and tracking hazardous waste for shipment, treatment, and disposal. They can also be involved in coordinating hazardous waste programs for both private industry and the public sector. They can be employed by waste recycling and treatment facilities or with large companies, packaging and shipping their hazardous waste. Hazardous waste technicians have specialized training on how to safely handle and dispose of chemical, biohazard, and radioactive wastes.

For more information or to look up other environmental careers go to www.eco.ca.

MONTREAL'S WASTE MANAGEMENT PLAN

Waste management is becoming a growing concern for industrial societies as waste amounts continually increase and landfills are saturated beyond capacity for facility expansion. Waste generation and disposal must be reduced by implementing new recycling facilities and programs. Waste management and recycling facilities often have a hard time finding a suitable location, however, as they receive considerable local opposition from stakeholders who do not want such facilities in their neighbourhoods.

To combat this problem, Montreal has implemented a Metropolitan Waste Management Master Plan (MWMMP), created by Dessau Construction and its partners. In the fall of 2003, the MWMMP was submitted for public consultation by the Montreal Metropolitan Community (MMC). The MWMMP covers 64 municipalities, which includes close to half of the population in the province of Quebec. The plan outlines reduction at the source, recovery, reuse, and disposal of waste along with increased citizen participation and responsibility in waste management. The plan aims to recover and reuse 60 percent of the waste managed by the municipalities by locating and building new eco-centres, recovery centres, composting centres, disposal sites, and waste transfer stations in the industrial employment areas of Montreal.

In April 2005, the city of Montreal launched its integrated waste management master plan citing the government's commitment to sustainable development. The Integrated Urban Waste Management Model (IUWMM), a sophisticated mixed integer linear programming model, is used to help decision makers with the long-term planning for waste management activities. The model is used to evaluate different scenarios considering various variables and constraints to establish solutions and minimize total system costs.

In 2005, Montreal diverted only 20 percent of the overall waste from landfill sites in the form of compost and recycling programs. By 2008, it had been hoped that 60 percent of Montreal's waste would be diverted, but in 2007, the city was struggling to divert even a third of its waste from landfills to be recycled or reused. Meanwhile, other cities, such as Vancouver, Edmonton, and Halifax, are diverting between 52 and 60 percent from their landfills.

Part of Montreal's waste management plan includes increasing recycling.

SUMMARY

1 Solid Waste

1. **Municipal solid waste** consists of solid materials discarded by homes, office buildings, stores, restaurants, schools, hospitals, prisons, libraries, and other facilities. **Nonmunicipal solid waste** consists of solid waste generated by industry, agriculture, and mining.

2. The **sanitary landfill** is the most common method of solid waste disposal; waste is compacted and buried under a shallow layer of soil. Layers of compacted clay and plastic sheets prevent leachate (liquid waste) from seeping into groundwater. Newer landfills possess a double liner system and collect leachate and gases that form during decomposition. Problems include the potential for methane gas to seep out and cause explosions, the accidental leaking of toxic leachate, a lack of existing landfill space, and resistance to new landfills near homes and businesses.

3. A **mass burn incinerator** is a large furnace that burns all solid waste except for unburnable items such as appliances. Problems associated with incineration of solid waste include the potential for air pollution, difficulties

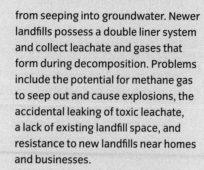

in disposing of the toxic ash produced, the high costs of the process, and the difficulties in choosing incinerator sites.

4. In composting, yard waste, food scraps, and other organic wastes are transformed by microbial action into a material that, when added to soil, improves its condition.

2 Reducing Solid Waste

1. In **source reduction**, products are designed and manufactured in ways that decrease the volume of solid waste and the amount of hazardous waste in the solid waste stream.

2. The volume of solid waste produced can be reduced through source reduction, reusing products, and recycling materials. Recycling conserves natural resources and is more environmentally benign than landfill disposal but requires a market for the recycled goods.

3. **Integrated waste management** is a combination of the best waste management techniques into a consolidated program to deal effectively with solid waste.

3 Hazardous Waste

1. **Hazardous waste** is any discarded chemical that threatens human health or the environment. The chemicals may be solids, liquids, or gases and include a variety of acids, dioxins, abandoned explosives, heavy metals, infectious waste, nerve gas, organic solvents, polychlorinated biphenyls (PCBs), pesticides, and radioactive substances.

2. Dioxins are hazardous chemicals formed as unwanted by-products during the combustion of many chlorine compounds. Polychlorinated biphenyls (PCBs) are hazardous, oily, industrial chemicals composed of carbon, hydrogen, and chlorine.

4 Managing Hazardous Waste

1. **Hazardous wastes** are waste substances whose disposal in the environment could potentially pose hazards to human health, jeopardize natural or agricultural resources, or interfere with other amenities. Disposal of hazardous wastes is now carried out in such a manner that the associated threats to people, resources, and amenities are acceptable and minimal through environmental legislation and regulations.

2. The most effective approach to managing hazardous waste is source reduction, reducing the amount and toxicity of hazardous materials used in industrial processes. Source reduction relies on **environmental chemistry**, a subdiscipline of chemistry in which commercially important chemical processes are redesigned to reduce environmental harm.

KEY TERMS

- **municipal solid waste** p. 443
- **nonmunicipal solid waste** p. 443
- **sanitary landfill** p. 444

- **mass burn incinerator** p. 448
- **source reduction** p. 451
- **integrated waste management** p. 456

- **hazardous waste** p. 458
- **environmental chemistry** p. 463

CRITICAL AND CREATIVE THINKING QUESTIONS

1. Compare the advantages and disadvantages of disposing of waste in sanitary landfills and by incineration.

2. How could source reduction efforts reduce the volume of waste that arises from abandoned automobiles?

3. List what you think are the best ways to treat each of the following types of solid waste, and explain the benefits of the processes you recommend: paper, plastic, glass, metals, food waste, and yard waste.

4. How do organizations such as Goodwill, which accepts donations of clothing, appliances, and furniture for resale, affect the volume of solid waste?

5. What are dioxins, and how are they produced? What harm do they cause?

6. Suppose hazardous chemicals were suspected to be leaking from an old dump near your home. Outline the steps you would take to:

    (1) have the site evaluated to determine if there is a danger, and

    (2) mobilize the local community to get the site cleaned up.

7. What is integrated waste management? Why must a sanitary landfill always be included in any integrated waste management plan?

8–10. In an effort to reduce municipal solid waste, many communities have required customers to pay for garbage collection according to the amount of garbage they generate, an approach termed unit pricing or "pay as you throw." The figure below illustrates the effects of unit pricing in one community on garbage sent to landfills and on wastes diverted through recycling and through separation of yard wastes.

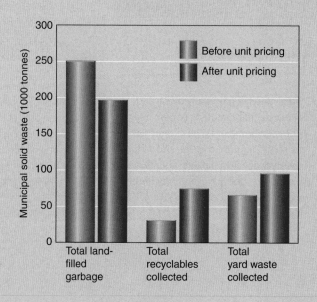

8. How did the implementation of unit pricing affect the amount of garbage sent to landfills?

9. How did the implementation of unit pricing affect the quantity of materials recycled or of yard wastes collected?

10. Which component of municipal solid wastes pictured in the graph was most affected by the implementation of unit pricing? Why do you suppose this was the result?

What is happening in this picture ?

This papermaker is converting denim waste into paper the colour of faded jeans.

How would this process reduce solid waste?

What type of industrial cooperation would be needed for this type of papermaking?

What are some other industries that might use these recycled materials?

Achieving Our Preferred Future Environment

13

DEFINING YOUR OWN "PERFECT" ENVIRONMENT

What do you think of as your own "perfect" environment? A boreal shield lake, a sun-drenched beach, an old-growth forest, a prairie wetland, the snow-covered arctic tundra, a mountainous retreat, a country home surrounded by trees and gardens, a suburban bungalow in a new subdivision, a downtown condo redeveloped from historic warehouse space? The list of possible preferred environments is endless.

Almost anyone who spends some time thinking about their preferences should be able to come up with their own definition of a "favourite" place: a place where they would like to spend the rest of their life, free from other obligations and responsibilities. A broader definition of this place might well be our own planet Earth—the place we call our global home.

As with many things in life, a preferred environment can mean many different things to different people. And the beauty of this reality is that no one can be right and no one can be wrong. Personal preferences are our own expressions of the nature of life as we see it.

But without question, as society moves forward to plan and manage our future environments, whatever we might choose those to be, there is an overriding need for maintaining the diversity and complexity of the factors that have produced the intricacies of the natural world as we know it today. To retain that wonderful complexity, through the use of paradigms such as sustainable development, ecosystem management, and adaptive management, is paramount to the future. To lose the meaning of "life on Earth" is to admit defeat in the face of adversity, and that is something that our species has never been able to accept. So go forward, and as Paul Hawken suggests on the first few pages of this chapter, take the opportunity and "run as if your life depends on it". . . since it probably does.

NATIONAL GEOGRAPHIC

CHAPTER OUTLINE

471

You Are Brilliant, and the Earth Is Hiring

The title of this section comes from a 2009 commencement speech given by Paul Hawken, a renowned entrepreneur, visionary environmental activist, author, and founder of WiserEarth (a nongovernmental organization dedicated to working for social justice, indigenous rights, and environmental stewardship), when he received an honorary doctorate of humane letters from the University of Portland. Portions of his speech are included below, but you can view the entire commencement address at www.charityfocus.org/blog/view.php?id=2077.

There is invisible writing on the back of the diploma you will receive, and in case you didn't bring lemon juice to decode it, I can tell you what it says: YOU ARE BRILLIANT, AND THE EARTH IS HIRING. The Earth couldn't afford to send any recruiters or limos to your school. It sent you rain, sunsets, ripe cherries, night blooming jasmine, and that unbelievably cute person you are dating. Take the hint. And here's the deal: Forget that this task of planet-saving is not possible in the time required. Don't be put off by people who know what is not possible. Do what needs to be done, and check to see if it was impossible only after you are done.

When asked if I am pessimistic or optimistic about the future, my answer is always the same: If you look at the science about what is happening on Earth and aren't pessimistic, you don't understand data. But if you meet the people who are working to restore this earth and the lives of the poor, and you aren't optimistic, you haven't got a pulse. What I see everywhere in the world are ordinary people willing to confront despair, power, and incalculable odds in order to restore some semblance of grace, justice, and beauty to this world. The poet Adrienne Rich wrote, "So much has been destroyed I have cast my lot with those who, age after age, perversely, with no extraordinary power, reconstitute the world." There could be no better description. Humanity is coalescing. It is reconstituting the world, and the action is taking place in schoolrooms, farms, jungles, villages, campuses, companies, refugee camps, deserts, fisheries, and slums. . . . This is the largest movement the world has ever seen. . . .

This extraordinary time when we are globally aware of each other and the multiple dangers that threaten civilization has never happened, not in a thousand years, not in ten thousand years. Each of us is as complex and beautiful as all the stars in the universe. We have done great things, and we have gone way off course in terms of honoring creation. You are graduating to the most amazing,

The gifts of the Earth FIGURE 13.1

In his commencement speech, Paul Hawken said "Ralph Waldo Emerson once asked what we would do if the stars only came out once every thousand years. No one would sleep that night, of course. The world would become religious overnight. We would be ecstatic, delirious, made rapturous by the glory of God. Instead the stars come out every night, and we watch television." What point is Hawken making here?

challenging, stupefying challenge ever bequested to any generation. The generations before you failed. They didn't stay up all night. They got distracted and lost sight of the fact that life is a miracle every moment of your existence. Nature beckons you to be on her side. You couldn't ask for a better boss. The most unrealistic person in the world is the cynic, not the dreamer. Hopefulness only makes sense when it doesn't make sense to be hopeful. This is your century. Take it and run as if your life depends on it.

SUSTAINABLE ECOSYSTEM MANAGEMENT

Near the beginning of his speech, Paul Hawken says, "This planet came with a set of operating instructions, but we seem to have misplaced them. Important rules like don't poison the water, soil, or air, and don't let the Earth get overcrowded, and don't touch the thermostat have been broken." As described throughout the previous chapters of this textbook, our planet Earth has undergone considerable change since first developing from a cosmic flash through millions of years of geological history and ecological evolution to the significant impacts that human societies have wreaked upon it. But the Earth is a resilient planet and continues to bounce back so that its potential future remains bright. In order to fully achieve all of the attributes that our future generations might require from it, a framework for action is needed so that Earth and the species that depend on its very existence can continue to flourish, evolve, and develop in ways that we today are likely not even capable of considering.

In Chapter 1, we indicated that this book would introduce the major environmental problems that humans have created, but more importantly, it would consider useful ways to address these issues and minimize human impact on our planet. We noted that we cannot afford to ignore the environment because our lives, as well as those of future generations, depend on it. As a wise proverb says, "We have not inherited the world from our ancestors; we have only borrowed it from our children."

And as we complete our task, we believe that this book has described the theoretical basis of our environment—using scientific and traditional knowledge as our foundation—and introduced you to a series of environmental issues and viable solutions. You should recall that most of these issues or problems are human-based, but note as well that the solutions or management strategies have similarly been established through human ingenuity, since humans possess an amazing set of skills, abilities, and knowledge. In fact, virtually nothing is impossible to achieve if we set our minds to it. So it is with this belief that we end our text and hopefully leave each of you with the undeniable realization that our Earth is perhaps reeling, but not down for the count.

To provide the most useful summary of this book, in this chapter we intertwine the key facts from the previous chapters into a real-world case study that can be used as an example of making a difference. We use a cookbook-like approach with a list of requirements and series of steps to "make it happen." The case study presents a somewhat hypothetical example of how a community (Royalwood, a real suburban subdivision in south Winnipeg, Manitoba; FIGURE 13.2 on page 474) can make a difference in preserving the environment in a way that can be emulated by others and by future generations wishing to accomplish the goals of sustainability. We lay out a model for sustainability of the Royalwood urban environment, and suggest that this approach can be used in any environment, any location, of any size.

The approach that we believe is most useful in achieving our preferred future environment is that of **sustainable ecosystem management**. This approach is useful because it empowers humans to take part in decisions that affect themselves and their use of the resources upon which we all depend. This collective decision-making strategy can be used to consider the needs of the present without compromising the ability of future generations to meet their needs—the paradigm of sustainable development.

The sustainable ecosystem management approach not only considers environmental well-being, but also economic and social requirements in

sustainable ecosystem management An approach to achieving our preferred future environment by meeting societal goals and objectives for the biotic communities and their abiotic components in a defined geographic area that takes into account economic, social, and environmental requirements.

Park setting in Royalwood FIGURE 13.2

Royalwood is a subdivision in south Winnipeg that is used as a case study to show how environmental sustainability can be implemented in real life to create a preferred future environment.

an intricately balanced and interconnected system of checks and balances. In the commencement speech quoted above, Paul Hawken has this to say about economics and environmentalism:

> *We have tens of thousands of abandoned homes without people and tens of thousands of abandoned people without homes. We have failed bankers advising failed regulators on how to save failed assets. Think about this: We are the only species on this planet without full employment. Brilliant. We have an economy that tells us that it is cheaper to destroy Earth in real time than to renew, restore, and sustain it. You can print money to bail out a bank but you can't print life to bail out a planet. At present we are stealing the future, selling it in the present, and calling it gross domestic product. We can just as easily have an economy that is based on healing the future instead of stealing it. We can either create assets for the future or take the assets of the future. One is called restoration and the other exploitation. And whenever we exploit the Earth we exploit people and cause untold suffering. Working for the Earth is not a way to get rich, it is a way to be rich.*

Sustainable ecosystem management is a way to marry economic richness with environmental protection. Furthermore, it is relevant in any location in the world and for any environmental consideration, whether on land or in water, concerning agriculture, wildlife, fisheries, forestry, land use, mining, energy, parks and protected areas, urban planning, or other considerations. The concept is being increasingly utilized in a diverse array of locations and for a wide range of decisions.

Thus, the process that is outlined in this chapter can be used by any person, organization, government, or other entity to achieve their collective goals—in this case, for the environment, but in other cases, to solve other important issues.

CONCEPT CHECK STOP

What is sustainable ecosystem management?

What are the benefits of this approach?

Defining and Creating Our Preferred Future Environment

LEARNING OBJECTIVES

Define *sustainable development.*

Describe how ecosystem management helps achieve the goals of sustainable development.

Each of us has a particular **worldview**—that is, a commonly shared perspective based on a collection of our basic values that helps us make sense of the world, understand our place and purpose in it, and determine right and wrong behaviours. These worldviews lead to behaviours and lifestyles that may or may not be compatible with environmental sustainability. (See Chapter 2 for more on the various environmental worldviews.)

In 1987, the Brundtland Commission put forward the worldview, or paradigm, of **sustainable development** (also known as

> **sustainable development**
> Economic growth that meets the needs of the present without compromising the ability of future generations to meet their needs.

environmental sustainability) as the preferred choice for preserving humanity's well-being into the future. This concept is best defined as the ability to meet humanity's needs without compromising the ability of future generations to meet their needs. The concept acknowledges that human actions can affect the environment, that Earth's resources are finite, that humans need to understand the impact of their consumption of environmental resources, and that there is a shared responsibility to make all of this happen. In short, everyone is part of the problem and everyone must be part of the solution.

Sustainable development considers not only environmental parameters, but also social and economic attributes. As **FIGURE 13.3** indicates, decision making that is sound in all areas is best able to maintain the future environment that we wish to attain.

As societies began to embrace the concept of sustainable development, it became necessary to identify a process by which

Sustainable development FIGURE 13.3

Three factors—environmentally sound decisions, economically viable decisions, and socially equitable decisions—interact to promote sustainable development.

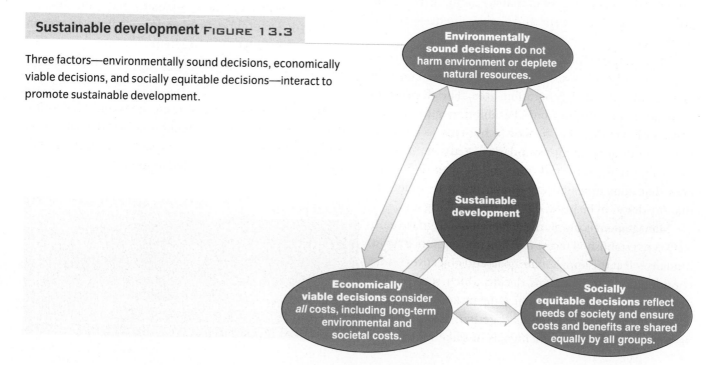

Environmentally sound decisions do not harm environment or deplete natural resources.

Sustainable development

Economically viable decisions consider *all* costs, including long-term environmental and societal costs.

Socially equitable decisions reflect needs of society and ensure costs and benefits are shared equally by all groups.

What is an ecosystem? FIGURE 13.4

What constitutes an ecosystem is somewhat arbitrary and is a decision made by humans. Depending on their objectives, some people might consider this picture of Medicine Lake in Jasper, Alberta, to be one ecosystem or several interacting ecosystems.

the principles of environmental well-being, social equity, and economic viability could best be achieved. The approach that most countries of the world have adopted is that of (sustainable) ecosystem management, a strategy that is directly connected to the concepts of sustainable development. **Ecosystem management** is based on the integration of environmental or ecological, social, and economic considerations in decision making with respect to the environment. This approach is increasingly being used to enhance environmental decision making by governments at all levels, as well as by private and nongovernmental organizations. In order to describe the term ecosystem management, its two component terms—ecosystem and management—need to be defined.

In its simplest form, an **ecosystem** is the biotic communities and their associated abiotic components that interact in a defined geographic area. The decision of what biotic communities and what abiotic components to consider is a choice that is made by humans, as is the decision about geographic boundaries (**FIGURE 13.4**). And there really is no right or wrong way to define the area that constitutes an ecosystem, but it is important that the decision be justifiable by those making it.

Management is simply the achievement of goals or objectives as established by the decision makers. Once again, humans select the appropriate goals and objectives under consideration, and humans decide which stakeholders should be part of the decision-making process.

So to put the definitions together, *ecosystem management* is a process by which goals of stakeholders in a

> **ecosystem management**
> An evolving concept that recognizes the need to use ecological components in environmental decision making.

defined geographic area are achieved for the interacting biotic communities and their abiotic components. Note that humans are active participants not only in the process but also in the ecosystem itself.

Ecosystem management can be implemented by following these five steps:

Step 1: Determine the main stakeholders, define the ecosystem area, and develop the relationship between them.

Step 2: Characterize the structure and function of the ecosystem, and set in place mechanisms to manage and monitor it.

Step 3: Identify the important economic issues that will affect the ecosystem and its inhabitants.

Step 4: Determine the likely impact of the ecosystem on adjacent ecosystems.

Step 5: Decide on long-term goals and flexible ways of reaching them.

We will address each of these steps and how to implement them in the next section, which describes one example of how sustainable ecosystem management has been successfully implemented in south Winnipeg.

CONCEPT CHECK STOP

What are the goals of sustainable development?

How does ecosystem management attempt to achieve the goals of sustainable development?

The Royalwood Urban Ecosystem: A Case Study

LEARNING OBJECTIVES

Outline the five steps in the ecosystem management approach.

Explain why Royalwood is a good example of sustainable ecosystem management.

R oyalwood is a real suburban development of about seven square kilometres in south Winnipeg, Manitoba, that serves as a case study to explore how the five steps of sustainable ecosystem management can be implemented in a real-world setting. The area was developed in the early 2000s in a creative and novel way by Ladco Company Limited. Ladco achieved a new standard for urban environmental decision making by using an ecosystem management–like approach in partnership with landscape architects Scatliff Miller Murray, and Native Plant Solutions, a subsidiary of Ducks Unlimited Canada.

Although many cities in Europe, Australia, and elsewhere in the world began practising novel approaches to urban planning and management long ago, North American cities have been slower to respond to the requirements of moving forward with sustainable development principles. Some cities in the United States, however, such as Minneapolis, Denver, and Burlington, have been innovators in using ecosystem management principles in planning urban landscapes.

Overall, Royalwood is an example of an innovative application in water and land management in an urban setting. Further, the project can be described as an urban ecosystem management exercise where, as our working definition of ecosystem management suggests, societal goals are achieved in a defined geographic area for the biotic communities and their associated abiotic components.

STEP 1: DETERMINING THE STAKEHOLDERS AND DEFINING THE ECOSYSTEM AREA

Often the most difficult step is to effectively identify the appropriate stakeholders and define the area for which management will be considered. This step is best achieved by working simultaneously on its two component parts, and it will take considerable time and effort to ensure that a "good fit" is achieved.

Stakeholders When identifying stakeholders, all of the key persons with interests in the area must be considered. Stakeholders can be weighted as primary, secondary, or tertiary, and their views can be assessed in that light. **Primary stakeholders** are those who are most dependent on the area and who are most likely to take an active part in its management. **Secondary** and **tertiary stakeholders** may by representatives of governments or private interest groups who may or may not have a strong personal affinity to the area under consideration and are generally assigned their position (as secondary or tertiary stakeholders) by the primary stakeholders. Regardless of the type, it is critical to build a strong relationship among the various stakeholders as early as possible in the process, and this is often effectively done through public forums and regular meetings.

The stakeholders for Royalwood could include, but are not limited to, the local developer and partners; city, provincial, and federal government representatives; and current residents and their community groups (**FIGURE 13.5** on page 478). It would be important for all groups to work together, attend public meetings, and establish general goals and objectives for the area. Some examples of possible goals and objectives that might be considered include the following:

- Create a naturally functioning wetland
- Preserve some urban forest lands
- Maintain a riparian buffer along river and stream systems
- Enhance overall biological diversity of the area

Since Royalwood is a subdivision, one of the primary stakeholders would be the population of residents who live there. Other stakeholders include the developer, various levels of government, and community or environmental groups active in the area.

Ecosystem area Determining the geographic boundaries of Royalwood is quite easy, since the developer had the rights to develop only a prescribed area within the city of Winnipeg. However, defining the *ecosystem area* for Royalwood must take into consideration many different factors other than just geography. Among the biotic components would be humans, pets, and other wildlife in the area; the abiotic components would include the structures that are built. In an urban ecosystem, humans influence ecological factors (plants, air, soil, animals), and human decisions (where to build houses, parks, highways, schools) are influenced by ecological factors.

STEP 2: CHARACTERIZING THE ECOSYSTEM STRUCTURE, FUNCTION, AND MANAGEMENT OPTIONS

The next step in the ecosystem management approach is to characterize the ecosystem structure and function with the stakeholders. The management goals and objectives that arise from this endeavour should be determined by using the best available knowledge from all stakeholders. This strategy not only uses the optimal data available, but also builds trust among all

involved. It is important that all parties be realistic in their expectations since they might need to settle for what is possible under time and budget constraints rather than what is theoretically ideal.

An important consideration in defining the management goals and objectives is to use the concept of **adaptive resource management (ARM)**. This approach admits that considerable uncertainty exists in most situations involving the establishment of environmental goals and objectives (**FIGURE 13.6**). But as managers, we cannot wait to make decisions until all of the data have been collected. Therefore, every management decision should be treated as a management experiment, in which the outcomes are monitored and future decisions are revised based on the results.

adaptive resource management (ARM) An approach to management that acknowledges uncertainty and the need for managers to learn from system responses while they manage.

Several approaches could be used to help identify appropriate management goals and objectives for an ecosystem. One appropriate measure and level of organization is ecosystem diversity. **Ecosystem diversity** characterizes the array of ecosystems in an overall area and plans to maintain or restore a level of functional representation

Adaptive resource management FIGURE 13.6

Adaptive resource management involves creating a plan based on uncertain or incomplete information, but then analyzing the results of that plan to learn from it and adapt the plan to account for new information that has been obtained. The process never ends because variations in the plan will always result in new information.

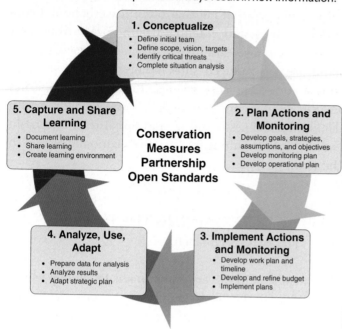

Historical range of variability FIGURE 13.7

HRV, in effect, defines the ecological sphere of ecosystem management.

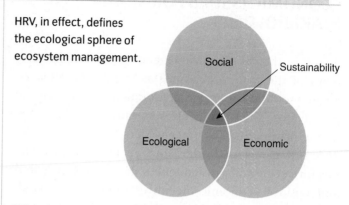

for each; therefore, using ecosystem diversity as a measure of goals and objectives best achieves the principles of sustainable development.

To actually implement a measure as broad as ecosystem diversity, the **historical range of variability (HRV)** approach is the most realistic for managers to consider. The assumption of this approach is that biodiversity (and habitat for species of concern) will be provided by the diverse array of ecosystems that result. The HRV approach indicates that the historical range of conditions that existed in an area provides the best scientific basis for understanding landscape dynamics and conditions and provides a reference point for biodiversity. In a sense, HRV defines the "natural environment" (**FIGURE 13.7**). Ecological sustainability can be met with a representation of historical conditions for every ecosystem.

The situation of Royalwood So let's consider the situation of Royalwood in the area of south Winnipeg. The historical condition of south

Winnipeg could generally be characterized in the following manner:

- Grassland dominated
- Wetland complexes to retain and filter water
- Shrubs and trees interspersed
- Disturbance factors of fire, floods, and bison
- High level of genetic, species, community, and ecosystem diversity
- First Nations and Metis peoples used the area for cultural and spiritual needs
- *Dynamic, fully functioning "natural" ecosystem*

Contrast that to the current condition found in most parts of south Winnipeg today:

- Suburban communities
- Manicured residential properties
- Little retention of water
- Plantings of exotic vegetative species
- Disturbance from human activities
- Relatively low level of biodiversity
- *Highly altered urban ecosystem*

As Steps 1 and 2 are implemented for a community like Royalwood, the stakeholders would need to become active in defining the extent to which the historical range of conditions can be achieved. In some situations, more "natural" landscapes might be preferred, but in others, a more typical urban situation might arise. But in either case, the stakeholders need to take the initiative to determine the choices that they are making and accepting with respect to the environment.

STEP 3: IDENTIFYING THE IMPORTANT ECONOMIC ISSUES FOR THE STAKEHOLDERS

Step 3 in the ecosystem management process involves many of the same procedures as Step 2, but from an economic rather than an environmental or ecological perspective. This step involves identifying the key economic drivers that the stakeholders believe will influence the management decisions. As much as possible, negative economic consequences should be eliminated and positive economic incentives should be created, such as giving stakeholders subsidies, awards, rewards, reduced taxes, or increased values for their particular land holdings.

For Royalwood, economic factors played a considerable role for the developer and partners, and also for the residents themselves. The approach taken involved developing naturalized stormwater retention basins (FIGURE 13.8; see Chapter 9 for more information on stormwater retention basins). These basins use natural wetland vegetation, submerged shelves throughout the ponds, and native tall grass prairie species on the upland fringes. These naturalized systems offer significantly reduced construction costs and a degree of water treatment, and they virtually eliminate the major issues identified by the city of Winnipeg of traditional-style ponds, namely shoreline erosion, submerged island shallows, and regular and costly removal of aquatic vegetation and algae. The natural shoreline vegetation is much taller than that found around traditional ponds, which eliminates the problem of attracting giant flocks of Canada geese in fall that often leave behind a considerable mess and potential health hazard from their waste droppings. And finally, the need to mow the upland areas is reduced substantially, resulting in additional long-term cost savings. Residential properties in this subdivision increased in value compared with similar residences found in traditional retention basin subdivisions, resulting in a positive economic benefit to homeowners.

STEP 4: DETERMINING THE IMPACT OF THE ECOSYSTEM

The next step in the ecosystem management approach is to identify the likely impact of the proposed management plan on adjacent ecosystems. The plan would then need to be adapted to ensure that benefits to all

Natural versus traditional stormwater retention ponds FIGURE 13.8

The stormwater retention pond in A is a picture taken from Royalwood. The pond in B is a more traditional retention pond. The tall grasses and natural vegetation around the pond in A provide many economic benefits to all stakeholders involved, including the developer, the city, and the residents.

SAVE OUR SEINE

The Seine River is increasingly threatened with urban development. Save Our Seine is just one example of the effects community-based environmental groups can have on preserving and restoring a natural landscape.

Save our Seine (SOS) is a community-based stewardship group located in Winnipeg, Manitoba. The Seine River (see picture) is a small tributary to the Red River and connects to many rural communities in southeast Manitoba. Within the city of Winnipeg, urban development is expanding toward the Seine River corridor, choking the natural area and threatening its continued survival.

Prior to settlement, the area provided an important source of game for Aboriginal hunters and supported a healthy ecosystem. Since settlement, however, the river has been exposed to ditches, bridges, agriculture, and urban development, all of which have resulted in changes to the ecosystem and its associated flora and fauna, the river basin, and the actual flow of the river itself. The most dramatic of these changes were observed following the construction of the Red River Floodway in 1967.

SOS's goals are to protect, preserve, and enhance the Seine River environment within the city of Winnipeg. The Seine River flows through areas now dominated by new housing developments such as the Royalwood community. Working closely with the community through public education, leadership, and participatory events, SOS attempts to develop a community-based ethic of environmental stewardship.

In April 2007, a collaborative effort between SOS, the city of Winnipeg, and local stakeholders developed a vision and action plan for a portion of the Seine River corridor known as the Seine River Greenway that stretches from the Red River to Happyland Park. The action plan identifies major goals aimed at protecting and enhancing the Seine River corridor for present and future generations by preventing exploitation and destruction of the natural environment.

There are many managerial challenges that face the Seine River Greenway initiatives. Beaver populations continue to grow in the absence of natural predators and take a heavy toll on mature trees, altering the tree canopy, water flow, and ecosystem balance. The presence of invasive species that are difficult to remove impact the successful survival of native species, and Dutch Elm disease is a major environmental concern for the greenway. There is also the added challenge of maintaining suitable water levels on the Seine River. Traditionally, water diversion has occurred for economic reasons and local flood protection, and efforts to change the existing diversion requires the support of the provincial government. Through these collaborative efforts, the Seine River Greenway plan hopes to encourage community-based participation and stewardship to protect and enhance the Seine River ecosystem.

For more information on the Save Our Seine Organization and their activities, please visit their website at www.saveourseine.com.

---

ecosystems are being realized. In particular, **synergistic benefits** could accrue to neighbouring ecosystems if plans are jointly developed. This step is therefore important not only to the ecosystem under consideration, but also to the larger landscape or jurisdictional area.

For Royalwood, stakeholders were particularly concerned about the impacts that their plan might have on the adjoining Seine River and riparian forest zone (see the Making a Difference feature for information on Save Our Seine, a community-based group that had a large impact on developing strategies to preserve the Seine River area). Not only was this location considered biologically unique within the city of Winnipeg, it was also recognized as an important cultural and spiritual site for the Metis people of this part of Manitoba. The riparian forest is known to them as *Les Bois des Spirits*, or Spirit Woods, and its preservation was considered paramount by area residents and the developer.

As well, other remnant forest blocks provided habitat to great horned owls. The owls appeared to have a unique place in the Royalwood system, had apparently nested in the location for several years, and were likely being somewhat aided by the disturbances caused by the subdivision development that resulted in their prey being easier to capture. Thus, the stakeholder groups built a stronger preservation requirement for these urban forest blocks into the management plan to provide preferred habitat for nesting owls as well as habitat for other bird species.

STEP 5: USING ADAPTIVE MANAGEMENT OVER TIME

The final step in the ecosystem management process is to create long-term goals and flexible ways of reaching them. The stakeholders need to continue to meet regularly after the plan is developed and implementation begins so that they can identify any unforeseen issues that might affect plan goals. In addition, alternative strategies should be considered to maximize the overall success of the plan. This phase should also identify methods to ensure that regular, science-based monitoring is undertaken and completed. This step helps to build trust among stakeholders in the plan and is regularly revisited and altered as necessary.

For Royalwood, the longer-term interests of stakeholders of various types has been considered since the beginning of development. For instance, Royalwood is now serving as a research platform for Ducks Unlimited in investigating carbon sequestration in urban retention ponds. Newer developments by Ladco are not only drawing upon the results achieved at Royalwood, but are also moving forward with ponds connected by meandering channels and areas designed to support specific vegetation types. And although Native Plant Solutions has noted that the naturalized basins were not constructed for wildlife habitat, a succession of amphibians and passerine bird species now make use of the maturing plant communities. So a win-win situation has been created for all concerned.

SUMMARY OF ROYALWOOD ACCOMPLISHMENTS

We have used Royalwood as an example of how to implement sustainable ecosystem management in part because of the success of this development.

Here are some of the results to date of the Royalwood development:

- Two fully functioning wetlands have been created, which filter toxic substances and chemical contaminants.
- Biodiversity in the area has increased.
- Property values in the subdivision have increased.
- Residents have a connection with the natural environment.
- A 10-hectare urban forest has been preserved, which includes preserved habitat for great horned owls.
- Cost savings to the development company have been realized.

All in all, the management plan objectives for Royalwood are being achieved, which has led to benefits to all stakeholders involved. Perhaps more importantly, the historic condition of the ecosystem is being restored while at the same time allowing for urban development.

Clearly, ecosystem management is a viable approach to enhancing decision making that ensures that more sustainable environmental practices are achieved. Whether this approach is possible or even necessary for all management areas is a question that is still under debate, but currently Royalwood is serving as an example of sustainable urban ecosystem management that is being studied and expanded on in new sustainable endeavours.

OTHER EXAMPLES OF HOW TO CREATE OUR PREFERRED FUTURE ENVIRONMENT

TABLE 13.1 offers additional real-world examples of how a preferred future environment can be defined and created. Each is important to review in its own right, as there are always new and better ways that humanity develops to improve upon past performance. TABLE 13.2 lists provincial and territorial government websites where you can access information on how your government is addressing sustainable development. In all cases, however, the important elements are the process steps that have been created to set an individual or organization on the right path to achieve success—in this case, for sustainability of our future environments.

Selected examples of agencies using ecosystem management in Canada TABLE 13.1

Agriculture and Agri-Foods Canada	www.agr.gc.ca
Agriculture and Agri-Foods Canada—Agriculture Environmental Services Branch	www4.agr.gc.ca/AAFC-AAC/display-afficher .do?id=1187362338955
B.C. Ministry of Forest and Range—Silviculture Treatment and Ecosystem Management	www.for.gov.bc.ca/hre/stems/
Canadian Environmental Assessment Agency	www.ceaa-acee.gc.ca
Canadian Food Inspection Agency	www.inspection.gc.ca
Canadian Nuclear Safety Commission	www.nuclearsafety.gc.ca
Canadian Polar Commission	www.polarcom.gc.ca
Canadian Wildlife Service	www.cws-scf.ec.gc.ca
David Suzuki Foundation	www.davidsuzuki.org
Department of Fisheries and Oceans Canada	www.dfo-mpo.gc.ca
Department of Fisheries and Oceans Canada— New Scientific Ecosystem Management Framework	www.dfo-mpo.gc.ca/science/publications/ecosystem/index-eng.htm
Ducks Unlimited Canada	www.ducks.ca
Environment Canada	www.ec.gc.ca
Environment Canada—Basin Management	www.pnr-rpn.ec.gc.ca/nature/ecosystems/nrei-iern/dg00s09.en.html
Environment Canada—Convention on Biological Diversity—Ecosystem Management Approach	www.cbin.ec.gc.ca/enjeux-issues/ecosystemique-ecosystem.cfm?lang=eng
Farm Credit Canada	www.fcc-fac.ca
Farm Products Council of Canada	www.fpcc-cpac.gc.ca
Fisheries Resource Conservation Council	www.frcc-ccrh.ca
Foreign Affairs and International Trade Canada	www.dfait-maeci.gc.ca
Freshwater Fish Marketing Corporation	www.freshwaterfish.com
Great Bear Ecosystem	www.savethegreatbear.org/
Great Lakes Net—Ecosystem Management in the Great Lakes Region	www.great-lakes.net/envt/air-land/ecomanag.html
Greening Canada's Government Operations	www.greeninggovernment.gc.ca/default.asp?lang=En
Greening Government	www.tpsgc-pwgsc.gc.ca/ecologisation-greening/index-eng.html
Health Canada	www.hc-sc.gc.ca
Manitoba Department of Conservation	www.gov.mb.ca/conservation/wildlife/
NAFTA Secretariat—Canadian Sector	www.nafta-alena.gc.ca
National Energy Board	www.neb-one.gc.ca
Natural Resources Canada	www.nrcan-rncan.gc.ca
Natural Resources Canada—Ecosystem Management and Climate Change	www.sst-ess.rncan-nrcan.gc.ca/ercc-rrcc/sector/eco_e.php
Natural Resources Canada—Ecosystem Management in Boreal Forests	www.cfs.nrcan.gc.ca/news/645

Selected examples of agencies using ecosystem management in Canada TABLE 13.1 (CONTINUED)

Nature Conservancy of Canada	www.natureconservancy.ca/site/
North American Agreement on Environmental Cooperation	www.naaec.gc.ca
Ontario Ministry of Natural Resources	www.mnr.gov.on.ca/en/index.html
Parks Canada	www.pc.gc.ca
Parks Canada—Ecosystem Management (in parks and fire management)	www.pc.gc.ca/eng/progs/np-pn/eco/eco5.aspx
World Wildlife Fund—Ecosystem Management in Marine Fisheries	www.assets.wwf.ca/downloads/wwf_northwestatlantic_ implementationofecosystem_ basedmanagement.pdf

Provincial and territorial government environmental agencies TABLE 13.2

Alberta—www. alberta.ca

- Agriculture and Rural Development
- Energy
- Sustainable Resource Development
- Tourism, Parks and Recreation

British Columbia—www.gov.bc.ca

- Environmental Assessment Office
- Environmental Appeal Board
- Forest Renewal BC
- Forest Appeals Commission
- Ministry of Agriculture, Fisheries and Food
- Ministry of Environment, Lands and Parks
- Ministry of Forests

Manitoba—www.gov.mb.ca

- Agriculture, Food and Rural Initiatives
- Conservation
- Innovation, Energy and Mines
- Water Stewardship

New Brunswick—www.gnb.ca

- Agriculture and Aquaculture
- Energy
- Environment
- Fisheries
- Natural Resources
- Tourism and Parks

Newfoundland and Labrador—www.gov.nf.ca

- Environment and Conservation
- Office of Climate Change, Energy Efficiency and Emissions Trading
- Task Force on Adverse Health Events
- Fisheries and Aquaculture
- Natural Resources
- Tourism, Culture and Recreation

Northwest Territories—www.gov.nt.ca

- Environment and Natural Resources
- Industry, Tourism and Investment

Nova Scotia—www.gov.ns.ca

- Agriculture
- Energy
- Environment
- Fisheries and Aquaculture
- Natural Resources

Nunavut—www.gov.nu.ca

- Department of the Environment

TABLE 13.2 (CONTINUED)

Ontario—www.ontario.ca

- Agriculture, Food and Rural Affairs
- Economic Development and Trade
- Energy and Infrastructure
- Environment
- Natural Resources
- Northern Development, Mines and Forestry

Prince Edward Island—www.gov.pe.ca

- Agriculture
- Environment, Energy and Forestry
- Fisheries, Aquaculture and Rural Development

Quebec—www.gouv.qc.ca

- Ministry of Natural Resources and Wildlife
- Ministry of Sustainable Development, Environment and Parks

Saskatchewan—www.gov.sk.ca

- Agriculture
- Energy and Resources
- Farm Land Security Board
- Saskatchewan Crop Insurance Corporation
- Saskatchewan Watershed Authority
- SaskEnergy Incorporated
- SaskPower
- SaskWater
- Tourism, Parks, Culture and Sport
- Water Appeal Board

Yukon—www.gov.yk.ca

- Energy, Mines and Resources
- Environment
- Yukon Energy Corporation

GETTING INVOLVED YOURSELF!

So how can you become involved in the future with these kinds of interesting and unique opportunities? It is really not that difficult, but it does take some initiative on your part. There is an old adage that says those who make a difference get involved in things (**FIGURE 13.9** on page 486)!

Some ways to do this include joining an environmental organization or club at your university or college or in your local community. But joining is just the first step. You have to attend meetings, take on a duty, or volunteer for an activity, and not be shy about bringing your ideas forward in a public setting. It may be difficult at first, but the more you speak up about things, the easier it gets. Practise in classroom settings first where there is less pressure to say or do the "right things."

But don't just stop by getting yourself involved. Encourage your classmates, friends, relatives, and family members to take an active role in the future of our world. Who could possibly not be interested in joining YOU as an advocate for societal change and action? And be sure to practise what you preach—if you are advocating for urban transit, be sure you are using it! And if buying local products is part of your mantra, make sure that's what you do.

Another important requirement for making a difference is to keep current on environmental matters. It is not enough to take one or a few courses in environmental studies or science, or even to get a degree in that discipline. You must continue to be active in *lifelong learning*—an approach to continually upgrade yourself in your respective area of interest and expertise. A professor at Colorado State University, Dale Hein, who used to teach one of the authors of this text used to say, "You will become obsolete in your discipline within six months of graduating unless you become involved with your professional organization." So find your professional organization. If you don't know what it is, just enter some key words into a search engine and see what comes up on the Internet. And don't just join but get involved! You *can* do it, you *need* to do it, and the Earth will thank you.

Joining an environmental organization that shares your passion and beliefs is one of the best ways to get involved.

CONCEPT CHECK STOP

What are the five steps in the ecosystem management approach?

Why is Royalwood a good example of successful ecosystem management?

ECO CANADA **CAREER FOCUS**

SUSTAINABLE ARCHITECT

Sustainable architects create attractive, affordable, and comfortable buildings that do not harm the environment, either during their construction or their lifetime.

Imagine you are standing amid a group of onlookers admiring the magnificent new city hall building. You are a sustainable architect and this classic structure is your latest creation. Two years ago, members of city council asked you to design a replacement for the old city hall, which was becoming too small. The council asked you to design not only a larger space but an environmentally friendly building to demonstrate the city's commitment to conserving and protecting the environment. After many revisions to the plans, many meetings with structural engineers and builders, and many months of construction, the building is finished. Based on your innovative design, the new city hall is a model for sustainable building.

As a sustainable architect, you had to fulfill a number of objectives when designing the new city hall—it was a challenging project. To begin, council needed more space but didn't want to build an imposing megastructure that dominated the landscape, so one of your design priorities was a building that had a positive impact on its environment while meeting council's space needs.

Next, you focused on resource efficiency: You tried to use only materials produced locally, to reduce dependency on truck transportation and the pollution trucks contribute. You selected durable materials that won't need to be replaced often and avoided anything from nonrenewable resources. You also concentrated on improving energy efficiency and providing healthy indoor air. The new building relies primarily on natural ventilation and has huge windows that capitalize on natural light. You incorporated solar energy, water-conserving bathrooms, and the most energy-efficient climate control system on the market. You used every opportunity to conserve energy and water and reduce emissions—with spectacular results.

Sustainable architects oversee the design and construction of buildings with a focus on the role a structure will play in its environment. For example, sustainable architects try to use only construction materials that won't add to landfills and make certain all the building's systems and equipment have high energy-efficiency ratings. A sustainable architect's primary goal is to create attractive, affordable, and comfortable buildings that do not harm the environment, either during their construction or their lifetime.

For more information or to look up other environmental careers go to www.eco.ca.

GREEN ROOF TOPS AT TRENT UNIVERSITY

Now-retired professor Tom Hutchinson from Trent University has spent over a decade supervising the development of the Trent University Environmental and Resource Sciences Vegetable Garden Project. This project saw the creation of an elaborate vegetable garden on the roof of the Environmental Sciences building at Trent University (see photo).

The green rooftop project set out with two main purposes. The first was to monitor ozone levels and conduct environmental research. The second was to raise 30 to 40 different crops of produce to be donated to the university's vegetarian café, local community groups, missions, the YMCA, and numerous kitchen volunteers. The green rooftop, measuring close to 30 by 90 metres, provides a forum for students and researchers to learn more about the effects of air pollution and smog on agricultural crops, provides a test region for tropospheric ozone, and illustrates how unused spaces can become incredibly productive. Coupled with these uses, the green, or "living," rooftop absorbs rainwater,

provides insulation, lowers urban air temperatures, and creates urban habitat for wildlife. The addition of green rooftops mitigates the negative energy effects of buildings and reduces the energy consumption of buildings. Green rooftops absorb water and offset overburden stormwater flow into lakes and rivers. Many avian and invertebrate species thrive in green rooftop environments.

Rooftop conditions can, however, present challenges for plant survival. Vegetation growing on rooftops are often moisture stressed and experience drought, elevated temperatures, high light intensity, and high winds. Plants growing in these environments need to withstand these conditions. Green roofs like that at Trent illustrate the economic, social, educational, and environmental benefits of creating these urban ecosystems in otherwise unproductive spaces.

For more information on the Trent University Environmental and Resource Sciences Vegetable Garden Project, please visit Trent University's website at www.trentu.ca.

The rooftop garden at Trent University has managed to successfully grow a variety of vegetables and has created an urban ecosystem in an otherwise unproductive space.

SUMMARY

1 You Are Brilliant and the Earth Is Hiring

1. **Sustainable ecosystem management** is an approach to achieving our preferred future environment by meeting societal goals and objectives for the biotic communities and their abiotic components in a defined geographic area that takes into account economic, social, and environmental requirements. This approach allows us to create a framework for action so that Earth and the species that depend on its very existence can continue to flourish, evolve, and develop in ways that we today are likely not even capable of considering.

2 Defining and Creating Our Preferred Future Environment

1. **Sustainable development** is best defined as the ability to meet humanity's needs without compromising the ability of future generations to meet their needs. The concept acknowledges that human actions can affect the environment, that Earth's resources are finite, that humans need to understand the impact of their consumption of environmental resources, and that there is a shared responsibility to make all of this happen. In short, everyone is part of the problem and everyone must be part of the solution.

2. **Ecosystem management** is a way to marry economic richness with environmental protection. It helps achieve the goals of sustainable development by taking into account many perspectives on and determining the best course of action for achieving a long-term solution that serves all stakeholders—both current and future.

3 The Royalwood Urban Ecosystem: A Case Study

1. Ecosystem management can be implemented by following these five steps:

   Step 1: Determine the main stakeholders, define the ecosystem area, and develop the relationship between them.
   Step 2: Characterize the structure and function of the ecosystem, and set in place mechanisms to manage and monitor it.
   Step 3: Identify the important economic issues that will affect the ecosystem and its inhabitants.

   Step 4: Determine the likely impact of the ecosystem on adjacent ecosystems.
   Step 5: Decide on long-term goals and flexible ways of reaching them.

2. Royalwood is a real suburban development in south Winnipeg, Manitoba, that serves as a case study to explore how the five steps of sustainable ecosystem management can be implemented in a real-world setting. Although not without encountering some problems along the way, Royalwood is an example of what can happen when sustainable ecosystem management is undertaken.

The project took into account all of the various needs of all the stakeholders, both environmental and economic, and anticipated and planned for the impact the community would have on surrounding ecosystems.

KEY TERMS

- sustainable ecosystem management p. 473

- sustainable development p. 475
- ecosystem management p. 476

- adaptive resource management (ARM) p. 478

CRITICAL AND CREATIVE THINKING QUESTIONS

1. Find a photo of your own preferred future environment and compare it with the photos shown on the opening pages of this chapter. What is different in your photo? What is the same?

2. Define the features of your own preferred future environment. What makes it so special to you?

3. Is society capable of meeting the requirements for environmental sustainability that Paul Hawken has identified? What will be the most difficult aspects to respond to?

4. Paul Hawken notes the following: "You are graduating to the most amazing, challenging, stupefying challenge ever bequested to any generation. The generations before you failed. . . . They got distracted and lost sight of the fact that life is a miracle every moment of your existence. Nature beckons you to be on her side. You couldn't ask for a better boss." Do you think that you are able to respond to the challenge that Hawken sets forward? What are some ways that you feel you can respond?

5. Is sustainable ecosystem management the best approach for achieving society's preferred future environment? Why or why not?

6. Briefly describe the five steps in the ecosystem management approach. Are all steps equally important, or are some more critical than others? Explain.

7. Describe an ecosystem of your own choosing with which you are familiar. Try to implement the five steps in the ecosystem management approach on your ecosystem. Which step(s) will be most difficult to achieve and why?

8. Is adaptive resource management (ARM) a necessary and sufficient condition to achieve environmental sustainability? Discuss.

9. If Royalwood were located in downtown Winnipeg rather than in a suburban area, would the management plan for its development have changed? Explain how.

10. Do you have a development like Royalwood in your own community? Do you have a development that should or could have followed the Royalwood development philosophy? What can you do about it?

What is happening in this picture ?

- Traditional stormwater retention ponds like the one shown here are havens for flocks of Canada geese. But geese are not always welcome inhabitants in a community. What problems can Canada geese cause in suburban communities?

- Newer stormwater retention ponds like the ones built in Royalwood are not as preferred by Canada geese. How does altering the design of stormwater retention ponds reflect an ecosystem management approach?

GLOSSARY

Aboriginal worldview   A worldview that emphasizes the interconnectedness among all living and nonliving things. The belief that humans need to live in harmony with each other and with nature is integral to this worldview.

acid mine drainage   Pollution caused when sulphuric acid and dangerous dissolved materials such as lead, arsenic, and cadmium wash from coal and metal mines into nearby lakes and streams.

acid neutralizing capacity (ANC)   The ability of water or soil to neutralize acids that have been added.

acid precipitation   A group of secondary air pollutants that include sulphuric, nitrous, and nitric acids formed by reactions of sulphur dioxide and nitrogen oxides with water in the atmosphere (wet deposition) or following contact with water (dry deposition).

active solar heating   A system of putting the sun's energy to use in which a series of collectors absorb the solar energy, and pumps or fans distribute the collected heat.

acute toxicity   Adverse effects that occur within a short period after high-level exposure to a toxicant.

adaptive resource management (ARM)   An approach to management that acknowledges uncertainty and the need for managers to learn from system responses while they manage.

age structure   The number and proportion of people at each age in a population.

air pollution   Various chemicals (gases, liquids, or particulates) present in the atmosphere in high enough concentrations to harm humans, ecosystems, or materials.

albedo   The extent to which an object diffusely reflects light from the sun.

aquaculture   The growing of aquatic organisms (fishes, shellfish, and seaweeds) for human consumption.

aquifer depletion   The removal of groundwater faster than it can be recharged by precipitation or melting snow.

aquifers   Underground reservoirs of permeable rock from which groundwater can be extracted.

atmosphere   The gaseous envelope surrounding Earth composed primarily of nitrogen and oxygen. Other gases are found in trace amounts.

atmospheric pressure   The force exerted by the column of air molecules located above a surface.

benthic environment   The ocean floor, which extends from the intertidal zone to the deep ocean trenches.

biochemical oxygen demand (BOD)   The amount of oxygen that microorganisms need in respiration to decompose biological wastes into carbon dioxide, water, and minerals.

biodiversity hotspots   Relatively small areas of land that contain an exceptional number of endemic species and that are at high risk from human activities.

biogeochemical cycles   The various cycles that move matter from one part of an ecosystem to another. They include the carbon, hydrologic, nitrogen, and phosphorus cycles.

biological control   A method of pest control that involves the use of naturally occurring disease organisms, parasites, or predators to control pests.

biological diversity   The variety of life in all forms, at all levels, and in all combinations in a defined area. It can be considered at the level of genes, species, populations, communities, ecosystems, and landscapes. The area is defined by humans, and can vary in size from something as large as the Earth to a country, province, region, municipality, or a puddle in your backyard.

biological magnification   The increase in toxicant concentrations as a toxicant passes through successive levels of the food chain.

biomass   Plant and animal material used as fuel.

biome   A large, relatively distinct terrestrial region with similar climate, soil, plants, and animals regardless of where it occurs in the world.

biosphere   The layer of Earth that contains all living organisms.

biotic potential   The maximum rate at which a population could increase under ideal conditions.

boreal forest   A region of coniferous forest (such as pine, spruce, and fir) in the northern hemisphere; located just south of the tundra. Also called taiga.

carbon management   Ways to separate and capture the CO_2 produced during the combustion of fossil fuels and then sequester (store) it.

carcinogen   Any substance (for example, chemical, radiation, virus) that causes cancer.

carrying capacity (K)   The largest population that can be sustained over the long term given that there are no changes in that environment.

catalytic reaction   Following the destructive reactions of ozone with reactive chlorine (or bromine), the halogen is reformed and continues to be available to undergo reactions. Ultimately one chlorine molecule can destroy hundreds of ozone molecules in the process.

chaparral   A biome with mild, moist winters and hot, dry summers; vegetation is typically small-leafed evergreen shrubs and small trees.

chronic toxicity   Adverse effects that occur after a long period of low-level exposure to a toxicant.

clear-cutting   A logging practice in which all the trees in a stand of forest are cut, leaving just the stumps.

climate   The average weather conditions that occur in a place over a period of years.

climate change   A change in the long-term weather patterns observed over a period from decades to millions of years.

cogeneration (CHP)   An energy technology that involves recycling "waste" heat.

community   A natural association that consists of all the populations of different species that live and interact together within an area at the same time.

compact development   Design of cities in which tall, multiple-unit residential buildings are close to shopping and jobs, and all are connected by public transportation.

competition   The interaction among organisms that vie for the same resources in an ecosystem (such as food or living space).

conservation The sensible and careful management of natural resources.

conservation biology The scientific study of how humans impact organisms and of the development of ways to protect biological diversity.

conservation tillage A method of cultivation in which residues from previous crops are left in the soil, partially covering it and helping to hold it in place until the newly planted seeds are established.

consumers Organisms that can't make their own food and use the bodies of other organisms as a source of energy and bodybuilding materials.

consumption overpopulation A situation in which each individual in a population consumes too large a share of resources.

contour plowing Plowing that matches the natural contour of the land.

cost-benefit diagram A diagram that helps policy makers make decisions about costs of a particular action and benefits that would occur if that action were implemented.

critical load The maximum amount of pollution that an ecosystem can tolerate without being damaged.

crop rotation The planting of a series of different crops in the same field over a period of years.

cultural eutrophication Enrichment of an aquatic ecosystem with nutrients such as ammonia and phosphorus from human activities such as agriculture, detergents, and discharge from sewage treatment plants.

decommission To dismantle a nuclear power plant after it closes.

decomposers Microorganisms that break down dead organic material and use the decomposition products to supply themselves with energy.

deep ecology worldview A worldview based on harmony with nature, a spiritual respect for life, and the belief that humans and all other species have an equal worth.

deforestation The temporary or permanent clearance of large expanses of forest for agriculture or other uses.

demographic transition The process whereby a country moves from relatively high birth and death rates to relatively low birth and death rates.

demographics The applied branch of sociology that deals with population statistics.

desert A biome in which the lack of precipitation limits plant growth; deserts are found in both temperate and tropical regions.

desertification Degradation of once-fertile rangeland or tropical dry forest into nonproductive desert.

dose-response curve In toxicology, a graph that shows the effects of different doses on a population of test organisms.

ecological niche The totality of an organism's adaptations, its use of resources, and the lifestyle to which it is fitted.

ecological succession The process of community development over time, which involves species in one stage being replaced by different species.

ecology The study of the interactions among organisms and between organisms and their abiotic environment.

ecosystem Biotic communities and their associated abiotic components that interact in a defined geographic area.

ecosystem goods and services Environmental benefits, such as food, fibre, clean air, clean water, and fertile soil, provided by ecosystems.

ecosystem management An evolving concept that recognizes the need to use ecological components in environmental decision making.

ecosystem services Important environmental benefits, such as clean air to breathe, clean water to drink, and fertile soil in which to grow crops, that the natural environment provides.

electromagnetic spectrum The entire range of electromagnetic radiation including gamma, x-ray, ultraviolet, visible, infrared, microwave, and radio.

El Niño-Southern Oscillation (ENSO) A periodic, large-scale warming of surface waters of the tropical eastern Pacific Ocean that temporarily alters both ocean and atmospheric circulation patterns.

endangered species A species that is facing imminent extinction or extirpation.

endemic species Organisms that are native to or confined to a particular region.

energy The capacity or ability to do work.

energy budget The gains and losses of solar radiation to Earth. When in equilibrium, there are no changes in overall energy amount for the planet.

energy conservation Using less energy by reducing energy use and waste, for example.

energy efficiency Using less energy to accomplish a given task by using new technology, for example.

energy flow The passage of energy in a one-way direction through an ecosystem.

enhanced greenhouse effect The additional warming that results from increased levels of CO_2 and other gases that absorb infrared radiation. Carbon dioxide is the gas of most concern in causing the enhanced warming influence.

enrichment The process by which uranium ore is refined after mining to increase the concentration of fissionable U-235.

environmental chemistry A subdiscipline of chemistry in which commercially important chemical processes are redesigned to reduce environmental harm.

environmental ethics A field of applied ethics that considers the moral basis of environmental responsibility.

environmental hazard Situations and events that pose threats to the surrounding environment.

environmental justice The right of every citizen to adequate protection from environmental hazards.

environmental science and environmental studies The interdisciplinary fields of humanity's relationship with other organisms and the physical environment. Environmental science relies more heavily on the scientific method. Environmental studies deals more with social issues, which often are value-laden. But both "science" and "studies" deal with the all-encompassing and interdisciplinary aspects of the environment.

environmental sustainability The ability to meet humanity's current needs without compromising the ability of future generations to meet their needs.

environmental worldview A worldview based on how the environment works, our place in the environment, and right and wrong environmental behaviours.

estuary A coastal body of water, partly surrounded by land, with access to the open ocean and a large supply of fresh water from a river.

eutrophication An increase in the concentration of chemical nutrients in an ecosystem where the enrichment stimulates an increase in photosynthetic productivity by plants and algae and alters the dynamics involving dissolved oxygen concentrations.

evolution The cumulative genetic changes in populations that occur during successive generations.

exponential population growth The population growth that occurs when environmental resources are not limited and there is a constant rate of reproduction. With exponential growth, the birth rate alone controls how fast (or slow) the population grows.

external cost A harmful environmental or social cost that is borne by people not directly involved in selling or buying a product.

extinction The elimination of a species from Earth.

extirpated species A species that is no longer present in the wild in Canada.

feedback loop A chain of cause-and-effect responses where the output from an event influences the same event in the future to form a loop.

first law of thermodynamics A physical law which states that energy cannot be created or destroyed, although it can change from one form to another.

fission The splitting of an atomic nucleus into two smaller fragments, accompanied by the release of a large amount of energy.

flowing-water ecosystem A freshwater ecosystem such as a river or stream in which water flows in a current.

food insecurity The condition in which people live with chronic hunger and malnutrition.

freshwater wetlands Lands that shallow fresh water covers for at least part of the year; wetlands have a characteristic soil and water-tolerant vegetation.

fuel cell A device that directly converts chemical energy into electricity without producing steam that runs a turbine and generator; the fuel cell requires hydrogen and oxygen from the air.

fusion The joining of two lightweight atomic nuclei into a single, heavier nucleus, accompanied by the release of a large amount of energy.

genetic engineering The manipulation of genes (for example, taking a specific gene from one species and placing it into an unrelated species) to produce a particular trait.

geothermal energy The use of energy from Earth's interior for space heating or generation of electricity.

global warming potential (GWP) How much a given mass of a greenhouse gas contributes to global warming over a given time period compared with the same mass of carbon dioxide.

greenhouse gases Gases that absorb infrared radiation, including water vapour, carbon dioxide, methane, nitrous oxide, chlorofluorocarbons, and tropospheric ozone.

groundwater The supply of fresh water under Earth's surface that is stored in underground aquifers.

growth rate (r) The rate of change (increase or decrease) of a population's size, which is typically expressed in percentage per year.

gyres Large, circular ocean current systems that often encompass an entire ocean basin.

habitat fragmentation The breakup of large areas of habitat into small, isolated patches.

hazardous waste Any discarded chemical that threatens human health or the environment.

high-level radioactive wastes Radioactive solids, liquids, or gases that initially give off large amounts of ionizing radiation.

highly developed countries Countries with complex industrialized bases, low rates of population growth, and high per person incomes.

hydropower A form of renewable energy that relies on flowing or falling water to generate electricity.

industrialized agriculture Modern agricultural methods, which require a large capital input and less land and labour than traditional methods.

industrial smog A whitish-grey haze resulting from sulphur-rich coal combustion. Sulphuric acid, particulate matter, carbon dioxide, and carbon monoxide are major end products of the reaction.

infant mortality rate The number of deaths of infants under age 1 per 1000 live births.

integrated pest management (IPM) A combination of pest control methods that, if used in the proper order and at the proper times, keep the size of a pest population low enough to prevent substantial economic loss.

integrated waste management A combination of the best waste management techniques into a consolidated program to deal effectively with solid waste.

intertidal zone The area of shoreline between low and high tides.

invasive species Foreign species that spread rapidly in a new area if free of predators, parasites, or resource limitations that may have controlled their population in their native habitat.

keystone species A species that has a disproportionate effect on its environment relative to its biomass.

land degradation Natural or human-induced processes that decrease the future ability of the land to support crops or livestock.

landscape A region that may include several interacting ecosystems.

landscape ecology The study of the relationship between spatial patterns and ecological processes on a number of different landscape scales and organizational levels.

less developed countries Countries with low levels of industrialization, very high rates of population growth, very high infant mortality rates, and very low per person incomes compared with highly developed countries.

low-level radioactive wastes Solids, liquids, or gases that give off small amounts of ionizing radiation.

marginal cost of pollution The added cost of an additional unit of pollution.

marginal cost of pollution abatement The added cost of reducing one unit of a given type of pollution.

mass burn incinerator A large furnace that burns all solid waste except for unburnable items such as refrigerators.

microirrigation A type of irrigation that conserves water by piping it to crops through sealed systems.

minerals Elements or compounds of elements that occur naturally in Earth's crust.

moderately developed countries Countries with medium levels of industrialization and per person incomes lower than those of highly developed countries.

municipal solid waste Solid materials discarded by homes, offices, stores, restaurants, schools, hospitals, prisons, libraries, and other facilities.

national income accounts A measure of the total income of a nation's goods and services for a given year.

natural capital Earth's resources and processes that sustain living organisms, including humans; includes minerals, forests, soils, groundwater, clean air, wildlife, and fisheries.

natural resources Naturally forming substances that are considered valuable in their relatively unmodified or "natural" form.

natural selection The tendency of better-adapted individuals—those with a combination of genetic traits best suited to environmental conditions—to survive and reproduce, increasing their proportion in the population.

neritic province The part of the pelagic environment that overlies the ocean floor from the shoreline to a depth of 200 metres.

nonmunicipal solid waste Solid waste generated by industry, agriculture, and mining.

nonpoint source pollution Pollutants that enter bodies of water over large areas rather than being concentrated at a single point of entry.

nonrenewable resources Natural resources that are present in limited supplies and are depleted as they are used.

nuclear energy The energy released by nuclear fission or fusion.

nuclear reactor A device that initiates and maintains a controlled nuclear fission chain reaction to produce energy for electricity.

ocean conveyor belt (thermohaline gradient) Deep-water circulation of the oceans due to a temperature salinity gradient. The flow of sea water transports vast amounts of absorbed energy from the equator to higher latitudes.

oceanic province The part of the pelagic environment that overlies the ocean floor at depths greater than 200 metres.

optimum amount of pollution The amount of pollution that is economically most desirable.

overgrazing When too many grazing animals consume the plants in a particular area, leaving the vegetation destroyed and unable to recover.

ozone Three oxygen atoms bonded together through reactions driven by the presence of ultraviolet radiation in the stratosphere. Ozone occurs naturally in the stratosphere as the "ozone layer" shielding the Earth's surface from UV radiation.

ozone depletion The destruction of stratospheric ozone by human-produced chlorine- and bromine-containing chemicals.

passive solar heating A system of putting the sun's energy to use without requiring mechanical devices to distribute the collected heat.

pathogen An agent (usually a microorganism) that causes disease.

people overpopulation A situation in which there are too many people in a given geographic area.

persistent organic pollutants (POPs) Persistent toxicants that bioaccumulate in organisms and travel through air and water to contaminate sites far from their source.

pesticide Any toxic chemical used to kill pests.

photochemical smog A brownish-orange haze formed by reactions of nitrogen oxides, volatile organic compounds, and particulate matter, typically on a sunny day. Tropospheric ozone and fine particulate matter are major end products of the reaction.

photosynthesis The biological process that captures solar energy and transforms it into the chemical energy of organic molecules, which are manufactured from carbon dioxide and water.

photovoltaic (PV) solar cell A wafer or thin film of solid state materials, such as silicon or gallium arsenide, that are treated with certain metals in such a way that the film generates electricity—that is, a flow of electrons—when solar energy is absorbed.

point source pollution Water pollution that can be traced to a specific spot.

population A group of organisms of the same species that live and interact together in the same area at the same time.

population ecology The branch of biology that deals with the number of individuals of a particular species found in an area and why those numbers increase or decrease over time.

poverty A condition in which people are economically unable to meet their basic needs for food, water, clothing, shelter, education, and health services.

precautionary principle A practice that involves making decisions about adopting a new technology or chemical product by assigning the burden of proof of its safety to its developers.

predation The consumption of one species (the prey) by another (the predator).

preservation Setting aside undisturbed areas, maintaining them in a pristine state, and protecting them from human activities.

primary air pollutants Harmful chemicals that enter directly into the atmosphere from either a human or natural source.

primary treatment Treatment of waste water that involves removing suspended and floating particles by mechanical processes.

producers An organism that manufactures large organic molecules from a simple inorganic substance.

radiative forcing The imbalance in Earth's energy budget that results when either the gains through energy absorbed or losses through the amount of energy radiated to outer space is changed through either natural or human influences. Positive forcing results in a warming influence and negative forcing leads to a cooling influence.

rangeland Land that is not intensively managed and is used for grazing livestock.

relative humidity The amount of moisture present in the atmosphere in relation to the amount the atmosphere can hold given the present temperature.

renewable resources Resources that are replaced by natural processes and that can be used forever, provided they are not overexploited in the short term.

replacement-level fertility The number of children a couple must produce to "replace" themselves.

restoration ecology The study of the historical condition of a human-damaged ecosystem, with the goal of returning it as close as possible to its former state.

risk The probability of harm (such as injury, disease, death, or environmental damage) occurring under certain circumstances.

risk assessment The use of statistical methods to quantify risks so they can be compared and contrasted.

runoff The movement of fresh water from precipitation and snowmelt to rivers, lakes, wetlands, and the ocean.

salinization The gradual accumulation of salt in soil, often as a result of improper irrigation methods.

salt water intrusion The movement of sea water into a freshwater aquifer near the coast when the water table drops due to withdrawal rates that exceed groundwater recharge.

sanitary landfill The most common method of disposing of solid waste, by compacting it and burying it under a shallow layer of soil.

savanna A tropical grassland with widely scattered trees or clumps of trees.

scientific method The way a scientist approaches a problem, by formulating a hypothesis and then testing it.

second law of thermodynamics A physical law which states that when energy is converted from one form to another, some of it is degraded into heat, a less useable form that disperses into the environment.

secondary air pollutants Harmful chemicals that form when primary air pollutants react in the atmosphere.

secondary treatment Biological treatment of waste water to decompose suspended organic material; secondary treatment reduces the water's biochemical oxygen demand.

sediment pollution A type of water pollution caused by the transportation of silt, clay, and sand particles into waterways.

sewage Waste water from drains or sewers (from toilets, washing machines, and showers); includes human wastes, soaps, and detergents.

shelterbelt A row of trees planted as a windbreak to reduce soil erosion of agricultural land.

sick building syndrome Eye irritations, nausea, headaches, respiratory infections, depression, and fatigue caused by indoor air pollution.

soil The uppermost layer of Earth's crust, which supports terrestrial plants, animals, and microorganisms.

soil erosion The wearing away or removal of soil from the land.

soil horizons The horizontal layers into which many soils are organized, from the surface to the underlying parent.

solar thermal electric generation A means of producing electricity in which the sun's energy is concentrated by mirrors or lenses onto a fluid-filled pipe; the heated fluid is used to generate electricity.

source reduction An aspect of waste management in which products are designed and manufactured in ways that decrease the amount of solid and hazardous waste in the solid waste stream.

special concern species A species that is particularly sensitive to human activities or natural events.

species dominance The most abundant species in a community.

species evenness The relative number of species present in a community.

species richness The number of different species in a community.

spent fuel Used fuel elements that were irradiated in a nuclear reactor.

standing-water ecosystem A body of fresh water surrounded by land and whose water does not flow; a lake or a pond.

stormwater retention pond A pond designed to manage stormwater runoff by diverting the water into an artificial lake with vegetation around the perimeter, and including a permanent pool of water in its design.

subsistence agriculture Traditional agricultural methods, which depend on labour and a large amount of land to produce enough food to feed oneself and one's family.

subsurface mining The extraction of mineral and energy resources from deep underground deposits.

surface mining The extraction of mineral and energy resources near Earth's surface by first removing the soil, subsoil, and overlying rock strata.

surface water Precipitation that remains on the surface of the land and does not seep down through the soil.

sustainable agriculture Agricultural methods that maintain soil productivity and a healthy ecological balance while having minimal long-term impacts.

sustainable consumption The use of goods and services in a way that satisfies basic human needs and improves the quality of life but that also minimizes resource use and preserves resources for the use of future generations.

sustainable development Economic growth that meets the needs of the present without compromising the ability of future generations to meet their needs.

sustainable ecosystem management An approach to achieving our preferred future environment by meeting societal goals and objectives for the biotic communities and their abiotic components in a defined geographic area that takes into account economic, social, and environmental requirements.

sustainable forestry The use and management of forest ecosystems in an environmentally balanced and enduring way.

sustainable soil use The wise use of soil resources without a reduction in the amount or fertility of soil so it is productive for future generations.

sustainable water use The wise use of water resources without harming the essential functioning of the hydrologic cycle or the ecosystems on which present and future humans depend.

symbiosis An intimate relationship or association between members of two or more species; includes mutualism, commensalism, and parasitism.

temperate deciduous forest   A forest biome that occurs in temperate areas where annual precipitation ranges from about 75 to 126 centimetres.

temperate grassland   A grassland with hot summers, cold winters, and less rainfall than is found in the temperate deciduous forest biome.

temperate rainforest   A coniferous biome with cool weather, dense fog, and high precipitation.

temperature inversion   A layer of cold air temporarily trapped near the ground by a warmer upper layer.

tertiary treatment   Advanced wastewater treatment methods that are sometimes employed after primary and secondary treatments.

threatened species   A species that has a high probability of becoming endangered.

tidal energy   A form of renewable energy that relies on the ebb and flow of the tides to generate electricity.

total fertility rate (TFR)   The average number of children born to each woman.

toxicology   The study of toxicants, chemicals with adverse effects on health.

tropical rainforest   A lush, species-rich forest biome that occurs where the climate is warm and moist throughout the year.

tundra   The treeless biome in the far north that consists of boggy plains covered by lichens and mosses; it has harsh, cold winters and extremely short summers.

urbanization   A process whereby people move from rural areas to densely populated cities.

utilitarian conservationist   A person who values natural resources because of their usefulness to humans but uses them sensibly and carefully.

water consumption   The removal of water for human use without any return of water to its original source.

water pollution   A physical or chemical change in water that adversely affects the health of humans and other organisms.

water withdrawal   The removal of water from a source such as a lake or river where a portion of this water is returned to the source and is available to be used again.

weather   The current temperature and precipitation conditions for a location at a given point in time.

Western worldview   A worldview based on human superiority over nature, the unrestricted use of natural resources, and economic growth to manage an expanding industrial base.

wildlife corridor   A protected zone that connects isolated unlogged or undeveloped areas.

wind energy   Electric energy obtained from surface air currents caused by the solar warming of air.

zero population growth   The state in which the population remains the same size because the birth rate equals the death rate.

VISUALIZING THE ENVIRONMENT
TEXT, TABLE, AND LINE ART CREDITS

Chapter 1

Figure 1.2: Data from Population Reference Bureau; **Environmental Economics section (beginning on p. 21):** Main source: Levin, J. "The Economy and the Environment: Revising the National Accounts." *IMF Survey* (June 4, 1990); **Graph in Creative and Critical Thinking Questions:** Data from *Science* Vol. 315 (February 16, 2007), p. 913.

Chapter 2

Excerpt from the "Wilderness Essay" (pp. 39–40): A letter written by Wallace Stegner to David Pesonen of the U. of California's Wildland Research Center; **Eight principles of deep ecology (pp. 49–50):** A. Naess and D. Rothenberg. *Ecology, Community and Lifestyle.* Cambridge, UK: Cambridge University Press (2001). Reprinted with permission of Cambridge University Press; **Five recommendations in section on Strategies for Sustainable Living:** Brown, L. *Plan B 2.0: Rescuing a Planet under Stress and a Civilization in Trouble.* New York: W.W. Norton (2006); **Graphs in Creative and Critical Thinking Questions:** Data from U.S. National Climate Assessment.

Chapter 3

Figure 3.5B: After V.C. Scheffer. "The Rise and Fall of a Reindeer Herd." *Science Monthly,* Vol. 73 (1951); **Figures 3.6, 3.13, 3.14, and 3.15 and graph for Critical and Creative Thinking Question:** Data from Population Reference Bureau; **Figure 3.8:** Data from *World Population Prospects, The 2004 Revision,* United Nations Population Division; **Figure 3.16:** Data from Statistics Canada, 2006, Census of Population, Statistics Canada catalogue no. 97-551-XCB2006009; **Table 3.1:** "Urban Agglomerations 2005," U.N. Department of Economic and Social Affairs, Population Division.

Chapter 4

Chapter introduction inset: E.A. Guilette, M.M. Meza, M.G. Aguilar, A.D. Soto, and I.E. Garcia. "An Anthropological Approach to the Evaluation of Preschool Children Exposed to Pesticides in Mexico." *Environmental Health Perspectives* (May 1998); **Figure 4.1:** Data from National Safety Council; art design from Health feature in *National Geographic* (August 2006); **Figure 4.2:** Adapted from *Science and Judgment in Risk Assessment.* Washington, D.C.: National Academy Press (1994); **Tables 4.1 and 4.2:** Physicians for a Smoke-Free Canada, "Tobacco and the Health of Canadians" www.Smoke-free.ca/ health/pscissues_health.htm. Accessed May 12, 2009; **Figure 4.9B:** Data from J.W. Grier, "Ban of DDT and Subsequent Recovery of Reproduction in Bald Eagles" Copyright 1982, American Association for the Advancement of Science; **Figure 4.10B:** Data from G.M. Woodwell, C.F. Worster, Jr. and P.A. Isaacson, "DDT Residues in an East Coast Estuary: A Case of Biological Concentration of a Persistent Insecticide. *Science.* Vol 156 (May 12, 1967); **Figure 4.12:** Philippe Rekacewicz, UNEP/GRID-Arendal, *UNEP/GRID-Arendal Maps and Graphics Library,* http://maps.grida.no/go/graphic/ long-range-transport-of-air-pollutants-to-the-arctic. Accessed July 7, 2010; **Table 4.5:** Josten, M.D. and J.L. Wood. *World of Chemistry,* 2nd edition. Philadelphia: Saunders College Publishing (1996).

Chapter 5

Figures 5.9, 5.10, 5.11, and 5.13: Values are from W.H. Schlesinger, *Biogeochemistry: An Analysis of Global Change,* 2nd edition. Academic Press, San Diego (1997) and based on several sources; **What a Scientist Sees: Resource Partitioning:** Adapted from R.H. MacArthur "Population Ecology of Some Warblers of Northeastern Coniferous Forests." *Ecology,* Vol. 39 (1958); **EnviroDiscovery: The Ecosystem Approach:** Adapted from Gill Shepherd, *The Ecosystem Approach: Five Steps to Implementation.* IUCN, Gland, Switzerland and Cambridge, UK, p. 2 (2004). Available online at http://data.iucn.org/dbtw-wpd/edocs/CEM-003.pdf; **Case Study: Human Appropriation of Net Primary Productivity (HANPP):** Image by Jesse Allen, based on data from NASA's Socioeconomic Data Center.

Chapter 6

Figure 6.1: Based on data from the World Wildlife Fund; **Figure 6.2:** Based on L. Holdridge, *Life Zone Ecology.* Tropical Science Center, San Jose, Costa Rica (1967); **Data for all climate graphs (in Figures 6.4–6.12):** www.worldclimate.com; **Figure 6.14:** Adapted from Figure 14-1 in G. Karleskint, *Introduction to Marine Biology.* Philadelphia: Harcourt College Publishers (1998); **Figure 6.15A:** Adapted from J.E. Marangos, M.P. Crosby, and J.W. McManus, "Coral Reefs and Biodiversity: A Critical and Threatened Relationship." *Oceanography,* Vol. 9 (1996); **Figure 6.24:** Adapted from Figure 14.11 on p. 428 in B.W. Murck, B.J. Skinner, and D. Mackenzie. *Visualizing Geology,* Hoboken, NJ: John Wiley & Sons, Inc. (2008); **Figures 6.25B and 6.25C:** Adapted from Figure 15.13 on p. 244 and Figure 15.19 on p. 247, respectively, in S.A. Alters and B. Alters *Biology: Understanding Life,* Hoboken, NJ: John Wiley & Sons, Inc. (2006).

Chapter 7

Data in Measuring Biological Diversity at the Species Level section (p. 203): Data from M.L. Reaka-Kudla, D.E. Wilson, and E.O. Wilson. *Biodiversity II.* Washington, D.C.: Joseph Henry Press (1997); **Figure 7.2B:** After M.L. Cody and J.M. Diamond, eds., *Ecology and Evolution of Communities.* Harvard University, Cambridge, (1975); **Table 7.1:** Adapted from page 527 of *Climate Change Impacts on the United States: A report of the National Assessment Synthesis Team,* U.S. Global Change Research Program, Cambridge University Press (2001); **What a Scientist Sees: Where Is Declining Biological Diversity the Greatest Problem:** Based on a map by Conservation International; **graph in Critical and Creative Thinking Questions:** Data from The Nature Conservancy.

Chapter 8

Table 8.1: Natural Resources Canada, GeoAccess Division. Retrieved August 9, 2009, from www40.statcan.ca/l01/cst01/phys01-eng.htm; **Figure 8.2:** Adapted from G.H. Heichel, "Agricultural Production and Energy Resources." *American Scientist,* Vol. 64 (January/February 1976); **Figure 8.7:** Worldwatch Institute, *World Watch* magazine, www. worldwatch.org; **Figure 8.10:** Reproduced with the permission of Natural Resources Canada 2010, courtesy Atlas of Canada; **Figure 8.1:** Values are from W.H. Schlesinger, *Biogeochemistry: An Analysis of Global Change,* 2nd edition. Academic Press, San Diego (1997) and based on several sources; **Figure 8.15A:** Based on Goode Base Map (S.R. Eyre 1968); **Figure 8.26:** Reproduced with the permission of Natural Resources Canada 2010, courtesy Atlas of Canada; **Figure 8.33:** *One Planet, Many People: Atlas of Our Changing Environment.* United Nations Environment Program (2005)— Part A based on graph on page 30 (FAO 2000), Part B based on table on page 29 (*World Atlas of Desertification*).

Chapter 9

Figure 9.3: Reproduced with the permission of Natural Resources Canada 2010, courtesy Atlas of Canada; **Figure 9.7:** *Control of Water Pollution from Urban Runoff.* Paris: Organization for Economic Cooperation and Development (1986); **Figure 9.12:** P. Raven and L.R. Berg, *Environment,* 6th Edition, Figure 22.4 on p. 511. New York: John Wiley & Sons, Inc. (2008); **Figure 9.13:** Adapted from M.D. Joesten and J.L. Wood. *World of Chemistry,* 2nd edition. Philadelphia: Saunders College Publishing (1996); **Figure 9.22:** A.Y. Hoekstra and A.K. Chapagain, "Water Footprints of Nations: Water Use by People as a Function of Their Consumption Pattern," *Water Resource Management,* Vol. 21 (2007), p. 41; **Figure 9.26:** Copyright 2010, City of Winnipeg; **Figure 9.29:** After B.S. Halpern et al. "A Global Map of Human Impact on Marine Ecosystems." *Science,* Vol. 319, No. 5865, pp. 948–952 (February 15, 2008); **Figure 9.30:** Adapted from Figure 6.31 on p. 148 in A.F. Arbogast. *Discovering Physical Geography.* Hoboken, NJ: John Wiley & Sons, Inc. (2007); **What a Scientist Sees: Ocean Warming and Coral Bleaching, Figure B:** Adapted from National Assessment Synthesis Team, *Climate Change Impacts on the United States: The Potential Consequences of Climate Variability and Change* (Report for the U.S. Global Change Research Program). Cambridge, UK: Cambridge University Press (2001).

Chapter 10

Figure 10.8: Adapted from Figure 6.11 in A.F. Arbogast *Discovering Physical Geography.* Hoboken, NJ: John Wiley & Sons, Inc. (2007); **Figure 10.9:** NASA. Accessed at http://www.jpl.nasa.gov/news/news.cfm?release=2010-101, July 29, 2010; **Figures 10.11 and 10.12:** Data from Environment Canada; **Figure 10.20:** Based on "Acid Rain and pH Scale." Environmental Protection Agency; **Figure 10.23:** P. Raven and L.R. Berg, *Environment,* 6th edition, Figure 20.19b on p. 474. Hoboken, NJ: John Wiley & Sons, Inc. (2008); **Figure 10.28c:** Data from Dave Keeling and Tim Whorf, Scripps Institution of Oceanography, La Jolla, California; **EnviroDiscovery: Your Carbon Footprint:** Based on data from Carbon Footprint; **EnviroDiscovery: ArcticNet:** NASA Earth Observatory. Available at http://earthobservatory.nasa.gov/IOTD/view.php?id=7679; **Figure 10.35:** Richard Feely and Dana Greeley, NOAA Pacific Marine Environmental Laboratory, as shown on page 111 in *National Geographic* (November 2007); **Figure 10.38:** Based on data from United Nations Environment Programme.

Chapter 11

Unless noted otherwise, all energy facts cited and data used to create figures in this chapter were obtained from the Energy Information Administration (EIA), the statistical agency of the U.S. Department of Energy (DOE). **Figure 11.1B:** Adapted from Figure 13-3a in J.M. Harris, *Environmental and Natural Resource Economics: A Contemporary Approach,* 2nd edition. Houghton Mifflin (2006) and based on data from the World Bank; **Figure 11.2:** Based on data from the World Coal Institute; **Table 17.2:** Adapted from two sources: R.A. Hinrichs and M. Kleinbach. *Energy: Its Use and the Environment,* 3rd edition. Philadelphia: Harcourt College Publishers (2002), and *Science for Democratic Action,* Vol. 8, No. 3 (May 2000); **Table 18.1:** Data from D.M. Kammen, "The Rise of Renewable Energy." *Scientific American* (September 2006); **Figure 11.24B:** Adapted from R.A. Hinrichs, *Energy: Its Use and the Environment,* 2nd ed. Philadelphia: Saunders College Publishing (1996); **Figure 11.27C:** Worldwatch Institute, *Vital Signs* 2005, www.worldwatch.org; **Case Study: Hydrogen Economy:** © PolicalConnections.net and PoliticalBull.net; www.politicalbull.net.

Chapter 12

Figure 12.3: Data from EPA; **Figure 12.5:** Data from Ecobalance, Inc., and Integrated Waste Services Association; **Figure 12.10A:** Adapted from T. Zeller, Jr., "Recycling: The Big Picture," *National Geographic* (January 2008); **graph in Critical and Creative Thinking Questions:** Data from EPA.

Chapter 13

Excerpt from Paul Hawken's University of Portland commencement speech: Reprinted courtesy of Paul Hawken; **Figure 13.6:** Based on Figure 1 in *Open Standards for the Practice of Conservation, Version 2.0,* p. 4 (2007) by Conservation Measures Partnerships.

PHOTO CREDITS

Chapter 1

Page 2–3: National Geographic Stock; page 5 (top): Earth Imaging/ Stone/Getty Images; page 5 (bottom): National Geographic Stock; page 6 (left and right): Peter Menzel; page 7: Peter Manzel; page 8: © Ali Abbas/Corbis; page 10: Oleksiy Maksymenko/Alamy; page 12: Raymond Gehman/National Geographic Stock; page 14: Mario Beauregard/CPI/ The Canadian Press; page 15 (left): Peter Ryan/National Geographic Stock; page 15 (top right): Medford Taylor/National Geographic Stock; page 15 (bottom right): Sarah Leen/National Geographic Stock; page 16 (top left): Gene Carl Feldman/Sea WIFs/NASA; page 16 (top left): Lara Hansen, Adam Markham/WWF/NG Maps; page 16 (top right): Photodisc/ Getty Images; page 16 (top right): Global Forest Watch/WRI/NG Maps; page 16 (bottom left): Caroline Rogers/USGS; page 16 (bottom left): NOAA/NMFS/UNEP-WCMC/NG Maps; page 16 (bottom right): Bruce Dale/National Geographic Society; page 16 (bottom right): USDA Global Desertification Map/National Geographic Society; page 19 (left): Peter Essick/National Geographic Stock; page 19 (right): National Geographic Stock; page 20: Carlos Osorio/Getstock; page 21 (top left): Creatas/ SUPERSTOCK; page 21 (top right): age fotostock/SUPERSTOCK; page 22: Randy Olson/National Geographic Stock; page 24 (top left): Tim Fitzharris/Minden/National Geographic Stock; page 24 (bottom left): CP PHOTO/Kevin Frayer; page 24 (right): JAUBERT IMAGES/ Alamy; page 26: Jim Richardson/National Geographic Stock; page 28: Nicole Duplaix/National Geographic Stock; page 28 (bottom): Courtesy of Matthew Price; page 29: Courtesy Eco Canada; page 30: CP PHOTO/ Winnipeg Free Press-Ken Gigliotti; page 31: Medford Taylor/National Geographic Stock; page 33: Courtesy Sierra Club of Canada, BC Chapter.

Chapter 2

Page 34–35: Sam Abell/National Geographic Stock; page 36: Robert E Peary/National Geographic Stock; page 37: Courtesy Library of Congress; page 38: Courtesy Conservation Force; page 39 (top): Photo A37421 appears courtesy of Provincal Archives of Alberta; page 39 (bottom): Photo by Tom Coleman/ Courtesy Aldo Leopold Foundation, Baraboo, WI.; page 40: Erich Hartmann/Magnum Photos, Inc; page 41: CP PHOTO/ Frank Gunn; page 42: Roger Rosentreter/iStockphoto; page 44: NG Maps; page 211 (top left): Shutterstock; page 211 (top right): National Archives of Canada/PA-51347; page 211 (bottom): Mark Edwards/Peter Arnold, Inc.; page 49 (top): Sam Abell/National Geographic Stock; page 49 (center): Secret Sea Visions/Peter Arnold, Inc.; page 49 (bottom): Alan and Sandy Cary/Peter Arnold, Inc.; page 50: Priit Vesilind/National Geographic Stock; page 51: © Robert Wagenhoffer/Corbis; page 52: Jenny Hager/ The Image Works; page 54 (top left): Donna DeCesare; page 54 (center left): Emory Kristof/National Geographic Stock; page 54 (bottom left): © Bertrand Rieger/Hemis/Corbis; page 54 (top right): Rafael Macia/ Photo Researchers, Inc.; page 54 (bottom right): © Christian Shallert/ Corbis; page 55: Jodi Cobb/National Geographic Stock; page 57 (top): Nick Nichols/National Geographic Stock; page 57 (bottom): Fred Hoogervorst/Panos Pictures; page 59: Peter Treanor/Alamy; page 60: Courtesy Eco Canada; page 61: Tim Laman/National Geographic Stock; page 65: Liane Cary/Age Fotostock.

Chapter 3

Page 66–67: Lou Linwei/Alamy; page 66 (insert): Sidney Hastings/ National Geographic Stock; page 68 (left): Wolcott Henry/National Geographic Stock; page 68 (right): David Pluth/National Geographic Stock; page 71: CNRI/Science Photo Library/Photo Researchers; page 72: Yvona Momatiuk and John Eastcott/Photo Researchers; page 73:

© DLILLC/Corbis; page 76–77: NGS Maps; page 168 (bottom): NG Maps; page 76: NG Maps; page 77 (bottom right): NG Maps; page 79 (left): Annie Griffiths Belt/National Geographic Stock; page 79 (right): Karen Kasmauski/National Geographic Stock; page 84 (left): Pablo Corral Vega/National Geographic Stock; page 84 (right): Larry C. Price; page 85: Kenneth Macleish/National Geographic Stock; page 86: Sean Spague/ Alamy; page 87: Karen Kasmauski/National Geographic Stock; page 89: Richard Nowitz/National Geographic Stock; page 90: mediacolor's/ Alamy; page 91: Urban Redevelopment of Pittsburgh; page 92 (top): Tara Smith; page 92 (bottom): Nathan Waite - globalairphotos.com; page 95: Courtesy of Eco Canada; page 96 (left): Time & Life Pictures/Getty Images; page 96 (right): © COLLART HERVE/CORBIS SYGMA; page 99: Lynn Johnson/National Geographic Stock.

Chapter 4

Page 100–101: David Alan Harvey/National Geographic Stock; page 103: Sisse Brimberg/National Geographic Stock; page 104: Scott Camazine/ Alamy; page 106: Shutterstock; page (top): SPL/Photo Researchers, Inc.; page 107 (bottom left): SPL/Photo Researchers,Inc.; page 107 (bottom right): Courtesy Millipore Corp; page 108: Bruce Dale/National Geographic Stock; page 109 (top): Justin Guariglia/National Geographic Stock; page 109 (bottom): Carsten Koall/Getty Images; page 111 (top): Roy Toft/National Geographic Stock; page 111 (bottom): Rod Planck/ Photo Researchers, Inc.; page 114: Courtesy Experimental Lakes Area; page 114: Andy Crump/Photo Researchers, Inc.; page 116: © Wolfgang Flamish/zefa/Corbis; page 118: Courtesy Centers for Disease Control; page 119: Daniel LeClair/Reuters/Landov; page 121 (top): Raymond Gehman/National Geographic Stock; page 121 (bottom): Courtesy of Eco Canada; page 122: Stuart Bauer; page 125: James L. Amos/National Geographic Stock.

Chapter 5

Page 126–127: Harmut Schwarzbach/Peter Arnold, Inc.; 126 (insert): Robert Caputo/National Geographic Stock; page 129 (top): Norbert Rosing/National Geographic Stock; page 129 (bottom): Geroge F. Mobley/National Gegraphic Stock; page 130 (top): Joel Sartore/National Geographic Stock; page 130 (bottom): Paul Nicklen/National Geographic Stock; page 132 (left and right): Romeo Gacad/AFP/Getty Images; page 133 (left): Raymond Gehman/National Geographic Stock; page 133 (right): Paul Nicklen/National Geographic Stock; page 134 (top left): John Eastcott and Yva Momatiuk/National Geographic Stock; page 134 (bottom left): Roy Toft/National Geographic Stock; page 134 (top right): Tim Laman/National Geographic Stock; page 134 (bottom right): Norbert Rosing/National Geographic Stock; page 141: © Roger Ressmeyer/ CORBIS; page 144 (left): Dr Jeremy Burgess/Photo Researchers; page 144 (right): Dr. Robert Calentine/Visuals Unlimited; page 147: Jason Edwards/National Geographic Stock; page 148: Rob and Ann Simpson/ Visuals Unlimited; page 150 (top left): Jack Jeffrey Photography; page 150 (bottom left): Tim Laman/National Geographic Stock; page 150 (top right): Minden Pictures/Getty Images; page 150 (bottom right): Darlyne A. Murawski/National Geographic Stock; page 151: Chris Johns/National Geographic Stock; page 152 (left): Des & Jen Bartlett/National Geographic Stock; page 152 (right): Robert Sisson/National Geographic Stock; page 153: Raymond Gehman/National Geographic Stock; page 154 (left): Raymond Gehman/National Geographic Stock; page 154 (right): Joel Sartore/National Geographic Stock; page 155: Courtesy of Don Sexton; page 157: Courtesy Eco Canada; page 158: NASA; page 158: Darlyne A. Murawski/National Geographic Stock.

Chapter 6

Page 162–163: © Mike Grandmaison/All Canada Photos; page 166: Paul Davis/The Global Land Cover Facility, University of Maryland Institute for Advanced Computer Studies; page 166: Brand X/SUPERSTOCK; page 166: Hemis. Fr/SUPERSTOCK; page 166: Photodisc/SUPERSTOCK; page 166: age fotostock/SUPERSTOCK; page: 166: Polka Dot Images/SUPERSTOCK; page 166: Michael McCoy/Photo Researchers, Inc.; page 166: Corbis/SUPERSTOCK; page 166: age fotostock/SUPERSTOCK; page 167: age fotostock/SuperStock, Inc.; page 167: James Steinberg/Photo Researchers, Inc.; page 167: Corbis/SUPERSTOCK; page 167: Imagemore/SUPERSTOCK; page 167: Steve Coleman/SUPERSTOCK; page 167: Ernest Manewal/SUPERSTOCK; page 168: Paul Nicklen/National Geographic Stock; page 169: Maria Stenzel/National Geographic Stock; page 170: Mike Dobel/Alamy; page 171: Taylor S. Kennedy/National Geographic Stock; page 172: Tom/Vezo/Minden/National Geographic Stock; page 173: Peter Ryan/National Geographic Stock; page 174: Robert Hynes/National Geographic Stock; page 174: Ken Lucas/Visuals; page 176: Mitsuaki Iwago/National Geographic Stock; page 174: Medford Taylor/National Geographic Stock; page 179: (top left): Burt Curtsinger/NG Image Collection; page 179 (top right): Stuart Westmorland/Photo Researchers; page 179 (bottom left): Alex Kerstitch/Visuals Unlimited; page 179 (bottom right): Paul Zahl/NG Image Collection; page 180: Tim Laman/National Geographic Stock; page 181 (top): Raul Touzon/National Geographic Stock; page 181 (bottom): Norbert Wu/Peter Arnold; page 182: (top) Hugh Rose/Danita; page 182 (bottom): Joe Stancampiano/National Geographic Stock; page 186: Dante Fenolio/Photo Researchers, Inc.; page 183: Kathleen Revis/National Geographic Stock; page 185: Frans Lanting/National Geographic Stock; page 186: © Mike Grandmaison/All Canada Photos; page 187 (top): Photo Canada Digital; page 187 (bottom): Joe Stancampiano/National Geographic Stock; page 189: Nicole Duplaix/National Geographic Stock; page 190: (top right): NG Maps; page 190: Ralph Lee Hopkins/NG Image Collection; page 190: © J. Dunning/VIREO; page 190: © Gerald & Buff Corsi/Visuals Unlimited; page 190: © Fritz Polking/Visuals Unlimited; page 190: © Joe & Mary Ann McDonald/Visuals Unlimited; page 190: Tierbild Okapia/Photo Researchers,Inc.; page 190: Eric Hosking/Photo Researchers, Inc.; page 191: O. Louis Mazzatenta/National Geographic Stock; page 193 (top left): Wolfgang Kaehler; page 193 (center left): Glenn N. Oliver/Visuals; page 193 (right): Wolfgang Kaehler; page 195: Courtesy Eco Canada; page 196 (left): CP PHOTO/Richard Lam; page 196 (right): CP PHOTO/Richard Lam; page 199 (bottom): Jodi Cobb/National Geographic Stock.

Chapter 7

Page 200–201: George Grall/National Geographic Stock; page 200 (inset): Frans Lanting/Minden Pictures, Inc.; page 206 (left): Sumia Harda/Minden/National Geographic Stock; page 206 (right): David Cavagnaro; Page 207: Doug Wechsler; page 211: Joel Sartore/National Geographic Stock; page 212: Joel Sartore/National Geographic Stock; page 213 (top left): Frans Lanting/Minden Pictures, Inc.; page 213 (bottom): Chris Johns/National Geographic Stock; page 213 (top right): W. Robert Moore/National Geographic Stock; page 215 (top): Dan McCoy/Science Faction/Getty Images; page 215 (bottom left): Kenneth R. Law, USDA APHIS PPQ, Bugwood.org; page 215 (bottom right): Pennsylvania Department of Conservation and Natural Resources - Forestry Archive, Bugwood.org; page 216: Andrew Sacks/Time Life Pictures/Getty Images; page 218 (left): USDA/National Geographic Stock; page 218 (right): USDA/National Geographic Stock; page 219 (top left): Flip Nicklin/National Geographic Stock; page 219 (bottom left): Tom Verzo/Minden/National Geographic Stock; page 219 (top right): Konrad Wothe/Minden/National Geographic Stock; page 219 (bottom right): Tom Verzo/Minden/National Geographic Stock; page 220: Terry Whitaker/Alamy; page 222: Rolf Hicker Photography/Alamy; page 223:

CP PHOTO/Calgary Herald-Lorraine Hjalte/ The Canadian Press; page 224: Courtesy Ducks Unlimited Canada; page 225: Happy Valley Forest (Ontario) in spring; photo by NCC; page 226 (top left): Steve Nexbitt, Florida Fish and Wildlife Conservation; page 226 (top right): Bryan & Cherry Alexander Photography/Alamy; page 226 (bottom): Richard O. Bierregaard,Jr.; page 227: Courtesy Eco Canada; page 228: Index Stock/Alamy; page 231: Michael Nichols/National Geographic Stock.

Chapter 8

Page 232–233: Norbert Rosing/National Geographic Stock; page 235 (left): Steven May/Alamy; page 235 (right): Nick Norman/National Geographic Stock; page 237: Biosphoto/NouN/Peter; page 238: David Keith Jones/Alamy; page 239: Sarah Leen/National Geographic Stock; page 241 (top left): Courtesy NASA; page 241 (top center): Lawrence Migdale/Photo Researchers, Inc.; page 241 (top right): John Elk III/Alamy; page 241 (bottom left): © AP/Wide World Photos; page 357 (bottom right): Courtesy Peggy Greb/USDA; page 244 (bottom): Steve Smith/SUPERSTOCK; page 246: Bruce Dale/National Geographic Stock; page 247: © John E. Marriott/All Canada Photos/Corbis; page 248: Reproduced with the permisson of Natural Resources Canada 2010, courtesy Atlas of Canada; page 250: Danita Delimont/Alamy; page 251: Kirtley-Perkins/Visuals Unlimited; page 252: Courtesy Forest Stewardship Council—US; page 254: Stephen Sharnoff/National Geographic Stock; page 255: Peter Essick/National Geographic Stock; page 256 (top left): NRSC/Photo Researchers, Inc.; page 256 (bottom left): NRSC/Photo Researchers, Inc.; page 256 (right): Marin Hervey/Natural History Photographic Agency; page 259: CP PHOTO/Adrian Wyld; page 260: Peter Essick/National Geographic Stock; page 291 (top): R. Ashley/Visuals Unlimited; page 291 (bottom): David Hiser/Stone/Getty Images; page 262: Peter Essick/National Geographic Stock; page 263: Courtesy U.S. Dept. of Agriculture; page 264: Photo by Ray Weil, courtesy Martin Rabenhorst; page 266: Grant Heilman/Grant Heilman Photography; page 267: Courtesy U. S. Department of Agriculture; page 268: Reproduced with the permission of Natural Resources Canada 2010, courtesy of the Atlas of Canada.; page 269 (top left): Tim Fitzharris/Minden/National Geographic Stock; page 269 (bottom left): Peter Arnold, Inc./Alamy page 269 (top right): John Sylvester/Alamy; page 269 (bottom right): Carr Clifton/Minden/National Geographic Stock; page 270: © Thomas Kitchin & Victoria Hurst/All Canada Photos; page 272: age fotostock/SuperStock, Inc.; page 272: Digital Vision/SUPERSTOCK; page 272: Ingram Publishing/SUPERSTOCK; page 272: Prisma/SUPERSTOCK; page 272: SuperStock, Inc.; page 272: Pacific Stock/SUPERSTOCK; page 272: NG Maps; page 273 (top): Bruce Dale/NG Image Collection; page 273 (bottom): James P. Blair/NG Image Collection; page 279 (top left): Courtesy U. S. Department of Agriculture; page 279 (bottom left): Blaine Harrington III/Alamy; page 279: © Kevin Flaming/Corbis' page 280: Courtesy Eco Canada; page 281: Carl Pedersen/Alamy; page 281 (bottom): Kalundborg, Denmark; page 285: Lyroky/Alamy.

Chapter 9

Page 286–287: Brian J. Skerry/National Geographic Stock; page 286 (inset): Jose Cort/Courtesy NOAA; page 288: Jodi Cobb/National Geographic Stock; page 289 (right): Reproduced with the permisson of Natural Resources Canada 2010, courtesy Atlas of Canada; page 295 (left): CP PHOTO 1997 (stf-Tom Hanson)th; page 295 (right): CP PHOTO 1996 (stf/Jacques Boissinot) jqb; page 296: Courtesy NASA; page 297: © Gary Fiegehen/All Canada Photos; page 298: Courtesy U. S. Department of Agriculture; page 302 (left): Rich Buzzelli/Tom Stack & Associates; page 302 (right): W.A. Banaszewski/Visuals Unlimited; page 302: Courtesy Experimental Lakes Area; page 304: Department of Biological Sciences, University of Alberta; page 305: Steve Winter/National Geographic Stock; page 307 (left): Dennis Frates/alamy; page 307 (right): Dennis Frates/alamy; page 309: NG Maps; page 311 (top): Courtesy U. S. Dept. of

Energy; page 311 (bottom): © Raymond G. Barnes; page 312: Hoekstra, A.Y. and Chapagain, A.K. (2007) Water footprints of nations: water use by people as a function of their consumption pattern, Water Resources Management 21(1): 35–48; page 214 (left): James L. Stanfield/National Geographic Stock; page 314 (right): © AP/Wide World Photos; page 315: Rama; page 317: Copyright 2010, City of Winnipeg; page 318: Ladco Company Limited's Royalwood; page 319: © Ted Shreshinsky/Corbis Images; page 327: Guy Hoppen/Accent Alaska; page 328: Gary Murray/All Canada Photos; page 329: Brian J. Skerry /National Geographic Stock; page 330: Flip Nicklin /Getty Images; page 331: Peter W. Glynn/National Geographic Stock; page 332: Science VU/Visuals Unlimited; page 333: Courtesy Eco Canada; page 334: Courtesy of Greg McCullough; page 337: Ken Gillespie Photography/Alamy.

Chapter 10

Page 338–339: Norbert Rosing/National Geographic Stock; page 341: Roger Harris/Photo Researchers, Inc.; page 342 (left): Kenneth Garrett/NG Image Collection; page 342 (right): Jack Finch/Photo Researchers; page 351: Sarah Leen/National Geographic Stock; page 352 (top): David Cavagnaro/Peter Arnold, Inc.; page 352 (center): Courtesy (INSTAAR) University of Colorado Boulder; page 352 (bottom): Accent Alaska.com/Alamy; page 352 (top left): Shutterstock; page 352 (bottom left): Emory Kristof/National Geographic Stock; page 352 (right): Mario Tama/Getty Images; page 358: Michael S. Yamashita/National Geographic Stock; page 360: David Wei/Alamy; page 361: James P. Blair/National Geographic Stock; page 363: Al Petteway/National Geographic Stock; page 364 (top): blickwinkel/Alamy; page 364 (bottom): FLPA/Alamy; page 366: Joerg Boethling/Peter Arnold, Inc.; page 372 (left): George F. Mobley/National Geographic Stock; page 372 (right): University of Bern (Switzerland), W. Berner, 1978; page 374: Stock Trek/PhotoDisc Green/Getty; page 376 (left and right): NASA; page 377: Courtesy of David Barber; page 378: © Tom Nevesely/All Canada Photos; page 379: Michael Nichols/National Geographic Stock; page 380 (left): FLPA/Alamy; page 380 (center): Peter Scoones /Photo Researchers, Inc.; page 380 (right): Thomas R. Fletcher/Alamy; page 382: David R. Frazier Photolibrary/Alamy; page 386: David Austen/National Gegraphic Stock; page 387: Courtesy Eco Canada; page 391: Thomas Nebbia/National Geographic Stock.

Chapter 11

Page 392-393: CP PHOTO/Paul Chiasson; page 396: Peter Essick/National Geographic Stock; page 397 (left): Kristin Finnegan Photograhy; page 397 (right): John Shaw Photography; page 424: Jim Richardson/NG Image Collection; page 424: Mark Burnett/Photo Researchers, Inc.; page 402: Courtesy National Renewable Energy Lab; page 402: John Mead/Photo Researchers, Inc.; page 402: Photo Researchers, Inc.; page 402: U.S. Department of Energy/USGS Mineral Resources Program; page 402: The Geothermal Energy Association/Barry Soloman/Michigan

Technological University/Sandia National Laboratories/Windpower Monthly News Magazine; page 402: NG Image Collection/BP Statistical View of World Energy; page 404: John Shaw Photography; page 405: Chris Wilkins/AFP/Getty Images; page 406 (left): NASA; page 406 (right): AP Images/AP Images; page 409 (left): Courtesy Westinghouse Electric Corp., Commercial Nuclear Fuel Division; page 409 (right): Jayson Mellom Photography; page 410: Photo provided courtesy of Ontario Power Generation; page 412 (top): Novosti; page 412 (bottom): © Caroline Penn/Corbis Images; page 414 (top): Peter Essick/Aurora Photos; page 414 (bottom): Howells David/Gamma-Presse, Inc.; page 415: Charles Mason/Black Star; page 417: Otis Imboden/National Geographic Stock; page 418: Manor Photography/Alamy; page 419: Courtesy of Uni-Solar Energy Conservation Devices, Discover magazine Magazine; page 419 (inset): United Solar Systems; page 420: Hank Morgan /Getty Images; page 421 (left): Pasquale Sorrentino/Photo Researchers, Inc.; page 421 (right): Steve McCurry/National Geographic Stock; page 422: Prof. David Hall/Photo Researchers, Inc.; page 424: Justin Sullivan /Getty Images; page 426: Bill Brooks/Alamy; page 427: Arcticphoto/Alamy; page 428: China Images/Alamy; page 429: Phil Degginger/Alamy; page 421: Courtesy Ontario Hydro; page 434: Courtesy Eco Canada; page 439: Sarah Leen/National Geographic Stock.

Chapter 12

Page 440-441: © Perry Mastrovito/First Light/Corbis; page 442: Mario Beauregard/CPI/The Canadian Press; page 446: Ashley Cooper/Visuals Unlimited, Inc. /Getty Images; page 448: Robert McGouey/Alamy; page 449: Ian Wood/Alamy; page 452: Michael Harvey/Alamy; page 454 (left): Frances Roberts/Alamy; page 544 (right): page 545 (right): Shari Lewis/© AP/Wide World Photos; page 455: Johnny Greig LSL/Alamy; page 456: Patrick Durand/ABACAPRESS.COM; page 458: Andy Levin/Photo Researchers, Inc.; page 460: Chromosohm/Photo Researchers, Inc.; page 461: DOE/Science Source; page 462 (top): Gary Milburn/Tom Stack & Associates; page 462 (bottom): Courtesy USDA; page 465: Courtesy Eco Canada; page 466: REUTERS/Christinne Muschi; page 469: © AP/Wide World Photos.

Chapter 13

Page 470–471: © Chris Harris/All Canada Photos; page 470 (top): Rolf Hicker Photography/Alamy; page 470 (bottom): travelstock44/Alamy; page 471 (top): Danita Delimont/Alamy; page 471 (bottom): Nick Norman/National Geographic Stock; page 472: Shutterstock; page 474: Ladco Company Limited's Royalwood; page 476: Emily Riddell/Alamy; page 478: Ladco Company Limited's Royalwood; page 480 (left): Ladco Company Limited's Royalwood; page 480 (right): Courtesy Rick Baydack; page 481: Courtesy Bill Rutherford; page 486 (top): Jim West/Alamy; page 486 (bottom): Courtesy Eco Canada; page 487: Robert Loney; page 489: Terrance Klassen/Alamy.

INDEX

Note: *f* indicates figures; *t* indicates tables

A

abiotic, 128, 165

Aboriginal peoples
  Aboriginal worldview, 50–51
  drinking water, 320
  impacts on forests, 36
  mercury contamination, compensation
    for, 112
  treaty rights, 235

Aboriginal worldview, 50–51

Acadian flycatcher, 172

acid deposition, 397

acid mine drainage, 261, 261*f*, 396

acid neutralizing capacity (ANC), 365

acid precipitation, 361–366

acid rain, 361–366

acidification, 332, 380*f*

acidity, 362

active management, 156

active solar heating, 416–417, 416*f*

acute toxicity, 105, 116

adaptation, 188

adaptive ecosystem management, 155–156

adaptive resource management (ARM), 478,
  479*f*, 482

additivity, 119

Advance Very High Resolution Radiometer
  (AVHRR), 167

aerosols, 373

aesthetic value of organisms, 207

Africa, 212, 253

African elephant, 212, 220–221

age structure, 80–82, 81*f*, 82*f*

Agri-Environmental Services Branch
  (AESB), 245

agriculture
  agricultural water consumption, reduction
    of, 311–313
  agroecosystem, 275
  alternative agriculture, 274
  in Canada, 236
  challenges of, 238–244
  climate change, effects of, 376–378
  conservation tillage, 277–278
  contour plowing, 278
  control of agricultural pests, 242, 243*t*
  crop rotation, 277–278
  crop yields, 239, 240
  domesticated plant and animal varieties,
    decline in, 239
  environmental impacts, 240–242, 241*f*
  green revolution, 239–240
  high-input agriculture, 236
  industrialized agriculture, 236–237
  integrated pest management, 242–243

  integrated pest management (IPM), 243*f*
  irrigation, 266, 295
  loss of agricultural land, 238–239
  low-input agriculture, 274
  microirrigation, 311–313, 314*f*
  organic agriculture, 274–275
  organisms, agricultural importance of,
    206–207
  principal types of, 236–238
  rangelands for livestock production, 244–247
  salination of irrigated soil, 297–298, 298*f*
  shelterbelts, 279
  slash-and-burn agriculture, 237, 257
  soil conservation and regeneration, 277–280
  strip cropping, 278
  subsistence agriculture, 237–238
  sustainable agriculture, 271–275
  terracing, 278

Agriculture and Agri-Food Canada (AAFC),
  236, 245, 298, 307

agroecosystem, 275

air pollution
  acid precipitation, 361–366
  Canada's response, 356–357
  and children, 119
  defined, 348
  emissions trading, 355
  hazardous air pollutants, 351
  health effects of major air pollutants,
    354–355, 354*t*
  indoor air pollution, 366–367, 367*f*
  industrial smog, 357–358
  as intercontinental or transboundary
    issue, 338
  long-distance transport of, 338
  long-range transport of air pollutants
    (LRTAP), 112–113
  in Los Angeles, 361
  major classes of air pollutants, 349–352
  management, 355–356
  natural sources, 352*f*
  photochemical smog, 358–359, 359*f*,
    360–361
  primary air pollutants, 348–349, 349*f*
  radon, 366–367
  secondary air pollutants, 348–349, 349*f*
  smog, 357–361
  sources of air pollutants, 352*f*, 353, 353*f*
  urban air pollution, effects of, 357–367

air quality, 46

air toxics, 351

Alaska Pipeline Project, 401

albedo, 374*f*, 375, 375*f*

Alberta tar sands, 400

alpine tundra, 168

alternative agriculture, 274

Alternative Land Use Services (ALUS), 158, 280

aluminum cans, 451

American Warbler, 148

ammonification, 144

amoebic dysentery, 107*t*

Amphibian Ark, 200

amphibians, 200

Anan, Kofi, 13

ancylostomiasis, 107*t*

androgens, 122

antagonism, 119

Antarctica, 375–376

anthropocentric, 50

aquaculture, 57–58, 329–330

aquatic ecosystems
  benthic environment, 178–181, 179*f*
  brackish ecosystems, 187–188
  classification of, 178–188
  estuaries, 187–188
  flowing-water ecosystems, 185
  freshwater ecosystems, 183–185
  freshwater wetlands, 186
  intertidal zone, 178, 179*f*
  major marine life zones, 178–183
  neritic province, 178, 182
  oceanic province, 178, 183
  standing-water ecosystems, 183–185

aquatic resources. *See* the ocean; water

aquifer depletion, 296

aquifers, 289–292

Aral Sea, 309, 309*f*

Arcata Marsh and Wildlife Sanctuary, 319

the Arctic
  climate change, effects of, 375–376
  dominating air currents, and
    pollution, 113*f*
  ecosystems, 112
  mercury concentration, 114
  natural gas reserves, 401
  warming Arctic, 17*f*

Arctic Goose Joint Venture (AGJV), 73

arctic tundra, 168–169

ArcticNet, 377

asbestos, 355

Aspen Face, 351

assimilation, 143

atmosphere
  air pollution. *See* air pollution
  air quality, 348
  climate change. *See* climate change
  composition, 340, 340*f*
  cooling of, 373–374
  defined, 130–131, 340
  electromagnetic spectrum, 341–343